GREEN SUSTAINABLE PROCESS FOR CHEMICAL AND ENVIRONMENTAL ENGINEERING AND SCIENCE

GREEN SUSTAINABLE PROCESS FOR CHEMICAL AND ENVIRONMENTAL ENGINEERING AND SCIENCE

Supercritical Carbon Dioxide as Green Solvent

Edited by

INAMUDDIN
Department of Chemistry, King Abdulaziz University, Saudi Arabia

ABDULLAH M. ASIRI
Department of Chemistry, King Abdulaziz University, Saudi Arabia

ARUN M. ISLOOR
Department of Chemistry, National Institute of Technology
Karnataka, India

ELSEVIER

Elsevier
Radarweg 29, PO Box 211, 1000 AE Amsterdam, Netherlands
The Boulevard, Langford Lane, Kidlington, Oxford OX5 1GB, United Kingdom
50 Hampshire Street, 5th Floor, Cambridge, MA 02139, United States

Library of Congress Cataloging-in-Publication Data
A catalog record for this book is available from the Library of Congress

British Library Cataloguing-in-Publication Data
A catalogue record for this book is available from the British Library

ISBN: 978-0-12-817388-6

For information on all Elsevier publications visit
our website at https://www.elsevier.com/books-and-journals

Publisher: Susan Dennis
Acquisition Editor: Kostas KI Marinakis
Editorial Project Manager: Michael Lutz
Production Project Manager: Vignesh Tamil
Cover Designer: Christian Bilbow

Typeset by SPi Global, India

Contents

Contributors

Poonam Aggarwal
Department of Food Science and Technology, Punjab Agricultural University, Ludhiana, India

Mukta Agrawal
Rungta College of Pharmaceutical Sciences and Research, Bhilai, Chhattisgarh, India

Mudasir Ahmad
Department of Food Science and Technology, University of Kashmir, Srinagar, India

Ajazuddin
Rungta College of Pharmaceutical Sciences and Research, Bhilai, Chhattisgarh, India

Sumia Akram
University of Education, Bank Road Campus, Lahore, Pakistan

Amit Alexander
Rungta College of Pharmaceutical Sciences and Research, Bhilai, Chhattisgarh, India

Franco Rico Amado
Materials and Environment Laboratory (LAMMA), State University of Santa Cruz—UESC, Ilheus, Brazil

Andrew N. Amenaghawon
Department of Chemical Engineering, Faculty of Engineering, University of Benin, Benin City, Nigeria

Nanjangud V Anil Kumar
Department of Chemistry, Manipal Institute of Technology, Manipal Academy of Higher Education, Manipal, India

Chinedu L. Anyalewechi
Department of Chemical Engineering, Faculty of Engineering, University of Benin, Benin City, Nigeria

Raouf Aslam
Department of Processing and Food Engineering, Punjab Agricultural University, Ludhiana, India

Mojhdeh Baghbanbashi
Department of Chemical Engineering, Amirkabir University of Technology (Tehran Polytechnic), Tehran, Iran

Ming Bao
State Key Laboratory of Fine Chemicals, Dalian University of Technology, Ganjingzi District, Dalian China; School of Petroleum and Chemical Engineering, Dalian University of Technology, Liaodongwan New District, Panjin, China

Daniel Assumpção Bertuol
Environmental Processes Laboratory (LAPAM), Chemical Engineering Department, Federal University of Santa Maria—UFSM, Santa Maria, Brazil

F.W.F. Bezerra
Technology Institute, Program of Post-Graduation in Food Science, Technology and Engineering, Federal University of Pará, Belém, Brazil

P.N. Bezerra
Technology Institute, Program of Post-Graduation in Food Science, Technology and Engineering, Federal University of Pará, Belém, Brazil

R.N. Carvalho Junior
Technology Institute, Program of Post-Graduation in Food Science, Technology and Engineering; Technology Institute, Program of Post-Graduation in Amazon Natural Resources Engineering, Federal University of Pará, Belém, Brazil

Rajesh Chandra
Bioenergy Research Laboratory, Department of Polymer & Process Engineering, Indian Institute of Technology Roorkee (Saharanpur Campus), Saharanpur, India

A.M.J. Chaves Neto
Technology Institute, Program of Post-Graduation in Amazon Natural Resources Engineering; Laboratory of Preparation and Computation of Nanomaterials, Federal University of Pará, Belém, Brazil

Ahmad Cheikhyoussef
University of Namibia, Windhoek, Namibia

Natascha Cheikhyoussef
Ministry of Higher Education, Training and Innovation, Windhoek, Namibia

Gun-Hean Chong
Faculty of Food Science and Technology, Universiti Putra Malaysia, Serdang, Selangor, Malaysia

R.M. Cordeiro
Technology Institute, Program of Post-Graduation in Amazon Natural Resources Engineering, Federal University of Pará, Belém, Brazil

Estevan Dorneles Cruz
Environmental Processes Laboratory (LAPAM), Chemical Engineering Department, Federal University of Santa Maria—UFSM, Santa Maria, Brazil

V.M.B. Cunha
Technology Institute, Program of Post-Graduation in Food Science, Technology and Engineering, Federal University of Pará, Belém, Brazil

W.A. da Costa
Technology Institute, Program of Post-Graduation in Amazon Natural Resources Engineering, Federal University of Pará, Belém, Brazil

J.N. da Cruz
Laboratory of Preparation and Computation of Nanomaterials, Federal University of Pará, Belém, Brazil

M.S. de Oliveira
Technology Institute, Program of Post-Graduation in Food Science, Technology and Engineering, Federal University of Pará, Belém, Brazil

Ana Lúcia Barbosa de Souza
University Center SENAI CIMATEC, Health Institute of Technologies (CIMATEC ITS), National Service of Industrial Learning—SENAI, Salvador, Bahia, Brazil

Noor U Din Reshi
Department of Chemistry, University of Kashmir, Srinagar, India

Janice Izabel Druzian
Federal University of Bahia (UFBA), Faculty of Pharmacy, Salvador, Bahia, Brazil

Sunil Kumar Dubey
Department of Pharmacy, Birla Institute of Technology and Science, Pilani, Rajasthan, India

Adil Gani
Department of Food Science and Technology, University of Kashmir, Srinagar, India

Sandra Gonçalves
Faculty of Sciences and Technology, MeditBio, University of Algarve, Faro, Portugal

Naghmeh Hadidi
Department of Clinical Research and Electronic Microscope, Pasteur Institute of Iran, Tehran, Iran

Noor Hadzuin Nik Hadzir
Faculty of Food Science and Technology, Universiti Putra Malaysia, Serdang, Selangor, Malaysia

Shahryar Jafarinejad
Department of Chemical Engineering, College of Engineering, Tuskegee University, Tuskegee, AL, United States

Junaid Khan
University Teaching Department (Pharmacy), Sant Gahira Guru University, Sarguja, Ambikapur, Chhattisgarh, India

Heri Septya Kusuma
Department of Chemical Engineering, Faculty of Industrial Technology, Institut Teknologi Sepuluh Nopember, Surabaya, Indonesia

Wan-Jun Lee
Faculty of Food Science and Technology, Universiti Putra Malaysia, Serdang, Selangor, Malaysia

Xin Liu
State Key Laboratory of Fine Chemicals, Dalian University of Technology, Ganjingzi District, Dalian China; School of Petroleum and Chemical Engineering, Dalian University of Technology, Liaodongwan New District, Panjin, China

Bruna Aparecida Souza Machado
University Center SENAI CIMATEC, Health Institute of Technologies (CIMATEC ITS), National Service of Industrial Learning—SENAI, Salvador, Bahia, Brazil

Mahfud Mahfud
Department of Chemical Engineering, Faculty of Industrial Technology, Institut Teknologi Sepuluh Nopember, Surabaya, Indonesia

Leonidas Matsakas
Biochemical Process Engineering, Division of Chemical Engineering, Department of Civil, Environmental and Natural Resources Engineering, Luleå University of Technology, Luleå, Sweden

Syed Kazim Moosvi
School Education Department, J&K Government, Srinagar, India

Muhammad Mushtaq
Department of Chemistry, GC University, Lahore, Pakistan

Ravish Patel
Ramanbhai Patel College of Pharmacy, Charotar University of Science and Technology, CHARUSAT Campus, Changa, Ta-Petlad, Anand, Gujarat, India

Alok Patel
Biochemical Process Engineering, Division of Chemical Engineering, Department of Civil, Environmental and Natural Resources Engineering, Luleå University of Technology, Luleå, Sweden

Gholamreza Pazuki
Department of Chemical Engineering, Amirkabir University of Technology (Tehran Polytechnic), Tehran, Iran

Fernando Luiz Pellegrini Pessoa
University Center SENAI CIMATEC, Health Institute of Technologies (CIMATEC ITS), National Service of Industrial Learning—SENAI, Salvador, Bahia, Brazil

R.H.H. Pinto
Technology Institute, Program of Post-Graduation in Food Science, Technology and Engineering, Federal University of Pará, Belém, Brazil

Jagbir Rehal
Department of Food Science and Technology, Punjab Agricultural University, Ludhiana, India

João Henrique de Oliveira Reis
Federal University of Bahia (UFBA), Faculty of Pharmacy, Salvador, Bahia, Brazil

Masood Ahmad Rizvi
Department of Chemistry, University of Kashmir, Srinagar, India

Anabela Romano
Faculty of Sciences and Technology, MeditBio, University of Algarve, Faro, Portugal

Shailendra Saraf
University Institute of Pharmacy, Pt. Ravishankar Shukla University, Raipur, Chhattisgarh, India

Swarnlata Saraf
University Institute of Pharmacy, Pt. Ravishankar Shukla University, Raipur, Chhattisgarh, India

Km Sartaj
Molecular Microbiology Laboratory, Biotechnology Department, Indian Institute of Technology (IIT-R), Roorkee, India

Sabahuddin Siddique
Patel College of Pharmacy, Madhyanchal Professional University, Bhopal, Madhya Pradesh, India

M.P. Silva
Technology Institute, Program of Post-Graduation in Food Science, Technology and Engineering, Federal University of Pará, Belém, Brazil

Emma Suali
Faculty of Engineering, Jalan UMS, Universiti Malaysia Sabah, Kota Kinabalu, Sabah, Malaysia

Norhidayah Suleiman
Faculty of Food Science and Technology, Universiti Putra Malaysia, Serdang, Selangor, Malaysia

Dhanya Sunil
Department of Chemistry, Manipal Institute of Technology, Manipal Academy of Higher Education, Manipal, India

Eduardo Hiromitsu Tanabe
Environmental Processes Laboratory (LAPAM), Chemical Engineering Department, Federal University of Santa Maria—UFSM, Santa Maria, Brazil

Nader Vahdat
Department of Chemical Engineering, College of Engineering, Tuskegee University, Tuskegee, AL, United States

Wan-Hui Wang
State Key Laboratory of Fine Chemicals, Dalian University of Technology, Ganjingzi District, Dalian China; School of Petroleum and Chemical Engineering, Dalian University of Technology, Liaodongwan New District, Panjin, China

Pooja Yadav
Rungta College of Pharmaceutical Sciences and Research, Bhilai, Chhattisgarh, India

Mudasir Yaqoob
Department of Food Science and Technology, Punjab Agricultural University, Ludhiana, India

Giovani L. Zabot
Laboratory of Agroindustrial Processes Engineering (LAPE), Federal University of Santa Maria (UFSM), Cachoeira do Sul, Brazil

CHAPTER 1

Polymer production and processing using supercritical carbon dioxide

Pooja Yadav[a], Mukta Agrawal[a], Amit Alexander[a], Ravish Patel[b], Sabahuddin Siddique[c], Shailendra Saraf[d], Ajazuddin[a]
[a]Rungta College of Pharmaceutical Sciences and Research, Bhilai, Chhattisgarh, India
[b]Ramanbhai Patel College of Pharmacy, Charotar University of Science and Technology, CHARUSAT Campus, Changa, Ta-Petlad, Anand, Gujarat, India
[c]Patel College of Pharmacy, Madhyanchal Professional University, Bhopal, Madhya Pradesh, India
[d]University Institute of Pharmacy, Pt. Ravishankar Shukla University, Raipur, Chhattisgarh, India

Contents

1. Introduction

In the past few decades, polymers have become an important part of daily life. Their feasible synthesis and processing are needed for various applications. Instead of using conventional solvents, now the research has been shifted to supercritical fluids. The supercritical fluids are described as a material maintained at a condition of temperature and pressure more than its critical values. It holds the unique combination of viscosity (like gas) and density (like liquid) which makes a supercritical fluid an excellent solvent [1, 2]. Supercritical carbon dioxide (SCO_2) is a non-toxic solvent, which is the best alternative

Green Sustainable Process for Chemical and Environmental Engineering and Science
https://doi.org/10.1016/B978-0-12-817388-6.00001-5

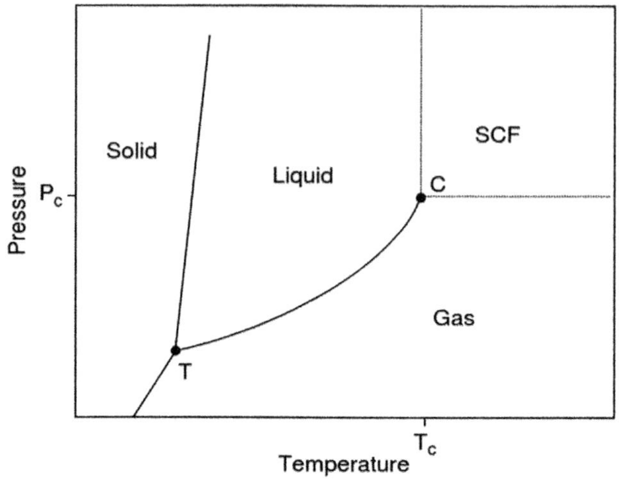

Fig. 1 Schematic representation of temperature–pressure phase diagram of the pure component, surrounded by triple point (T) and critical point (C). *(Adapted from S.P. Nalawade, F. Picchioni, L.P.B.M. Janssen, Supercritical carbon dioxide as a green solvent for processing polymer melts: processing aspects and applications, Prog. Polym. Sci. 31(1) (2006) 19–43.)*

for noxious organic solvents and chlorofluorocarbons. Owing to their immense physical properties such as non-flammability, chemically inertness, and cheapness, these find applications, not only in polymer synthesis, but also in polymer production. The supercritical conditions of carbon-dioxide (CO_2) can easily be achieved ($Tc = 304$ K and $P_c = 7.38$ MPa) (Fig. 1) and extracted out from the reaction by simply depressurizing the reactor. The high yield and increased product quality attract the usage of SCO_2 [4].

Moreover, CO_2 is a gas in ambient conditions, which makes it easy to remove from the polymeric product instead of using costly methods of solvent evaporation or drying. Mixtures of solvents are used to enhance the strength of chemical reaction during the synthesis and processing of polymers [5]. However, by incorporating SCO_2 into the response gives advantages over conventional solvents, such as:

- it possesses environmentally benign characteristics such as being non-toxic, non-flammable, and low cost;
- easy removal of the solvent after the reaction;
- faster rate of reaction at mild conditions; and
- higher selectivity and higher yield in chemical reactions (obtained due to temperature and pressure above critical values).

2. Properties of supercritical CO_2

SCO_2 has widely gained importance in recent years because of its assured diffusivity and density. These can change the glass transition temperature (Tg) of the polymers leading to their reduced viscosity [6]. Undeniably, even though CO_2, due to its distinct symmetrical structure, has no dipole moment, it can show inviolable quadruple moment (acting at a smaller distance than dipolar interaction), hence increasing the solubility [7, 8]. The CO_2 itself shows excellent solubilizing power and is therefore used as a solvent and antisolvent. CO_2 is preferred as a polymerization medium because:

- these are cost-effective, non-flammable, non-toxic, and are readily obtainable in pure form;
- solvent recovery becomes easy due to the use of CO_2;
- the high solubility of various polymers within CO_2 finds its application for their synthesis;
- for porous polymers to be synthesized, the supercritical fluid having CO_2 facilitates as non-solvating power and acts as an organic diluent; and
- carbon dioxide usually doesn't interact with the strong nucleophiles (e.g., alkoxides, primary amines, etc.), and hence it can be suggested polymerization in CO_2 is generally through the anionic mechanism.

The nonpolar molecules, having low molecular weight, can be readily solubilized in supercritical CO_2 rather than water and ionic compounds because supercritical CO_2 has a relatively small dielectric constant ($\varepsilon = 2$). SCO_2, owing to its tremendous physiochemical properties, is also cheap, non-toxic, and non-flammable, hence finding applications in the production and refinement of polymers. Owing to the high mass transfer, low viscosity, and low toxicity profile, the supercritical carbon dioxide can be potentially used in polymer processing, which also overcomes the Trommsdorff effect. Also, with supercritical CO_2, dry polymers can be easily made by merely depressurizing the reactor after polymerization [9, 10].

3. Applications of SCO_2 in polymer production and processing

3.1 Purification of polymers

A prior quantitative analysis of the extracted compound from the polymeric matrix by different conventional methods, like during synthesis or processing of polymers, is essential because there is a chance to obtain by-products or residual raw materials that have to be extracted. The conventional methods for purification of polymers include various methods such as solvent-intensive, Soxhlet extraction, or polymer dissolution method [11]. SCO_2 possesses the ability to extract out these by-products by firstly swelling the polymer and then permeating within the matrix where these impurities get solubilized into the SCO_2 and again upon reducing the pressure, SCO_2 is easily diffused out of the polymer. Interestingly, the solvating power of SCO_2 is regulated by slight modification in pressure and temperature. Hence, it is used in purification of polymers [12, 13]. Moreover, these do not leave any harmful residue, representing the non-toxic behavior of SCO_2. In addition, these can be easily recycled due to their volatile nature. Compared to the conventional extraction processes, if we use a dry membrane that is pre-treated with supercritical carbon dioxide, this can result in less shrinkage and improved water permeability [14, 15].

3.2 Impregnation and supercritical dyeing

CO_2 can be used as a solvent to introduce numerous dyes and metal complexes into the polymeric hosts. Hence, it may be applicable during dyeing. A doping solute behaving like a guest is introduced within the host polymer matrix during polymer impregnation and dyeing. The supercritical properties of SCO_2, including low surface tension, higher diffusivity, and easy recovery of solvent, assists the preparation of a new polymer [16, 17]. Impregnation is the technique to absorb the liquids within the polymer matrix. During impregnation using SCO_2, the therapeutic drug is dissolved into the solvent and then impregnated within the polymer matrix, which can be prepared for drug delivery. There are two mechanisms for SCO_2 infiltration as additives in the polymeric matrix. The first mechanism deals with the solute, which gets solubilized into the SCO_2 and then is passed through the polymer matrix. After reducing the pressure, CO_2 vacates the polymer matrix, and the solute particles becomes trapped within the polymer [4, 18]. The second mechanism deals with the partitioning of solute within the polymer matrix because of their low solubility in SCO_2. The high affinity of solute towards polymer is the primary factor for successful supercritical dyeing [19].

3.3 Particle production

Normal milling and grinding may deteriorate the polymer particles; therefore, administering SCO_2 as a suitable solvent, or antisolvent, could offer a more promising approach to reduce the particle size and control the morphological properties of the polymer. In addition to this, adjusting the process parameters such as pressure, temperature, rate of depressurization, and nozzle diameter can also result in different particle size [20]. Although a high pressure is required for the particle formation while using SCO_2 instead of using organic solvents, which tend to produce undesirable solvent residues, it is beneficial to use SCO_2 where no such toxic wastes are generated. The micron sized particles can be produced by low-melting polyester coatings, acrylic coating, and polyester-epoxy systems [21]. Either of the two methods can form polymeric particles, precipitation generally through a rapid expansion, which involves the use of SCO_2 or by using SCO_2 as an antisolvent.

3.4 Polymer modification

Monomers and initiators can get dissolved within the CO_2 leading to enhanced diffusion within the polymer matrix and can therefore easily modify the morphology of polymer. Grafting of chemical groups within the polymer substrate can be done through additional reactions [22].

The isopropyl-isocyanate grafting on ethylene-vinyl alcohol copolymers (EVOH) bare the advantage of SCO_2 providing a selective dispersion method in addition to frequent reactions of the monomer in amorphous form of EVOH maintaining their crystalline nature [23].

4. Polymer production

4.1 Step-growth polymerization

In this method, the functional group of difunctional monomers is condensed. By using SCO_2, polycondensation of components such as polyesters, polyamides (nylon), polyurethanes, and polyureas is possible (Fig. 2).

The name itself shows that the polymerization process will be stepwise from dimers to large molecules and then to polymer chains. The main problem associated with this process is tedious removal of high viscosity by-products, which can largely be overcome by supercritical CO_2, showing an excellent plasticizing effect that will decrease the viscosity of thaw influencing finer stirring. Predominantly, supercritical CO_2 permits extraction of by-products [8, 24, 25].

Polycarbonates can be synthesized by incorporating SCO_2 in the transesterification of Bisphenol A using diphenyl carbonate. Here, SCO_2 has a tendency to remove the phenol by-products and enhance the diffusivity of phenol during solid state polymerization process [8, 26]. Similarly, polyurethanes and polydimethylsiloxane (PDMS) were synthesized using SCO_2 as dispersant media at 60°C and above 200 bar pressure where ethylene glycol and tolylene-2,4-diisocyanate act as monomers and hydroxy- or isocyanate-terminated polydimethylsiloxane serve as a surfmer (Fig. 3) [8, 27].

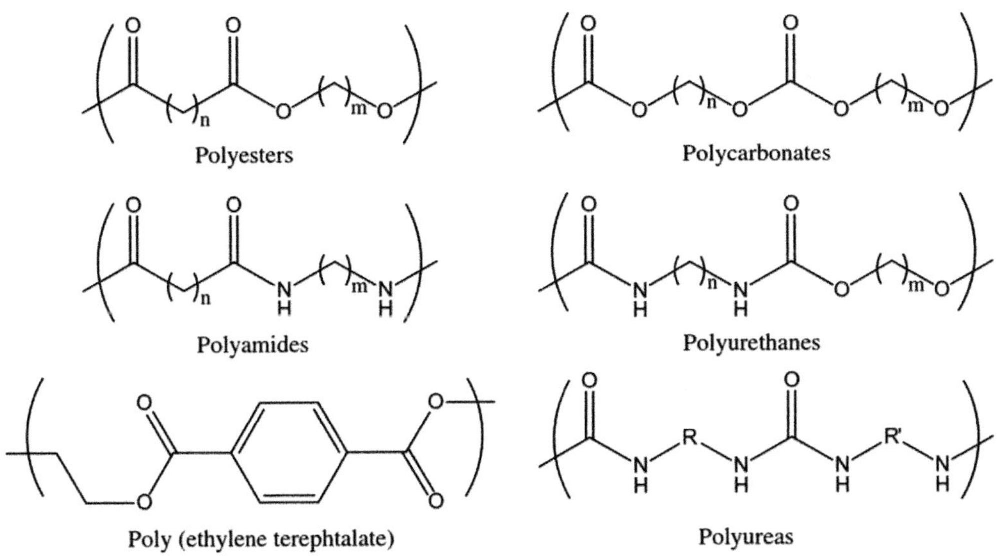

Fig. 2 Schematic representation of step-growth polymerization using SCO2. *(Adapted from C. Boyère, C. Jérôme, A. Debuigne, Input of supercritical carbon dioxide to polymer synthesis: an overview, Eur. Polym. J. 61 (2014) 45–63.)*

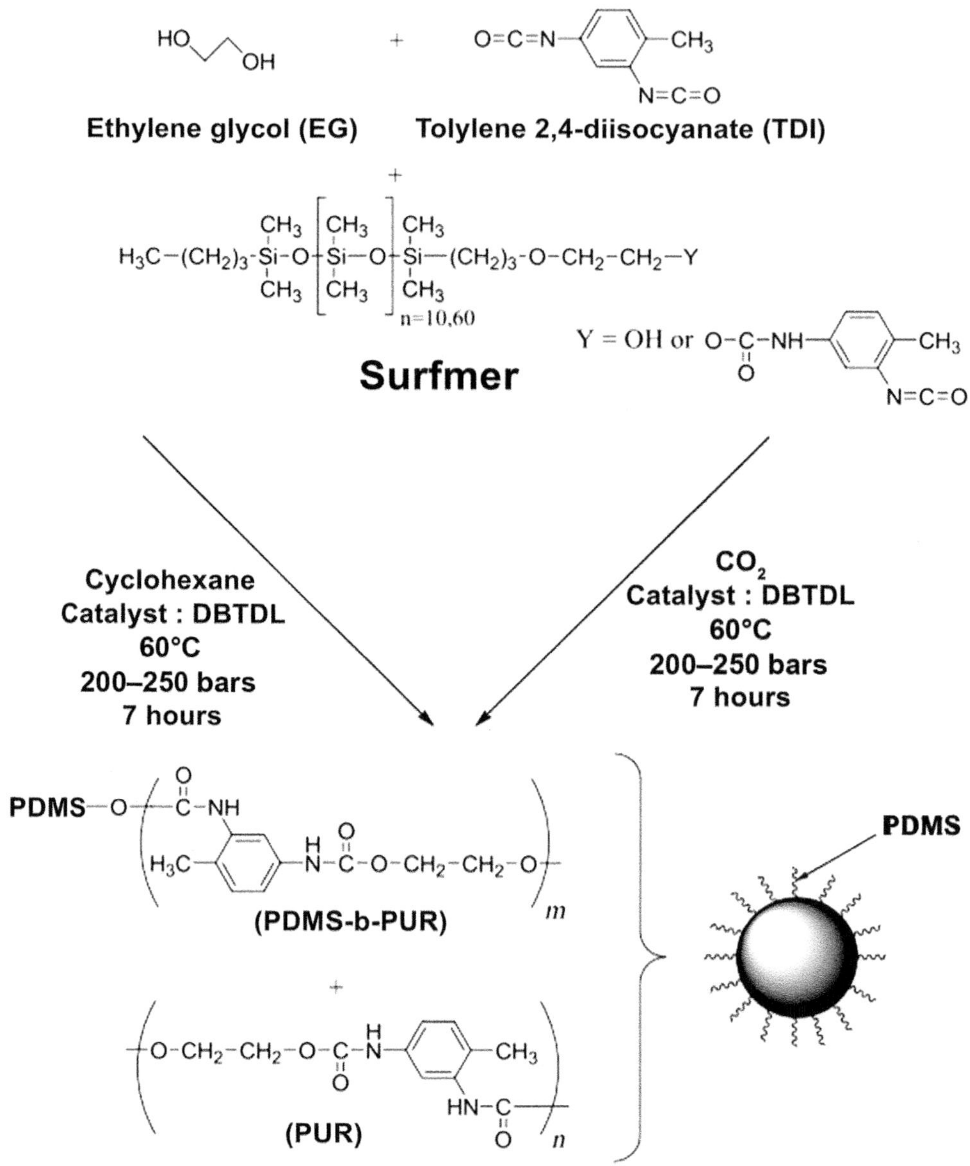

Fig. 3 Reaction between tolylene 2,4-diisocyanate, and ethylene glycol in the presence of surfmer. *(Adapted from P. Chambon, E. Cloutet, H. Cramail, T. Tassaing, M. Besnard, Synthesis of core-shell polyurethane–polydimethylsiloxane particles in cyclohexane and in supercritical carbon dioxide used as dispersant media: a comparative investigation, Polymer 46(4) (2005) 1057–1066.)*

4.2 Chain growth polymerization

Supercritical CO_2 is best suited for radical polymerization where all the three components, i.e., monomer, initiator, and control reagents, are effortlessly solubilized in supercritical CO_2 [28]. It acts as a green solvent for the synthesis of various fluorinated and non-fluorinated polymers by atom transfer radical polymerization (ATRP). Moreover, it plays a binary role by stabilizing the growing chains of PMMA and complexion of the copper salt in the synthesis of poly(methyl methacrylate) (PMMA) using ATRP [29, 30]. A one-pot process using dispersion ATRP and azide-alkyne 1,3-dipolar Huisgen's cycloaddition was employed for preparing functionalized polymers (Fig. 4) [31].

4.2.1 Homogeneous polymerization

The fluoropolymers can be synthesized by using SCO_2 via cationic polymerization or free radical polymerization [32]. The homogeneous polymerization of 1,1-dihydro perfluorooctyl acrylate (FOA) and azobisisobutyronitrile (AIBN) using SCO_2 yields various fluoropolymers associated with FOA [8, 33]. Furthermore, FOA can be *co*-polymerized using vinyl hydrocarbon monomers in the presence of SCO_2, which yields poly(1,1-dihydro perfluorooctyl acrylate) (PFOA) (Fig. 5).

Another co-polymer tetrafluoroethylene (TFE) can be used for the synthesis of various fluoropolymers; although TFE as a monomer causes rapid explosions, when combined with CO_2 it forms a "pseudo" azeotrope, which makes it easier to handle. Various TFE-based fluoropolymers have been synthesized using CO_2 such as fluorinated ethylene propylene (FEP), perfluoroalkoxy alkanes (PFA), ethylene tetrafluoroethylene (ETFE), TFE/vinyl acetate, Nafion type materials, and Teflon-AF-type materials (Fig. 6) [35].

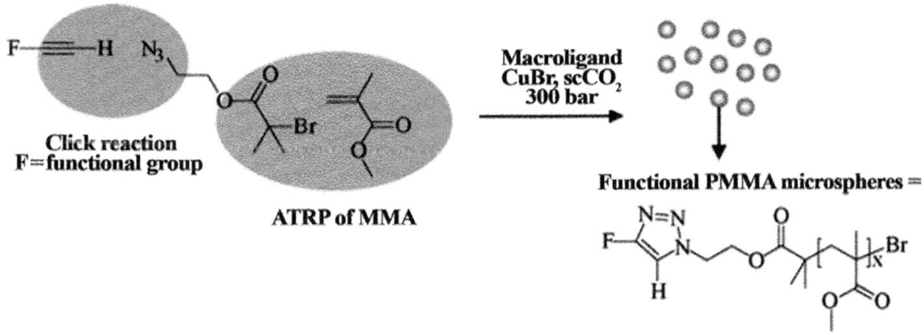

Fig. 4 Reaction between dispersion ATRP and click reaction containing functional group for the formation of functional PMMA microspheres. *(Adapted from B. Grignard, C. Calberg, C. Jerome, C. Detrembleur, "One-pot" dispersion ATRP and alkyne-azide Huisgen's 1,3-dipolar cycloaddition in supercritical carbon dioxide: towards the formation of functional microspheres, J. Supercrit. Fluids 53(1) (2010) 151–155.)*

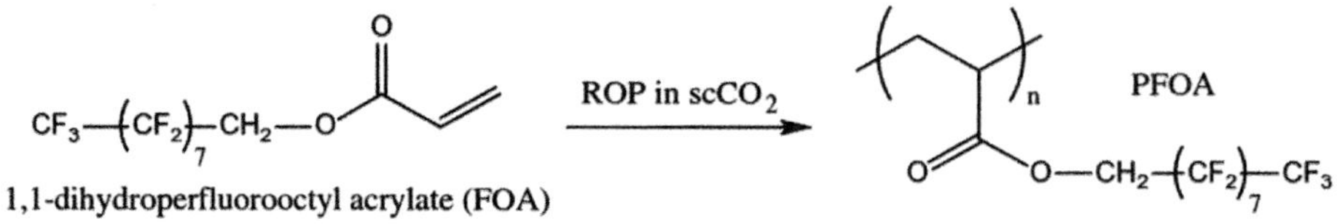

Fig. 5 Homogeneous polymerization PFOA from FOA monomer using SCO2. *(Adapted from C. Boyère, C. Jérôme, A. Debuigne, Input of supercritical carbon dioxide to polymer synthesis: an overview, Eur. Polym. J. 61 (2014) 45–63.)*

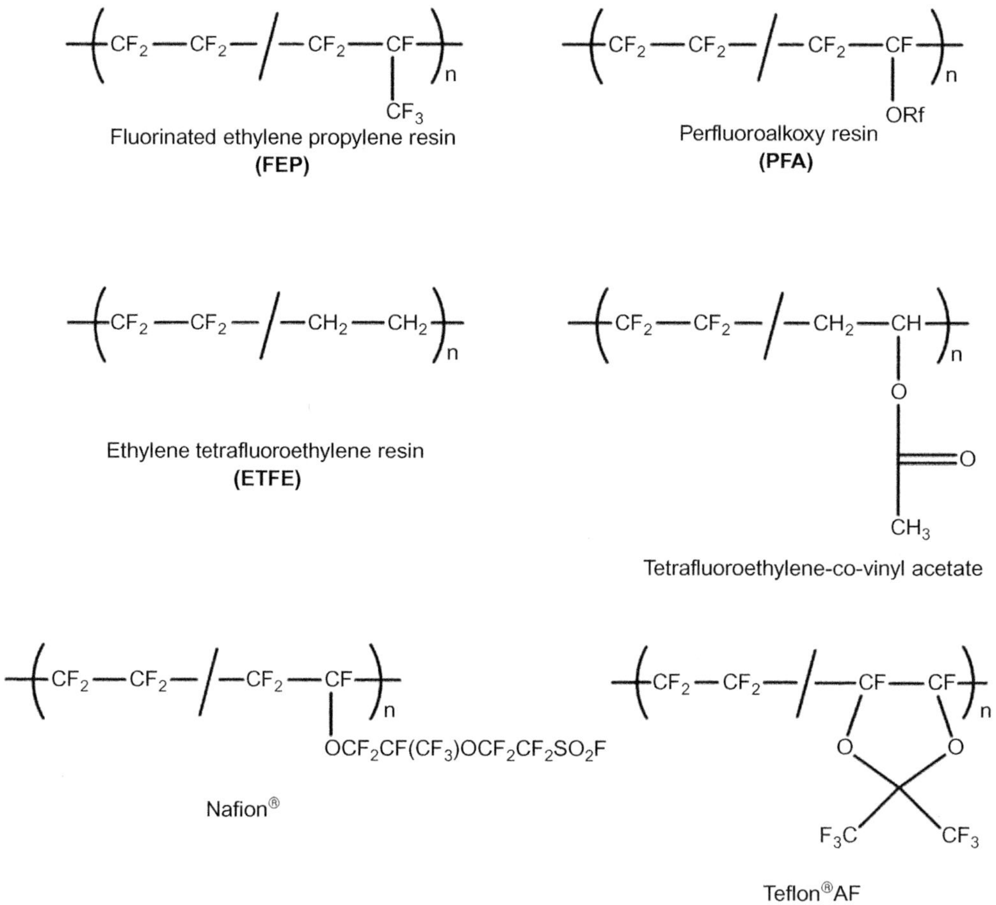

Fig. 6 Different Tetrafluoroethylene-based fluoropolymers synthesized by using carbon dioxide *(Adapted from C.D. Wood, A.I. Cooper, J.M. DeSimone, Green synthesis of polymers using supercritical carbon dioxide, Curr. Opin. Solid State Mater. Sci. 8(5) (2004) 325–331.)*

However, TFE and SCO_2 based fluoropolymers produce high molecular weight polymers due to the presence of much less acidic end groups. Hence, it becomes necessary to add further chain transfer agents, which will decrease molecular weight and optimize melt processability of the polymer [36]. SCO_2 plays a dual role here by facilitating the monomer into the polymer phase (because of the plasticizing property of CO_2) and favoring cross propagation due to decreased temperature [34, 37].

4.2.2 Precipitation polymerization

As such, vinyl monomers are readily soluble in CO_2 but their respective polymers after synthesis precipitate out from the solution. The use of SCO_2 here signifies that upon simple depressurization of the reactor, the precipitated polymers will crystallize out in dry form with no solvent leftover behind [38].

Semi-crystalline fluoropolymers can be treated using the method of precipitation polymerization, which yields high molecular weight polymers. By using TFE and perfluoro (propyl vinyl ether) copolymers, high molecular weight polymers were produced and confirmed by using FT-IR analysis where the role of SCO_2 was also to eliminate the unwanted end groups [8, 39]. Polymers that are thermoresponsive were also synthesized by using N-isopropylacrylamide (NIPAM), which reacted with acrylic acid (AA). These were subjected to hydration-dehydration behavior to analyze the temperature response [40]. Similar work was done by using 2-hydroxyethyl methacrylate (HEMA) [41] and vinylidene fluoride (VDF) [42]. Insoluble aliphatic polyesters were produced from feedstocks of non-petrochemical compounds by ring opening polymerization (ROP) of lactones [43]. The reaction resulted in deleterious environmental hazards in the presence of Lewis acid catalyst. Therefore, the ROP was performed using CO_2. The aliphatic polyesters were insoluble in CO_2. Hence, the latter was polymerized using precipitation polymerization (Fig. 7) [8, 44].

4.2.3 Dispersion polymerization

In dispersion polymerization, the polymer is obtained by polymerizing the monomer and initiator into CO_2 by using a stabilizer, which tends to stabilize the dispersed polymer

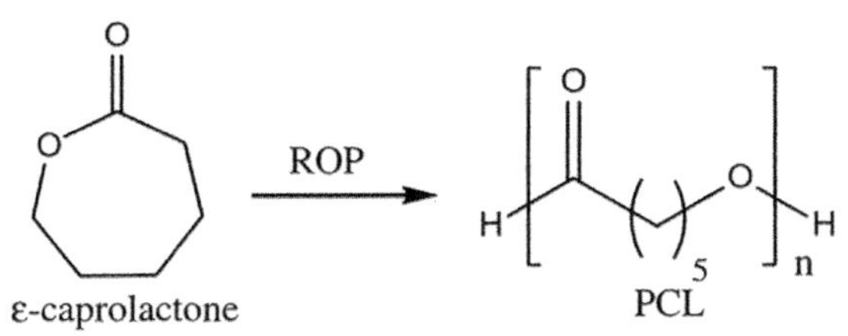

Fig. 7 ROP of e-caprolactone acting as monomer and its corresponding polycaprolactone (PCL). *(Adapted from C. Boyère, C. Jérôme, A. Debuigne, Input of supercritical carbon dioxide to polymer synthesis: an overview, Eur. Polym. J. 61 (2014) 45–63.)*

particles while there is a change in the reaction medium from homogeneous to heterogeneous [45]. The stabilizer acts by inducing steric repulsion between the growing particles. However, nowadays the research has shifted towards non-fluorinated steric stabilizer, which is more beneficial in polymerization involving dispersion using SCO_2. Recently, hydrocarbon block copolymers were employed as stabilizers but, due to lack of solubility in CO_2, resulted in agglomerated polymer particles [46]. Interestingly, some polyvinyl esters like polyvinyl acetate and polyvinyl pivalate were a better replacement for the conventional fluorinated surfactants [47–50].

4.2.4 Suspension polymerization

In opposition to dispersion polymerization, here these form a heterogeneous media where neither monomer nor polymer are soluble in the media. The monomer forms drops when the initiator is added to it with the aid of mechanical stirring. Suspension polymerization yielded submicron-sized water-soluble polymers by N-ethyl acrylamide. The size of the polymer particles varies widely, depending upon monomer concentration, stabilizer, initiator, and, more importantly, upon varying the pressure and temperature [8, 51]. The poly(3-O-methacryloyl-D-glucopyranose)-b-PFOMA[a] is used as an emulsifier for the suspension polymerization, which is an amphiphilic di-block copolymer [52].

Recently, various polymers such as poly(acrylic acid) (PAA) [53], poly(lactide-co-dioxanone) (PLDO) [41], and copolymers of N-isopropylacrylamide (NIPA) [54] were also synthesized by suspension polymerization in SCO_2.

4.2.5 Emulsion polymerization

Emulsion polymerization deals with biphasic liquids/organic solvent mixtures containing stabilizers, which will modify the structure of polymers after polymerization. This method of polymerization yields high molecular weight polymers because these generally do not depend upon the viscosity; rather, the elongating chains remain within the particles at low concentrations. The stable emulsion can be formed by using surfactants having CO_2-phillic and CO_2-phobic parts attached, which will determine the nature of emulsion formed, i.e., CO_2-in-water or water-in-CO_2 type emulsion. If we have hydrophilic monomers undergoing water-in-CO_2 (W/C) polymerization, then water-soluble polymers will form and vice versa.

In the case of a W/C system, surfactant tends to form micelle with a hydrophilic monomer, which upon polymerization becomes solubilized within the aqueous phase and results in suspended particles within SCO_2 (Fig. 8).

[a] Poly(3-O-methacryloyl-D-glucopyranose)-b-1,1-dihydroperfluorooctyl methacrylate.

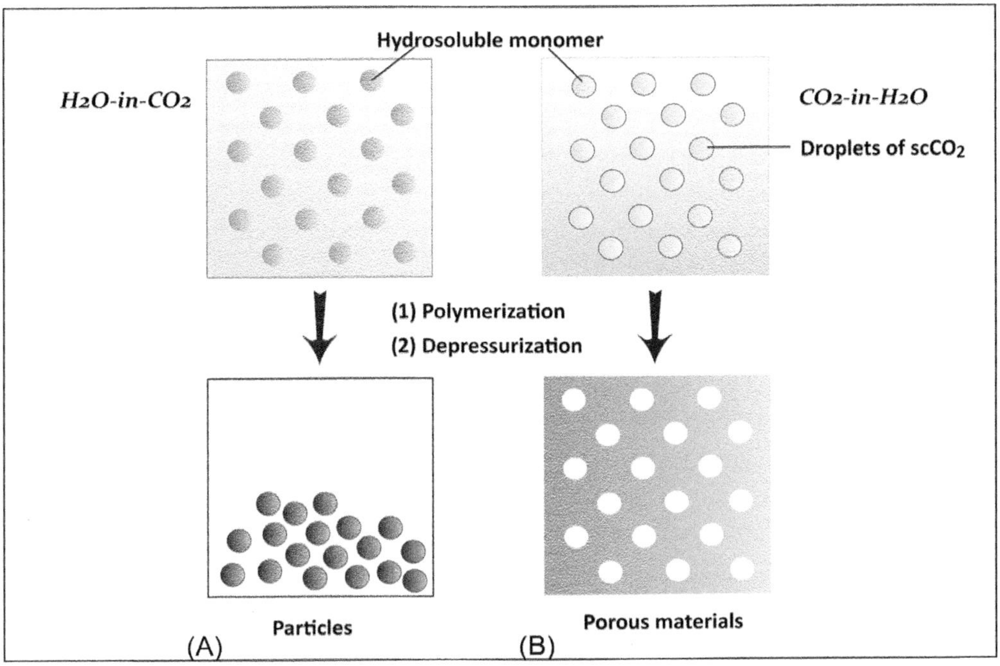

Fig. 8 Graphical illustration of heterogeneous microemulsion polymerization technique using (A) W/C and (B) CO_2-in-water (C/W) systems, which results in formation of solid particles and porous material, respectively. *(Adapted and modified from C. Boyère, C. Jérôme, A. Debuigne, Input of supercritical carbon dioxide to polymer synthesis: an overview, Eur. Polym. J. 61 (2014) 45–63.)*

5. Polymer processing

5.1 Plasticization of polymers

Supercritical CO_2 has the potential to lower the glass transition temperature (Tg) of glassy polymers. These polymers tend to absorb supercritical CO_2 and henceforth swell and reform their physical and mechanical strength. The interaction of a functional group of polymers and supercritical SCO_2 tend to decrease the chain-chain interaction and elevate the mobility of polymers [55, 56]. Plasticization of polymers using supercritical SCO_2 can be measured by different methods such as gas sorption, permeability [57, 58], and polymer swelling [59, 60].

5.2 Viscosity reduction

High molecular weight polymers are difficult to process due to the high viscosity solvents used during the processing. However, viscosity can be reduced if we apply a high temperature to the system, as we know that by increasing the temperature, due to the increase in kinetic energy, molecular vibrations increases, and the viscosity decreases. Indeed, the increase in temperature can result in polymer degradation. Therefore, a suitable solvent such as SCO_2 can be used to reduce the viscosity at low temperature [61]. SCO_2 has a good plasticization effect on polymers. CO_2 is readily solubilized in polymers, decreasing their Tg, and thus the decrease in the melting point of polymers also reduces the viscosity.

Due to the elongation of polymer molecules during synthesis, CO_2 could not escape from the fibers; hence, after solidification, in addition to reduced viscosity, we get a low-density polymer [3].

5.3 Microcellular foam

A microcellular foamed polymer consists of a cell diameter of fewer than 10 μm and a cell density of more than 100 cells/cm^3 [17]. Microcellular foam can be formed by a single-phase solution where after depressurizing the SCO_2, supersaturation occurs, leading to nucleation of cells. Drug entrapment within the polymer foam using supercritical CO_2 has become an interesting topic for research. In a recent investigation, hydrophilic drug gentamicin and hydrophobic drug curcumin have been loaded as microcellular foam using poly(lactic-*co*-glycolic acid) (PLGA) polymer with the aid of SCO_2. The entrapment efficiency was found to be 75% by using the drug: a polymer ratio of 75:25 [62]. The foaming process is generally completed in two basic steps: (a) dissolution of CO_2 in the polymeric matrix under an optimum pressure forming a polymer/gas solution; and (b) nucleation and growth of the particles upon a decrease in pressure or an increase in temperature [12].

5.4 Polymer blending

For processing and blending, interactions associated with the supercritical media and the functional group of polymers have to be kept in mind. Their ratio of the viscosities will determine the size of the phases. The supercritical media behaves differently with every component of the blend because of their varying molecular structures, which will have a different effect on the Tg of each element. Therefore, the plasticizing effect of each component will be different. During extrusion using supercritical fluid, the solubility of CO_2 plays an important role upon varying temperature, pressure, and shear rate. After employing SCO_2, the shear thinning is decreased and a fine dispersion is obtained. Therefore, one can easily manipulate the morphology of polymer blend and tune its size easily [63]. By blending the polymer with initiator and SCO_2, a polymer composite can also be formed.

6. Future prospects

During the manufacture of fluoropolymers, SCO_2 shows increased safety as well as polymer properties. Therefore, these can find application during handling of explosive monomers. SCO_2 can be found advantageous when considering the drawbacks associated with the scaffold fabrication method. These drawbacks not only can manipulate scaffold porosity but can also be used to make biocomposites and preserve protein activity. As discussed above, CO_2 not only is a better alternative for polymer production, but in many cases is the most needed one. A wide range of applications associated with SCO_2 suggest it to be a promising solvent, but further investigations related to the interaction of the polymer and SCO_2 system are required.

7. Challenges ahead

A large volume of SCO_2 is needed during the encapsulation of drugs into polymeric hosts which proved to be limited. Another challenge is the high solubility of monomers required for polymer processing and synthesis. Highly SCO_2 soluble polymers show significant polymer processing ability. Overall, it has become clear that SCO_2 has moved ahead from scientific curiosity, but it still needs to be more commercialized concerning polymer processing. A key challenge is to make a large number of drugs to be processed in SCO_2. However, using SCO_2 as an antisolvent with other additional solvents overcomes this problem. Hence, SCO_2 will play an essential role in future drug delivery applications.

8. Conclusion

SCO_2 is an exciting substitute for various organic/non-polar solvents in multiple applications such as extraction, purification, impregnation of polymers, and supercritical dyeing. It can also be used in different polymerization techniques including homogeneous polymerization, dispersion, suspension, precipitation, and emulsion polymerizations. Various processing strategies have emerged that depend on supercritical CO_2 because of its plasticizing and antisolvent properties. In summary, supercritical fluids offer a tremendous opportunity in research associated with polymers, and provide great scope for development of more promising and sustainable technologies to the polymeric industry.

References

[1] M. Champeau, J.M. Thomassin, T. Tassaing, C. Jerome, Drug loading of polymer implants by supercritical CO2 assisted impregnation: a review, J. Control. Release 209 (2015) 248–259.

[2] A. Aguiar-Ricardo, V.D. Bonifacio, Supercritical carbon dioxide design strategies: from drug carriers to soft killers, Philos. Trans. A Math. Phys. Eng. Sci. 373 (2057) (2015).

[3] S.P. Nalawade, F. Picchioni, L.P.B.M. Janssen, Supercritical carbon dioxide as a green solvent for processing polymer melts: processing aspects and applications, Prog. Polym. Sci. 31 (1) (2006) 19–43.

[4] A.I. Cooper, S.G. Kazarian, M. Poliakoff, Supercritical fluid impregnation of polyethylene films, a new approach to studying equilibria in matrices; the hydrogen bonding of fluoroalcohols to (η5-C5Me5)Ir (CO)2 and the effect on C-H activation, Chem. Phys. Lett. 206 (1) (1993) 175–180.

[5] O.R. Davies, A.L. Lewis, M.J. Whitaker, H. Tai, K.M. Shakesheff, S.M. Howdle, Applications of supercritical CO2 in the fabrication of polymer systems for drug delivery and tissue engineering, Adv. Drug Deliv. Rev. 60 (3) (2008) 373–387.
[6] J. Zhou, Application of supercritical fluid in nanolithographic processes, Recent Pat. Nanotechnol. 4 (2) (2010) 78–84.
[7] K. MF, Supercritical carbon dioxide for sustainable polymer processes, in: M.T. Kemmere MF (Ed.), Supercritical Carbon Dioxide: in Polymer Reaction Engineering, FRG: Wiley-VCH Verlag GmbH & Co, Weinheim, 2006.
[8] C. Boyère, C. Jérôme, A. Debuigne, Input of supercritical carbon dioxide to polymer synthesis: an overview, Eur. Polym. J. 61 (2014) 45–63.
[9] P. Sheth, H. Sandhu, D. Singhal, W. Malick, N. Shah, M.S. Kislalioglu, Nanoparticles in the pharmaceutical industry and the use of supercritical fluid technologies for nanoparticle production, Curr. Drug Deliv. 9 (3) (2012) 269–284.
[10] M. Bhamidipati, A.M. Scurto, M.S. Detamore, The future of carbon dioxide for polymer processing in tissue engineering, Tissue Eng. Part B Rev. 19 (3) (2013) 221–232.
[11] K. Khosravi-Darani, E. Vasheghani-Farahani, Application of supercritical fluid extraction in biotechnology, Crit. Rev. Biotechnol. 25 (4) (2005) 231–242.
[12] E. Di Maio, E. Kiran, Foaming of polymers with supercritical fluids and perspectives on the current knowledge gaps and challenges, J. Supercrit. Fluids 134 (2018) 157–166.
[13] E. Kiran, Supercritical fluids and polymers—The year in review—2014, J. Supercrit. Fluids 110 (2016) 126–153.
[14] G. Brunner, Calculation of phase equilibria and their relation to separation with supercritical fluids, J. Supercrit. Fluids 134 (2018) 2–11.
[15] Y. Hiraga, A. Duereh, R.L. Smith, Aspects of solvent polarity and solvent properties in developing efficient systems for processing biomass with ionic liquid mixtures and supercritical CO2, J. Supercrit. Fluids 134 (2018) 12–20.
[16] M. Tang, T.-Y. Wen, T.-B. Du, Y.-P. Chen, Synthesis of electrically conductive polypyrrole–polystyrene composites using supercritical carbon dioxide: II. Effects of the doping conditions, Eur. Polym. J. 39 (1) (2003) 151–156.
[17] C.L. Higginbotham, J.G.L. Yons, J.E. Kennedy, 13-Polymer processing using supercritical fluids, in: S. Thomas, Y. Weimin (Eds.), Advances in Polymer Processing, Woodhead Publishing, 2009, pp. 384–401.
[18] M. Miya, R. Iwamoto, S. Mima, FT-IR study of intermolecular interactions in polymer blends, J. Polym. Sci. Polym. Phys. Ed. 22 (1984) 1149–1151.
[19] J. von Schnitzler, R. Eggers, Mass transfer in polymers in a supercritical CO2-atmosphere, J. Supercrit. Fluids 16 (1) (1999) 81–92.
[20] T.J. Yoon, Y.-W. Lee, Current theoretical opinions and perspectives on the fundamental description of supercritical fluids, J. Supercrit. Fluids 134 (2018) 21–27.
[21] E. Weidner, M. Petermann, K. Blatter, V. Rekowski, Manufacture of powder coatings by spraying of gas-enriched melts, Chem. Eng. Technol. 24 (5) (2001) 529–533.
[22] G.Y. Friedmann, J.M. Catala, Modification chimique de polymeres en milieu CO2 supercritique Greffage de groups isocyanate-isopropyle sur une chaine de poly(ethylene-coalcool vinylique), Eur. Polym. J. 36 (2000) 13–20.
[23] A. Alexander, J. Ajazuddin, S. Khan, S. Saraf, S. Saraf, Poly(ethylene glycol)-poly(lactic-co-glycolic acid) based thermosensitive injectable hydrogels for biomedical applications, J. Control. Release 172 (3) (2013) 715–729.
[24] O. Ihata, Y. Kayaki, T. Ikariya, Synthesis of Thermoresponsive polyurethane from 2-Methylaziridine and supercritical carbon dioxide, Angew. Chem. Int. Ed. 43 (6) (2004) 717–719.
[25] B. Tan, H.M. Woods, P. Licence, S.M. Howdle, A.I. Cooper, Synthesis and CO2 solubility studies of poly(ether carbonate)s and poly(ether ester)s produced by step growth polymerization, Macromolecules 38 (5) (2005) 1691–1698.
[26] S.M. Gross, G.W. Roberts, D.J. Kiserow, J.M. DeSimone, Crystallization and solid-state polymerization of poly(bisphenol a carbonate) facilitated by supercritical CO2, Macromolecules 33 (1) (2000) 40–45.

[27] P. Chambon, E. Cloutet, H. Cramail, T. Tassaing, M. Besnard, Synthesis of core-shell polyurethane–polydimethylsiloxane particles in cyclohexane and in supercritical carbon dioxide used as dispersant media: a comparative investigation, Polymer 46 (4) (2005) 1057–1066.
[28] P.B. Zetterlund, F. Aldabbagh, M. Okubo, Controlled/living heterogeneous radical polymerization in supercritical carbon dioxide, J. Polym. Sci. A Polym. Chem. 47 (15) (2009) 3711–3728.
[29] B. Grignard, C. Calberg, C. Jérôme, W. Wang, S. Howdle, C. Detrembleur, Supported ATRP of fluorinated methacrylates in supercritical carbon dioxide: preparation of $scCO_2$ soluble polymers with low catalytic residues, Chem. Commun. (44) (2008) 5803–5805.
[30] A. Alexander, M. Ajazuddin, M. Swarna, M. Sharma, D. Tripathi, Polymers and permeation enhancers: Specialized components of mucoadhesives, Stamford J. Pharmaceut. Sci. 4 (1) (2011) 91–95.
[31] B. Grignard, C. Calberg, C. Jerome, C. Detrembleur, "One-pot" dispersion ATRP and alkyne-azide Huisgen's 1,3-dipolar cycloaddition in supercritical carbon dioxide: towards the formation of functional microspheres, J. Supercrit. Fluids 53 (1) (2010) 151–155.
[32] A. Alexander, J. Ajazuddin, S. Khan, S. Saraf, S. Saraf, Formulation and evaluation of chitosan-based long-acting injectable hydrogel for PEGylated melphalan conjugate, J. Pharm. Pharmacol. 66 (9) (2014) 1240–1250.
[33] J.M. DeSimone, Z. Guan, C.S. Elsbernd, Synthesis of Fluoropolymers in supercritical carbon dioxide, Science (New York, N.Y.) 257 (5072) (1992) 945–947.
[34] C.D. Wood, A.I. Cooper, J.M. DeSimone, Green synthesis of polymers using supercritical carbon dioxide, Curr. Opin. Solid State Mater. Sci. 8 (5) (2004) 325–331.
[35] A. Alexander, J. Ajazuddin, S. Khan, S. Saraf, S. Saraf, Polyethylene glycol (PEG)-poly(N-isopropylacrylamide) (PNIPAAm) based thermosensitive injectable hydrogels for biomedical applications, Eur. J. Pharm. Biopharm. 88 (3) (2014) 575–585.
[36] K. Takahashi, Polymer analysis by supercritical fluid chromatography, J. Biosci. Bioeng. 116 (2) (2013) 133–140.
[37] A.I. Cooper, Polymer synthesis and processing using supercritical carbon dioxide, J. Mater. Chem. 10 (2) (2000) 207–234.
[38] A. Alexander, S. Saraf, S. Saraf, Understanding the role of Poloxamer 407 based Thermoreversible in situ gelling hydrogel for delivery of PEGylated Melphalan conjugate, Curr. Drug Deliv. 13 (4) (2016) 621–630.
[39] T.J. Romack, J.M. DeSimone, T.A. Treat, Synthesis of Tetrafluoroethylene-based, nonaqueous Fluoropolymers in supercritical carbon dioxide, Macromolecules 28 (24) (1995) 8429–8431.
[40] L. Cao, L. Chen, W. Lai, Precipitation copolymerization of N-isopropylacrylamide and acrylic acid in supercritical carbon dioxide, J. Polym. Sci. A Polym. Chem. 45 (5) (2007) 955–962.
[41] X.-H. Wang, L.-Q. Cao, L.-J. Zhang, J.-D. Wang, Precipitation polymerization of 2-hydroxyethyl methacrylate in supercritical carbon dioxide, Polym. Adv. Technol. 23 (3) (2012) 529–533.
[42] J. Liu, H. Tai, S.M. Howdle, Precipitation polymerisation of vinylidene fluoride in supercritical CO2 and real-time calorimetric monitoring, Polymer 46 (5) (2005) 1467–1472.
[43] C. Jerome, P. Lecomte, Recent advances in the synthesis of aliphatic polyesters by ring-opening polymerization, Adv. Drug Deliv. Rev. 60 (9) (2008) 1056–1076.
[44] A.-F. Mingotaud, F. Cansell, N. Gilbert, A. Soum, Cationic and anionic ring-opening polymerization in supercritical CO2. Preliminary results, Polym. J. 31 (1999) 406.
[45] E.J. Park, A.P. Richez, N.A. Birkin, H. Lee, N. Arrowsmith, K.J. Thurecht, S.M. Howdle, New vinyl ester copolymers as stabilisers for dispersion polymerisation in $scCO_2$, Polymer 52 (24) (2011) 5403–5409.
[46] H. Lee, E. Terry, M. Zong, N. Arrowsmith, S. Perrier, K.J. Thurecht, S.M. Howdle, Successful dispersion polymerization in supercritical CO2 using Polyvinylalkylate hydrocarbon surfactants synthesized and anchored via RAFT, J. Am. Chem. Soc. 130 (37) (2008) 12242–12243.
[47] Y. Wang, L. Hong, D. Tapriyal, I.C. Kim, I.-H. Paik, J.M. Crosthwaite, A.D. Hamilton, M.C. Thies, E.J. Beckman, R.M. Enick, J.K. Johnson, Design and evaluation of Nonfluorous CO2-soluble oligomers and polymers, J. Phys. Chem. B 113 (45) (2009) 14971–14980.
[48] E. Girard, T. Tassaing, S. Camy, J.-S. Condoret, J.-D. Marty, M. Destarac, Enhancement of poly(vinyl ester) solubility in supercritical CO2 by partial fluorination: The key role of polymer–polymer interactions, J. Am. Chem. Soc. 134 (29) (2012) 11920–11923.

[49] M. Altarsha, F. Ingrosso, M.F. Ruiz-Lopez, A new glimpse into the CO2-Philicity of carbonyl compounds, ChemPhysChem 13 (14) (2012) 3397–3403.
[50] B. Tan, J.-Y. Lee, A.I. Cooper, Ionic hydrocarbon surfactants for emulsification and dispersion polymerization in supercritical CO2, Macromolecules 39 (22) (2006) 7471–7473.
[51] W. Ye, J.M. DeSimone, Emulsion polymerization of N-Ethylacrylamide in supercritical carbon dioxide, Macromolecules 38 (6) (2005) 2180–2190.
[52] W. Ye, J.M. DeSimone, Synthesis of sugar-containing Amphiphiles for liquid and supercritical carbon dioxide, Ind. Eng. Chem. Res. 39 (12) (2000) 4564–4566.
[53] Y.A. Hussain, T. Liu, G.W. Roberts, Synthesis of cross-linked, partially neutralized poly(acrylic acid) by suspension polymerization in supercritical carbon dioxide, Ind. Eng. Chem. Res. 51 (35) (2012) 11401–11408.
[54] P. O'Connor, R. Yang, W.M. Carroll, Y. Rochev, F. Aldabbagh, Facile synthesis of thermoresponsive block copolymers of N-isopropylacrylamide using heterogeneous controlled/living nitroxide-mediated polymerizations in supercritical carbon dioxide, Eur. Polym. J. 48 (7) (2012) 1279–1288.
[55] S.G. Kazarian, M.F. Vincent, F.V. Bright, C.L. Liotta, C.A. Eckert, Specific intermolecular interaction of carbon dioxide with polymers, J. Am. Chem. Soc. 118 (7) (1996) 1729–1736.
[56] S.G. Kazarian, K.L. Chan, Micro- and macro-attenuated total reflection Fourier transform infrared spectroscopic imaging, Plenary Lecture at the 5th International Conference on Advanced Vibrational Spectroscopy, 2009, Melbourne, Australia, Appl. Spectrosc. 64 (5) (2010) 135a–152a.
[57] E.S. Sanders, S.M. Jordan, R. Subramanian, Penetrant-plasticized permeation in polymethylmethacrylate, J. Membr. Sci. 74 (1) (1992) 29–36.
[58] J.-S. Wang, Y. Naito, Y. Kamiya, Effect of penetrant-induced isothermal glass transition on sorption, dilation, and diffusion behavior of polybenzylmethacrylate/CO2, J Polym Sci B 34 (12) (1996) 2027–2033.
[59] R.G. Wissinger, M.E. Paulaitis, Glass transitions in polymer/CO2 mixtures at elevated pressures, J Polym Sci B 29 (5) (1991) 631–633.
[60] Y. Han, H. Zheng, X. Jing, L. Zheng, Swelling behavior of polyester in supercritical carbon dioxide, J. CO2 Utiliz. 26 (2018) 45–51.
[61] L.J. Gerhardt, C.W. Manke, E. Gulari, Rheology of polydimethylsiloxane swollen with supercritical carbon dioxide, J Polym Sci B 35 (3) (1997) 523–534.
[62] Y.X.J. Ong, L.Y. Lee, P. Davoodi, C.-H. Wang, Production of drug-releasing biodegradable microporous scaffold using a two-step micro-encapsulation/supercritical foaming process, J. Supercrit. Fluids 133 (2018) 263–269.
[63] S.G. Kazarian, Applications of FTIR spectroscopy to supercritical fluid drying, extraction and impregnation, Appl. Spectrosc. Rev. 32 (4) (1997) 301–348.

CHAPTER 2

Extraction of lipids from algae using supercritical carbon dioxide

Alok Patel*, Leonidas Matsakas*, Km Sartaj†, Rajesh Chandra‡
*Biochemical Process Engineering, Division of Chemical Engineering, Department of Civil, Environmental and Natural Resources Engineering, Luleå University of Technology, Luleå, Sweden
†Molecular Microbiology Laboratory, Biotechnology Department, Indian Institute of Technology (IIT-R), Roorkee, India
‡Bioenergy Research Laboratory, Department of Polymer & Process Engineering, Indian Institute of Technology Roorkee (Saharanpur Campus), Saharanpur, India

Contents

1. Introduction

The rapid increase in the world's population and developing new technologies has resulted in high energy demands, along with rapidly declining of current energy resource reserves [1–3]. Moreover, sky-rocketing development, including industrialization, urbanization, and our modern way of living, makes the transportation sector a field of major energy demands, which lead to long-term unavoidable deprivation in the availability of petroleum products [3, 4]. According to the current scenario,

Green Sustainable Process for Chemical and Environmental Engineering and Science
https://doi.org/10.1016/B978-0-12-817388-6.00002-7

environmental pollution and energy crises are predicted to be major challenges in the near future [5]. Therefore, the world's scientific communities and policymakers have focused on the development of alternative fuels derived from renewable sources so that the energy sector can overcome the problems associated with depleting reserves of crude oils [3, 6].

Although there are various forms of renewable sources, biomass derived from plant materials or biological matters, including both flora and fauna, are gaining more interest day by day and are recognized as a major world renewable energy source [7]. In general, lignocellulosic biomass is composed of cellulose (linear polymer of glucopyranose), hemicelluloses (a polymerized form of glucose, mannose, and xylose), lignin (amorphous phenolic polymer), and minute amount of other organics depending on the plant species [8, 9]. Lignocellulosic biomass can be converted into various forms of energy, such as liquid fuels including bio-oils, which can serve as a promising candidate to replace petroleum fuels [3, 8, 10].

Pyrolysis is one way in which biomass can be converted directly into bio-oils, and even though it can be produced by all types of available biomasses such as woods, agricultural residues, organic wastes, aquatic plants, and algae. Pyrolysis oil, biofuel oil, pyrolytic oil, and liquid wood are different synonyms of bio-oils [3]. However, the bio-oil produced by pyrolysis is totally different from the microbial oils. Microalgae showed their potential to produce microbial oil in a sustainable manner due to their fast growth rate in comparison to terrestrial photosynthetic plants [11]. Microalgal cells usually have a short doubling time of just 24 h; however, this time reduces to 3.5 h at the peak of exponential growth [11]. Moreover, microalgae have various other advantages including the ability to grow in saline water and exploit low-quality agricultural land, use of carbon dioxide as an inorganic carbon source, mitigation of greenhouse gases, higher photosynthesis efficiency, being less water-dependent than plants, wide range of tolerance to environmental stresses, and tolerance to shear force with high biomass productivity. All these factors make microalgae a suitable option for microbial oil production [12–20].

Along with these advantages, the most important factor which makes microalgae more promising for oil production is having no overlap with food supply, which completely overcomes the "food versus fuel" debate [21–24]. Microalgae is characterized as an extraordinary specific group of photosynthetic microorganisms, which consist mainly (90% of dry weight) of carbohydrate, protein, and oil [17], and in terms of morphology, microalgae come under the extremely diverse group from unicellular to multicellular organisms like plants [25]. Unicellular algae consist of chloroplast for photosynthetic reactions, pyrenoids for energy storage, and contractile vacuoles for osmotic balance [25]. Some unique features of high growth rate and accumulation of the extensive amount of lipids make microalgae a very promising alternative source of energy, which has become one of the hottest topics for the world scientific community working on renewable resources for oil production [26, 27]. Due to their photosynthetic activity, microalgae confine solar energy and fix atmospheric CO_2 for their growth and produce a variety of nutritional and bioactive compounds such as phycobiliproteins, carotenoids, fatty acids, and natural oxidants [16, 27].

Microalgal oils can be processed into biodiesel through a series of upstream and downstream processes including growth on various inorganic or organic carbon sources, harvesting of cultures, drying of biomass or processing wet biomass, pretreatment of biomass for the disruption of cells, lipid extraction by different methods, purification of lipids (triacylglycerols), and finally transesterification to convert lipids into ethyl or methyl esters forms of which biodiesel consists [24]. Lipid extraction from oleaginous microorganisms involves high cost and labor-intensive processes that hinder commercialization of biodiesel production [28]. The lipids are synthesized intracellularly, which creates a problem in the downstream process in lab- or large-scale production [29]. The cellular integrity must be interrupted for the enhanced lipid extraction, which can be proceeded by various pretreatment techniques of cellular biomass followed by organic solvent extraction of lipids with the lysed biomass [30]. However, it is not easy to disintegrate the cellular structure in wet conditions; a prior dewatering/drying step is required, which is again an energy-intensive and expensive procedure that makes this task difficult on a large scale [31]. Bligh & Dyer and Folch methods for lipid extraction are considered conventional methods that only work at lab scale using organic solvents such as chloroform and methanol [32]. Different issues related to these techniques should likewise be considered for the improvement of cell disruption [32, 33].

At present, different mechanical, chemical, and enzymatic pretreatment methods such as high-speed homogenization, microwave, ultrasonication, bead beating, osmolysis, and oxidative destruction are utilized to disintegrate the cells of oleaginous microorganisms in laboratories [34, 35]. However, the efficiency of all these methods are not yet reported on an industrial scale [36]. Some common methods, such as expeller press and bead beating, are applied to break the cellular structure of microorganisms before solvent extraction on an industrial scale; however, the microbial biomass should be free from moisture. The organic solvent cannot be applied to wet biomass for the lipid extraction purpose because the presence of moisture or water creates surface charges on the cells that interfere in the establishment of the contact between cells and organic solvents [37].

Drying of biomass is an expensive step for industrial-scale lipid production, hence, it is necessary to develop a technique appropriate for lipid extraction from wet biomass [38]. Moreover, lipid extraction, along with other techniques, also has some disadvantages regarding efficiency and energy intensiveness. The most effective alternative of solvent extraction is supercritical fluid extraction, where supercritical CO_2 is used as a primary solvent which is considered as non-toxic and fireproof solvent that makes a certain extent of adjustable selectivity [39–42]. Its operating conditions are very flexible to operate in low temperature (30–40°C), making it appropriate for thermosensitive lipids, while its moderate pressure (72.9 bar) does not increase the compression cost [43–45]. CO_2 has been a very promising fluid in supercritical fluid extraction due to its low toxic, nonflammable, and unreactive nature [46, 47].

2. Lipid accumulation in microalgae

Oleaginous microalgae are considered promising microorganisms as they can utilize both organic and inorganic carbon sources through four different modes of nutrition: autotrophic, mixotrophic, heterotrophic, and photoheterotrophic [17, 36, 48, 49]. Sub-cellular compartments in microalgae are the main organelles where synthesis of triacylglycerols takes place as a result of multiple enzymatic reactions [50]. The three major steps of lipid synthesis in oleaginous microalgae involve the fatty acid accumulation in chloroplasts, the assembly of glycerolipids in the endoplasmic reticulum, and finally, the deposition of triacylglycerols in the form of lipid droplets in the cellular compartment of microalgae [51].

It has been reported that various types of physical, chemical, or physicochemical stresses to the oleaginous microorganisms support the accumulation of large amounts of lipid droplets inside the cells [52]. Under stress conditions with excess amounts of carbon source, algae alter their biosynthetic pathway towards the accumulation of neutral lipids in the form of triacylglycerol, which serves as a storage form of carbon and energy [26, 53]. Microalgae are well known for their characteristic de novo lipid accumulating feature, which involves a series of events starting from the chloroplast, in which CO_2 fixes into sugars, which are further processed to form acetyl CoA, a precursor component for fatty acid synthesis. A number of enzymatic reactions and multiple pathways participate to maintain the pool of acetyl CoA [16, 54]. Fatty acid synthesis also requires a high amount of energy and reducing power in form of ATP, NADH, and NADPH [52, 55, 56]. Photosynthetic reaction plays a characteristic role in the formation of reducing power (NADH and NADPH) as well as the fixation of inorganic carbon source [57]. Previous studies on fatty acid synthesis in microalgae show that the final acyl chain length in most of the algae at the time of coming out from chloroplast varies from 16 to 18 carbon.

Finally, the formation triacylglycerol occurs in the endoplasmic reticulum with the help of a group of enzymes acyltransferases that catalyze the consecutive acylation of synglycerol-3-phosphate (G3P) backbone with three acyl-CoA [58]. Synthesized triacylglycerol forms oil droplets, also known as oleosomes, lipid droplets, and oil bodies, which are generally separated from the endoplasmic reticulum membrane and considered as distinct organelles, which are dispersed in the cytoplasm by a single layer of phospholipids with hydrophilic head groups on the surface [57]. The lipid droplets are usually composed of an inner lipid core surrounded by a protein-coated lipid monolayer [59, 60].

3. Existing methods for lipid extraction from microalgae

Sustainable production of microbial oil entirely depends upon efficient extraction of lipids from microalgal cells [61]. The presence of thick and robust cell walls in microalgae is a major drawback, making lipids recovery challenging and more complicated, which in turn restricts the uses of microalgae biomass as a raw material on an industrial scale due to the high cost and energy demands [19]. To overcome these challenges, there are plenty of methods for lipid extraction which make the process easier and can meet the essential requirements of being eco-friendly, less time consuming, energy efficient, cost-effective, environmentally friendly, and efficient for industrial-scale production [62, 63].

Lipid extraction from oleaginous microorganisms can be carried out by mechanical, chemical, thermal, electromagnetic, or biological methods, which have their own advantages and disadvantages [64]. Apart from this, there are two broad routes in which processing of dry and wet cell biomass for lipid extraction occurs [65, 66]. Compared to dry biomass methods, the wet route seems to be a better way of lipid extraction due to lower cost and energy demands, which can also mitigate the toxicity arising from the use of organic solvents, making the process more feasible on an industrial scale as a result of eliminating the drying process prior to lipid extraction [26, 67].

3.1 Folch and Bligh & Dyer extraction methods

The Folch method and the Bligh & Dyer method are the oldest and most widely practiced techniques for lipid extraction and are based on solvent extraction by using chloroform-methanol mixtures (2:1 by volume) [68, 69]. The Folch method can process a large number of samples within a short time; however, its low sensitivity over the latest technologies is a major disadvantage, whereas the simplicity of Bligh & Dyer method over the Folch method makes it more applicable for pilot and large-scale extraction processes [29].

3.2 Superior solvents extraction method

Although Folch and Bligh & Dyer methods are widely practiced, low lipid recovery, together with the use of toxic chemicals such as chloroform, methanol, and hexane, which affect the environment and human health adversely, is a major drawback. Therefore, the use of less toxic and less effective solvents or a mixture of solvents would be a better option for extraction purposes. As a solvent, 2-ethoxyethanol (2-EE) is very effective and considered environmentally safer in comparison to traditional solvents [29]. Mixtures of polar and non-polar solvents can be a good option to enhance lipid yield as well as FAME recovery by up to 50%. Microalgae accumulate lipids in the form of TAGs that are non-polar compounds, hence the mixture of polar and non-polar solvents can positively affect the extraction and lipid recovery [70]. In a mixture, polar solvents are responsible for separating the lipids from their protein-lipid complexes and facilitate their dissolution in non-polar solvent [71]. Moreover, this strategy can be applied to wet microalgal biomass, in which polar solvent can cross the water barrier and facilitate the lipid extraction of the non-polar solvent [72, 73].

3.3 Expeller press and bead beating

Expeller press and bead beating are among the oldest mechanical techniques applied to enhance lipid extraction from microalgae [19, 74]. The expeller press uses mechanical crushing in an oil press, while bead beating relies on a grinding mechanism. Shaking vessels and agitated beads are types of beads that break cells by shaking the culture vessel [75–80]. High hardness and density of zirconia-silica, zirconium oxide, or titanium carbide beads improve the disruption rate and extraction efficiency [81, 82]. However, in an expeller press, special care should be taken when choosing the working pressure as under certain values the excretion of pigments alongside lipid extraction can be enhanced, resulting in an increase of the overall downstream processing cost [83]. On the downside, generation of high heat, low lipid recovery, and machine choking problems are the main demerits of the bead beating technique, which restricts its use to extraction processes [29, 84].

3.4 Microwave-assisted extraction (MAE)

Microwave-assisted extraction (MAE) offers a green, safe, rapid, and economical way of lipid extraction, which also reduces the cost associated with drying of microalgal biomass [33, 61, 76, 85, 86]. Cost-effectiveness, low energy requirement, rapid procedure, high lipid recovery, accompanied with purity and avoidance of utilization of hazardous substances, are the major advantages of this method [61, 87]. However, the maintenance cost on large-scale production is a significant issue with this technique [88]. Microwaves create an oscillating electric field that generates heat when in contact with a polar substance such as water, due to which cells are ruptured as a result of the formation of water vapor inside the cells; the effect is similar to the electroporation effect, which is responsible for leakages in the cell membrane and release of intracellular contents [33, 76, 85, 86, 89–92]. Implementation of MAE along with ultrasonication can be a good choice for lipid extraction from oleaginous microalgae (*Chlorella vulgaris*) and yeast (*Rhodosporidium kratochvilovae* HIMPA1) [87]. Factors like microalgal species, temperature, the power of microwave, solvent properties, and volume used are the key factors which determine the efficiency of microwave-assisted extraction [93].

3.5 Ultrasound-assisted extraction (UAE)

UAE is another simple, eco-friendly, and time efficient alternative to extract lipids from oleaginous microorganisms that overcome the problems associated with the conventional methods. It can be operated under mild temperatures and pressures and it does not require any chemicals. Cavitation and acoustic streaming are two different phenomena that are generated when ultrasound is applied to cells. Cavitation is the factor through which ultrasonication ruptures the cell walls and at the same time produces microbubbles in presence of liquid cultures [77, 94–96]. These bubbles eventually collapse and emit a shockwave, due to which cell walls are shattered and release the intracellular content in the corresponding medium [65, 89, 93, 97–99]. The mixture of polar and non-polar solvents, high frequencies, along with operation time can be optimized for enhanced lipid extraction [93, 100]. UAE has several advantages such as low operating temperature, low set-up costs, short processing time, and high purity of the final product, making this process useful for lipid extraction from microbial biomass [93].

3.6 Osmotic shock

This method works on the principle of osmotic balance by varying the salt concentration and causing hyper and hypo-osmotic conditions [73, 101]. Among them, hypo-osmotic conditions play a significant role in lipid removal from microalgae due to the high intracellular salt concentration, causing water or fluids to move into the cellular compartment which in turn results in cell swelling and bursting [102–104]. Osmotic shock treatment, using a mixture of polar and non-polar solvents on wet biomass of *Chlamydomonas reinhardtii,* showed two-fold increase in lipid recovery as compared to other processes [73, 104]. In other microalgae such as *Botryoccus* sp. and *Chlorella sorokiniana,* it also showed positive results [105]. Lipid extraction through this method is dependent on the cell wall properties and other species specificities, making this process more complicated, which limits its application [99, 106–108]. Although osmotic shock-assisted lipid extraction has been successfully used at lab scale; however, the feasibility of technology and pilot-scale production requires intensive research in this field [65, 73].

3.7 Oxidative stress

The technique of oxidative stress involves pre-treatment of algal biomass with oxidative agents like free nitrous acid (FNA) that enhance the extraction efficiency [109]. This can be a most efficient technique in terms of low production cost, and from an environmental point of view because FNA is a green and renewable chemical. According to the studies, the cultures treated with FNA show a 2.4 fold increase in lipid yield (up to 2.19 mg HNO2-N/L) [109]. Other than this, UV is also a good oxidative agent for the pretreatment method to improve the lipid extraction [110].

3.8 Electroporation

Electroporation is a technique that is diversely used in molecular biology research to enable transportation of chemical, drug, and foreign DNA products into the cell [111, 112]. Nowadays, this technique is also used for lipid recovery from microalgal cells [102, 111, 112]. The applied electrical field on wet algal biomass creates aqueous pores in the cell walls, which enhances both membrane permeability and conductivity, hence mass transfer across the cell membrane is also increased [113–115]. It has been reported that lipid recovery can be enhanced with this method without affecting the composition and quality of lipids, however, more studies are required to prove its effectiveness on various types of microorganisms [63, 113, 116].

3.9 Isotonic extraction method

Isotonic extraction methods include different solvent-free extraction techniques [29, 76, 117]. These techniques use ionic liquids instead of toxic organic solvents for lipid recovery [117]. It is relatively easy to design solvents with specific polarity, hydrophobicity, conductivity, and solubility as a result of the high synthetic flexibility of ionic liquids. However, there are only a few studies that have been performed in this field and it is too early to conclude whether this method will be applicable [65, 73].

3.10 Enzymatic disruption

This method involves the use of different enzymatic cocktails for cell wall disruption prior to recovery of lipids. Cellulase, neutral protease, alkaline protease, papain, and lysozyme are the main enzymes used in this process [116, 118]. In contrast to other mechanical disruption techniques, like sonication, this makes the extraction process easy and low time consuming [116, 118]. Although enzymatic disruption is a poorly studied method in the algal cell, the selectivity of reactions and lack of any side effects on the final products leads to excellent lipid recovery [119]. According to Fu et al., an increment of approximately 14% was observed in lipid extraction efficiency when the cell wall of *Chlorella* was disrupted by enzymatic hydrolysis, compared to unhydrolyzed microalgae [119a]. High selectivity, optimal operating temperature, and no corrosion are the main advantages that make this process more accessible over physical and chemical methods of lipid extraction [116].

4. Problems associated with currently available methods

The above discussion highlighted the different techniques and methods that are involved in lipid extraction. Production of microbial oil or other products totally depends on the yield of the product or scalable recovery, making the downstream process the foremost step during pilot-scale production [120]. Success of any method or technique is strongly dependent on the cost involved in the process, easiness of the protocol, time requirements, toxicity levels of the compounds used in the process, eco-friendly nature, viability of process, minimum dependence on conditions such as temperature and pressure, and worldwide acceptability [33, 65, 117, 121]. However, a number of techniques, as discussed earlier, are used for extraction purposes, although all these methods have problems at some stage or level. For example, physicochemical methods involve the use of toxic organic solvents like chloroform, methanol, and hexane, which are not only harmful to the environment but also create hazardous impacts on human health unless carefully controlled [89].

While physical methods such as crushing, grinding, and pressure-assisted disruption of cells are enormously efficient techniques, their high energy demands and the need for skilled labor increase the overall cost and makes them impractical for industrial applications. Generation of high amounts of heat, and strain dependence are also some common problems with these methods, which affect the quality of products and restrict its uses. Apart from this, the thick cell wall and non-porosity of microalgae cells prevent the release of intracellular lipids; hence, a pretreatment step is required to disrupt the indehiscent cell wall of microalgae, which ultimately increases the overall production cost.

5. Hydrothermal liquefaction (HTL)

Hydrothermal liquefaction (HTL), also referred to as the Environment-Enhancing Energy (E2 -Energy) system, is a process which tackles the direct conversion of wet biomass into fuel energy in an energy-efficient manner that results in high bio-oil yield [8, 122–128]. The process of biomass conversion into bio-oils in HTL is achieved with high temperature (280–374°C) and moderate pressure (10–25 MPa), which decreases the strength of hydrogen bond and lowers the dielectric constant of water, due to which, solubility of lower polarity molecules is enhanced [123, 126]. High temperature and pressure also help in the polymerization of small molecules that further act as a precursor for bio-oil [123]. At the same time, a mixture of several reactions like acid catalysis and hydrolysis also performs fundamental roles in the decomposition of macromolecules present in algal biomass such as lipids, proteins, carbohydrates, and algaenans [8, 122, 127].

Properties like conversion of proteins and carbohydrates into bio-oil, recycling of nutrients essential for algal growth, i.e., nitrogen, phosphorus, magnesium, iron, calcium, potassium, etc. proves the HTL process to be more promising in the field of bio-oil production in comparison to other process, i.e., pyrolysis and gasification [8, 91, 127]. Although the HTL process provides a superior quality of bio-oil in terms of thermal and storage stability, it cannot, however, be used as a transportation fuel because of the presence of high amounts of O and N, which still require subsequent upgrading before use [8, 122, 123, 129, 130]. High energy consumption due to the utilization of high temperature and pressure is another demerit of this method; however, it can be advanced by using continuous-scale subcritical water reactors [126].

6. Supercritical fluid extraction

Although supercritical fluid extraction (SFE) has been known of for >100 years for its solvent abilities; however, its commercial application was not feasible due to the lack of focused research in this area [131]. Incessant research in the field of extraction evolved this technique to a new level with lipid extraction from microalgae in the presence of supercritical fluids like ethylene, ethane, methanol, benzene, ethanol, toluene, CO_2, and water [107, 121].

The basic principle behind SFE is to achieve a condition in which the meniscus separating the liquid and vapor phases disappears, which is obtained beyond the critical point of a fluid [132]. Among the different choices of supercritical fluids, CO_2 is gaining increasing attention in the extraction of pharmaceutical and health-related products [133]. Due to the non-polar nature of CO_2, this shows high selectivity towards neutral lipids, especially TAGs, which together with its inability to solubilize phospholipids, makes the process highly specific towards TAGs [47, 107]. Along with these properties, low critical temperature (31.1°C) of SFE provides an excellent solution to extract thermally sensitive lipids avoiding thermal degradation [43, 134]; however, supercritical CO_2 (SC-CO_2) extracted products also contain a small amount of impurities of free fatty acids, sterol, and pigments [135].

The efficiency of the supercritical CO_2 extraction processes depends on the chosen operating conditions, in which different conditions of several physical parameters such as pressure (20–60 MPa), temperature (303.15–333.15 K), and CO_2 flow rate (0.06–30 g/min) have been tested, and it has been suggested that up to 100% extraction yield can be achieved only by increasing the pressure at a constant temperature [134]. Often the extraction efficiency can present variations between dried and wet microalgal biomass and it has been proposed that high extraction yield can be obtained with biomass of low moisture content, hence drying of biomass is required prior to SC-CO_2 extraction to enhance the lipid yield [136]. As such, this technique can serve as a suitable alternative for the extraction technique.

6.1 Major advantages of supercritical fluid extraction

The incredible properties of supercritical CO_2 eventually increase the levels of lipid extraction compared to other conventional methods. This process overcomes one of the biggest disadvantages of other methods, which is the degradation of extracts, by providing a non-oxidizing environment and the low critical temperature (around 31°C) also prevents the thermal degradation of extract [47]. Other advantages include non-toxicity, simple and easy downstream process as a result of the easy separation of CO_2, and the high diffusivity and low surface tension increase the penetration of pores that are too small for chemical solvents [137]. SFE can be a promising technique with several distinct properties:

(1) Prior discussion highlights that SFE involves different parameters like temperature and pressure that are easy to modify and provide high selectivity, which allows easy extraction of complex samples [138].

(2) Extraction and quantification process of highly volatile compounds is another advantage of SFE, which becomes possible due to the direct coupling with a chromatographic method [137].

(3) It can be applied on a wide range of sample amounts ranging from a few grams to kilograms and even up to tons, which is very helpful for both lab and pilot levels of production [139].

(4) Quick and selective extraction reduces the separation cost [140].

(5) SFE is the latest technique to provide more information about extraction and purification. One can use this information for optimization purposes and to evaluate the extraction efficiency [135].

(6) Low viscosity and high diffusivity enhance porosity, which enables penetration into solid material very efficiently, resulting in faster, quantitative, and complete extraction [47].

(7) Proper elimination of polluting organic solvents and flexibility of process encourages its worldwide application [39].

(8) Supercritical CO_2 is more advantageous as compared to other organic solvents, it is considered as a safe and non-flammable solvent [40, 42, 82, 132, 141, 142].

(9) SC-CO_2 demonstrates enormous capabilities to separate less volatile and high molecular weight compounds; as pressure rises, more polar molecules also can be separated.

(10) Use of CO_2 as compared to other solvents makes the extraction cheaper because of the reusable properties of CO_2. Its easy availability and contamination-free products are the major advantages of the process [40].

6.2 Extraction of lipids from microalgae using supercritical carbon dioxide

A list of microalgal species that have been tested for lipids extraction with SC-CO_2 under optimal conditions and their advantages and disadvantages are listed in Table 1. Tang et al. [143] isolated lipids from the microalga *Schizochytrium limacinum* by using SC-CO_2 and under the optimum pressure (35 MPa) and temperature (40°C) conditions; 33.9% of lipid yield was achieved while high-purity of docosahexenoic acid (DHA) was processed by a urea complexation method [143]. In another study, it was revealed that bead milling of microbial biomass enhances the lipid extraction from *Chlorella vulgaris* when using SC-CO_2 [82]. The lipids and pigments from the oleaginous microalga *Nannochloropsis* sp. were extracted with SC-CO_2 at the optimized conditions of pressure and temperature, while ethanol was used as co-solvent [144]. It was reported that a maximum of 45 g of lipids per 100 g dry microalgal biomass and 70% of the total pigment can be extracted after optimized condition by this method [144].

In a study, various extraction techniques were evaluated for the fatty profiles of marine microalga *Tetraselmis* sp. M8, where it was reported that SC extraction method was the most effective method among all tested techniques for the lipid extraction especially for long-chain unsaturated fatty acids [145]. Solana et al. [136] compared three microalgae, *Scenedesmus obliquus*, *Chlorella protothecoides*, and *Nannochloropsis salina* for the oil rich in α-linolenic acid as essential fatty acids by supercritical fluid extraction [136]. The highest extraction yield was reported at 60°C and 30 MPa with 0.4 kg/h of CO_2 and 5% of ethanol as co-solvent. Among all tested microalgae strains, *Scenedesmus*

Table 1 List of lipids extraction from different microalgae using SC-CO_2 under optimum conditions

Microalgae species	SC-CO_2 condition for lipid extraction e.g. pressure (MPa), temp (°C), CO_2 flow rate and/or co-solvents	Lipid concentration, LC (%, *w/w*) or extraction yield, EY (%)	Remarks	References
Schizochytrium limacinum	35 MPa, 40°C, 95% v/v ethanol as a co-solvent	LC, 33.9%	High DHA purity and superb product quality along with total lipid extraction	[143]
Chlorella vulgaris	60 MPa, 59.85°C, 5% ethanol	LC, 11.43%	Low toxicity	[82]
Scenedesmus sp.	50 MPa, 53.85°C, 1.9 g/min CO_2 flow rate	LC, 7.34%	Totally free of solvents	[44]
Nannochloropsis sp.	30 MPa, 39.85°C	LC, 33%	No heavy metals are present in CO_2 or the equipment; Energy-intensive due to use of high pressure	[144]
Nannochloropsis sp.	30 MPa, 40°C, 20% ethanol	LC, 45%	No extra unit operations needed, and yield of useful material is very high	[144]
Tetraselmis sp	35 MPa, 39.85°C, CO_2 + ethanol	LC, 10.88%	CO_2 recycling avoids greenhouse effect	[145]
Commercial algae	30 MPa, 29.85°C, propanol	EY, 90.56%		[147]
S. obliquus	20 MPa, 65°C	LC, 18.15% and EY 24.67%	Lower extraction time	[136]
Pavlova sp.	35 MPa, 40°C	LC, 10.4% and EY 34.0%	Mass transfer equilibrium could be favorable; High equipment and operational cost	[148]

Table 1 List of lipids extraction from different microalgae using SC-CO_2 under optimum conditions—cont'd

Microalgae species	SC-CO_2 condition for lipid extraction e.g. pressure (MPa), temp (°C), CO_2 flow rate and/or co-solvents	Lipid concentration, LC (%, *w*/w) or extraction yield, EY (%)	Remarks	References
Chlorella protothecoides	30 MPa, 70°C	LC, 21%	High efficiency	[45]
Nannochloropsis	55 MPa, 55°C	LC, 44%	CO_2 is highly selective and no chance of polar substances forming polymers exists	[146]
Hypnea charoides	37.9 MPa, 50°C	LC, 6.7%		[149]
Chlorella vulgaris	35 MPa, 55°C	LC, 13%	Recycling of CO_2 minimizing waste generation	[150]
Crypthecodinium cohnii	30.0 MPa, 49.85°C	LC, 50%	Selective extraction of the specific compound; Energy-intensive due to use of high pressure	[40]
Botryococcus braunii	25 MPa, 50°C	LC, 17.6%	Ideal technique to study thermally labile compounds	[43]
Arthrospira (Spirulina) maxima	25 MPa, 49.95°C	LC, 40%		[151]
Chlorella vulgaris	35.0 MPa, 54.95°C	LC, 52%		[151]

obliquus showed the highest amount of total ω-3 fatty acid and α-linolenic acid compared to the other species [136]. Andrich et al. [146] used supercritical CO_2 for the extraction of bioactive lipids with a high proportion of polyunsaturated fatty acids (PUFA) from unicellular microalga *Nannocloropsis* sp. [146].

6.3 Application of supercritical fluid extraction

SFE is one of the oldest methods for lipid extraction and, on the analytical and preparative scale, it is one of the most widely used techniques, which enables the efficient lipid recovery from not only microalgae but also allows the smooth and obstacle-free extraction of essential oils from rosemary, fennel, coffee seeds, and anise [140]. Nowadays, refining is the main method for removing undesired components from crude vegetable oils and has several advantages in the food industry, but it also eradicates some of the valuable compounds that also affect the quality of oils. SFE is an excellent method to overcome these problems as inactive CO_2 does not interact with the desired product and, due to its gaseous nature, it is easy to separate, allowing a good product quality [152].

Wheat germ oil, rice bran oil and crude palm oil, essential oil, fatty acids, and bioactive compounds can be extracted by SC-CO_2 in a better way as compared to other techniques [141]. Along with these properties, it is also important for the fractionation of fish oils and their conversion into omega-3 fatty acids. It also plays a key role in the pharmaceutical industry for the purification and quantification of active enantiomer, among several other applications. This technique not only offers an excellent solution in the field of food and pharmaceuticals but also is advantageous in environmental research such as for the removal of heavy metals and reduction of secondary waste generation [41].

7. Conclusions and future perspectives

Increasing population, alongside global industrialization, has increased the consumption rate of petroleum oils, making the sufficient supply of petroleum challenging in the long run. Hence, we must move to the use of sustainable and renewable sources of oils to overcome the imminent shortage problems and reduce the negative impact of the use of fossil resources. In this regard, microbial oils can be considered as an appropriate substitute for petroleum oils.

Microbial oils in the form of lipid droplets are usually synthesized intracellularly and lipid extraction must be carried out after the disintegration of the indehiscent microbial cellular framework. Although various physical, chemical, physicochemical, and biological methods have been introduced to facilitate lipid extraction from oleaginous microorganisms, this presently has certain limitations. Achieving an extraction efficiency close to 100% is a crucial step aiming to enhance the lipids yield and meet the current demands of oils. Recent research advances have proved that SFE can be an excellent alternative technique over other available traditional methods of extraction, avoiding pretreatment of the microbial cells prior to its application. Although this is the oldest technique, the process has been evolved over time and certain improvements have been made in the field of lipids extraction from microalgae using SC-CO_2.

There are some other important factors needed to improve the further use of this technique, e.g., its low polarity problem during operation, which can be solved by using co-solvents as a polar modifier. However, this cannot solve the problem at a significant level, and the other important issue is the critical point that creates a phase equilibrium situation because overall process is highly sensitive to change in parameters of the reaction or change in the operating conditions (theoretically, the solubility of the compounds in SC-CO_2 as a function of temperature and pressure) where phase engineering plays a key role, therefore, process optimization requires a lot of studies that can improve the technology in an efficient manner. Finally, the lack of realistic economic studies on the SC-CO_2 extraction of lipids also hinders its industrial-scale application.

References

[1] M.I. Al Irsyad, A. Halog, R. Nepal, Renewable energy projections for climate change mitigation: an analysis of uncertainty and errors, Renew. Energy 130 (2019) 536–546, https://doi.org/10.1016/j.renene.2018.06.082.

[2] Y. Ma, Z. Gao, Q. Wang, Y. Liu, Biodiesels from microbial oils: opportunity and challenges, Bioresour. Technol. (2018), https://doi.org/10.1016/j.biortech.2018.05.028.

[3] S. Xiu, A. Shahbazi, Bio-oil production and upgrading research: a review, Renew. Sustain. Energy Rev. (2012), https://doi.org/10.1016/j.rser.2012.04.028.

[4] M. Paul Abishek, J. Patel, A. Prem Rajan, Algae oil: a sustainable renewable fuel of future, Biotechnol. Res. Int. 2014 (2014) 1–8, https://doi.org/10.1155/2014/272814.

[5] Y. Lu, J. Yuan, X. Lu, C. Su, Y. Zhang, C. Wang, X. Cao, Q. Li, J. Su, V. Ittekkot, R.A. Garbutt, S. Bush, S. Fletcher, T. Wagey, A. Kachur, N. Sweijd, Major threats of pollution and climate change to global coastal ecosystems and enhanced management for sustainability, Environ, Pollut. 239 (2018) 670–680, https://doi.org/10.1016/j.envpol.2018.04.016.

[6] A. Patel, N. Arora, K. Sartaj, V. Pruthi, P.A. Pruthi, Sustainable biodiesel production from oleaginous yeasts utilizing hydrolysates of various non-edible lignocellulosic biomasses, Renew. Sustain. Energy Rev. 62 (2016) 836–855, https://doi.org/10.1016/j.rser.2016.05.014.

[7] J.S. Lim, Z. Abdul Manan, S.R. Wan Alwi, H. Hashim, A review on utilisation of biomass from rice industry as a source of renewable energy, Renew. Sustain. Energy Rev. 16 (2012) 3084–3094, https://doi.org/10.1016/j.rser.2012.02.051.

[8] J. Akhtar, N.A.S. Amin, A review on process conditions for optimum bio-oil yield in hydrothermal liquefaction of biomass, Renew. Sustain. Energy Rev. (2011), https://doi.org/10.1016/j.rser.2010.11.054.

[9] L. Matsakas, V. Raghavendran, O. Yakimenko, G. Persson, E. Olsson, U. Rova, L. Olsson, P. Christakopoulos, Lignin-first biomass fractionation using a hybrid organosolv—Steam explosion pretreatment technology improves the saccharification and fermentability of spruce biomass, Bioresour, Technol. 273 (2019) 521–528, https://doi.org/10.1016/j.biortech.2018.11.055.

[10] I. Ambat, V. Srivastava, M. Sillanpää, Recent advancement in biodiesel production methodologies using various feedstock: a review, Renew. Sustain. Energy Rev. 90 (2018) 356–369, https://doi.org/10.1016/j.rser.2018.03.069.

[11] Y. Chisti, Biodiesel from microalgae beats bioethanol, Trends Biotechnol. 26 (2008) 126–131, https://doi.org/10.1016/j.tibtech.2007.12.002.

[12] A.L. Ahmad, N.H.M. Yasin, C.J.C. Derek, J.K. Lim, Microalgae as a sustainable energy source for biodiesel production: a review, Renew. Sustain. Energy Rev. 15 (2011) 584–593, https://doi.org/10.1016/j.rser.2010.09.018.
[13] M. Al-Qasmi, N. Raut, S. Talebi, S. Al-Rajhi, T. Al-Barwani, A review of effect of light on microalgae growth, Proc. World Congr. Eng. I (2012) 8–10.
[14] Z. Baicha, M.J. Salar-garcía, V.M. Ortiz-martínez, F.J. Hernández-fernández, A.P.D.L. Ríos, A critical review on microalgae as an alternative source for bioenergy production: a promising low cost substrate for microbial fuel cells, Fuel Process. Techology 154 (2016) 104–116, https://doi.org/10.1016/j.fuproc.2016.08.017.
[15] L.L. Beer, E.S. Boyd, J.W. Peters, M.C. Posewitz, Engineering algae for biohydrogen and biofuel production, Curr. Opin. Biotechnol. (2009), https://doi.org/10.1016/j.copbio.2009.06.002.
[16] M.A. Borowitzka, Algal biotechnology products and processes—matching science and economics, J. Appl. Phycol. 4 (1992) 267–279, https://doi.org/10.1007/BF02161212.
[17] G.M. Elrayies, Microalgae: prospects for greener future buildings, Renew. Sustain. Energy Rev. 81 (2018) 1175–1191, https://doi.org/10.1016/j.rser.2017.08.032.
[18] D. Hernández, B. Riaño, M. Coca, M. Solana, A. Bertucco, M.C. García-gonzález, Microalgae cultivation in high rate algal ponds using slaughterhouse wastewater for biofuel applications, Chem. Eng. J. 285 (2016) 449–458, https://doi.org/10.1016/j.cej.2015.09.072.
[19] M. Mubarak, A. Shaija, T.V. Suchithra, A review on the extraction of lipid from microalgae for biodiesel production, Algal Res. 7 (2015) 117–123, https://doi.org/10.1016/j.algal.2014.10.008.
[20] T. Naumann, Z. Çebi, B. Podola, M. Melkonian, Growing microalgae as aquaculture feeds on twin-layers: a novel solid-state photobioreactor, J. Appl. Phycol. 25 (2013) 1413–1420, https://doi.org/10.1007/s10811-012-9962-6.
[21] H.H. Mardhiah, H.C. Ong, H.H. Masjuki, S. Lim, H.V. Lee, A review on latest developments and future prospects of heterogeneous catalyst in biodiesel production from non-edible oils, Renew. Sustain. Energy Rev. 67 (2017) 1225–1236, https://doi.org/10.1016/j.rser.2016.09.036.
[22] M. Saber, B. Nakhshiniev, K. Yoshikawa, A review of production and upgrading of algal bio-oil, Renew. Sustain. Energy Rev. 58 (2016) 918–930, https://doi.org/10.1016/j.rser.2015.12.342.
[23] S.A. Shields-Menard, M. Amirsadeghi, W.T. French, R. Boopathy, A review on microbial lipids as a potential biofuel, Bioresour. Technol. 259 (2018) 451–460, https://doi.org/10.1016/j.biortech.2018.03.080.
[24] S.K. Yellapu, Bharti, R. Kaur, L.R. Kumar, B. Tiwari, X. Zhang, R.D. Tyagi, Recent developments of downstream processing for microbial lipids and conversion to biodiesel, Bioresour. Technol. 256 (2018) 515–528, https://doi.org/10.1016/j.biortech.2018.01.129.
[25] S.A. Khan, Rashmi, M.Z. Hussain, S. Prasad, U.C. Banerjee, Prospects of biodiesel production from microalgae in India, Renew. Sustain. Energy Rev. 13 (2009) 2361–2372, https://doi.org/10.1016/j.rser.2009.04.005.
[26] Q. Hu, M. Sommerfeld, E. Jarvis, M. Ghirardi, M. Posewitz, M. Seibert, A. Darzins, Microalgal triacylglycerols as feedstocks for biofuel production: perspectives and advances, Plant J. 54 (2008) 621–639, https://doi.org/10.1111/j.1365-313X.2008.03492.x.
[27] L.D. Zhu, Z.H. Li, E. Hiltunen, Strategies for lipid production improvement in microalgae as a biodiesel feedstock, Biomed Res. Int. (2016), https://doi.org/10.1155/2016/8792548.
[28] S.Y. Lee, J.M. Cho, Y.K. Chang, Y.K. Oh, Cell disruption and lipid extraction for microalgal biorefineries: a review, Bioresour. Technol. 244 (2017) 1317–1328, https://doi.org/10.1016/j.biortech.2017.06.038.
[29] R. Ranjith Kumar, P. Hanumantha Rao, M. Arumugam, Lipid extraction methods from microalgae: a comprehensive review, Front. Energy Res. 2 (2015) 1–9, https://doi.org/10.3389/fenrg.2014.00061.
[30] T. Dong, E.P. Knoshaug, P.T. Pienkos, L.M.L. Laurens, Lipid recovery from wet oleaginous microbial biomass for biofuel production: a critical review, Appl. Energy 177 (2016) 879–895, https://doi.org/10.1016/j.apenergy.2016.06.002.
[31] C.R. Soccol, C.J. Dalmas Neto, V.T. Soccol, E.B. Sydney, E.S.F. da Costa, A.B.P. Medeiros, L.P.S. Vandenberghe, Pilot scale biodiesel production from microbial oil of Rhodosporidium

toruloides DEBB 5533 using sugarcane juice: Performance in diesel engine and preliminary economic study, Bioresour. Technol. 223 (2017) 259–268, https://doi.org/10.1016/j.biortech.2016.10.055.
[32] R. Halim, M.K. Danquah, P.A. Webley, Extraction of oil from microalgae for biodiesel production: a review, Biotechnol. Adv. 30 (2012) 709–732, https://doi.org/10.1016/j.biotechadv.2012.01.001.
[33] A. Ranjan, C. Patil, V.S. Moholkar, Mechanistic assessment of microalgal lipid extraction, Ind. Eng. Chem. Res. 49 (2010) 2979–2985, https://doi.org/10.1021/ie9016557.
[34] E. D'Hondt, J. Martín-Juárez, S. Bolado, J. Kasperoviciene, J. Koreiviene, S. Sulcius, K. Elst, L. Bastiaens, Cell disruption technologies, in: Microalgae-Based Biofuels and Bioproducts, Elsevier, 2017, pp. 133–154, https://doi.org/10.1016/B978-0-08-101023-5.00006-6.
[35] A. Patel, F. Mikes, L. Matsakas, An overview of current pretreatment methods used to improve lipid extraction from oleaginous microorganisms, Molecules (2018) 23, https://doi.org/10.3390/molecules23071562.
[36] Y. Chisti, Biodiesel from microalgae, Biotechnol. Adv. 25 (2007) 294–306, https://doi.org/10.1016/j.biotechadv.2007.02.001.
[37] J. Kim, G. Yoo, H. Lee, J. Lim, K. Kim, C.W. Kim, M.S. Park, J.W. Yang, Methods of downstream processing for the production of biodiesel from microalgae, Biotechnol. Adv. 31 (2013) 862–876, https://doi.org/10.1016/j.biotechadv.2013.04.006.
[38] Y.H. Chen, T.H. Walker, Fed-batch fermentation and supercritical fluid extraction of heterotrophic microalgal Chlorella protothecoides lipids, Bioresour. Technol. 114 (2012) 512–517, https://doi.org/10.1016/j.biortech.2012.03.026.
[39] R.M. Couto, J. Fernandes, M.D.R.G. da Silva, P.C. Simões, Supercritical fluid extraction of lipids from spent coffee grounds, J. Supercrit. Fluids 51 (2009) 159–166, https://doi.org/10.1016/j.supflu.2009.09.009.
[40] R.M. Couto, P.C. Simões, A. Reis, T.L. Da Silva, V.H. Martins, Y. Sánchez-Vicente, Supercritical fluid extraction of lipids from the heterotrophic microalga Crypthecodinium cohnii, Eng. Life Sci. 10 (2010) 158–164, https://doi.org/10.1002/elsc.200900074.
[41] M. Herrero, J.A. Mendiola, A. Cifuentes, E. Ibáñez, Supercritical fluid extraction: recent advances and applications, J. Chromatogr. A 1217 (2010) 2495–2511, https://doi.org/10.1016/j.chroma.2009.12.019.
[42] A. Mouahid, C. Crampon, S.A.A. Toudji, E. Badens, Supercritical CO2 extraction of neutral lipids from microalgae: Experiments and modelling, J. Supercrit. Fluids 77 (2013) 7–16, https://doi.org/10.1016/j.supflu.2013.01.024.
[43] A. Santana, S. Jesus, M.A. Larrayoz, R.M. Filho, Supercritical carbon dioxide extraction of algal lipids for the biodiesel production, Procedia Eng. 42 (2012) 1755–1761, https://doi.org/10.1016/j.proeng.2012.07.569.
[44] H. Taher, S. Al-Zuhair, A. AlMarzouqi, Y. Haik, M. Farid, S. Tariq, Supercritical carbon dioxide extraction of microalgae lipid: process optimization and laboratory scale-up, J. Supercrit. Fluids 86 (2013) 57–66, https://doi.org/10.1016/j.supflu.2013.11.020.
[45] M. Viguera, A. Marti, F. Masca, C. Prieto, L. Calvo, The process parameters and solid conditions that affect the supercritical CO_2 extraction of the lipids produced by microalgae, J. Supercrit. Fluids 113 (2016) 16–22, https://doi.org/10.1016/j.supflu.2016.03.001.
[46] J. Harris, K. Viner, P. Champagne, P.G. Jessop, Advances in microalgal lipid extraction for biofuel production: a review, Biofuels, Bioprod. Biorefining (2018) 1118–1135, https://doi.org/10.1002/bbb.1923.
[47] F. Sahena, I.S.M. Zaidul, S. Jinap, A.A. Karim, K.A. Abbas, N.A.N. Norulaini, A.K.M. Omar, Application of supercritical CO_2 in lipid extraction—a review, J. Food Eng. 95 (2009) 240–253, https://doi.org/10.1016/j.jfoodeng.2009.06.026.
[48] S. Fon Sing, A. Isdepsky, M.A. Borowitzka, N.R. Moheimani, Production of biofuels from microalgae, Mitig. Adapt. Strateg. Glob. Chang. 18 (2013) 47–72, https://doi.org/10.1007/s11027-011-9294-x.
[49] A. Guldhe, F.A. Ansari, P. Singh, F. Bux, Heterotrophic cultivation of microalgae using aquaculture wastewater: a biorefinery concept for biomass production and nutrient remediation, Ecol. Eng. 99 (2017) 47–53, https://doi.org/10.1016/j.ecoleng.2016.11.013.

[50] O. Perez-Garcia, F.M.E.E. Escalante, L.E. de-Bashan, Y. Bashan, Heterotrophic cultures of microalgae: metabolism and potential products, Water Res. 45 (2011) 11–36, https://doi.org/10.1016/j.watres.2010.08.037.
[51] J. Chen, J. Li, W. Dong, X. Zhang, R.D. Tyagi, P. Drogui, R.Y. Surampalli, The potential of microalgae in biodiesel production, Renew. Sustain. Energy Rev. 90 (2018) 336–346, https://doi.org/10.1016/j.rser.2018.03.073.
[52] W.L. Yu, W. Ansari, N.G. Schoepp, M.J. Hannon, S.P. Mayfield, M.D. Burkart, Modifications of the metabolic pathways of lipid and triacylglycerol production in microalgae, Microb. Cell Fact. 10 (2011) 91, https://doi.org/10.1186/1475-2859-10-91.
[53] I.A. Guschina, J.L. Harwood, Lipids and lipid metabolism in eukaryotic algae, Prog. Lipid Res. 45 (2006) 160–186, https://doi.org/10.1016/j.plipres.2006.01.001.
[54] J. Ohlroggeav, J. Browseb, Lipid biosynthesis, Plant Cell 7 (1995) 957–970.
[55] S. Papanikolaou, G. Aggelis, Lipids of oleaginous yeasts. Part I: Biochemistry of single cell oil production, Eur. J. Lipid Sci. Technol. 113 (2011) 1031–1051, https://doi.org/10.1002/ejlt.201100014.
[56] C. Ratledge, J.P. Wynn, The biochemistry and molecular biology of lipid accumulation in oleaginous microorganisms, Adv. Appl. Microbiol. 51 (2002) 1–51.
[57] J. Fan, C. Andre, C. Xu, A chloroplast pathway for the de novo biosynthesis of triacylglycerol in Chlamydomonas reinhardtii, FEBS Lett. 585 (2011) 1985–1991, https://doi.org/10.1016/j.febslet.2011.05.018.
[58] A. Mühlroth, K. Li, G. Røkke, P. Winge, Y. Olsen, M.F. Hohmann-Marriott, O. Vadstein, A.M. Bones, Pathways of lipid metabolism in marine algae, co-expression network, bottlenecks and candidate genes for enhanced production of EPA and DHA in species of chromista, Mar. Drugs 11 (2013) 4662–4697, https://doi.org/10.3390/md11114662.
[59] L.A. Garay, K.L. Boundy-Mills, J.B. German, Accumulation of high-value lipids in single-cell microorganisms: a mechanistic approach and future perspectives, J. Agric. Food Chem. 62 (2014) 2709–2727, https://doi.org/10.1021/jf4042134.
[60] Y. Guo, K.R. Cordes, R.V. Farese, T.C. Walther, Lipid droplets at a glance, J. Cell Sci. 122 (2009) 749–752, https://doi.org/10.1242/jcs.037630.
[61] R. Kapoore, T. Butler, J. Pandhal, S. Vaidyanathan, Microwave-assisted extraction for microalgae: from biofuels to biorefinery, Biology (Basel) 7 (2018) 18, https://doi.org/10.3390/biology7010018.
[62] D. Dvoretsky, S. Dvoretsky, M. Temnov, E. Akulinin, E. Peshkova, Enhanced lipid extraction from microalgae chlorella vulgaris biomass: experiments, modelling, optimization, Chem. Eng. Trans. 49 (2016) 175–180, https://doi.org/10.3303/CET1649030.
[63] F. Ghasemi Naghdi, L.M. González González, W. Chan, P.M. Schenk, Progress on lipid extraction from wet algal biomass for biodiesel production, Microb. Biotechnol. 9 (2016) 718–726, https://doi.org/10.1111/1751-7915.12360.
[64] R.P. Rastogi, A. Pandey, C. Larroche, D. Madamwar, Algal Green Energy—R&D and technological perspectives for biodiesel production, Renew. Sustain. Energy Rev. 82 (2018) 2946–2969, https://doi.org/10.1016/j.rser.2017.10.038.
[65] J.Y. Lee, C. Yoo, S.Y. Jun, C.Y. Ahn, H.M. Oh, Comparison of several methods for effective lipid extraction from microalgae, Bioresour. Technol. 101 (2010) 2009–2011, https://doi.org/10.1016/j.biortech.2009.03.058.
[66] D. Ramesh, Lipid identification and extraction techniques, in: F. Bux (Ed.), Biotechnological Applications of Microalgae: Biodiesel and Value-Added Products, CRC Press, Boca Raton, FL, 2013, pp. 89–97.
[67] N. Uduman, Y. Qi, M.K. Danquah, G.M. Forde, A. Hoadley, Dewatering of microalgal cultures: A major bottleneck to algae-based fuels, J. Renew. Sustain. Energy (2010) 2, https://doi.org/10.1063/1.3294480.
[68] E. Bligh, W.J. Dyler, A rapid method of total lipid extraction and purification, Can. J. Biochem. Physiol. 37 (1959) 911–917.
[69] J. Folch, M. Lees, G. Sloane Stanley, A simple method of isolation and purification of total lipids from animal tissues, J. Biol. Chem 226 (1957) 497–509.
[70] D.Y. Kim, D. Vijayan, R. Praveenkumar, J.I. Han, K. Lee, J.Y. Park, W.S. Chang, J.S. Lee, Y.K. Oh, Cell-wall disruption and lipid/astaxanthin extraction from microalgae: Chlorella and Haematococcus, Bioresour. Technol. 199 (2016) 300–310, https://doi.org/10.1016/j.biortech.2015.08.107.

[71] M. Khot, D. Ghosh, Lipids of *Rhodotorula mucilaginosa* IIPL32 with biodiesel potential: oil yield, fatty acid profile, fuel properties, J. Basic Microbiol. (2017) 1–8, https://doi.org/10.1002/jobm.201600618.

[72] R.B. Levine, T. Pinnarat, P.E. Savage, Biodiesel production from wet algal biomass through in situ lipid hydrolysis and supercritical transesterification, Energy and Fuels 24 (2010) 5235–5243, https://doi.org/10.1021/ef1008314.

[73] G. Yoo, W.-K.K. Park, C.W. Kim, Y.-E.E. Choi, J.-W.W. Yang, Direct lipid extraction from wet Chlamydomonas reinhardtii biomass using osmotic shock, Bioresour. Technol. 123 (2012) 717–722, https://doi.org/10.1016/j.biortech.2012.07.102.

[74] N.S. Topare, S.J. Raut, V.C. Renge, S.V. Khedkar, Y.P. Chavan, S.L. Bhagat, Extraction of oil from algae by solvent extraction and oil expeller method, Int. J. Chem. Sci. 9 (2011) 1746–1750.

[75] M. Follows, P.J. Hetherington, P. Dunnill, M.D. Lilly, Release of enzymes from bakers' yeast by disruption in an industrial homogenizer, Biotechnol. Bioeng. 13 (1971) 549–560, https://doi.org/10.1002/bit.260130408.

[76] E. Hounslow, J. Noirel, D.J. Gilmour, P.C. Wright, Lipid quantification techniques for screening oleaginous species of microalgae for biofuel production, Eur. J. Lipid Sci. Technol. 119 (2017) 1–24, https://doi.org/10.1002/ejlt.201500469.

[77] M. Kwak, S.G. Kang, W.K. Hong, J.I. Han, Y.K. Chang, Simultaneous cell disruption and lipid extraction of wet aurantiochytrium sp. KRS101 using a high shear mixer, Bioprocess Biosyst. Eng. (2018) 1–8, https://doi.org/10.1007/s00449-018-1901-8.

[78] J.N. Liu, J.N. Liu, M. Yuan, Z.H. Shen, K.M. Peng, L.J. Lu, X.F. Huang, Bioconversion of volatile fatty acids derived from waste activated sludge into lipids by Cryptococcus curvatus, Bioresour. Technol. 211 (2016) 548–555, https://doi.org/10.1016/j.biortech.2016.03.146.

[79] M.M. Mendes-Pinto, M.F.J. Raposo, J. Bowen, A.J. Young, R. Morais, Evaluation of different cell disruption process on encysted cells of Haematococcus pluvialis, J. Appl. Phycol. 13 (2001) 19–24, https://doi.org/10.1023/A:1008183429747.

[80] Y.Y. Sere, M. Regnacq, J. Colas, T. Berges, A Saccharomyces cerevisiae strain unable to store neutral lipids is tolerant to oxidative stress induced by α-synuclein, Free Radic. Biol. Med. 49 (2010) 1755–1764, https://doi.org/10.1016/j.freeradbiomed.2010.09.007.

[81] P.R. Postma, T.L. Miron, G. Olivieri, M.J. Barbosa, R.H. Wijffels, M.H.M. Eppink, Mild disintegration of the green microalgae Chlorella vulgaris using bead milling, Bioresour. Technol. 184 (2015) 297–304, https://doi.org/10.1016/j.biortech.2014.09.033.

[82] C. Safi, S. Camy, C. Frances, M.M. Varela, E.C. Badia, P.Y. Pontalier, C. Vaca-Garcia, Extraction of lipids and pigments of Chlorella vulgaris by supercritical carbon dioxide: influence of bead milling on extraction performance, J. Appl. Phycol. 26 (2014) 1711–1718, https://doi.org/10.1007/s10811-013-0212-3.

[83] P.R. Postma, E. Suarez-Garcia, C. Safi, G. Olivieri, G. Olivieri, R.H. Wijffels, R.H. Wijffels, Energy efficient bead milling of microalgae: effect of bead size on disintegration and release of proteins and carbohydrates, Bioresour. Technol. 224 (2017) 670–679, https://doi.org/10.1016/j.biortech.2016.11.071.

[84] M. Al Hattab, A. Ghaly, A. Hammouda, Microalgae harvesting methods for industrial production of biodiesel: critical review and comparative analysis, J. Fundam. Renew. Energy Appl. 05 (2015), https://doi.org/10.4172/2090-4541.1000154.

[85] M. Athenaki, C. Gardeli, P. Diamantopoulou, S.S. Tchakouteu, D. Sarris, A. Philippoussis, S. Papanikolaou, Lipids from yeasts and fungi: physiology, production and analytical considerations, J. Appl. Microbiol. 124 (2018) 336–367, https://doi.org/10.1111/jam.13633.

[86] T.H. Lee, J.S. Chang, H.Y. Wang, Current developments in high-throughput analysis for microalgae cellular contents, Biotechnol. J. 8 (2013) 1301–1314, https://doi.org/10.1002/biot.201200391.

[87] A. Patel, N. Arora, V. Pruthi, P.A. Pruthi, A novel rapid ultrasonication-microwave treatment for total lipid extraction from wet oleaginous yeast biomass for sustainable biodiesel production, Ultrason. Sonochem. (2018), https://doi.org/10.1016/j.ultsonch.2018.05.002.

[88] D. Boldor, A. Kanitkar, B.G. Terigar, C. Leonardi, M. Lima, G.A. Breitenbeck, Microwave assisted extraction of biodiesel feedstock from the seeds of invasive chinese tallow tree, Environ. Sci. Technol. 44 (2010) 4019–4025, https://doi.org/10.1021/es100143z.

[89] M. Cooney, G. Young, N. Nagle, Extraction of bio-oils from microalgae, Sep. Purif. Rev. 38 (2009) 291–325, https://doi.org/10.1080/15422110903327919.
[90] A.F. Ferreira, A.P.S. Dias, C.M. Silva, M. Costa, Effect of low frequency ultrasound on microalgae solvent extraction: analysis of products, energy consumption and emissions, Algal Res. 14 (2016) 9–16, https://doi.org/10.1016/j.algal.2015.12.015.
[91] A.R.K. Gollakota, N. Kishore, S. Gu, A review on hydrothermal liquefaction of biomass, Renew. Sustain. Energy Rev. 81 (2018) 1378–1392, https://doi.org/10.1016/j.rser.2017.05.178.
[92] A. Patel, L. Matsakas, A comparative study on de novo and ex novo lipid fermentation by oleaginous yeast using glucose and sonicated waste cooking oil, Ultrason. Sonochem. (2018), https://doi.org/10.1016/j.ultsonch.2018.12.010.
[93] M. Gong, A. Bassi, Carotenoids from microalgae: a review of recent developments, Biotechnol. Adv. 34 (2016) 1396–1412, https://doi.org/10.1016/j.biotechadv.2016.10.005.
[94] J. Cheng, J. Sun, Y. Huang, J. Zhou, K. Cen, Fractal microstructure characterization of wet microalgal cells disrupted with ultrasonic cavitation for lipid extraction, Bioresour. Technol. 170 (2014) 138–143, https://doi.org/10.1016/j.biortech.2014.07.090.
[95] A. Patel, M. Pravez, F. Deeba, V. Pruthi, R.P. Singh, P.A. Pruthi, Boosting accumulation of neutral lipids in Rhodosporidium kratochvilovae HIMPA1 grown on hemp (Cannabis sativa Linn) seed aqueous extract as feedstock for biodiesel production, Bioresour. Technol. (2014), https://doi.org/10.1016/j.biortech.2014.03.142.
[96] A. Talebian-Kiakalaieh, N.A.S. Amin, H. Mazaheri, A review on novel processes of biodiesel production from waste cooking oil, Appl. Energy 104 (2013) 683–710, https://doi.org/10.1016/j.apenergy.2012.11.061.
[97] A. Abghari, S. Chen, Yarrowia lipolytica as an oleaginous cell factory platform for production of fatty acid-based biofuel and bioproducts, Front. Energy Res. 2 (2014) 1–21, https://doi.org/10.3389/fenrg.2014.00021.
[98] N.N.A.N. Yusuf, S.K. Kamarudin, Z. Yaakub, Overview on the current trends in biodiesel production, Energy Convers. Manag. 52 (2011) 2741–2751, https://doi.org/10.1016/j.enconman.2010.12.004.
[99] X. Zhang, S. Yan, R.D. Tyagi, P. Drogui, R.Y. Surampalli, Ultrasonication assisted lipid extraction from oleaginous microorganisms, Bioresour. Technol. 158 (2014) 253–261, https://doi.org/10.1016/j.biortech.2014.01.132.
[100] A.H. Metherel, A.Y. Taha, H. Izadi, K.D. Stark, The application of ultrasound energy to increase lipid extraction throughput of solid matrix samples (flaxseed), Prostaglandins Leukot. Essent. Fat. Acids 81 (2009) 417–423, https://doi.org/10.1016/j.plefa.2009.07.003.
[101] D.Y. Rahman, F.D. Sarian, A. van Wijk, M. Martinez-Garcia, M.J.E.C. van der Maarel, Thermostable phycocyanin from the red microalga Cyanidioschyzon merolae, a new natural blue food colorant, J. Appl. Phycol. 29 (2017) 1233–1239, https://doi.org/10.1007/s10811-016-1007-0.
[102] E. Günerken, E. D'Hondt, M.H.M. Eppink, L. Garcia-Gonzalez, K. Elst, R.H. Wijffels, Cell disruption for microalgae biorefineries, Biotechnol. Adv. 33 (2015) 243–260, https://doi.org/10.1016/j.biotechadv.2015.01.008.
[103] J.-Y. Lee, C. Yoo, S.-Y. Jun, C.-Y. Ahn, H.-M. Oh, Comparison of several methods for effective lipid extraction from microalgae, Bioresour. Technol. 101 (2010) S75–S77, https://doi.org/10.1016/j.biortech.2009.03.058.
[104] P. Mercer, R.E. Armenta, Developments in oil extraction from microalgae, Eur. J. Lipid Sci. Technol. 113 (2011) 539–547, https://doi.org/10.1002/ejlt.201000455.
[105] S. Rakesh, D.W. Dhar, R. Prasanna, A.K. Saxena, S. Saha, M. Shukla, K. Sharma, Cell disruption methods for improving lipid extraction efficiency in unicellular microalgae, Eng. Life Sci. 15 (2015) 443–447, https://doi.org/10.1002/elsc.201400222.
[106] F. Chemat, Zill-E-Huma, M.K Khan, Applicationss of ultrasound in food technology: Processing, preservation and extraction, Ultrason. Sonochem. 18 (2011) 813–835, https://doi.org/10.1016/j.ultsonch.2010.11.023.
[107] C.-H.H. Cheng, T.-B.B. Du, H.-C.C. Pi, S.-M.M. Jang, Y.-H.H. Lin, H.-T.T. Lee, Comparative study of lipid extraction from microalgae by organic solvent and supercritical CO2, Bioresour. Technol. 102 (2011) 10151–10153, https://doi.org/10.1016/j.biortech.2011.08.064.

[108] J.R. McMillan, I.A. Watson, M. Ali, W. Jaafar, Evaluation and comparison of algal cell disruption methods: microwave, waterbath, blender, ultrasonic and laser treatment, Appl. Energy 103 (2013) 128–134, https://doi.org/10.1016/j.apenergy.2012.09.020.
[109] X. Bai, F. Ghasemi Naghdi, L. Ye, P. Lant, S. Pratt, Enhanced lipid extraction from algae using free nitrous acid pretreatment, Bioresour. Technol. 159 (2014) 36–40, https://doi.org/10.1016/j.biortech.2014.01.133.
[110] K. Sharma, Y. Li, P.M. Schenk, UV-C-mediated lipid induction and settling, a step change towards economical microalgal biodiesel production, Green Chem. 16 (2014) 3539–3548, https://doi.org/10.1039/c4gc00552j.
[111] E. Sakuradani, S. Shimizu, Single cell oil production by Mortierella alpina, J. Biotechnol. 144 (2009) 31–36, https://doi.org/10.1016/j.jbiotec.2009.04.012.
[112] P. Singh, S. Kumari, A. Guldhe, R. Misra, I. Rawat, F. Bux, Trends and novel strategies for enhancing lipid accumulation and quality in microalgae, Renew. Sustain. Energy Rev. 55 (2016) 1–16, https://doi.org/10.1016/j.rser.2015.11.001.
[113] T.Y. Tsong, Electroporation of cell membranes, Minireview. Biophys. J. 60 (1991) 297–306.
[114] J.C. Weaver, G.I. Harrison, J.G. Bliss, J.R. Mourant, K.T. Powell, Electroporation: high frequency of occurrence of a transient high-permeability state in erythrocytes and intact yeast, FEBS Lett. 229 (1988) 30–34, https://doi.org/10.1016/0014-5793(88)80791-9.
[115] A. Yousuf, M.R. Khan, M.A. Islam, Z.A. Wahid, D. Pirozzi, Technical difficulties and solutions of direct transesterification process of microbial oil for biodiesel synthesis, Biotechnol. Lett. 39 (2016) 1–11, https://doi.org/10.1007/s10529-016-2217-x.
[116] A. Ghaly, M.A. Hattab, Microalgae oil extraction pre-treatment methods: critical review and comparative analysis, J. Fundam. Renew. Energy Appl. 05 (2015), https://doi.org/10.4172/2090-4541.1000172.
[117] A.K. Lee, D.M. Lewis, P.J. Ashman, Disruption of microalgal cells for the extraction of lipids for biofuels: processes and specific energy requirements, Biomass Bioenergy 46 (2012) 89–101, https://doi.org/10.1016/j.biombioe.2012.06.034.
[118] K. Liang, Q. Zhang, W. Cong, Enzyme-assisted aqueous extraction of lipid from microalgae, J. Agric. Food Chem. 60 (2012) 11771–11776, https://doi.org/10.1021/jf302836v.
[119] M. Demuez, A. Mahdy, E. Tomás-Pejó, C. González-Fernández, M. Ballesteros, Enzymatic cell disruption of microalgae biomass in biorefinery processes, Biotechnol. Bioeng. 112 (2015) 1955–1966, https://doi.org/10.1002/bit.25644.
119a. C.C. Fu, T.C. Hung, J.Y. Chen, C.H. Su, W.T. Wu, Hydrolysis of microalgae cell walls for production of reducing sugar and lipid extraction, Bioresour. Technol. 101 (2010) 8750–8754.http.//doi.org/10.1016/j.biortech.2010.06.100.
[120] A. Patel, P.A. Pruthi, V. Pruthi, Oleaginous yeast—a promising candidatea for high quality biodiesel production, Adv. Biofeedstocks Biofuels Prod. Technol. Biofuels 2 (2016) 107–128, https://doi.org/10.1002/9781119117551.ch4.
[121] D. Hernández, M. Solana, B. Riaño, M.C. García-González, A. Bertucco, Biofuels from microalgae: lipid extraction and methane production from the residual biomass in a biorefinery approach, Bioresour. Technol. 170 (2014) 370–378, https://doi.org/10.1016/j.biortech.2014.07.109.
[122] W.T. Chen, Y. Zhang, J. Zhang, G. Yu, L.C. Schideman, P. Zhang, M. Minarick, Hydrothermal liquefaction of mixed-culture algal biomass from wastewater treatment system into bio-crude oil, Bioresour. Technol. 152 (2014) 130–139, https://doi.org/10.1016/j.biortech.2013.10.111.
[123] Y. Guo, T. Yeh, W. Song, D. Xu, S. Wang, A review of bio-oil production from hydrothermal liquefaction of algae, Renew. Sustain. Energy Rev. 48 (2015) 776–790, https://doi.org/10.1016/j.rser.2015.04.049.
[124] D.C. Hietala, J.L. Faeth, P.E. Savage, A quantitative kinetic model for the fast and isothermal hydrothermal liquefaction of Nannochloropsis sp, Bioresour. Technol. 214 (2016) 102–111, https://doi.org/10.1016/j.biortech.2016.04.067.
[125] D.C. Hietala, C.K. Koss, A. Narwani, A.R. Lashaway, C.M. Godwin, B.J. Cardinale, P.E. Savage, Influence of biodiversity, biochemical composition, and species identity on the quality of biomass and biocrude oil produced via hydrothermal liquefaction, Algal Res. 26 (2017) 203–214, https://doi.org/10.1016/j.algal.2017.07.020.

[126] M. Lavanya, A. Meenakshisundaram, S. Renganathan, S. Chinnasamy, D.M. Lewis, J. Nallasivam, S. Bhaskar, Hydrothermal liquefaction of freshwater and marine algal biomass: a novel approach to produce distillate fuel fractions through blending and co-processing of biocrude with petrocrude, Bioresour. Technol. 203 (2016) 228–235, https://doi.org/10.1016/j.biortech.2015.12.013.

[127] T. Michael, T. Nonaka, S. Fukuda, H. Kujiraoka, Y. Kumagai, H. Inomata, R. Lee Jr., Nutrient recovery from municipal sludge for microalgae cultivation with two-step hydrothermal liquefaction, ALGAL 18 (2016) 61–68, https://doi.org/10.1016/j.algal.2016.06.009.

[128] J.D. Sheehan, P.E. Savage, Modeling the effects of microalga biochemical content on the kinetics and biocrude yields from hydrothermal liquefaction, Bioresour. Technol. 239 (2017) 144–150, https://doi.org/10.1016/j.biortech.2017.05.013.

[129] J. Jiang, P.E. Savage, Influence of process conditions and interventions on metals content in biocrude from hydrothermal liquefaction of microalgae, Algal Res. 26 (2017) 131–134, https://doi.org/10.1016/j.algal.2017.07.012.

[130] D. Xu, P.E. Savage, Effect of temperature, water loading, and Ru/C catalyst on water-insoluble and water-soluble biocrude fractions from hydrothermal liquefaction of algae, Bioresour. Technol. 239 (2017) 1–6, https://doi.org/10.1016/j.biortech.2017.04.127.

[131] K. Kitada, S. Machmudah, M. Sasaki, M. Goto, Y. Nakashima, S. Kumamoto, T. Hasegawa, Supercritical CO2 extraction of pigment components with pharmaceutical importance from Chlorella vulgaris, J. Chem. Technol. Biotechnol. 84 (2009) 657–661, https://doi.org/10.1002/jctb.2096.

[132] R. Sawangkeaw, K. Bunyakiat, S. Ngamprasertsith, A review of laboratory-scale research on lipid conversion to biodiesel with supercritical methanol (2001–2009), J. Supercrit. Fluids 55 (2010) 1–13, https://doi.org/10.1016/j.supflu.2010.06.008.

[133] M. Ota, H. Watanabe, Y. Kato, M. Watanabe, Y. Sato, R.L. Smith, H. Inomata, Carotenoid production from Chlorococcum littorale in photoautotrophic cultures with downstream supercritical fluid processing, J. Sep. Sci. 32 (2009) 2327–2335, https://doi.org/10.1002/jssc.200900154.

[134] H.W. Yen, S.C. Yang, C.H. Chen, Jesisca, J.S. Chang, Supercritical fluid extraction of valuable compounds from microalgal biomass, Bioresour. Technol. 184 (2015) 291–296, https://doi.org/10.1016/j.biortech.2014.10.030.

[135] Q. Lang, C.M. Wai, Supercritical fluid extraction in herbal and natural product studies—a practical review, Talanta 53 (2001) 771–782, https://doi.org/10.1016/S0039-9140(00)00557-9.

[136] M. Solana, C.S. Rizza, A. Bertucco, Exploiting microalgae as a source of essential fatty acids by supercritical fluid extraction of lipids: comparison between Scenedesmus obliquus, Chlorella protothecoides and Nannochloropsis salina, J. Supercrit. Fluids 92 (2014) 311–318, https://doi.org/10.1016/j.supflu.2014.06.013.

[137] M.K. Lam, K.T. Lee, Microalgae biofuels: a critical review of issues, problems and the way forward, Biotechnol. Adv. 30 (2012) 673–690, https://doi.org/10.1016/j.biotechadv.2011.11.008.

[138] E. Reverchon, M.C. Volpe, G. Caputo, Supercritical fluid processing of polymers: composite particles and porous materials elaboration, Curr. Opin. Solid State Mater. Sci. 7 (2003) 391–397, https://doi.org/10.1016/j.cossms.2004.01.002.

[139] E. Reverchon, G. Donsi, L.S. Osséo, Modeling of supercritical fluid extraction from herbaceous matrices, Ind. Eng. Chem. Res. 32 (1993) 2721–2726, https://doi.org/10.1021/ie00023a039.

[140] M. Raventos, S. Duarte, R. Alarcon, Application and possibilities of supercritical CO_2 extraction in food processing industry: an overview, Food Sci. Technol. Int. 8 (2002) 269–284, https://doi.org/10.1177/1082013202008005451.

[141] M. Eisenmenger, N.T. Dunford, Bioactive components of commercial and supercritical carbon dioxide processed wheat germ oil, JAOCS, J. Am. Oil Chem. Soc. 85 (2008) 55–61, https://doi.org/10.1007/s11746-007-1163-0.

[142] H.M. Shirazi, J. Karimi-Sabet, C. Ghotbi, Biodiesel production from spirulina microalgae feedstock using direct transesterification near supercritical methanol condition, Bioresour. Technol. 239 (2017) 378–386, https://doi.org/10.1016/j.biortech.2017.04.073.

[143] S. Tang, C. Qin, H. Wang, S. Li, S. Tian, Study on supercritical extraction of lipids and enrichment of DHA from oil-rich microalgae, J. Supercrit. Fluids 57 (2011) 44–49, https://doi.org/10.1016/j.supflu.2011.01.010.

[144] B.P. Nobre, F. Villalobos, B.E. Barragán, A.C. Oliveira, A.P. Batista, P.A.S.S. Marques, R.L. Mendes, H. Sovová, A.F. Palavra, L. Gouveia, A biorefinery from Nannochloropsis sp.

microalga—Extraction of oils and pigments. Production of biohydrogen from the leftover biomass, Bioresour. Technol. 135 (2013) 128–136, https://doi.org/10.1016/j.biortech.2012.11.084.
[145] Y. Li, F. Ghasemi Naghdi, S. Garg, T.C. Adarme-Vega, K.J. Thurecht, W.A. Ghafor, S. Tannock, P.M. Schenk, A comparative study: the impact of different lipid extraction methods on current microalgal lipid research, Microb. Cell Fact. 13 (2014) 1–9, https://doi.org/10.1186/1475-2859-13-14.
[146] G. Andrich, U. Nesti, F. Venturi, A. Zinnai, R. Fiorentini, Supercritical fluid extraction of bioactive lipids from the microalga Nannochloropsis sp, Eur. J. Lipid Sci. Technol. 107 (2005) 381–386, https://doi.org/10.1002/ejlt.200501130.
[147] K.T. Chen, C.H. Cheng, Y.H. Wu, W.C. Lu, Y.H. Lin, H.T. Lee, Continuous lipid extraction of microalgae using high-pressure carbon dioxide, Bioresour. Technol. 146 (2013) 23–26, https://doi.org/10.1016/j.biortech.2013.07.017.
[148] C.H. Cheng, T.B. Du, H.C. Pi, S.M. Jang, Y.H. Lin, H.T. Lee, Comparative study of lipid extraction from microalgae by organic solvent and supercritical CO2, Bioresour. Technol. 102 (2011) 10151–10153, https://doi.org/10.1016/j.biortech.2011.08.064.
[149] P.C.K. Cheung, Temperature and pressure effects on supercritical carbon dioxide extraction of n-3 fatty acids from red seaweed, Food Chem. 65 (1999) 399–403, https://doi.org/10.1016/S0308-8146(98)00210-6.
[150] R.L. Mendes, J.P. Coelho, H.L. Fernandes, I.J. Marrucho, J.M.S. Cabral, J.M. Novais, A.F. Palavra, Applications of supercritical CO2 extraction to microalgae and plants, J. Chem. Technol. Biotechnol. 62 (1995) 53–59, https://doi.org/10.1002/jctb.280620108.
[151] R.L. Mendes, B.P. Nobre, M.T. Cardoso, A.P. Pereira, A.F. Palavra, Supercritical carbon dioxide extraction of compounds with pharmaceutical importance from microalgae, Inorganica Chim. Acta 356 (2003) 328–334, https://doi.org/10.1016/S0020-1693(03)00363-3.
[152] S.P. Jeevan Kumar, G. Vijay Kumar, A. Dash, P. Scholz, R. Banerjee, Sustainable green solvents and techniques for lipid extraction from microalgae: a review, Algal Res. 21 (2017) 138–147, https://doi.org/10.1016/j.algal.2016.11.014.

CHAPTER 3

Extraction of catechins from green tea using supercritical carbon dioxide

Mukta Agrawal[a], Sunil Kumar Dubey[b], Junaid Khan[c], Sabahuddin Siddique[d], Ajazuddin[a], Swarnlata Saraf[e], Shailendra Saraf[e], Amit Alexander[a]
[a]Rungta College of Pharmaceutical Sciences and Research, Bhilai, Chhattisgarh, India
[b]Department of Pharmacy, Birla Institute of Technology and Science, Pilani, Rajasthan, India
[c]University Teaching Department (Pharmacy), Sant Gahira Guru University, Sarguja, Ambikapur, Chhattisgarh, India
[d]Patel College of Pharmacy, Madhyanchal Professional University, Bhopal, Madhya Pradesh, India
[e]University Institute of Pharmacy, Pt. Ravishankar Shukla University, Raipur, Chhattisgarh, India

Contents

Green Sustainable Process for Chemical and Environmental Engineering and Science
https://doi.org/10.1016/B978-0-12-817388-6.00003-9

1. Introduction

Solvent extraction has been the most common method of separation of desired constituents from the raw materials. The process extensively utilizes organic solvent for the extraction process. Such vast application of organic solvents by various industries globally creates a serious hazard to our environment [1, 2]. Hence, to restrict the manufacturing and use of such ozone-depleting solvents (like chlorofluorocarbon), in 1987, the Montreal Protocol was introduced, which encouraged production and supply of different kinds of solvents for the extraction process that are less hazardous to the environment [3]. Since then, various countries (>170) all over the globe have agreed to be a part of United Nations Environment Program and several amendments were made in the Montreal Protocol. Subsequently, the industries throughout the world were encouraged to adopt the new methodology of extraction that utilizes environmentally safe solvent system [4]. In this context, the concept of supercritical fluid was introduced in the late 1970s for the extraction of natural compounds. Initially, its application by the industries was limited, but now, after the development of various equipment and advanced processes, the industries have shown much more interest in supercritical fluids [5].

Supercritical fluid extraction (SFE) is a process of separation of active constituent/s from the natural material by using the supercritical fluid as extracting solvent. In recent years, the SFE has gained great popularity among scientists and researchers for extraction of plant actives and essential oils from plants and herbs [6, 7]. It is a novel, emerging technique, utilizing the supercritical fluid, which exhibits the unique properties of liquid and gases above the critical point. Carbon dioxide (CO_2), with or without a co-solvent (methanol/ethanol), is the most popular supercritical fluid used for industrial extraction processes [8]. The supercritical CO_2 ($SCCO_2$) is applied above the critical conditions, i.e., at 31°C temperature and 74bar pressure. It is a highly pure, colorless, odorless, non-flammable, non-toxic, safe, recyclable, and cost-effective gas. Another advantage of $SCCO_2$ is it possesses critical points as low as 31.4°C critical temperature and 74 bars critical pressure. In the critical pressure condition, below the critical temperature, the gas compresses and condenses into a dense liquid state. At this point, both the phases are in equilibrium. Then, boiling above the critical temperature, the boundaries between the liquid and gas region disappear and mingle to form a single supercritical phase (Fig. 1). At the supercritical point, various changes in the physical and chemical properties of the CO_2, like density, viscosity, and solvent properties have been observed. Owing to this unique behavior, $SCCO_2$ is widely used in research [9].

In general, the SFE is used for: (a) the collection of desired substances like essential oil from the plant material; (b) separation or removal of unwanted compounds such as

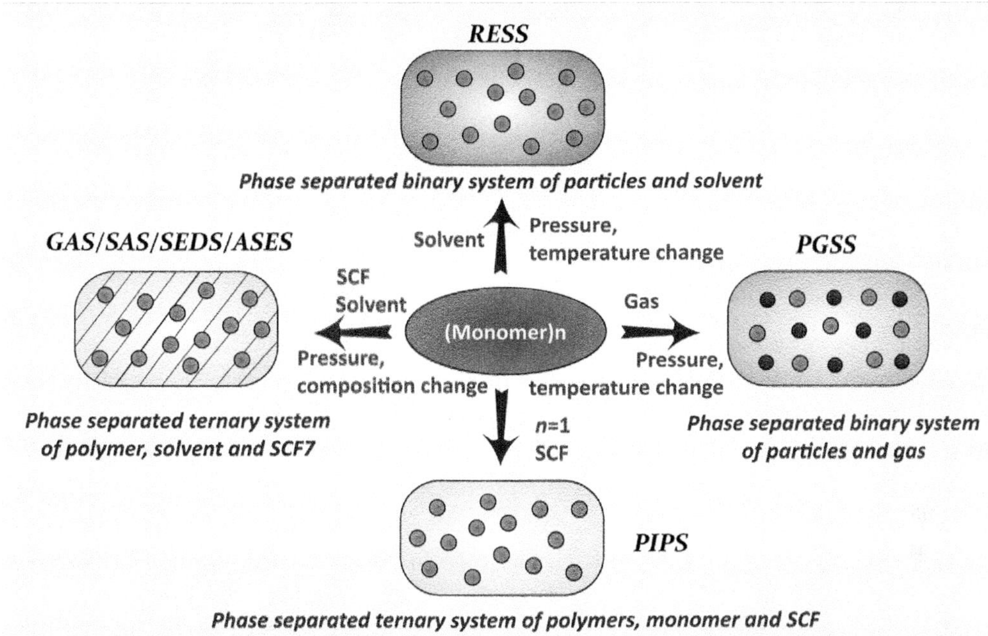

Fig. 1 Phase transition of carbon dioxide into supercritical CO_2 at critical temperature and pressure condition (>31.4°C and 74 bars). *(Adopted and modified from P. Girotra, S.K. Singh, K. Nagpal, Supercritical fluid technology: a promising approach in pharmaceutical research, Pharm. Dev. Technol. 18(1) (2013) 22–38.)*

decaffeination; or (c) sample preparation for the analytical process. The $SCCO_2$ is highly sensitive to very slight changes in the experimental conditions. Thus, a simple modification in the experimental temperature and pressure is sufficient to obtain the desired product [10, 11].

2. Green solvent

Green solvents or bio solvents are natural and environmentally friendly solvent systems, derived from different crops. In general, the organic solvents used in various chemical processes are petrochemical based, which are hazardous for our environment and exert ozone depletion effect [12]. Thus, the Montreal Protocol identified the necessity of modification in the chemical processes to make them environment-friendly. This evolved the concept of green solvents, which can be primarily produced by using either renewable natural resources (such as crops), environmentally harmless supercritical fluids (like supercritical carbon dioxide), or less volatile ionic liquids. Each green solvent has its unique properties, and there is no ideal solvent that suits all the chemical processes. Hence, the selection of a suitable green solvent needs a thorough knowledge and expertise in green chemistry [13]. An ideal green solvent must have the following properties:

(i) It should be non-corrosive.

(ii) It should be utterly biodegradable.

(iii) It can be easily recycled.
(iv) It should be non-carcinogenic.
(v) It should be non-ozone depleting.
(vi) It should be derived from natural and renewable sources.

Some commonly used green solvents are ethyl acetate, bioethanol, terpene, polyether, dibasic ester, siloxane polymer, and supercritical carbon dioxide.

3. Carbon dioxide as a green solvent

$SCCO_2$ is a very popular green solvent used in various industrial and chemical processes. It fulfills almost all the ideal characteristics of a green solvent such as recyclability, biodegradability, non-toxicity, it does not contribute to smog and global warming, and is also easy to remove from the product. The CO_2 produced in various industrial procedures like the production of fertilizer, cement, and other manufacturing units is collected, purified, compressed, and cooled to the liquid state. This liquid CO_2 is then stored in an insulated chamber and used as a solvent in various chemical processes instead of hazardous organic solvents [14].

The critical temperature of CO_2 is near ambient (32.1°C) which makes it suitable for temperature sensitive material, while the critical pressure is 73.7739 bar. The supercritical region of CO_2 is shown in Fig. 2.

It offers a good solvent system for most of the non-polar and some polar, low molecular weight substances. It is not suitable for most of the polar, high molecular weight compounds. These compounds are made soluble in $SCCO_2$ by the application of high process pressure and the addition of small amounts of non-polar or polar co-solvent. Together, some specific surfactants and ligands can also be used to improve the solubility of high molecular weight polar compounds in the $SCCO_2$. Owing to its environmentally friendly and less toxic nature, it is widely used in the production of various bioproducts, extraction of plant constituents, plant processing, and food, as well as the cosmetics industry [15, 16].

4. Green tea composition and bioactives

Tea is one of the most popular and most consumed beverages throughout the world. It is obtained from shoots and leaves of the plant *Camelia sinensis*, which belongs to the family *Theaceae*. Owing to the pleasant taste and aroma and CNS stimulant effect, it is very popular as a daily brew all over the globe. It also exerts some health benefits like reducing the risk of heart disease, improves bone health, weight loss, and acts as an anti-oxidant [17]. Depending on the different methods of processing (enzymatic conversion of constituents), the variety of tea brands are available these days such as black tea, green tea, oolong

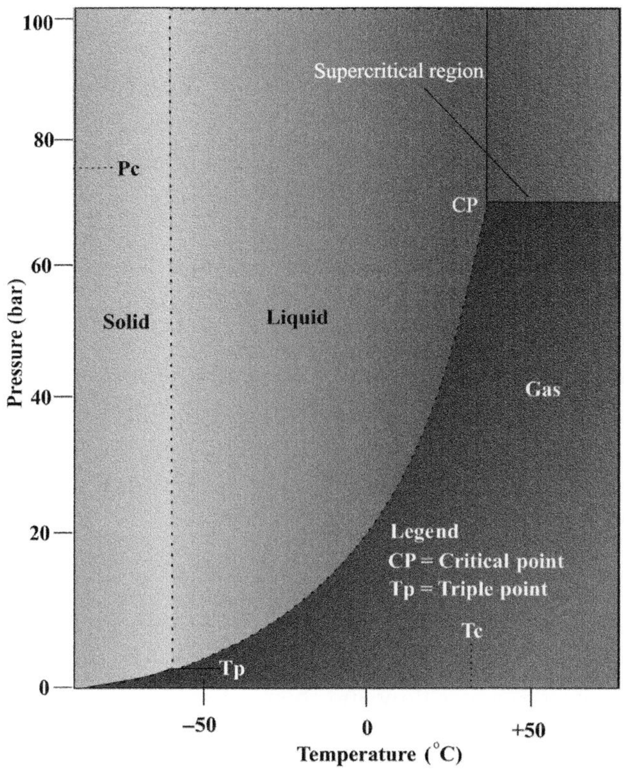

Fig. 2 The phase diagram of the supercritical region of carbon dioxide.

tea, white tea, and pu-erh tea. The organoleptic properties and composition of green tea differ from others because of minimum oxidation of the active constituents. Unlike the other fermented beverages, green tea is rich in polyphenolic compounds (30% of the total dry weight of leaves) [18]. It mainly consists of catechin (primary phenolic compound) and various other phenolic compounds such as flavonols, flavones, flavone glycosides, and phenolic acids. Among these compounds, some primary flavonols and flavone glycosides like quercetin, myricetin, luteolin, kaempferol, apigenin, and 5-O-galloylquinic acid are well-preserved from oxidation. However, the concentration of other phenolic compounds may change during processing [17]. Along with the phenolic compounds, green tea also contains some alkaloids including caffeine (2%–5% of dry weight), theophylline, and theobromine (<0.4%).

4.1 Decaffeination of green tea leaves

Owing to the CNS stimulant effect, anti-oxidant property, and other health benefits, caffeine is considered one of the most consumed alkaloids worldwide. It boosts brain functions like concentration, focus, endurance, and short-term memory, and increases body functions for a short duration [15]. Moreover, it reduces pain and relieves asthma. Apart from these benefits, the constant and long-term use of caffeinated products also causes severe side effects such as anxiety, tachycardia, high blood pressure, sleep deprivation, and miscarriage (in pregnant women), etc. [16] while acute caffeine intake may result in coronary diseases, insulin sensitivity, and high cholesterol level. Due to such adverse effects of caffeine, nowadays, decaffeinated tea has become popular and holds excellent commercial significance.

Commercially, the supercritical carbon dioxide and a liquid solvent both are used for decaffeination of green tea. Steaming and soaking in warm water are the typical steps of all the decaffeination processes, which facilitate the transfer of caffeine from plant material to the solvent system. Various commercial decaffeination processes are:

(i) Direct solvent decaffeination: This process involves the extraction or removal of caffeine from the plant material by keeping it directly in contact with the organic solvent. Traditionally, the chlorinated solvents have been popularly used but their high toxicity and harmful effect on the environment has restricted their application [19]. Apart from this, various other solvents, like ethyl acetate, are popular in commercial practice [20].

(ii) Indirect solvent decaffeination: In this process, the caffeine is moved from the plant material to the organic solvent through a water current, i.e., the extracted caffeine firstly comes into the water flow and then enters the organic solvent. Finally, the water is recycled for future use [21].

(iii) Swiss water process: This process only utilizes the water for caffeine extraction, not the organic solvent [22].

(iv) $SSCO_2$ extraction: This is the most advanced method of decaffeination. It is a more complicated method than liquid extraction but acceptable among consumers because it produces high-quality products.

4.2 Catechin

Catechin is a natural polyphenolic compound belonging to the flavonoid group, abundant in the full range of vegetables, fruits, and other plant-based materials. Catechin is not an essential nutritional requirement, but found effective in preventing various pathological conditions and improving human health [23]. Grape, cherries, pear, apples, and green tea are the primary sources of catechin. Additionally, it is also found in some legumes like green beans and beans. The catechin derivatives like gallocatechin (GC), catechin gallate (CG), gallocatechin gallate (GCG), epigallocatechin (EGC), epigallocatechin gallate (EGCG), epicatechin (EC), and epicatechin gallate (ECG) (Fig. 3) are the primary polyphenols found in fruit wine [24]. Various studies have demonstrated the effectiveness of catechin against coronary heart disease-associated mortality rate and in neurodegenerative disorders [25]. It possesses potential antibacterial activity against a wide range of

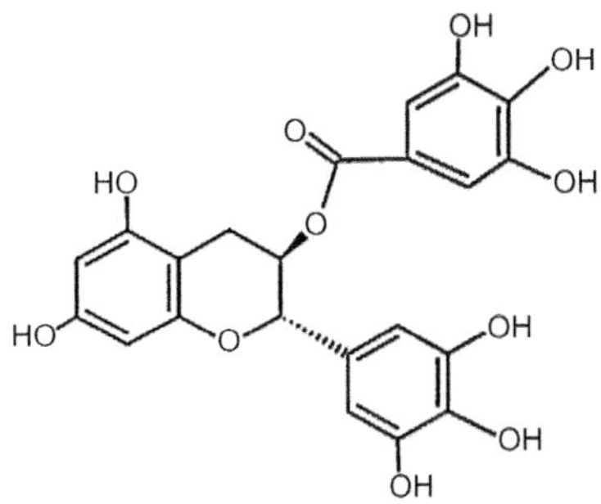

(–)-Gallocatechin gallate (GCG)

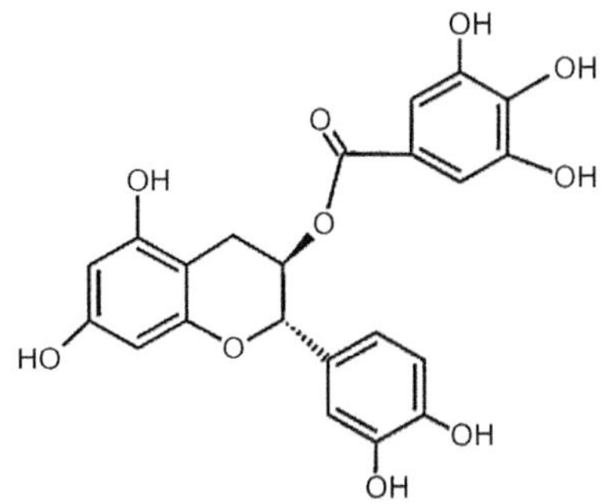

(–)-Catechin gallate (CG)

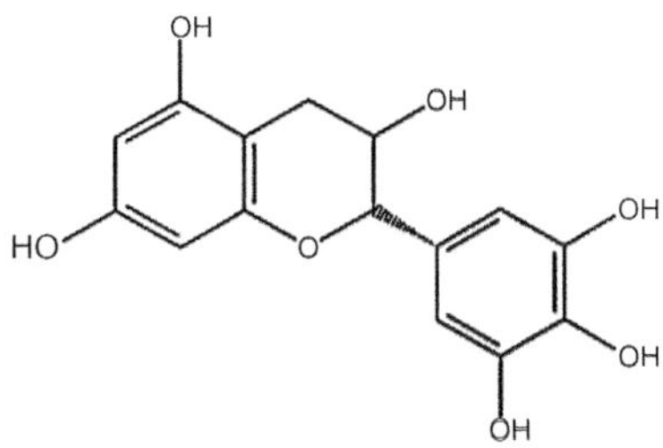

(+)-Gallocatechin (GC)

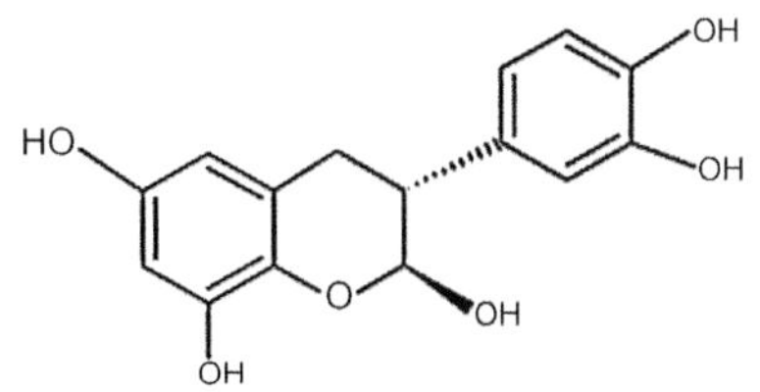

(+)-Catechin (C)

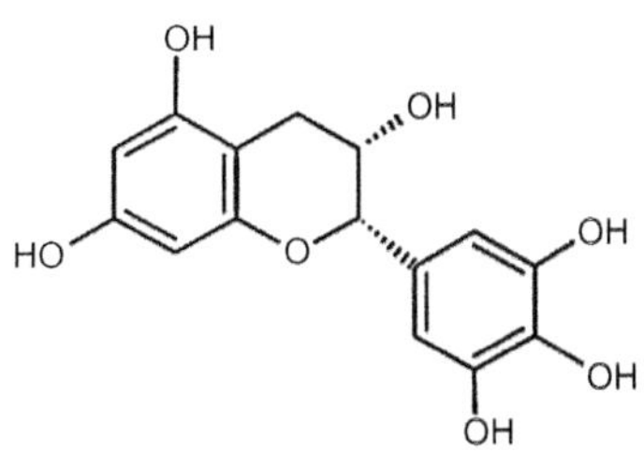

(–)-Epigallocatechin (EGC)

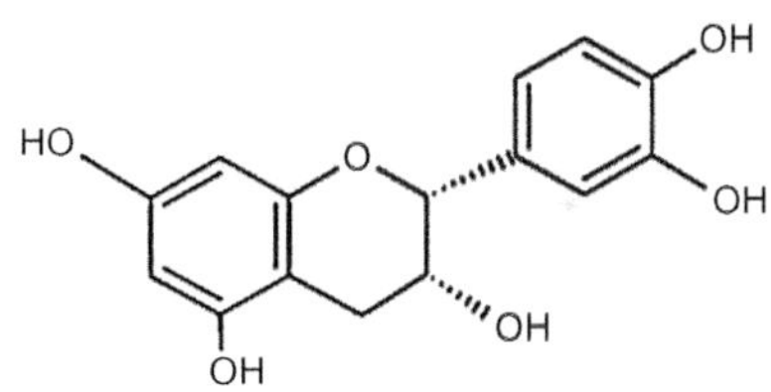

(–)-Epicatechin (EC)

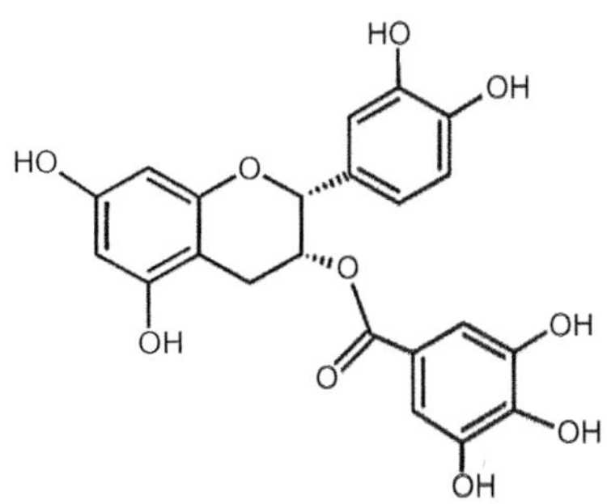

(–)-Epicatechin gallate (ECG)

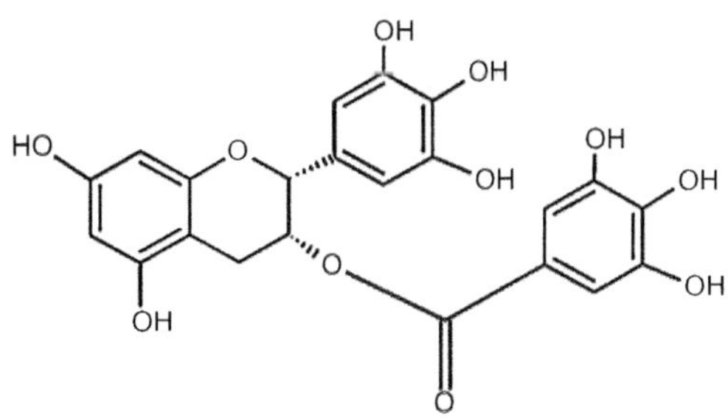

(–)-Epigallocatechin gallate (EGCG)

Fig. 3 Structures of different types of catechins. *(Adopted from P.V. Gadkari, M. Balarman, U.S. Kadimi, Polyphenols from fresh frozen tea leaves (Camellia assamica L.,) by supercritical carbon dioxide extraction with ethanol entrainer—application of response surface methodology, J. Food Sci. Technol. 52(2) (2015) 720–30.)*

Gram +ve and Gram-ve bacteria. Moreover, research shows that the green tea catechin and EGCG can potentially inhibit different kinds of cancer including lung, colon, intestine, esophagus, stomach, liver, bladder, prostate, breast, and skin cancers [26, 27]. In addition, the EGCG supplement significantly improves the memory and learning ability in an AD mouse model [28].

4.3 Physical properties of catechin

The basic physical properties of catechin derivatives are given in Table 1. Catechin is a colorless and crystalline material with an astringent, bitter taste. It is a hydrophilic compound, highly soluble in polar solvents like water and methanol. The melting point of catechin ranges from 175 to 177°C. Various extraction parameters such as time and temperature of extraction and nature of the solvent used may significantly affect the solubility of catechin. The amount of catechin extracted from green tea or other sources depends upon the temperature and time of extraction. The green tea contains EGCG at a greater extent than EGC, ECG, and EC. The ester derivatives (EGCG and ECG) of catechin are more bitter and have higher astringency than the non-esterified catechins (EGC and EC). The extracted catechin can be identified, characterized, and quantified by UV spectroscopic analysis and the absorption maxima of the catechin derivatives ranges typically from 210 to 280 nm [34]. However, the capillary electrophoresis and liquid chromatography are considered the most suitable methods for the separation and quantification of catechin [35].

4.4 Chemical properties of catechin

The chemical structures of some important catechins are shown in Fig. 3. The chemical composition of catechin consists of two aromatic rings with some hydroxyl groups. Based on the substitution, catechins are divided into two categories: a) free or non-esterified catechins and b) esterified catechins. Catechin, *epi*catechin, gallocatechin, and epigallocatechin are the free catechins while catechin gallate, epicatechin gallate, epigallocatechin gallate, and gallocatechin gallate are esterified catechins. In general, the catechins are unstable, and the epimerized form or epi-structure (like epicatechins) is highly unstable, which gets easily converted into non-epi form from the epi-structure upon change in pH and temperature. These are stable at pH <4 and are unstable at alkaline pH (from 4 to 8) [35]. The esterified catechins are more prone to interact with various enzymes and proteins and tend to get precipitated or have higher creaming ability than the non-esterified catechins [20].

4.5 Biological potential of catechin

Various in vitro and preclinical investigation have confirmed the potency of catechin as a protective agent against degenerative disorders. It can be used in immunological

Table 1 Basic physical properties of some essential catechin derivatives

Properties	Catechin	EC	EGC	ECG	EGCG	Reference
Molecular formula	$C_{15}H_{14}O_6$	$C_{15}H_{14}O_6$	$C_{15}H_{14}O_7$	$C_{22}H_{18}O_{10}$	$C_{22}H_{18}O_{11}$	[29]
Molecular weight (g/mol)	290.27	290	306	442	458	[29]
Absorption maxima (nm)	276	280	269	280	273	[29]
Melting Point (°C)	175–177	242	218	257	224	[30]
Solubility	Soluble in water, methanol, and ethanol; insoluble in chloroform					[31, 32]
Taste	Slightly astringent, not bitter	Bitter with a sweet taste	Bitter with a sweet taste	Astringent and bitter	Astringent and bitter	[33]
Appearance	Colorless solid	Yellowish to greenish powder	Light yellow powder	Light yellow powder	White to pink powder	[33]

Adapted from P.V. Gadkari, M. Balaraman, Catechins: Sources, extraction and encapsulation: A review, Food Bioprod. Process. 93 (2015) 122–138.

dysfunctioning as an immune modulator, as an anti-tumor agent, in the treatment of coronary heart disease [36], and many other chronic disorders [37]. EGCG diminishes the telomerase activity in the MCF-7 cell (human breast cancer cell) and induces cellular apoptosis. Hence, it could be used as a promising agent for the treatment of breast cancer [38]. The preclinical and clinical investigations in human and different animal models demonstrate that there are some specific receptors which mediate the anti-cancer activity of EGCG [39]. At the same time, epicatechin has been found useful in dementia, as the preclinical investigation evidenced the improvement in memory of laboratory animals [40]. Zhang et al. in 2012 showed that green tea catechin was capable of effectively reducing the visceral fat in obese Chinese adults after 12-weeks regular intake of green tea as a beverage [41]. Various other studies have also demonstrated the weigh losing ability of catechin and its derivatives in humans [42]. However, much further research is desired to establish the pharmacological potency of catechin.

5. Extraction techniques

Catechins appear in plant tissues in the chemically bounded form with the protein, sugar, or other biological components and are highly susceptible to light, oxidation, alkaline environment, and high temperature. Thus, the extraction of catechin is a difficult process [43]. Additionally, to get a stable form of catechin is also a challenge for researchers. Hence, scientists and industries are highly concerned about developing a proper technique for extraction, isolation, and stabilization of catechin. Infusion, maceration, exhalation, ultrasound extraction, microwave-assisted extraction, and supercritical CO_2 extraction are the common techniques of catechin extraction/isolation. However, pH, temperature, frequency and time of extraction, and the nature of the solvent used are major variables that significantly affect the extraction of catechin [44]. Some of the common extraction techniques of catechin from green tea leaves are discussed below.

5.1 Conventional extraction

Conventional extraction is based on the principle of solubility, where constituents present in a solid moiety (plant material, in this case) are solubilized and extracted out using a suitable solvent. The extraction of polyphenols using a combination of heat and a preferable solvent or a group of solvents is the most commonly used method. As the catechins are mostly considered to be water-soluble, polar solvents are preferred for their extraction [45].

Conventional methods of extraction include hot water extraction and soxhlet extraction in the case of organic solvents. The most straightforward technique is to suspend green tea leaves in hot/boiling water for a considerable period, usually between 2 and 8 h. While extracting catechins from tea leaves, separation of residual caffeine is also necessary. Dichloromethane is considered to be the most suitable solvent for the decaffeination process [46]. Bharadwaz and Bhattacharjee [47] have used ethyl acetate as a solvent to extract out polyphenols from the decaffeinated aqueous extract. Their experiments yielded the polyphenolic rich extract with a high content of polyphenolic compounds to about 19.33%, with only 2.56%, of caffeine. The tea to water ratio was another critical factor that affected the efficiency of concentration along with the pH [47]. Wati et al., in 2009, extracted the polyphenols from fresh tea leaves and found a tea to water ratio of 1:20 and a pH ranging from 4 to 5 as optimum for maximum yield [48]. In a similar study by Vuong et al. in 2011, pH <6 with a tea to water ratio of 1:50 (g/ml) showed maximum yield [49].

Extraction temperature, time of heating, and concentration of solvent were other critical factors influencing the yield and quality of the end product. In one study, Liang et al. (2007) reported that heating the water to 100°C brought about changes in the isomeric configuration of catechins and suggested a temperature of 80°C to maintain the integrity of catechins. Moreover, in the same study, the authors also concluded that 70% ethanol was better suited for the extraction of fresh leaves, whereas 50% ethanol gave better yield than water while extracting polyphenols from dried leaves [50].

In extraction through soxhlet technique, apart from the above-discussed parameters, factors like particle size of the sample, the solubility of polyphenols in the selected organic solvent/s, and pH of the medium affect the quality and yield of the extract. The extraction process can also be accelerated by using high temperatures resulting in an increase in efficiency with reduced time. Several studies have stressed the use of a proportionate mixture of different solvents to improve the polyphenolic extraction. The conventional extraction processes are easy to use and show high efficiency and extraction yields. However, there are certain limitations, like high temperature, which can destroy thermolabile constituents, and the need for drying and purification of contents [51].

5.2 Pressurized liquid extraction

To improve the efficiency and reduce the time of extraction via a conventional extraction process, the concept of pressurized liquid was introduced. It includes the use of high temperature—above the boiling point—and high-pressure conditions up to 3000 psi. The use of high pressure forces the solvent into the powdered sample and facilitates faster filling of the extraction cell. It also promotes the rate and extent of diffusion and increases desorption. The overall extraction time is drastically reduced. Studies have shown that pressurized extraction has improved the extraction of catechins and the method is more rapid than other methods like magnetic stirring and ultrasound-assisted extraction.

Experimental work on four different species of green tea has shown pressure values around 150 bars as optimum for extraction of catechins with a constant temperature around 75 ± 1°C for 60 min [52]. Extraction of catechins was also performed using ethyl lactate as a solvent and applying pressure around 10 MPa at 100°C followed by selective precipitation using the supercritical carbon dioxide- antisolvent technique. Since caffeine is more soluble in ethyl lactate, a combination of ethyl lactate with CO_2 was used for extracting out catechins as precipitate and removing caffeine as extracted in the solvent [53].

Pressurized extraction offers advantages like quick extraction with limited use of a solvent. The limited use of organic solvents makes the technique eco-friendly. By use of simple machinery, the process of pressurized extraction is easily programmed. The pressure conditions are usually varied across the liquid compressed region where the liquids are mostly uncompressed. The operation is carried out below the critical temperature to maintain the liquid state of the solvent and increase the rate of extraction. However, again the use of high temperature limits the use of this technique as thermolabile components are hardly able to sustain such high temperatures [54].

5.3 Microwave-assisted extraction

As the name suggests, this technique utilizes microwaves to improve solvent penetration and hence extraction efficiency and time. The energy consumption in microwave-assisted extraction is also low compared to conventional methods. This technique utilizes the essential moisture that is present in the dry plant material/s. The microwaves create zones of high temperatures, which facilitate the extrusion of internal moisture, thus rupturing the cell wall. Therefore, the desorption of constituents from cells is a classic feature of this extraction procedure. Another salient feature of this method is that there is no contact between the sample under study and the source of heat. The use of water or another suitable solvent further improves the yield of extraction by this method [55, 56].

Microwave-assisted extraction procedure has shown efficient extraction of catechins from green tea leaves within a short period of 4 min. Temperature range between 80°C and 100°C with microwaves for extraction has shown results better than conventional methods. Multiple exposures for short durations of a few seconds have resulted in excellent yields. The extraction can be carried either in a closed system with limited sample quantity and high temperature or in an open system with longer duration of time for extraction. In one study utilizing microwave-assisted and ultrasonic extraction of catechins from green tea leaves, the microwave assisted extraction was found to be significantly more efficient regarding yield and total polyphenolic contents. A temperature around 65°C for 7.8 min was found optimum for extraction, in this case. The duration of exposure to heat seems to be a key factor influencing extraction efficiency [57].

This statement was further supported by the study carried out by Albuqerque et al. in 2017, who demonstrated that the efficiency of catechins extraction from the fruits of *Arbutus unedo* L. was significantly influenced by exposure time and temperature. The parameters were optimized using response surface methodology, as represented in Fig. 4. The extraction was carried out in a closed system. The extraction efficiency was observed maximum

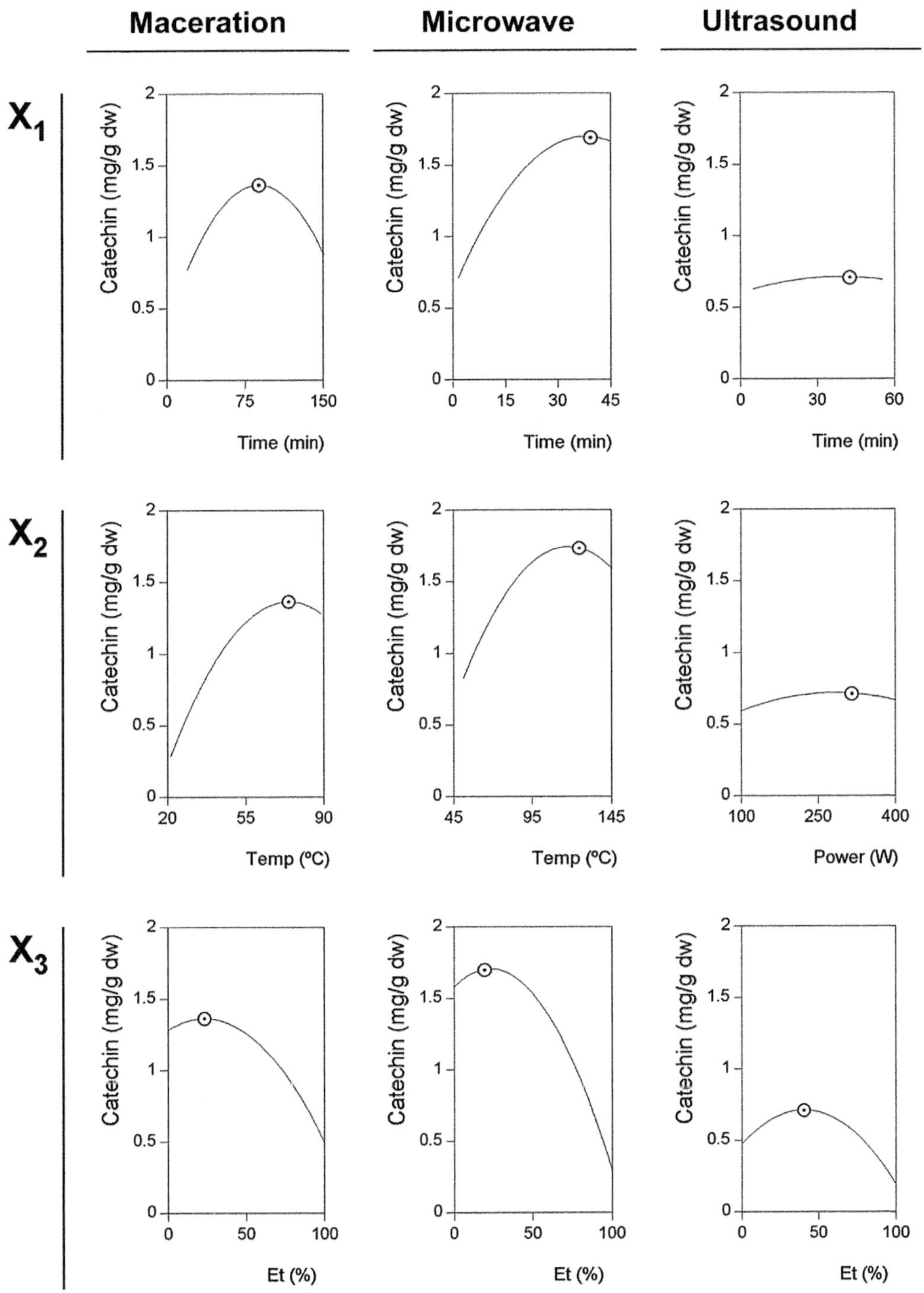

Fig. 4 2D responses for different variables viz. time, temperature and % ethanol (Et%) used. The dots represent the optimum value of each variable. *(Adopted from B.R. Albuquerque, M.A. Prieto, M.F. Barreiro, A. Rodrigues, T.P. Curran, L. Barros, I.C.F.R. Ferreira, Catechin-based extract optimization obtained from* Arbutus unedo *L. fruits using maceration/microwave/ultrasound extraction techniques, Ind. Crop Prod. 95 (2017) 404–415.)*

when the exposure time was around 42.2 ± 4.1 min, which was less compared to maceration (93.2 ± 3.7 min). Microwave-assisted extraction technique has shown better results in the extraction of catechins as compared to simple maceration and ultrasound assisted extraction. Microwave-assisted extraction provides extract with higher percentage purity within a short spell of time. The technique is also comfortable in operation, and the results are highly reproducible. Moreover, the use of solvent and consumption of energy are also decidedly less important compared to conventional techniques of extraction [58].

5.4 Solid phase extraction

This is another exciting technique used for extracting selective constituents from samples in gaseous and liquid states wherein the desired components are retained in the solid phase. It was developed as an alternative technique to liquid–liquid extraction. Initially, simple sorbents like silica and alumina were used, which were slowly replaced by polymer-based matrices. With the development of polymer science, a wide range of polymers are now being used as solid sorbents. Polymers offer advantages like broader pH range and ability to sustain the effect of different solvents and buffers that otherwise destroy the classical silica-based matrices. Polymers are mostly regular in shape and give rise to homogeneous packed beds. The use of molecularly imprinted polymers has further advanced the technique. These are selective polymers having marked sites for extracting constituents of interest [59].

Most of the MIPs are built using methacrylic acid-based monomers. Such methacrylic based MIPs show better adsorption of catechins than the non-template imprinted polymers as there are two regions of catechins adsorption in the MIPs, viz. template core and the non-template surface. Solid phase extraction of using catechins as templates with methacrylic acid-based monomer has shown higher adsorption of catechins from green tea leaves. Using epigallocatechin gallate as the model has shown the highest adsorption of catechins, followed by epicatechin and catechins [60].

5.5 Ultrasound-assisted extraction

Like microwaves, ultrasound waves have also been utilized for extraction of chemical constituents from plant materials. As per the Royal Society of Chemistry, ultrasound extraction is the process of transferring a substance from any matrix to an appropriate liquid phase, assisted by sound waves (>20 KHz in frequency) that propagate through the liquid media. Ultrasound waves show mechanical action through the pressure waves creating air bubbles or cavitation. The expanding bubbles create turbulence in the cellular structures resulting in disruption of the cell membrane [61]. This technique has been widely used for the extraction of polyphenols. Ultrasound waves can be used with any organic solvent for extraction, thus offering a broader scope of application. The application of ultrasound waves can be made in two ways, by using either probes or a bath [62].

Like other extraction techniques, extraction by ultrasonication is also affected by temperature and time. Ultrasonic radiation has been shown to increase the number of catechins extracted from green tea at low temperature with water [63]. Higher yields of catechins have been obtained with low levels, around 20% of ethanol as a solvent, with shorter extraction time, around 27 min at fixed temperature 24°C [64]. The low concentration of ethanol as the solvent also reduced the caffeine content of the extract. When compared with conventional extraction, the extraction of catechins by ultrasound intervention has been shown to reduce the caffeine content in the extract by 30%; thus, yielding a relatively pure form of extract. Moreover, the extraction by this method was rapid, with extraction time reduced to about 70%–78%. The solvent mixture containing 2% phosphoric acid and 40% ethanol was suitable for simultaneous extraction of catechins and caffeine from green tea leaves. In a study similar to the above-discussed one, Choung et al. in 2013 reported the optimal extraction conditions for extracting green tea catechins using ultrasound-assisted extraction method as 40% ethanol for 2 h at 40°C. Compared to reflux extraction and conventional extraction at room temperature, the ultrasonic method yielded the highest content of catechins [65].

The primary advantage of this extraction method was the higher yield at a lower temperature; thus, protecting the chemical integrity of the plant material in the extract. Researches have shown a better yield of polyphenolic contents at a lower temperature than higher temperature by ultrasonication [66].

5.6 Aqueous two-phase extraction

Aqueous two-phase extraction involves the use of immiscible polymers. Although this technique was primarily used for separating proteins in the biotech industry, its utility has now been extended to the extraction of phytoconstituents. The polymers are added into the water where two immiscible phases are formed. These two phases can then extract different components according to solubility [67]. There is also a third option for two-phase extraction system, which includes ionic liquids and short chain alcohols. Such a solvent system has shown efficient extraction of codeine and papaverine [68]. Polyphenols like curcuminoids have also been successfully extracted using ionic solvent 1,3-dimethylimidazole bis[(trifluoromethyl)sulfonyl]imide [69]. A two-phase extraction system comprising of PEG6000, water, and ammonium sulfate at pH 6.5, with a ratio of PEG 0.1037, g/g and ammonium sulfate 0.0925 g/g the extraction was found efficient enough to yield extraction efficiency of about 89.11%. This process has shown similar capability with significantly shorter time when extracted in a microextractor rather than macroextractor [70].

The polymer-based two-phase extraction technique offers advantages such as being eco-friendly, and resulting in sufficiently high extraction efficiency with limited use of time and energy. The yields are higher even at low temperatures—around room temperature—and thus it is also useful for extraction of thermolabile constituents. However, the technique suffers from certain limitations as well. The recovery of the solvent is difficult, and the cost of polymers is also higher than conventional solvents [71].

5.7 Supercritical carbon dioxide extraction

The supercritical fluid extraction uses supercritical solvents to extract constituents from the solid as well as liquid matrices. CO_2, with a critical temperature closer to room temperature, (31°C) is among the eco-friendliest solvents used for extraction. Regarding solvation efficiency, it is no less than the organic solvent, hexane. It is used either alone or in combination with other alcoholic solvents or extractant. CO_2 offers many advantages over toxic organic solvents and the products from the extraction are suitable for edible purposes. The extraction qualities like solubility and density of CO_2 can be easily modified as per requirement with minute changes in temperature and pressure. Another distinctive feature of CO_2 extraction is that there is almost no loss of solvent or extractives in the form of any residue [72]. Chang et al. (2000) successfully extracted green tea catechins using supercritical CO_2 and described specific equipment for the same (Fig. 5) [11].

Co-solvents like water and ethanol are, in some cases, helpful in improving the efficiency of supercritical CO_2 for extraction of green tea catechins. It is believed that these co-solvents enhance the hydrogen bonding and Van der Walls forces between the solvent and solutes. It is also suggested that water is not a good co-solvent with CO_2 for extraction of catechins [73]. One study has suggested optimum conditions of pressure 30 Mp,

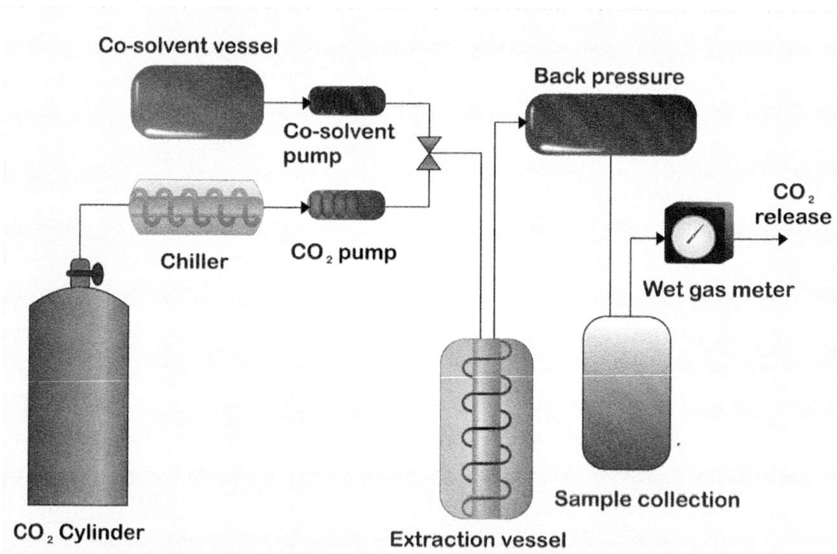

Fig. 5 Schematic representation of equipment used for supercritical CO_2 extraction of polyphenol from green tea.

temperature 47.99°C, with a duration of 40 min, using 50% ethanol as co-solvent for better extraction of green tea catechins. Apart from extraction, CO_2 is also used for decaffeination of green tea and obtaining relatively pure catechins extracts [74].

Temperature is the key factor in improving the efficiency of CO_2-based extraction of green tea polyphenols. Substantial improvement in extraction efficiency has been noted with temperature increase from 313 to 323 K. The extraction of polyphenols at a lower pressure, around 15–25 MPa, has resulted in getting polyphenol-rich extract, and hence such a pressure range has been recommended to prevent the loss of catechins [75].

6. Standardization of method

Standardization of the extraction process for catechin-like phenolic compounds is a multifarious mechanism. It involves chances of oxidation leading to degradation of active constituents. The possibility of using low-cost lignocellulosic adsorption has been described as a suitable method for large-scale extraction of tea components, despite large initial setup costs. It is essential to select an environment-friendly method for the extraction of phytoconstituents, which also preserves the physicochemical and biomedical properties of the extract. Supercritical fluid extraction is one such extraction technique. It offers excellent control over the operational parameters like pressure, temperature, co-solvent, etc., and represents an optimized technique, which enhances the extraction efficiency [8, 11, 76, 77]. Park et al. reported that the extraction parameters such as temperature, pressure, and co-solvents have a significant effect on the extraction of catechin from tea using SFE-CO_2. They revealed that extraction of 7 g of tea per 100 g of CO_2 with 95% v/v ethanol at pressure 300 bar and temperature 700°C for 120 min, extracted around 97% of caffeine. However, this method also results in the extraction of some undesirable components from the tea. Therefore, future research work is needed to derive the optimum SC-CO_2 conditions to achieve effective extraction of catechins in tea [76, 78].

7. Operating parameters

The quantity and quality of the final extract may vary according to various operational conditions like temperature, pressure, water content, and extraction time, etc. The effects of such parameters are explained below.

7.1 Effect of temperature and pressure

Pressure and temperature are the two crucial parameters controlling the SC-CO_2 extraction process in which solvent properties of supercritical fluids may be tuned by altering pressure and temperature conditions. The solubility of targeted compounds is mainly determined with the aid of SC-CO_2 density. Generally, for the extraction of a phenolic compound like catechin, SC-CO_2 density increases with pressure at constant temperature and decreases with temperature at constant or higher pressures. The majority of the studies indicate that an increase in the extraction pressure of supercritical carbon dioxide leads to an increase in the yield of obtained product. For instance, Bimakr and co-workers in 2011 studied the effect of operational conditions like pressure, temperature, and extraction time on the SC-CO_2 extraction of spearmint flavonoid. The study was performed using response surface methodology, full factorial completely randomized response model. The SC-CO_2 extraction was performed at a temperature of 60°C and 200 bars pressure for 60min and the results were compared with Soxhlet extraction data. The results demonstrated that the Soxhlet extraction results in higher crude content while the SC-CO_2 extraction results in higher amount of active compounds (flavonoids) [79]. [79]

7.2 Effect of flow rate

The speed of the supercritical fluid flowing through the cell has a strong influence on the extraction efficiency of the process. While extracting the phytoconstituents, the slower the fluid velocity, the deeper it penetrates the crude plant, thereby, enhancing productivity. Papamichail and co-workers in 2010 studied the effect of SC-CO_2 fluid on the extraction of milled celery seed oil, and they found that flow rate of solvent can be correlated with the amount of oil extracted with time, as an increase in flow rate enhanced the extraction of constituents of celery seeds [80]. In another study by Topal and co-workers in 2016, they tested different flow rates of SC-CO_2 ranging from 1.5 to 4.5mL/min and found that the yield of lycopene extract was maximum at a flow rate of 2.5mL/min, and a further increase in flow rate showed a decline in the extraction of lycopene [81].

7.3 Effect of organic modifier

Supercritical carbon dioxide is one of the known non-polar solvents that can be used for the extraction of non-polar compounds, and other low molecular weight volatile compounds. It can also be used for the extraction of polar compounds, but the yield during the extraction process is quite low. Nevertheless, the addition of small amounts of organic solvent or co-solvent modifier in the extraction process can enhance the efficiency of SC-CO_2 and ultimately improve the yield of phytoconstituents. Use of solvent modifiers such as methanol and ethanol may reduce the activation energy of analyte desorption in order to increase its movement towards the fluid. In an experiment, researchers studied the extraction of polyphenols using 15% water:SC-CO_2 followed by 15% ethanol: SC-CO_2; giving the higher yield of constituents up to 7.3% when compared with the SC-CO_2 alone [82]. Additionally, Hang et al. studied separation of catechins of green tea using SC-CO_2. He observed that content of catechins in extracts was increased 4.4-fold with the use of 95% ethanol as a co-solvent [80].

7.4 Extraction time

Extraction time is another crucial parameter that needs to be optimized for maximizing the yield of compound from plant matrices. To enhance the extraction efficiency of the SC-CO_2 process, it is necessary to prolong the contact time of the solvent with the plant matrix. Because the plant material has a complex physical structure, the penetration of the solvent and the diffusion of targeted compounds in the particles are very slow. Therefore, extraction time is usually limited to the quick extraction period since the yield recovered in the slow extraction period is near to negligible. Bimakr et al. (2013) studied the effect of extraction time at 60, 90, and 120 min on crucial phyto-compound recovery from Benincasahispida seeds. They found that the highest crude extraction yield (176.30 mg extract/g dried sample) was obtained at 97 min. As per their statement, 60 min of extraction time was not sufficient for its complete extraction of the bioactive compound, while thermal degradation was observed at 120 min of extraction time that led to dropped yields value [79]. Reducing the extraction time could also reduce costs as well as improve energy efficiency and can affect extraction efficiency.

Additionally, Bimakr et al. (2011) studied the effect of dynamic time on the SC-CO_2 extraction of bioactive compounds from *Mentha spicata* L. leaves. They observed that the solvent power of SC-CO_2 was dropped at 100 bar pressure due to the lower CO_2 density, but the maximum yield was obtained at 90 min dynamic extraction time. Application of higher pressures (200 and 300 bar) resulted in an increased extraction rate and enhanced yield up to 60 min of dynamic extraction time [76].

7.5 Particle size

Another critical operating parameter is the particle size of the raw material in the SC-CO_2 extraction processes. Larger particles size of plant material require long extraction times, which could directly influence yield value of bioactive compound. In contrast, a smaller sized particle is associated with a larger surface area and diminished intra-particle diffusion resistance. Besides, the extraction process can be enhanced by reducing the diffusion path, leading to increased extraction efficiency. For example, Özkal and Yener revealed that, by decreasing the particle size, the yield increased by 24% in the SC-CO_2 extraction of flax-seed oil. However, further reductions in particle size have caused drop in yields value; the possible reason being aggregation of the particles, i.e., strong cohesiveness of particles [83].

7.6 Drying time

Like extraction time, drying time also plays an essential role during the SC-CO_2 process. To enhance the extraction efficiency of the SC-CO_2 method, it is necessary to optimize drying time. An optimum drying is required to strengthen the stability of the catechin while minimizing the packaging requirements and reducing transport weight. Usually, sufficient drying time is used to remove the moisture through the simultaneous transfer of heat and humidity from the bioactive compound, which shows a significant impact on its cost and quality of the product. As the catechins are heat and oxygen sensitive, the selection of the drying method is a critical step. Hence, the optimal drying method like spray drying, rotary evaporation, vacuum drying, and freeze-drying must be energy efficient in order to maintain the natural physical and biochemical quality of the catechin [53]. Therefore, further desirable studies are required to identify the efficient methods and conditions for drying the catechin extracts.

7.7 The water content in the supercritical fluid extraction

Water content is one of the most significant characteristics in determining the quality of the output of extraction processes. Water is considered to be soluble at approximately 0.3% v/v in supercritical CO_2. The presence of water, however, may either assist in or be an impediment to the diffusion of supercritical carbon dioxide depending upon the type of compounds targeted [80]. In a study, Ivanovic and co-workers investigated the effects of moisture content on the extraction of essential oils from the flowers of *Helichrysum italicum* and they practically found that pre-soaking of the samples in water led to an increase in the yield value of about 40% and reduced CO_2 consumption with operating pressure and temperature at 100 bar and 40°C, respectively.

Moreover, preparation of plant materials is a critical step for the extraction process. Fresh plant matrices contain a high moisture content, which causes mechanical problems due to ice formation. Furthermore, the extraction efficiency will reduce due to the high water solubility of compounds dissolved in the aqueous phase, while the water solubility (0.3%) in SC-CO_2 is very low [84]. Therefore, it is necessary to control the water content of plant matrices by drying or mixing them with chemicals like sodium sulfate and silica gel. It has also been stated that when the moisture content of *Benincasahispida* seeds was increased from 5% to 20%, the crude extraction yield was decreased by up to 38% [83, 85]. Saldana and co-workers in 2014 found that 10% moisture content was adequate to reach high recovery of β-carotene from apricot pomace by SC-CO_2 extraction [83].

8. Qualitative assessment

Because tea extracts usually contain a combination of various types of phenolic compounds of different polarities, their extraction and isolation have been a critical challenge. For the identification and characterization of tea extract, several sophisticated qualitative techniques such as TLC, column chromatography, flash chromatography, HPLC, and LC–MS/MS have been used to achieve pure bio-compounds. The purified compounds have been used for the determination of their structures and biological activities. Moreover, non-chromatographic techniques such as immunoassay, microbial screening, and

Fourier-transform infrared spectroscopy (FTIR), have also been employed to facilitate the identification of the catechin-like phytochemical constituent [86, 87].

8.1 Microbial aspects

The health benefits of catechin extract, derived from the leaves of the *Camellia sinensis* plant, have been studied for many years. Recently, scientists have developed their interest in the use of catechin as an antimicrobial agent. Mostly four types of catechins (polyphenols) have been found in green tea extract. These are (−)-epicatechin (EC), (−)-epicatechin-3-gallate (ECG), (−)-epigallocatechin (EGC), and (−)-epigallocatechin-3-gallate (EGCG). Out of these, ECG, EGC, and EGCG have shown promising antimicrobial activity against various classes of organisms. Sirk and co-workers suggested that catechins can bind directly to the bacterial lipid bilayer cell membrane, causing damage to the bacterial cell membrane. This damage can then lead to a variety of related antimicrobial effects. Moreover, catechin-mediated photo-protection of human skin against bacterial infection and topical tea ointment for the treatment of impetigo have been reported [88–90].

9. Conclusion

Extraction is a process of separation of desired active substances from the plant material. The effective extraction of phytoconstituents depends upon the various factors including solubility, pH, extraction time, and temperature. The selection of extraction technique is based on the niche of the compound and the desired quality. The extraction of a polyphenolic compound like catechin mainly utilizes solvent extraction process due to its ease of applicability and excellent recovery. The solvent extraction techniques include heat flux solvent extraction (Soxhlet), microwave-assisted extraction, ultrasound extraction (sonication), and supercritical carbon dioxide extraction. Among these, supercritical fluid extraction provides an alternative non-hazardous solvent system.

CO_2 is not a suitable solvent for extraction of catechin because of its non polar nature. To overcome this, polar organic solvents such as ethanol, acetone, methanol, and water can be added to the primary solvent system before the extraction. There are some reports available on the extraction of catechin from green tea with the addition of polar co-solvents. $SSCO_2$ is a commonly used method for the extraction because it is easily recoverable from the extract due to high volatility; in fact, it does not leave any harmful residue in the extract. Despite the benefits of conserving the structural integrity of catechin, high capital investment is required to set up this method and equipment, and elevated pressure is also necessary for the operation of the equipment. Further, the compression of the solvent requires intense recycling measures to reduce the energy cost.

Acknowledgment

The author would like to acknowledge Rungta College of Pharmaceutical Sciences and Research, Kohka, Kurud Road, Bhilai, Chhattisgarh, India for providing necessary facilities for the compilation of this work.

Conflict of interest

None.

References

[1] M. Herrero, J.A. Mendiola, A. Cifuentes, E. Ibanez, Supercritical fluid extraction: recent advances and applications, J. Chromatogr. A 1217 (16) (2010) 2495–2511.
[2] D. Bhosle, A. Janghel, S. Deo, P. Raut, C. Verma, S.S. Kumar, M. Agrawal, N. Amit, M. Sharma, T. Giri, Emerging ultrasound assisted extraction (UAE) techniques as innovative green technologies for the effective extraction of the active phytopharmaceuticals, Res. J. Pharm. Technol. 8 (7) (2015) 963–970.
[3] S. Deo, A. Janghel, P. Raut, D. Bhosle, C. Verma, S.S. Kumar, M. Agrawal, N. Amit, M. Sharma, T. Giri, Emerging microwave assisted extraction (MAE) techniques as an innovative green technologies for the effective extraction of the active phytopharmaceuticals, Res. J. Pharm. Technol. 8 (5) (2015) 655.
[4] E. Ramsey, Q. Sun, Z. Zhang, C. Zhang, W. Gou, Mini-review: green sustainable processes using supercritical fluid carbon dioxide, J. Environ. Sci. (China) 21 (6) (2009) 720–726.
[5] G. Brunner, Supercritical fluids: technology and application to food processing, J. Food Eng. 67 (1) (2005) 21–33.
[6] M.M.R. de Melo, A.J.D. Silvestre, C.M. Silva, Supercritical fluid extraction of vegetable matrices: applications, trends and future perspectives of a convincing green technology, J. Supercrit. Fluids 92 (2014) 115–176.
[7] P. Raut, D. Bhosle, A. Janghel, S. Deo, C. Verma, S.S. Kumar, M. Agrawal, N. Amit, M. Sharma, T. Giri, Emerging pressurized liquid extraction (PLE) techniques as an innovative green technologies for the effective extraction of the active phytopharmaceuticals, Res. J. Pharm. Technol. 8 (6) (2015) 800–810.
[8] W.-J. Kim, J.-D. Kim, J. Kim, S.-G. Oh, Y.-W. Lee, Selective caffeine removal from green tea using supercritical carbon dioxide extraction, J. Food Eng. 89 (3) (2008) 303–309.
[9] P. Girotra, S.K. Singh, K. Nagpal, Supercritical fluid technology: a promising approach in pharmaceutical research, Pharm. Dev. Technol. 18 (1) (2013) 22–38.
[10] J.M. Walsh, G.D. Ikonomou, M.D. Donohue, Supercritical phase behavior: the entrainer effect, Fluid Phase Equilib. 33 (3) (1987) 295–314.
[11] C.J. Chang, K.-L. Chiu, Y.-L. Chen, C.-Y. Chang, Separation of catechins from green tea using carbon dioxide extraction, Food Chem. 68 (1) (2000) 109–113.
[12] S.P.J. Kumar, S.R. Prasad, R. Banerjee, D.K. Agarwal, K.S. Kulkarni, K.V. Ramesh, Green solvents and technologies for oil extraction from oilseeds, Chem. Cent. J. 11 (2017) 9.
[13] K. Häckl, W. Kunz, Some aspects of green solvents, C. R. Chim. 21 (6) (2018) 572–580.
[14] J. Peach, J. Eastoe, Supercritical carbon dioxide: a solvent like no other, Beilstein J. Org. Chem. 10 (2014) 1878–1895.
[15] S.P. Nalawade, F. Picchioni, L.P.B.M. Janssen, Supercritical carbon dioxide as a green solvent for processing polymer melts: Processing aspects and applications, Prog. Polym. Sci. 31 (1) (2006) 19–43.
[16] R. Tiwle, A. Ajaz, T.K. Giri, D.K. Tripathi, V. Jain, A. Alexander, An exhaustive review on solubility enhancement for hydrophobic compounds by possible applications of novel techniques, Trends Appl. Sci. Res. 7 (8) (2012) 596–619.

[17] D. Villanueva-Bermejo, G. Reglero, T. Fornari, Recent advances in the processing of green tea biomolecules using ethyl lactate: a review, Trends Food Sci. Technol. 62 (2017) 1–12.
[18] Q.-Y. Shi, V. Schlegel, Green tea as an agricultural based health promoting food: The past five to ten years, Agriculture 2 (4) (2012) 393.
[19] R. Wang, Y. Zhang, Q. Lan, T.R. Holford, B. Leaderer, S.H. Zahm, P. Boyle, M. Dosemeci, N. Rothman, Y. Zhu, Q. Qin, T. Zheng, Occupational exposure to solvents and risk of non-Hodgkin lymphoma in Connecticut women, Am. J. Epidemiol. 169 (2) (2009) 176–185.
[20] Q.V. Vuong, P.D. Roach, Caffeine in green tea: Its removal and isolation, Sep. Purif. Rev. 43 (2) (2014) 155–174.
[21] A. Farah, T. de Paulis, D.P. Moreira, L.C. Trugo, P.R. Martin, Chlorogenic acids and lactones in regular and water-decaffeinated arabica coffees, J. Agric. Food Chem. 54 (2) (2006) 374–381.
[22] K. Ramalakshmi, B. Raghavan, Caffeine in coffee: its removal. Why and how? Crit. Rev. Food Sci. Nutr. 39 (5) (1999) 441–456.
[23] S. Kondo, K. Tsuda, N. Muto, J.-E. Ueda, Antioxidative activity of apple skin or flesh extracts associated with fruit development on selected apple cultivars, Sci. Hortic. 96 (1) (2002) 177–185.
[24] M.G.L. Hertog, P.C.H. Hollman, B. van de Putte, Content of potentially anticarcinogenic flavonoids of tea infusions, wines, and fruit juices, J. Agric. Food Chem. 41 (8) (1993) 1242–1246.
[25] J.H. Stein, J.G. Keevil, D.A. Wiebe, S. Aeschlimann, J.D. Folts, Purple grape juice improves endothelial function and reduces the susceptibility of LDL cholesterol to oxidation in patients with coronary artery disease, Circulation 100 (10) (1999) 1050–1055.
[26] C.S. Yang, J.Y. Chung, G.Y. Yang, C. Li, X. Meng, M.J. Lee, Mechanisms of inhibition of carcinogenesis by tea, BioFactors (Oxford, England) 13 (1–4) (2000) 73–79.
[27] P. Shukla, A. Singh, S. Gawri, A. Alexander, S. Sonwane, In vitro propagation of Barleria prionitis Linn and its antibacterial activity, Int. J. Pharma Prof. Res. 2 (1) (2011) 170–172.
[28] A. Smith, B. Giunta, P.C. Bickford, M. Fountain, J. Tan, R.D. Shytle, Nanolipidic particles improve the bioavailability and alpha-secretase inducing ability of epigallocatechin-3-gallate (EGCG) for the treatment of Alzheimer's disease, Int. J. Pharm. 389 (1–2) (2010) 207–212.
[29] M. Ninomiya, L. Unten, M. Kim, Chemistry and application of green tea, in: T. Yamamoto, L.R. Juneja, D.C. Chu, M. Kim (Eds.), Yamamoto, CRC Press, Boca Raton, 1997, , pp. 23–36.
[30] I.M. Zijp, O. Korver, L.B. Tijburg, Effect of tea and other dietary factors on iron absorption, Crit. Rev. Food Sci. Nutr. 40 (5) (2000) 371–398.
[31] D. Labbé, A. Tremblay, L. Bazinet, Effect of brewing temperature and duration on green tea catechin solubilization: Basis for production of EGC and EGCG-enriched fractions, Sep. Purif. Technol. 49 (1) (2006) 1–9.
[32] J. Hu, D. Zhou, Y. Chen, Preparation and antioxidant activity of green tea extract enriched in epigallocatechin (EGC) and epigallocatechin gallate (EGCG), J. Agric. Food Chem. 57 (4) (2009) 1349–1353.
[33] R.L. Wickremasinghe, E.L. Wick, T. Yamanishi, Gas chromatographic-mass spectrometric analyses of "flavory" and "non-flavory" Ceylon black tea aroma concentrates prepared by two different methods, J. Chromatogr. 79 (1973) 75–80.
[34] V. Sharma, A. Gulati, S.D. Ravindranath, V. Kumar, A simple and convenient method for analysis of tea biochemicals by reverse phase HPLC, J. Food Compos. Anal. 18 (6) (2005) 583–594.
[35] V.K. Ananingsih, A. Sharma, W. Zhou, Green tea catechins during food processing and storage: A review on stability and detection, Food Res. Int. 50 (2) (2013) 469–479.
[36] C. Braicu, M.R. Ladomery, V.S. Chedea, A. Irimie, I. Berindan-Neagoe, The relationship between the structure and biological actions of green tea catechins, Food Chem. 141 (3) (2013) 3282–3289.
[37] J. Boyer, R.H. Liu, Apple phytochemicals and their health benefits, Nutr. J. 3 (2004) 5.
[38] A. Mittal, M.S. Pate, R.C. Wylie, T.O. Tollefsbol, S.K. Katiyar, EGCG down-regulates telomerase in human breast carcinoma MCF-7 cells, leading to suppression of cell viability and induction of apoptosis, Int. J. Oncol. 24 (3) (2004) 703–710.
[39] H. Tachibana, K. Koga, Y. Fujimura, K. Yamada, A receptor for green tea polyphenol EGCG, Nat. Struct. Mol. Biol. 11 (4) (2004) 380–381.
[40] H. van Praag, M.J. Lucero, G.W. Yeo, K. Stecker, N. Heivand, C. Zhao, E. Yip, M. Afanador, H. Schroeter, J. Hammerstone, F.H. Gage, Plant-derived flavanol (−)epicatechin enhances angiogenesis and retention of spatial memory in mice, J. Neurosci. 27 (22) (2007) 5869–5878.

[41] Y. Zhang, Y. Yu, X. Li, S. Meguro, S. Hayashi, M. Katashima, T. Yasumasu, J. Wang, K. Li, Effects of catechin-enriched green tea beverage on visceral fat loss in adults with a high proportion of visceral fat: A double-blind, placebo-controlled, randomized trial, J. Funct. Foods 4 (1) (2012) 315–322.
[42] D. Jin, Y. Xu, X. Mei, Q. Meng, Y. Gao, B. Li, Y. Tu, Antiobesity and lipid lowering effects of theaflavins on high-fat diet induced obese rats, J. Funct. Foods 5 (3) (2013) 1142–1150.
[43] N.S. Ncube, A.J. Afolayan, A.I. Okoh, Assessment techniques of antimicrobial properties of natural compounds of plant origin: current methods and future trends, Afr. J. Biotechnol. 7 (2008).
[44] C.D. Stalikas, Extraction, separation, and detection methods for phenolic acids and flavonoids, J. Sep. Sci. 30 (18) (2007) 3268–3295.
[45] M. Demeule, J. Michaud-Levesque, B. Annabi, D. Gingras, D. Boivin, J. Jodoin, S. Lamy, Y. Bertrand, R. Beliveau, Green tea catechins as novel antitumor and antiangiogenic compounds, Curr. Med. Chem. Anticancer Agents 2 (4) (2002) 441–463.
[46] S.T. Saito, G. Gosmann, C. Pungartnik, M. Brendel, Green tea extract-patents and diversity of uses, Recent Pat. Food Nutr. Agric. 1 (3) (2009) 203–215.
[47] A. Bharadwaz, C. Bhattacharjee, Extraction of polyphenols from dried tea leaves, Int. J. Sci. Eng. Res. 3 (5) (2012) 1–5.
[48] R. Wati, S. Sampanvejsobha, N. Punbusayakul, Effect of extraction conditions on Assam green tea, Agric. Sci. J. 40 (2009) 47–50.
[49] Q.V. Vuong, C.E. Stathopoulos, J.B. Golding, M.H. Nguyen, P.D. Roach, Optimum conditions for the water extraction of L-theanine from green tea, J. Sep. Sci. 34 (18) (2011) 2468–2474.
[50] H. Liang, Y. Liang, J. Dong, J. Lu, Tea extraction methods in relation to control of epimerization of tea catechins, J. Sci. Food Agric. 87 (2007) 1748–1752.
[51] G. Dominguez-Rodriguez, M.L. Marina, M. Plaza, Strategies for the extraction and analysis of non-extractable polyphenols from plants, J. Chromatogr. A 1514 (2017) 1–15.
[52] N.S. Jaiswal, SG, pressurized liquid extraction of antioxidant compounds from green tea, am, J. Food Technol. 12 (6) (2017) 358–366.
[53] D. Villanueva Bermejo, E. Ibáñez, G. Reglero, C. Turner, T. Fornari, I. Rodriguez-Meizoso, High catechins/low caffeine powder from green tea leaves by pressurized liquid extraction and supercritical antisolvent precipitation, Sep. Purif. Technol. 148 (2015) 49–56.
[54] D.P.C. Anandharamakrishnan, Techniques for extraction of green tea polyphenols: a review, Food Bioproc. Tech. 8 (5) (2015) 935–950.
[55] H.V. Chuyen, M.H. Nguyen, P.D. Roach, J.B. Golding, S.E. Parks, Microwave-assisted extraction and ultrasound-assisted extraction for recovering carotenoids from Gac peel and their effects on antioxidant capacity of the extracts, Food Sci. Nutr. 6 (1) (2018) 189–196.
[56] G. Sardar, E. Demir, S. Bayrak, M. Sokmen, New approaches for effective microwave assisted extraction of caffeine and catechins from green tea, Int. J. Second. Metabolite 3 (1) (2016) 3–13.
[57] V. Ghasemzadeh-mohammadi, B. Zamani, M. Afsharpour, A. Mohammadi, Extraction of caffeine and catechins using microwave-assisted and ultrasonic extraction from green tea leaves: an optimization study by the IV-optimal design, Food Sci. Biotechnol. 26 (2017) 1281–1290.
[58] B.R. Albuquerque, M.A. Prieto, M.F. Barreiro, A. Rodrigues, T.P. Curran, L. Barros, I.C.F. R. Ferreira, Catechin-based extract optimization obtained from Arbutus unedo L. fruits using maceration/microwave/ultrasound extraction techniques, Ind. Crop Prod. 95 (2017) 404–415.
[59] X. Zhu, J. Yang, Q. Su, J. Cai, Y. Gao, Selective solid-phase extraction using molecularly imprinted polymer for the analysis of polar organophosphorus pesticides in water and soil samples, J. Chromatogr. A 1092 (2) (2005) 161–169.
[60] Z.H. Tian, K.H. Row, Solid phase extraction of Catechin compounds from green tea by Catechin molecular imprinted polymers, Asian J. Chem. 24 (10) (2012) 4606–4610.
[61] M.M. Delgado-Povedano, M.D. Luque de Castro, Ultrasound-assisted extraction and in situ derivatization, J. Chromatogr. A 1296 (2013) 226–234.
[62] F. Chemat, N. Rombaut, A.G. Sicaire, A. Meullemiestre, A.S. Fabiano-Tixier, M. Abert-Vian, Ultrasound assisted extraction of food and natural products. Mechanisms, techniques, combinations, protocols and applications. A review, Ultrason. Sonochem. 34 (2017) 540–560.

[63] M.N. Koiwai, Extraction of Catechins from green tea using ultrasound, Jpn. J. Appl. Phys. 46 (7B) (2007) 4936–4938.
[64] L.S. Lee, N. Lee, Y.H. Kim, C.H. Lee, S.P. Hong, Y.W. Jeon, Y.E. Kim, Optimization of ultrasonic extraction of phenolic antioxidants from green tea using response surface methodology, Molecules (Basel, Switzerland) 18 (11) (2013) 13530–13545.
[65] H.Y. Choung, L.M.S. MG, J. Lee, S.T. Kang, T.H. Jun, Comparison of extraction and isolation efficiency of catechins and caffeine from green tea leaves using different solvent systems, Int. J. Food Sci. Technol. 49 (6) (2014) 1572–1578.
[66] T. Xia, S. Shi, X. Wan, Impact of ultrasonic-assisted extraction on the chemical and sensory quality of tea infusion, J. Food Eng. 74 (4) (2006) 557–560.
[67] P. Stanbury, A. Whitaker, S. Hall, Principles of Fermentation Technology, Elsevier, 2016.
[68] M. Iqbal, Y. Tao, S. Xie, Y. Zhu, D. Chen, X. Wang, L. Huang, D. Peng, A. Sattar, M.A. Shabbir, H. I. Hussain, S. Ahmed, Z. Yuan, Aqueous two-phase system (ATPS): an overview and advances in its applications, Biol. Proced. Online 18 (2016) 18.
[69] Y. Shu, M. Gao, X. Wang, R. Song, J. Lu, X. Chen, Separation of curcuminoids using ionic liquid based aqueous two-phase system coupled with in situ dispersive liquid-liquid microextraction, Talanta 149 (2016) 6–12.
[70] A. Salic, A.J. Tusek, D. Fabek, I. Rukavina, B. Zeli, Aqueous two- phase extraction of polyphenols using a microchannel system—Process optimization and intensification, Food Technol. Biotechnol. 49 (4) (2011) 495–501.
[71] Y.H. Goja, C.M. AM, C. Li, Aqueous two-phase extraction advances for bioseparation, Bioproces Biotechniq 4 (1) (2013) 140.
[72] G.N. Sapkale, S.M. Patil, U.S. Surwase, P.K. Bhatbhage, Supercritical fluid extraction, Int. J. Chem. Sci. 8 (2) (2010) 729–743.
[73] Q.L. Sun, S. Hua, J.H. Ye, J.L. Lu, X.Q. Zheng, Y.R. Liang, Decaffeination of green tea by supercritical carbon dioxide, J. Med. Plant Res. 4 (12) (2010) 1161–1168.
[74] B.A. James, S.M. Abdulkarim, F. Abas, Solvent and supercritical fluid extraction of Catechin from Camellia Sinensis (tea) leaves for utilization as functional food ingredient, Int. J. Eng. Technol. 16 (2) (2016) 21–33.
[75] S.M. Chacko, P.T. Thambi, R. Kuttan, I. Nishigaki, Beneficial effects of green tea: A literature review, Chinas Med. 5 (2010) 13.
[76] P.V. Gadkari, M. Balaraman, Catechins: Sources, extraction and encapsulation: A review, Food Bioprod. Process. 93 (2015) 122–138.
[77] S. Banerjee, J. Chatterjee, Efficient extraction strategies of tea (Camellia sinensis) biomolecules, J. Food Sci. Technol. 52 (6) (2015) 3158–3168.
[78] H.S. Park, N.G. Im, K.H. Kim, Extraction behaviors of caffeine and chlorophylls in supercritical decaffeination of green tea leaves, LWT- Food Sci. Technol. 45 (1) (2012) 73–78.
[79] R.A.R.M. Bimakr, F.S. Taip, L.T. Chuan, A. Ganjloo, L.M. Salleh, J. Selamat, A. Hamid, Supercritical carbon dioxide (SC-CO) extraction of Catechin, Epicatechin, Rutin and Luteolin from spearmint (*Mentha spicata* L.) Leaves, World Appl. Sci. J. 5 (4) (2008) 410–417.
[80] K.Y. Khaw, M.O. Parat, P.N. Shaw, J.R. Falconer, Solvent supercritical fluid technologies to extract bioactive compounds from natural sources: A review, Molecules (Basel, Switzerland) 22 (7) (2017).
[81] O. Wrona, K. Rafinska, C. Mozenski, B. Buszewski, Supercritical fluid extraction of bioactive compounds from plant materials, J. AOAC Int. 100 (6) (2017) 1624–1635.
[82] C. Da Porto, D. Decorti, A. Natolino, Water and ethanol as co-solvent in supercritical fluid extraction of proanthocyanidins from grape marc: a comparison and a proposal, J. Supercrit. Fluids 87 (2014) 1–8.
[83] S.M. Pourmortazavi, S.S. Hajimirsadeghi, Supercritical fluid extraction in plant essential and volatile oil analysis, J. Chromatogr. A 1163 (1–2) (2007) 2–24.
[84] P.V. Gadkari, M. Balarman, U.S. Kadimi, Polyphenols from fresh frozen tea leaves (Camellia assamica L.,) by supercritical carbon dioxide extraction with ethanol entrainer—application of response surface methodology, J. Food Sci. Technol. 52 (2) (2015) 720–730.
[85] P.V. Gadkari, M. Balaraman, Mass transfer and kinetic modelling of supercritical CO_2 extraction of fresh tea leaves (Camellia sinensis L.), Braz. J. Chem. Eng. 34 (03) (2017) 799–810.

[86] E. Susanti, Ciptati, R. Ratnawati, Aulanni'am, A. Rudijanto, Qualitative analysis of catechins from green tea GMB-4 clone using HPLC and LC-MS/MS, Asian Pac. J. Trop. Biomed. 5 (12) (2015) 1046–1050.
[87] R.S. Ahmad, M.S. Butt, N. Huma, M.T. Sultan, M.U. Arshad, Z. Mushtaq, F. Saeed, Quantitative and qualitative portrait of green tea Catechins (Gtc) through Hplc, Int. J. Food Prop. 17 (7) (2014) 1626–1636.
[88] A. Shirolkar, A. Gahlaut, A.K. Chhillar, R. Dabur, Quantitative analysis of catechins in Saraca asoca and correlation with antimicrobial activity, J. Pharm. Anal. 3 (6) (2013) 421–428.
[89] W.C. Reygaert, The antimicrobial possibilities of green tea, Front. Microbiol. 5 (2014) 434.
[90] D.A. Amit Alexander, T. Giri, A. Dulal, Unexplored areas and new findings in lipid emulsion serving as a potential drug carrier for lipophilic drugs: a review, Trends Med. Res. 7 (1) (2012) 1–24.

Further reading

[91] M.J. Glade, Caffeine-Not just a stimulant, Nutrition (Burbank, Los Angeles County, Calif.) 26 (10) (2010) 932–938.
[92] M. Giannelli, P. Doyle, E. Roman, M. Pelerin, C. Hermon, The effect of caffeine consumption and nausea on the risk of miscarriage, Paediatr. Perinat. Epidemiol. 17 (4) (2003) 316–323.

CHAPTER 4

Application of supercritical CO_2 for enhanced oil recovery

Sandra Gonçalves, Anabela Romano
Faculty of Sciences and Technology, MeditBio, University of Algarve, Faro, Portugal

Contents

1. Introduction

The plant kingdom is a repository of a great diversity of chemical compounds with complex structures designated as secondary metabolites that are responsible for the biological properties of plants [1]. Secondary metabolites include a vast range of compounds with medicinal properties that are biosynthesized through phenylpropanoid, mevalonate, 2-*C*-methyl-*D*-erythritol-4-phosphate, amino acid, glucose, acetate-malonate pathway, or a combination of pathways. These compounds have important physiological functions, mainly in the interaction between the plant and the environment, having a vital role in their existence and subsistence.

Plant bioactive compounds exhibit a broad range of properties useful for pharmaceutical, cosmetic, and food products and their demand is increasing nowadays. The advances in the identification and isolation techniques allowed the discovery of many plant compounds. Between 25% and 28% of current medicines and more than 60% of anticancer drugs are derived from higher plants [2–4]. The recent report of the British Broadcasting Corporation [5] indicates that drugs derived from plants will grow from $29.3 billion in 2017 to $39.2 billion by 2022 at an annual growth rate of 5.9%.

Green Sustainable Process for Chemical and Environmental Engineering and Science
https://doi.org/10.1016/B978-0-12-817388-6.00004-0

The extraction and identification of plant biologically active compounds is very important in food and pharmaceutical industries. The extraction procedure must be optimized according to the plant species and this process is limited by the solubilization or the diffusional step. Particularly, the extraction from the complex matrix of plant material, involving the breakage of plant cell walls, is a crucial step [6]. The conventional/traditional methods of extraction are generally simple but there are several drawbacks (e.g., usually are tedious to operate, use harmful processing solvents, many use heat leading to the degradation of thermolabile compounds, in some cases they are selective for a few types of compounds, usually have poor yield to the high energy input, etc.) [7]. Green extraction techniques appeared in the 1990s aiming at the reduction of energy consumption and the substitution of conventional solvents by environmentally friendly alternatives. Most of the limitations of the conventional extraction methods can be overcome by the use of supercritical fluid extraction (SFE) using CO_2 (SFE-CO_2). This technique offers the possibility of reducing processing energy and, in addition, uses alternative solvents [7].

Among natural products, essential oils (EOs) are very important due to their various biological properties [8] and have been used since ancient times as curative remedies and flavoring agents. Several reports show that EOs have a broad range of important biological properties, including antimicrobial, antioxidant, herbicidal, insecticidal, among others. The oil obtained by SFE is more correctly designated as volatile oil to distinguish it from the EO, which according to the European Pharmacopeia [9] is the extract obtained from aromatic plants by distillation processes (hydrodistillation and steam distillation); nevertheless, in literature EOs have been extensively used to designate the oil components isolated from plants, particularly from aromatic species (e.g., [10, 11]).

This chapter focuses on the importance of SFE-CO_2 for the recovery of oil components from plant matrices with emphasis on aromatic plant species. The different factors influencing oil recovery and the comparison of SFE with other extraction techniques will be discussed based on updated data, highlighting the main advantages of this technique.

2. Aromatic plants and EOs

Aromatic plants demand special attention within the plant kingdom because they synthesize a large diversity of bioactive compounds useful in many applications. The interest in this group of plants is continuously growing, mainly due to their contents in EOs. The aromatic plants are uniformly dispersed within the angiosperms of tropical and temperate habitats, including families from Gramineae to Rosaceae [12]. Several common and important aromatic plants are from the Lamiaceae, Apiaceae, Myrtaceae, Rutaceae, Verbenaceae, and Zingiberaceae families (Table 1). Lamiaceae is probably the most important family, including economically relevant plants such as oregano, thyme, sage, rosemary, etc. Many aromatic species have been extensively studied and their health

Table 1 Plant species of different families producing industrially important essential oils and their main components

Plant species	Common name	Family	Main compounds
Coriandrum sativum L.	Coriander	Apiaceae	Terpinene, *p*-cymene and pinene
Foeniculum vulgare Mill.	Fennel	Apiaceae	*trans*-Anethole and methyl chavicol
Pimpinella anisum L.	Anise	Apiaceae	Anethole
Lavandula sp.	Lavender	Lamiaceae	Linalool, linalyl acetate, and camphor
Mentha sp.	Mint	Lamiaceae	Mentol, menthone, and carvone
Origanum vulgare L.	Oregano	Lamiaceae	Carvacrol, thymol, β-fenchyl alcohol, and γ-cimene
Rosmarinus officinalis L.	Rosemary	Lamiaceae	1,8-Cineol, camphor, and borneol
Salvia officinalis L.	Sage	Lamiaceae	β-Caryophyllene, 4-terpineol, caryophyllene oxide, mannol, and linalool
Thymus sp.	Thyme	Lamiaceae	Thymol, *p*-cymene, γ-terpinene and carvacrol
Eucalyptus sp.	Eucalyptus	Myrtaceae	Cineole, pinene and limonene
Citrus limon L. (Osbeck)	Lemon	Rutaceae	D-Limonene and limonene
Verbena officinalis L.	Vervain	Verbenaceae	Ocymene, isobornyl formate, citral, carvacrol, and thymol
Zingiber officinale Rosc.	Ginger	Zingiberaceae	Terpenio, neral, and geranial

benefits demonstrated. For instance, the European Medicines Agency recommends rosemary EO to treat muscular pain, circulatory disorders, dyspepsia, and some digestive disorders [13]. Extracts and EOs from this plant can be used as a natural food preservative. Indeed, rosemary extract is approved as a food additive (E392) (Directive 2010/67/EU) [14]. In addition, the extract from lavender has antiseptic and anti-inflammatory properties for skin care. Menthol from mint is used to treat nasal congestion and thymol from thyme has a strong antimicrobial activity.

EOs are secondary metabolites produced with the involvement of different enzymatic reactions by two complex pathways [15]. They can be extracted from different parts of the plant such as leaves (e.g., mint, thyme, oregano), roots and rhizomes (e.g., ginger), flowers (e.g., rose, lavender, violet), fruits, and seeds (e.g., pepper, coriander, orange, lemon). EOs have been extensively investigated in recent years, due to their natural nature and because they have shown benefits in food preservation and human health [16]. They have been used for centuries as perfumes, in culinary applications, mainly as flavoring agents, and in folk medicine, and their industrial applications began in the 19th century [17]. More than 3000 EOs have industrial significance and are used mainly in the cosmetic and flavor industries [18]. Presently, EOs are mostly studied for their broad range of biological properties including antioxidant, antimicrobial, antitumor, analgesic, insecticidal, antidiabetic, and anti-inflammatory.

Plants can produce an amount of EO from 0.01% to 10% of the plant total weight. They are extremely complex mixtures composed of highly functionalized chemical substances from different chemical classes of volatile compounds and oxygenated derivatives [19], and also contain non-volatile compounds such as hydrocarbons, fatty acids, sterols, carotenoids, waxes, and flavonoids [20]. The composition of the oil depends on different factors such as the plant species, the part of the plant used, the cultivation and growing conditions including soil and climate, the harvesting time, etc.

Consumers' interest in the multiple biological properties of the EOS has led to an increase in the world production and consumption of these natural products. Usually, the main compounds present in the EOs are those responsible for their bioactivity, however the minor compounds also contribute for the activity and there is the possibility of synergism between the components, increasing the spectrum of action and improving the EO efficacy [21]. EOs contain many compounds with antimicrobial and antioxidant properties and, therefore, have preservative properties and can be used in food products to reduce or eliminate the presence of microorganisms and/or decrease the lipid oxidation phenomenon. The fact that EOs are generally recognized as safe (GRAS) by the Food and Drug Administration (FDA) [22] reinforces their importance for food applications.

3. Extraction methods

The choice of the extraction method and the optimization of extraction parameters is a critical step in the search for natural compounds. This can be a difficult task due to the great complexity of plant tissues and physicochemical properties of plant bioactive compounds. EOs occur in plants in small amounts and thus, the extraction process greatly affects the yield and quality of the recovered oil. It is important to use an extraction method that allows the highest yield and reduced changes in functional properties [11]. The extraction methods can enhance or reduce the biological properties of the EO obtained. In fact, the main traditional methods for EOs extraction are steam-distillation, hydrodistillation, and solvent extraction [10]. The most used method at commercial scale is steam-distillation and this method and hydrodistillation are the most commonly used at laboratorial scale. The method recommended by the European Pharmacopeia is hydrodistillation using a Clevenger apparatus. One of the main drawbacks of distillation methods is the high temperatures applied, which can damage some thermolabile compounds compromising the quality of the oil extracted [10].

The non-conventional or innovative methods of extraction aim to limit mass transfer, improve extraction yields in limited time, and reducing extraction solvents. The actual tendency is to use "green" techniques of extraction of natural compounds, which include the reduction of waste and carbon footprint, and the reuse of discarded byproducts [11]. These alternative extraction techniques must involve shorter time of extraction, fewer solvents, lower energy costs, and less CO_2 emissions [11]. Green extraction techniques must minimize the use of toxic solvents, reduce energy requirements, produce valuable products, reduce the time of extraction, and obtain a natural and safe extract [23].

The modern techniques include ultrasound-assisted extraction, microwave-assisted extraction, high voltage electrical discharges and subcritical fluid extraction and SFE. These techniques associated with green solvents allow sustainable production. Among these modern techniques, SFE is the most attractive and is used for oil recovery since it overcomes many of the limitations of traditional extraction methods, achieving high extraction yields and obtaining oil of superior quality in terms of composition [10].

4. Supercritical fluid extraction

SFE uses supercritical solvents that have singular physicochemical properties, namely, density, diffusivity, viscosity, and dielectric constant, and therefore, offer several operational advantages as compared with conventional extraction methods [10]. Supercritical fluids can give faster extraction rates since they have low viscosity and high diffusivity. The opportunity of changing fluid density by varying pressure and temperature is the main characteristic of this technique. Density is related to solubility, and therefore by changing the extraction pressure, the solvent force of the fluid can be modified. The use of GRAS solvents and the lower extraction periods are additional advantages of SFE [24]. Carbon dioxide (CO_2) is the most widely used solvent particularly because it is harmless to human health and to the environment, and its moderate critical temperature allows the preservation of extracted bioactive molecules [25]. After extraction and system decompression, it is recovered as an extract free of solvent since at ambient temperature carbon dioxide is a gas. It is less adequate to extract more polar components of the extract due to its low polarity nature and therefore it is frequent to use co-solvents to overcome this problem. CO_2 is extensively used for oil recovery from aromatic species and, since EOs contain lipophilic compounds, a co-solvent is not needed.

5. Factors affecting SFE

Basically, the SFE process involves the solubilization of the compounds and its separation from the solvent [26]. This process is affected by several variables, including factors related to the plant (plant species, part of the plant, plant location and harvesting time, etc.) and the extraction conditions used (e.g., pressure, temperature, CO_2 flow rate, type and quantity of co-solvent, extraction time, etc.) Furthermore, the pre-treatment also greatly affects yield and composition of the extracted products. In the following subsections will be addressed the main factors affecting the SFE process.

5.1 Plant matrix

The plant matrix is the key factor affecting SFE-CO_2 oil recovery. Studies using the same extraction conditions showed considerable variations in the extraction yields and kinetic behavior depending on the plant species studied [10, 27, 28]. In addition to the plant species, other factors related with the plant matrix can be considered such as the plant part, harvesting time, the water content, particle size, etc. The water content influenced the extraction by decreasing the extraction yields, mainly in the case of the most polar compounds as terpenes. The presence of water can induce hydrolysis of compounds. The water content must be lower than 12% and, therefore, it is recommended to dry the plant material before extraction [10]. The drying process can also influence the composition of the extracted oil [29].

Particle size is another factor that has been studied by several authors and, overall, results show that it greatly influences the extraction process. Particle size affects the length of the process and both are interrelated with CO_2 flow rate. The extraction is favored by reducing mass transfer resistance, shortening the extraction period [10]. Wilkinson et al. [30], working with soybean, observed that increasing the grinding time reduced the particle sizes and consequently enhanced the extraction yield. Grosso et al. [31] observed that, although volatiles composition from Italian coriander seeds was not considerably affected by particle size, the SFE yield increased when the particle size decreased, probably because more ducts were cracked with longer milling periods.

Shrigod et al. [32] investigated the effect of several SFE parameters (temperature, pressure, extraction time, and particle size) on extraction yield and carvone content in volatile oil of mint leaves. The authors observed that SFE yield was mostly affected by particle size, followed by pressure, temperature, and extraction time; on the other hand, carvone content in extract was more influenced by pressure, followed by particle size and extraction time. Hatami et al. [33] studied the effect of grinding time and mass of raw material in the mill on content of terpenes in fennel extract obtained by SFE. Authors observed that both factors must be taken into account as independent variables but, overall, both have substantial effects on anethole and fenchone content, the main components of volatile oil. In some cases, the grinding process can increase the temperature of the ground material, which can induce a loss of volatile oil and flavor constituents. Cryogenic grinding with liquid nitrogen has been successfully used to overcome this problem [34].

5.2 Extraction operational conditions

The selection of the SFE operational conditions is a key aspect to be considered in the extraction process. These are mainly selected based on the solute solubility in the solvent and on the global extraction efficiency (maximum amount of material extracted at specific conditions). The optimization of SFE process requires knowledge of thermodynamic and kinetic data [26]. There are several operational variables that can be optimized, such as temperature, pressure, flow rate of CO_2, extraction time, etc. The use of the optimal conditions can significantly improve the efficiency and recuperation of a target compound. Among the different variables, the extraction pressure and temperature are possibly the most important. Pressure can be used to adjust the selectivity of the solvent. In the case of thermolabile compounds—most of the EOs components—the selected temperature must be in the range of 35–50°C. Overall, EOs components can be extracted at moderate temperature and pressure, the use of high pressures induces the co-extraction of other substances. If high pressures are employed, fractionation with at least two separators must be used to separate the oil from the other co-extracted components [10].

There are several recent reports about the optimization of operational conditions to extract EOs by SFE-CO_2. In a study with allspice, EO was recovered by SFE-CO_2 at different temperatures and pressures [35]. Results showed that the maximum extraction yield (68.47%) was obtained after 6h of extraction at 45°C, 360bar, and the extraction greatly depended on temperature and pressure. Jerković et al. [36] also investigated the influence of different SFE extraction parameters (pressure, temperature, and CO_2 flow rate) on the extraction yield, chemical profile, and relative content of some oxygenated monoterpenes of *Lavandula angustifolia* Miller flowers. In addition, the authors determined the optimal extraction conditions by using response surface methodology (RSM). This model showed that the optimal conditions to achieve the greatest extraction yield and the higher amounts of oxygenated monoterpenes were 10MPa, 41°C and CO_2 flow rate 2.3kg/h. Indeed, several statistical models were used to optimize the SFE operating parameters including, mainly, pressure and temperature.

Recently, Priyanka and Khanam [37] used central composite design (CCD) under response surface methodology to optimize five operating parameters including pressure, temperature, solvent flow rate, particle size, and addition of co-solvent, for SFE of turmeric root oil. The effect of those parameters was evaluated on oil yield, total curcumin, and turmerone amounts. Results showed that pressure, temperature, flow rate, and co-solvent (ethanol) positively affect extraction yield, and this variable is very sensitive to particle size.

Shahsavarpour et al. [38] investigated the effects of pressure, temperature, CO_2 flow rate, mean particle size, and extraction time on the recovery of EOs from spearmint leaves, and observed that extraction yield is greatly affected by pressure. For temperature, this relation is not so direct. Both supercritical solvent density and sublimation pressure temperature affect EOs solubility. The best conditions were found to be 42.6°C and 119.7bar, 0.169g/min of CO_2 flow rate, particle size of 0.248mm, and 109.9min of static time of extraction. In another recent study, the influence of different extraction parameters (pressure, temperature and CO_2 flow rate) on extraction yield and chemical profile of extracts from *Salvia officinalis* L. was investigated [39]. Pressure was the factor that mostly affected the extraction yield and the amount of important compounds. The authors used RSM to establish the optimal extraction conditions for targeting the most abundant compounds (oxygenated monoterpenes, α- humulene, viridiflorol, and manool) as pressure 13 MPa, temperature 40°C, and CO_2 flow rate of 3 kg/h.

Reviewed data showed that EOs extraction using CO_2 are usually performed at moderate pressure and temperature and the mathematical models are valuable tools to plan and improve the extraction processes, and for their scale-up [40].

5.3 Strategies to improve extraction efficiency and selectivity

Many strategies can be applied to increase SFE extraction efficiency and selectivity. Biomass pretreatment, with several substances, namely enzymes, basic modifiers, or ionic liquids, and the combination with other approaches, namely ultrasound and enzymes, are some of the strategies that can be used to increase SFE efficiency [26]. Ultrasound assisted extraction can produce benefits in the extraction of plant metabolites by improving extraction rate and yield, and reducing the extraction temperature and time required. Due to cavitation, ultrasound causes turbulence, particle agglomeration, and cell disruption, breaking plant cell walls and improving the solvent penetration. The ultrasound assisted extraction in conventional solvent extraction is well established, however, the acoustic intensity needed to produce cavitation is a limitation when pressurized solvents are used [41].

Fractionation is a useful strategy to extract different families of compounds with different solubilities in supercritical CO_2. Basically, fractionation is the separation of the solutes in more than one fraction and can effectively improve selectivity of SFE. Fractionation can be used in the extraction and/or separation steps employing two or more modules, respectively. There are two different fractionation techniques available: the multi-step fractionation, in which the extraction occurs in successive steps, and on-line fractionation occurring in a cascade decompression system [10]. In multi-step fractionation, the conditions vary step by step in the extraction vessel, improving the CO_2 density to achieve the fractional extraction of the soluble components. The high soluble components are separated in the first fraction and components less soluble in the solvent are extracted in the successive steps. EOs components have a good solubility in supercritical CO_2 and, therefore, are generally recovered in the first fraction.

In on-line fractionation, the operating conditions in the extraction vessel are the same during the complete extraction procedure and different separators in series are used at distinct temperatures and declining pressures. The aim of this procedure is to provoke the selective precipitation of several compound families related to their distinct saturation conditions in the solvent. This practice was successfully used to extract volatile compounds [42].

6. Comparison of SFE-CO_2 with other extraction techniques for oil recovery

The comparison of SFE-CO_2 and hydrodistillation methods for oil recovered from plant matrices has been investigated for a vast multiplicity of plant species, including widely distributed and endemic species. In this section, this comparison will be discussed based on recent data of several aromatic species.

EOs obtained from Lamiaceae plant species are extensively studied due to their interest for cosmetic and food industries, mainly due to their antioxidant properties. Rosemary, *Rosmarinus officinalis* L. (Lamiaceae) is one of the most important Mediterranean aromatic plants. It is particularly used for its antioxidant properties, particularly due to its content in carnosol, rosmanol, isorosmanol, rosmadiol, carnosic acid, rosmarinic acid, and methyl carnosate [43]. The SFE of EOs from this species was investigated by several authors. Recently, Conde-Hernández et al. [44] analyzed the chemical composition and antioxidant activity of oil from rosemary samples from Mexico obtained by CO_2-SFE (temperatures of 40 and 50°C, pressure of 10.34 and 17.24 MPa, and particle size of 600 ± 50 mm), hydrodistillation, and steam distillation. Higher extraction yields and antioxidant activity were attained in oil recovered by CO_2-SFE at 40°C and 17.24 MPa, (Table 2), followed by steam distillation, and then by hydrodistillation.

Salvia species (Lamiaceae) have also been extensively used for their medicinal properties, particularly due to their anti-inflammatory effect. There are several studies reporting the isolation of bioactive compounds from *Salvia* species, particularly from *S. officinalis*, by conventional techniques and also by SFE (e.g., Arranz et al. [50]). Glisic et al. [51] made a comparative study of different techniques to extract diterpenes from Dalmatian *Salvia officinalis* L. From the different approaches tested, including SFE or a combination of SFE and ultrasound assisted extraction, the pretreatment of ultrasound assisted extraction with distilled water was the most adequate. Recently, Šulniūtė et al. [52] analyzed the composition of volatile compounds in ten *Salvia* species isolated by simultaneous distillation/extraction in a Likens-Nickerson apparatus (L-N) and SFE-CO_2. In this case, the authors observed that the number of volatile compounds isolated by SFE-CO_2 was inferior to that obtained by the L-N method, which is caused by the loss of volatiles with exhausting from the system CO_2 after depressurization. The results of these authors highlight the importance of the operational conditions used during the SFE process and its adequate optimization for each plant matrix.

Within Lamiaceae species, the EOs from oregano (*Origanum vulgare* L.) and marjoram (*Origanum majorana* L.) have been accepted as of the most importance all over the world due to their antimicrobial and antioxidant properties. A comparison between the chemical profile of marjoram and oregano EOs obtained by hydrodistillation and SFE-CO_2 was recently reported [53]. Higher extraction yields were obtained with SFE and qualitative differences on the chemical composition of the oil were observed.

Table 2 Selected examples of important bioactive oil components extracted by SFE-CO_2 from different plant species

Plant species	Common name	Main compound(s)	Bioactivity	Conditions	Results	Reference
Cordia verbendcea DC	Erva baleeira	α–Humulene and *trans*-caryophyllene	Antitumor	50°C, 300 bar, 0.3 kg/h of CO_2 and 270 min	Extraction yield 5.75%	[45]
Coriandrum sativum L.	Coriander	Linalool	Antioxidant	55°C, 100 bar, and 0.4 kg/h of CO_2	Highest yield of linalool 717.0 mg/g	[46]
Matricaria chamomilla L.	Chamomile	Matricine	Anti-inflammatory and anti-spasmodic	40°C and 250 bar	Extraction yield 3.81% and up to 28 mg/g of matricine	[47]
Mentha spicata L.	Spearmint	Carvone	Antioxidant and antimicrobial	48°C, 151 bar, 0.40 mm particle size and 37.5 min dynamic time, 150 min	Extraction yield 1.4% and 998 mg/100 g of carvone	[30]
Origanum vulgare L.	Oregano	*trans*-Sabinene hydrate, thymol, and carvacrol	Anti-inflammatory	40°C and 15 MPa S1, and 40°C and 2 MPa S2	S1: 45.21, 24.10 and 7.99% of *trans*-sabinene hydrate, thymol and carvavrol, respectively; S2: 45.05, 19.81, and 7.17%	[48]
Piper nigrum L.	Black-pepper	β-Caryophyllene, limonene, sabinene, 3-carene, β-pinene, and α-pinene.	Antioxidant	40°C, 300 bar, 2 mL/min, 30 min SP + 40 min DP	Extraction yield 2.05%	[49]

Rosmarinus officinalis L.	Rosemary	Camphor, eucalyptol, β-caryophyllene, and borneol acetate	Antioxidant	40°C, 17.24 MPa, 126.24 mL/min, 180 min	Extraction yield 2.53%	[43]
Salvia officinalis L.	Sage	Camphor, 1,8-cineole, and borneol	Anti-inflammatory	40°C, 100 bar S1, 300 bar S2, 50 g/min, 480 min	S1: 41.78, 13.34 and 7.28% of camphor, 1,8-cineole, and borneol, respectively; S2: 33.9, 4.64, and 9.56%, respectively	[44]
Thymus zygis L., *Thymus vulgaris* L., and *Thymus hyemalis* L.	Thyme	Thymol, borneol, carvacrol, camphor, and 1,8-cineole	Antiviral	40°C, 100 bar, 480 min	*T. zygis*: 70.86, 4.71, 4.20, 1.47 and 1.03% of thymol, borneol, carvacrol, camphor, and 1,8-cineole, respectively; *T. vulgaris*: 65.43, 3.35, 4.26, 1.49, and 1.29%, respectively; *T. hyemalis*: 10.05, 8.33, 0.64, 24.44, and 20.97%, respectively	[26]

S: separator.

Hydrodistillation allows a higher extraction of monoterpene hydrocarbons and SFE of oxygenated monoterpenes. Costa et al. [46] studied the chemical profiles of *Lavandula viridis* L'Hér EO obtained by hydrodistillation and SFE conducted at 40°C and pressures of 12 or 18 MPa in two different separators. The authors observed that more compounds were isolated by hydrodistillation but higher yields were obtained by SFE and, although oxygen-containing monoterpenes were the main compounds obtained by both methods, there were significant differences in the chemical profiles. The optimal SFE conditions for another commonly used Lamiaceae species, *Mentha spicata* L., obtained though numerical optimization, were 48°C, 151 bar, 0.40 mm, and 37.5 min. With these conditions, the authors obtained an extraction yield of 1.4% and 998 mg/100 g of carvone content [32].

Coriander, *Coriandrum sativum* L. (Apiaceae), is a Mediterranean medicinal and aromatic plant gaining importance due to its cooking applications and medicinal properties. EOs from this species, particularly from seeds, have been extensively studied in recent years. Some of those investigations are focused on the extraction of EOs by SFE and the effect of operational conditions on extraction yield and biological activity. The comparison of extraction yield and chemical composition of the extracts obtained by different extraction methods has also been described. Zeković et al. [54] extracted EOs by SFE from seeds and optimized the process by using Box–Behnken experimental design and response surface methodology. The three tested parameters (pressure, temperature, and CO_2 flow) influenced the extraction procedure but pressure showed the strongest effect. In terms of chemical composition, linalool, the main compound, was obtained in the greatest amount (Table 2) at 100 bar, 55°C and 0.4 kg/h of CO_2. Overall, it was observed that lower pressure, temperature, and CO_2 flow rate are the better conditions for acyclic and cyclic oxygenated monoterpenes extraction, while higher temperature was beneficial for the extraction of aromatic oxygenated monoterpenes and monoterpenes. In summary, distinctive requirements in extraction conditions are linked with differences in structure of monoterpene classes.

Fennel (*Foeniculum vulgare* Mill.) is another important aromatic plant from the Apiaceae family that shows a broad range of medicinal properties including anti-inflammatory, analgesic, expectorant, diuretic, laxative, and carminative, among others [55]. The main ingredients found in fennel volatile oil are anethole and fenchone. There are many studies reporting the extraction of volatile oil from this species using diverse extraction techniques, including SFE, and several variables affecting the process [33, 56, 57]. Data found in literature indicate that SFE is an efficient technique to extract bioactive ingredients from this species. Moura et al. [56] demonstrated that the yields were 5.2 times greater in SFE than in hydrodistillation. In addition to larger relative proportion of anethole and fenchone, the SFE extracts also contained other valuable compounds in the volatile oil. The importance of considering the harvesting time of the plant material was also highlighted [56]. Recently, Hatami et al. [33] concluded that grinding time and mass of raw material are also important factors to be considered in SFE of fennel, since these parameters have significant effects on anethole and fenchone contents.

Mouahid et al. [58] compared the extraction yield and oil composition of four Corsican endemic plants, including *Rosmarinus officinalis*, *Juniperus communis* ssp. *nana*, *Helichrysum italicum*, and *Pistacia lentiscus*, obtained by SFE-CO_2 and hydrodistillation. Although the authors observed that more components were obtained by hydrodistillation, regarding the major compounds, only limonene was not recovered by SFE-CO_2. The authors concluded that this technique is the most adequate method for the extraction of important compounds as verbenone, germacrene D, bornyl acetate, ferruginol, transcaryophyllene, elemol, γ-cadinene, geraniol, and β-eudesmol, in greater amounts.

Using a gray wolf optimization algorithm, Sodeifian et al. [59] determined the optimal operating conditions for SFE EO extraction from *Cleome coluteoides* Boiss, an endemic Iranian wild species with medicinal properties. The optimal conditions, which allow the maximum yield, were pressure 220 bar, temperature 308 K, particle size 0.58 mm and dynamic extraction time 135 min. The chemical composition of EOs obtained by SFE using these conditions and with hydrodistillation was compared. The compositions were similar but extraction yield of the SFE was superior showing that this procedure is more efficient than traditional distillation.

Eucalyptus is an extensively grown tree, renowned for the production of EO with high biological effects applied in various domains. Herzi et al. [47] compared the extraction yield and volatile composition of the EO from two *Eucalyptus* species (*E. cinerea* and *E. camaldulensis*) isolated by hydrodistillation and SFE-CO_2. SFE improved extraction yields by about 20% and 40% for *E. cinerea* and *E. camaldulensis*, respectively. The authors attribute this fact to the conditions used in SFE including low temperature and quickness (40°C, 30 min), which reduce the exposure to high temperature and therefore the thermal degradation occurring in hydrodistillation, where temperatures of around 100°C for long periods are applied. Authors also observed qualitative and quantitative differences in the chemical composition of the EOs obtained by SFE and hydrodistillation. As an example, 1,8-cineol is the major compound of the EO obtained with hydrodistillation in both species, but using SFE, this compound was detected at high concentration, without being the major compound. SFE extracts contain superior amounts of heavier molecular weight compounds than in the case of hydrodistillation. Lighter components present in the EO obtained by hydrodistillation were weakly detected in the SFE oil.

The major advantage of using SFE-CO_2 over conventional distillation methods for oil recovery from chamomile flower heads is the quality of the product obtained [60]. The oil obtained by SFE contains only matricine, the main compound of interest in this species, and chamazulene, a degradation product of matricine, is not present.

Conversely, steam distillation oil contains mainly chamazulene and trace amounts of matricine. This study highlights the effectiveness of SFE for the recovery of high-value products.

The optimal SFE conditions to isolate EO from the Chinese medicinal plant *Anoectochilus roxburghii* (Wall.) Lindl. (Orchidaceae) were determined using response surface methodology (37.6°C, 35 MPa, extraction time of 150 min and CO_2 flow rate of 8 L/h) [61]. The chemical composition of the oil extracted using these conditions was compared with those of EOs obtained by hydrodistillation and Soxhlet extraction with hexane. The results show no significant differences between the oils obtained by Soxhlet extraction and SFE-CO_2. Hexane and CO_2 are both non-polar solvents and may display similar extraction properties. However, the EO obtained by hydrodistillation has a different composition. Results also showed that oil extracted by SFE showed the better radical scavenging capacity.

There are other examples of improved biological activity by using SFE instead of conventional extraction techniques. EO obtained by SFE from *Echinophora platyloba* DC., an Iranian endemic plant with remarkable medicinal properties, showed higher antioxidant activity as well as higher extraction yield as compared with EO obtained by hydrodistillation [49]. Similar results were obtained for black-pepper [62]. Glisic et al. [45] observed that carrot EO isolated by SFE was more active against *Bacillus cereus* than that recovered by hydrodistillation. SFE extracts of *Cordia verbenacea* D.C produced an increase in cytotoxicity and antitumor activity in comparison with extracts produced by the conventional extraction technique, which is attributed to the high content in α-humulene and *trans*-caryophyllene [48]. Also, Santoyo et al. [28] showed that SFE represents a good alternative to obtain antiviral compounds from *Thymus* species. Furthermore, it has been observed that SFE-CO_2 extraction conditions considerably affected the biological activity of the EOs [62].

Some explanations for the good results obtained with SFE are that the temperatures usually used in this technique prevent the degradation of thermolabile constituents; moreover, this is a more selective technique producing EOs of superior activity, particularly antioxidant. Ocaña-Fuentes et al. [63] proved the anti-inflammatory activity of SFE extracts of oregano rich in *trans*-sabinene hydrate, thymol, and carvacrol. These authors observed no differences in terms of composition and activity between extracts obtained in different separators and thus the SFE fractional conditions applied allow recovering the fraction extracts with anti-inflammatory properties. Overall, the literature reviewed showed that SFE-CO_2 is an efficient approach to extract bioactive compounds from many aromatic species (Table 2). Among the different compounds with functional properties extracted by SFE from plant matrices, the antioxidant compounds stand out as the most investigated. Antioxidants have an important role in the food and pharmaceutical industries which justify the extensive search for new rich sources of these compounds and the optimization of extraction conditions for their recovery.

7. Conclusions

Plants produce a great diversity of bioactive molecules useful in many fields. EOs constituents from aromatic plants are some of the most important natural products with a broad range of applications in the food, pharmaceutical, perfume, and cosmetic industries. Several Mediterranean aromatic plants, mainly from Lamiaceae and Apiaceae families, have been subject of numerous investigations aiming at the extraction of bioactive constituents by SFE-CO_2 and evaluation of their biological properties. SFE is an innovative, effective, and environmental friendly technique for oil recovery from aromatic plants. Due to the lipophilic nature of EO components, they are highly soluble in CO_2 at moderate temperatures and pressures. The use o polar solvent is not needed and, therefore, a subsequent step of solvent elimination is not necessary. The low temperatures allow the recovery of high quality products with improved biological properties, and higher pressures also allow the extraction of compounds with lower solubility. Fractionation strategies allow obtaining products with distinct characteristics. Many variables have remarkable effects on the SFE process, such as pressure and temperature, CO_2 flow rate, extraction period, plant collection site and time, type of plant tissue, and pre-treatment, among others. These variables affect not only the extraction yields but also the composition of the recovered material. Several mathematical models have been successfully used in the optimization of operational conditions of the SFE process.

The reviewed literature indicates that SFE-CO_2 leads to higher extraction yield, improved oil quality, higher selectivity of the extracted components, and improved biological activity in many aromatic plant species. The reviewed data also highlights the importance of optimizing all the factors involved in the extraction process. SFE-CO_2 is an efficient approach to extract bioactive compounds from many aromatic species to be used in many applications (e.g., food products, nutraceuticals, pharmaceuticals, cosmetics).

Acknowledgments

This work was supported by National Funds—FCT (Fundação para a Ciência e a Tecnologia, I.P.), through the project UID/BIA/4325/2013 and to the project INTERREG—MD.Net: When Brand Meets People.

References

[1] M. Narayani, S. Srivastava, Elicitation: A stimulation of stress in *in vitro* plant cell/tissue cultures for enhancement of secondary metabolite production, Phytochem. Rev. 16 (2017) 1227–1252.
[2] G. Samuelsson, Drugs of Natural Origin: A Textbook of Pharmacognosy, fifth ed., Swedish Pharmaceutical Press, Stockholm, 2004.
[3] G.M. Cragg, D.J. Newman, Plants as a source of anticancer agents, J. Ethnopharmacol. 100 (2005) 72–79.
[4] M. Gordaliza, Natural products as leads to anticancer drugs, Clin. Transl. Oncol. 9 (2007) 767–776.

[5] British Broadcasting Corporation (BBC) Research, Plant-derived Drugs: Global Markets, 2017. Available from: https://www.bccresearch.com/market-research/biotechnology/botanical-and-plant-derived-drugs-global-markets-bio022h.html (Accessed 2018-11-08).
[6] M.S. Butler, F. Fontaine, M.A. Cooper, Natural product libraries: assembly maintenance and screening, Planta Med. 80 (2014) 1161–1170.
[7] K.-Y. Khaw, M.-O. Parat, P.N. Shaw, J.R. Falconer, Solvent supercritical fluid technologies to extract bioactive compounds from natural sources: a review, Molecules 22 (2017) 1186.
[8] S. Elshafie, I. Camele, An overview of the biological effects of some Mediterranean essential oils on human health, Biomed. Res. Int. 2017 (2017) (ID 9268468).
[9] Council of Europe-COE-European Directorate for Quality of Medicines, European Pharmacopeia, sixth ed., Council of Europe, Strasbourg, France, 2007.
[10] T. Fornari, G. Vicente, E. Vázquez, M.R. García-Risco, G. Reglero, Isolation of essential oil from different plants and herbs by supercritical fluid extraction, J. Chromatogr. A 1250 (2012) 34–48.
[11] J. Giacometti, D.B. Kovačević, P. Putnik, D. Gabrić, T. Bilušić, G. Krešić, V. Stulić, F.J. Barba, F. Chemat, G. Barbosa-Cánovas, A.R. Jambrak, Extraction of bioactive compounds and essential oils from Mediterranean herbs by conventional and green innovative techniques: a review, Food Res. Int. 113 (2018) 245–262.
[12] S. Öztekin, M. Martinov, Medicinal and Aromatic Crops: Harvesting, Drying, and Processing, Haworth Food & Agricultural Products Press, Binghamtion, New York, 2007.
[13] R. Ribeiro-Santos, D. Carvalho-Costa, C. Cavaleiro, H.S. Costa, T.G. Albuquerque, M.C. Castilho, F. Ramos, N.R. Melo, A. Sanches-Silva, A novel insight on an ancient aromatic plant: The rosemary (*Rosmarinus officinalis* L.), Trends Food Sci. Techn. 45 (2015) 355–368.
[14] F.S. Agency, European pharmacopoeia, sixth ed., Current EU approved additives and their E numbers, in: C.O. Europe (Ed.), Strasbourg, 2008, pp. 251.
[15] R. Rehman, M.A. Hanif, Z. Mushtaq, A.M. Al-Sadic, Biosynthesis of essential oils in aromatic plants: a review, Food Rev. Int. 32 (2016) 117–160.
[16] R. Ribeiro-Santos, M. Andrade, A. Sanches-Silva, N.R. Melo, Essential oils for food application: Natural substances with established biological activities, Food Bioprocess Tech. 11 (2018) 43–71.
[17] K.H.C. Baser, G. Buchbauer, Handbook of Essential Oils: Science, Technology, and Applications, CRC Press, Boca Raton, FL, 2009.
[18] F. Bakkali, S. Averbeck, D. Averbeck, M. Idaomar, Biological effects of essential oils—A review, Food Chem. Toxicol. 46 (2008) 446–475.
[19] N.S. Sangwan, A.H.A. Farooqi, F. Shabih, R.S. Sangwan, Regulation of essential oil production in plants, Plant Growth Regul. 34 (2001) 3–21.
[20] V.P.S. Rao, D. Pandey, Extraction of Essential Oils and its Applications. Orissa, National Institute of Technology Rourkela, India, 2006.
[21] S. Burt, Essential oils: their antibacterial properties and potential applications in foods-a review, Int. J. Food Microbiol. 94 (2004) 223–253.
[22] Food and Drug Administration (FDA), Code of Federal Regulations (CFR), Title 21: ood and Drugs. Chapter I—Food and Drug Administration, Department of Health and Human Services, Subchapter B—Food for Human Consumption (Continued), Part 182—Substances Generally Recognized as Safe (GRAS), Subpart —General Provisions, Subpart 182.20—Essential Oils, Oleoresins, and Natural Extractives, Office of the Federal Register, Washington, 2016.
[23] F. Chemat, N. Rombaut, A. Meullemiestre, M. Turk, S. Perino, A.S. Fabiano-Tixier, M. Abert-Vian, Review of green food processing techniques. Preservation, transformation, and extraction, Innov. Food Sci. Emerg. Technol. 41 (2017) 357–377.
[24] M. Herrero, A. Cifuentes, E. Ibanez, Sub- and supercritical fluid extraction of functional ingredients from different natural sources: plants, food-by-products, algae and microalgae, Food Chem. 98 (2006) 136–148.
[25] H.M.A. Barbosa, M.M.R. de Melo, M.A. Coimbra, C.P. Passos, C.M. Silva, Optimization of the supercritical fluid coextraction of oil and diterpenes from spent coffee grounds using experimental design and response surface methodology, J. Supercrit. Fluids 85 (2014) 165–172.

[26] R.P.F.F. Silva, T.A.P. Rocha-Santos, A.C. Duarte, Supercritical fluid extraction of bioactive compounds, Trends Analyt. Chem. 76 (2016) 40–51.
[27] T. Fornari, A. Ruiz-Rodriguez, G. Vicente, E. Vázquez, M.R. García-Risco, G. Reglero, Kinetic study of the supercritical CO_2 extraction of different plants from Lamiaceae family, J. Supercrit. Fluids 64 (2012) 1–8.
[28] S. Santoyo, L. Jaime, M.R. Garcia-Risco, M. Lopez-Hazas, G. Reglero, Supercritical fluid extraction as an alternative process to obtain antiviral agents from thyme species, Ind. Crop. Prod. 52 (2014) 475–480.
[29] E. Oca, A. Ibanez, G. Murga, S.L.D. Sebastian, J. Tabera, G. Reglero, Supercritical fluid extraction and fractionation of different preprocessed rosemary plants, J. Agric. Food Chem. 47 (1999) 1400.
[30] N. Wilkinson, R. Hilton, D. Hendry, C. Venkitasamy, W. Jacoby, Study of process variables in supercritical carbon dioxide extraction of soybeans, Food Sci. Technol. Int. 20 (2014) 63–70.
[31] C. Grosso, V. Ferraro, A.C. Figueiredo, J.G. Barroso, J.A. Coelho, A.M. Palavra, Supercritical carbon dioxide extraction of volatile oil from Italian coriander seeds, Food Chem. 111 (2008) 197–203.
[32] N.M. Shrigod, N.R. Swami Hulle, R.V. Prasad, Supercritical fluid extraction of essential oil from mint leaves (*Mentha spicata*): process optimization and its quality evaluation, J. Food Process Eng. 40 (2017) 1–9.
[33] T. Hatami, J.C.F. Johner, M.A.A. Meireles, Investigating the effects of grinding time and grinding load on content of terpenes in extract from fennel obtained by supercritical fluid extraction, Ind. Crop. Prod. 109 (2017) 85–91.
[34] M. Meghwal, T.K. Goswami, Cryogenic grinding of spices is a novel approach whereas ambient grinding needs improvement, Continental J. Food Sci. Technol. 4 (2010) 24–37.
[35] Y.Y. Andrade-Avila, J. Cruz-Olivares, C. Pérez-Alonso, C.H. Ortiz-Estrada, M.D.C. Chaparro-Mercado, Supercritical extraction process of allspice essential oil, J. Chem. 2017 (2017) (ID 6471684).
[36] I. Jerković, M. Molnar, S. Vidović, J. Vladić, S. Jokić, Supercritical CO_2 extraction of *Lavandula angustifolia* mill. Flowers: Optimisation of oxygenated monoterpenes, coumarin and herniarin content, Phytochem. Anal. 28 (2017) 558–566.
[37] P. Priyanka, S. Khanam, Influence of operating parameters on supercritical fluid extraction of essential oil from turmeric root, J. Clean. Prod. 188 (2018) 816–824.
[38] M. Shahsavarpour, M. Lashkarbolooki, M.J. Eftekhari, F. Esmaeilzadeh, Extraction of essential oils from *Mentha spicata* L. (Labiatae) via optimized supercritical carbon dioxide process, J. Supercrit. Fluids 130 (2017) 253–260.
[39] S. Jokić, M. Molnar, M. Jakovljević, K. Aladić, I. Jerković, Optimization of supercritical CO_2 extraction of *Salvia officinalis* L. leaves targeted on oxygenated monoterpenes, α-humulene, viridiflorol and manool, J. Supercrit. Fluids 133 (2018) 253–262.
[40] C. Grosso, A.C. Figueiredo, J. Burillo, A.M. Mainar, J.S. Urieta, J.G. Barroso, J.A. Coelho, A.M. Palavra, Composition and antioxidant activity of *Thymus vulgaris* volatiles: comparison between supercritical fluid extraction and hydrodistillation, J. Sep. Sci. 33 (2010) 2211–2218.
[41] J.L. Pasquel Reategui, A.P.D.F. Machado, G.F. Barbero, C.A. Rezende, J. Martinez, Extraction of antioxidant compounds from blackberry (*Rubus* sp.) bagasse using supercritical CO_2 assisted by ultrasound, J. Supercrit. Fluids 94 (2014) 223–233.
[42] C. Porto, D. Decorti, A. Natolino, Separation of aroma compounds from industrial hemp inflorescences (*Cannabis sativa* L.) by supercritical CO_2 extraction and on-line fractionation, Ind. Crop. Prod. 58 (2014) 99–103.
[43] E. Ibañez, A. Kubátová, F.J. Señoráns, S. Cavero, G. Reglero, S.B. Hawthorne, Subcritical water extraction of antioxidant compounds from rosemary plants, J. Agric. Food Chem. 51 (2003) 375–382.
[44] L.A. Conde-Hernández, J.R. Espinosa-Victoria, A. Trejo, J.A. Guerrero-Beltrán, CO_2-supercritical extraction, hydrodistillation and steam distillation of essential oil of rosemary (*Rosmarinus officinalis*), J. Food Eng. 200 (2017) 81–86.
[45] S.B. Glisic, D.R. Misic, M.D. Stamenic, I.T. Zizovic, R.M. Asanin, D.U. Skala, Supercritical carbon dioxide extraction of carrot fruit essential oil: Chemical composition and antimicrobial activity, Food Chem. 105 (2007) 346.

[46] P. Costa, C. Grosso, S. Gonçalves, P.B. Andrade, P. Valentão, M.G. Bernardo-Gil, A. Romano, Supercritical fluid extraction and hydrodistillation for the recovery of bioactive compounds from *Lavandula viridis* L'Hér, Food Chem. 135 (2012) 112–121.
[47] N. Herzi, J. Bouajila, S. Camy, S. Cazaux, M. Romdhane, J.S. Condoret, Comparison between supercritical CO_2 extraction and hydrodistillation for two species of eucalyptus: yield, chemical composition, and antioxidant activity, J. Food Sci. 78 (2013) 667–672.
[48] E.B. Parisotto, E.M.Z. Michielin, F. Biscaro, S.R.S. Ferreira, D.W. Filho, R.C. Pedrosa, The antitumor activity of extracts from *Cordia verbenacea* D.C. obtained by supercritical fluid extraction, J. Supercrit. Fluids 61 (2012) 101–107.
[49] G. Sodeifian, S.A. Sajadian, Investigation of essential oil extraction and antioxidant activity of *Echinophora platyloba* DC. Using supercritical carbon dioxide, J. Supercrit. Fluids 121 (2017) 52–62.
[50] E. Arranz, L. Jaime, M.C. Lopez de la Hazas, G. Vicente, G. Reglero, S. Santoyo, Supercritical sage extracts as anti-inflammatory food ingredients, Ind. Crop. Prod. 54 (2014) 159–166.
[51] S.B. Glisic, M. Ristic, D.U. Skala, The combined extraction of sage (*Salvia officinalis* L.): Ultrasound followed by supercritical CO_2 extraction, Ultrason. Sonochem. 18 (2011) 318–326.
[52] V. Šulniūtė, R. Baranauskienė, O. Ragažinskienė, P.R. Venskutonis, Comparison of composition of volatile compounds in ten *Salvia* species isolated by different methods, Flavour Fragr. J. 32 (2017) 254–264.
[53] C. Busatta, J. Barbosa, R.I. Cardoso, N. Paroul, M. Rodrigues, D. Oliveira, J.V. Oliveira, L.C. Rogério, Chemical profiles of essential oils of marjoram (*Origanum majorana*) and oregano (*Origanum vulgare*) obtained by hydrodistillation and supercritical CO_2, J. Essent. Oil Res. 29 (2017) 367–374.
[54] Z. Zeković, B. Pavlić, A. Cvetanović, S. Đurović, Supercritical fluid extraction of coriander seeds: process optimization, chemical profile and antioxidant activity of lipid extracts, Ind. Crop. Prod. 94 (2016) 353–362.
[55] B.N. Jahromi, A. Tartifizadeh, S. Khabnadideh, Comparison of fennel and mefenamic acid for the treatment of primary dysmenorrheal, Int. J. Gynaecol. Obstet. 80 (2003) 153–157.
[56] L.S. Moura, R.N. Carvalho, M.B. Stefanini, L.C. Ming, M.A. Meireles, Supercritical fluid extraction from fennel (*Foeniculum vulgare*): global yield, composition and kinetic data, J. Supercrit. Fluids 35 (2005) 212–219.
[57] R. Rodríguez-Solana, J.M. Salgado, J.M. Domínguez, S. Cortés-Diéguez, Estragole quantity optimization from fennel seeds by supercritical fluid extraction (carbon dioxide–methanol) using a Box–Behnken design. Characterization of fennel extracts, Ind. Crop. Prod. 60 (2014) 186–192.
[58] A. Mouahid, C. Dufour, E. Badens, Supercritical CO_2 extraction from endemic Corsican plants; comparison of oil composition and extraction yield with hydrodistillation method, J. CO_2 Util. 20 (2017) 263–273.
[59] G. Sodeifian, N.S. Ardestani, S.A. Sajadian, S. Ghorbandoost, Application of supercritical carbon dioxide to extract essential oil from *Cleome coluteoides* Boiss: experimental, response surface and grey wolf optimization methodology, J. Supercrit. Fluids 114 (2016) 55–63.
[60] P. Kotnik, M. Škerget, Z. Knez, Supercritical fluid extraction of chamomile flower heads: comparison with conventional extraction, kinetics and scale-up, J. Supercrit. Fluids 43 (2007) 192–198.
[61] Q. Shao, Y. Deng, H. Liu, A. Zhang, Y. Huang, G. Xu, M. Li, Essential oils extraction from *Anoectochilus roxburghii* using supercritical carbon dioxide and their antioxidant activity, Ind. Crop. Prod. 60 (2014) 104–112.
[62] H. Bagheri, M.Y.B.A. Manap, Z. Solati, Antioxidant activity of *Piper nigrum* L. essential oil extracted by supercritical CO_2 extraction and hydro-distillation, Talanta, 121 (2014) 220–228.
[63] A. Ocaña-Fuentes, E. Arranz-Gutiérrez, F.J. Señorans, G. Reglero, Supercritical fluid extraction of oregano (*Origanum vulgare*) essentials oils: anti-inflammatory properties based on cytokine response on THP-1 macrophages, Food Chem. Toxicol. 48 (2010) 1568–1575.

CHAPTER 5

Metal recovery using supercritical carbon dioxide

Daniel Assumpção Bertuol[a], Franco Rico Amado[b], Estevan Dorneles Cruz[a], Eduardo Hiromitsu Tanabe[a]
[a]Environmental Processes Laboratory (LAPAM), Chemical Engineering Department, Federal University of Santa Maria—UFSM, Santa Maria, Brazil
[b]Materials and Environment Laboratory (LAMMA), State University of Santa Cruz—UESC, Ilheus, Brazil

Contents

1. Introduction

The use of supercritical technology offers the possibility of obtaining new products with special characteristics, as well as the development of new processes that are ecologically acceptable and sustainable. The use of high pressure as a processing tool can eliminate problems associated with the use of conventional organic solvents, while also providing reaction rates that are much faster than in traditional processes. In some cases, it is possible to reduce the consumption of reagents, hence making processes much cleaner and more efficient [1].

Supercritical fluids are already used commercially in various established industrial processes, such as the extraction of solids and liquids [2], dry cleaning [3], high-pressure sterilization [4], jet cutting [5], deposition of thin films in microelectronics [6], separation of high added-value products from fermentation broths [7], and as solvents in many syntheses involving transition metals [8], among other uses [9].

This wide range of industrial-scale processes provides a strong case for expanding research in the field of metal recovery, with the objective of developing new technologies. It is clear that for some industrial applications the use of supercritical fluids may be more expensive, compared to conventional methods. Nonetheless, supercritical fluids can be used to obtain

Green Sustainable Process for Chemical and Environmental Engineering and Science
https://doi.org/10.1016/B978-0-12-817388-6.00005-2

high added-value products, with low environmental impacts and rapid extraction kinetics making the process attractive from the economic perspective.

A good example of an application of supercritical fluids that can provide substantial benefits is in the processing of waste electrical and electronic equipment (WEEE). This is because the composition of these wastes includes rare and strategically important metals, such as gold, rhodium, and rare earths, together with highly toxic components such as flame retardants based on halogenated compounds, or the metals lead and cadmium, among others. The large quantities of these wastes, together with the potential presence of materials harmful to humans and the environment, necessitate special treatment at the end of their useful life.

The main WEEE are goods such as televisions, computers, mobile phones, printers, and domestic appliances, which are considered to be electronic waste when they become obsolete or cease to function [10, 11]. The worldwide generation of WEEE in 2016 was around 44.7 million tons, with the quantity expected to increase to 52.2 million tons in 2021, at a rate of 3%–4% per year [12]. Of these 44.7 Mt., approximately 1.7 Mt. are likely to be incinerated or sent to landfill in developed countries. Globally, only 8.9 Mt. of e-waste is documented to be collected and recycled, corresponding to 20% of all the e-waste generated [13].

Mobile phones represent an increasingly important type of WEEE, due to the fact that they are now considered indispensable for facilitating daily life. In Brazil, according to data provided by the Brazilian Electrical and Electronic Industry Association (ABINEE) [14], the production of consumer goods in the electrical-electronic sector has increased rapidly, especially considering the sales of smartphones (around 48 million units) and notebook computers (3.5 million units), with increases of 10% and 21%, respectively, in 2017, compared to 2016 [14].

Printed circuit boards (PCBs) constitute one of the main components of WEEE. They contain metals including copper, lead, zinc, and tin, as well as precious metals such as gold, silver, and platinum [15]. Discarded PCBs therefore represent a useful and valuable raw material. In addition to metals and metal alloys, PCBs also contain various organic and inorganic compounds [16]. As a broad generalization, these boards can be considered to consist of 49% ceramic materials, glass, and oxides, 28% metals, 19% plastics, and 4% bromine. The actual composition depends on the origin of the PCB, as well as the type and age of the equipment [17].

Angerer et al. [18] described the differences in the compositions of PCBs from old computers, new computers, and video equipment. For example, the average concentrations of copper were found to be 29,800 ppm in old computers, 37,000 ppm in new computers, and 44,900 ppm in video equipment. The concentrations of gold were 1050 ppm in old computers, 750 ppm in new computers, and 870 ppm in video equipment. Considering the materials present in the composition of PCBs, metals constitute the most valuable part, justifying the recycling of these materials and making the process economically viable.

The supercritical fluid utilization is very promising for the treatment of waste PCBs as well as other types of technological scrap, including the recycling of indium from LCD screens [19], rare earths present in batteries, magnets, and fluorescent lamps, and the platinum group metals found in catalysts.

The processing of WEEE is complex, due to the heterogeneity of its composition, and the usual conventional methods of recycling can generate significant quantities of waste. As an alternative to these methods, supercritical fluids can be used to dissolve the materials. Compared to traditional solvents, the properties of supercritical fluids provide advantages in terms of enhanced mass transfer [20, 21]. While the idea of using supercritical fluids is not new, it is only more recently that their use has been reported for the recycling of technological wastes, evidencing that this is a new technology that is still in the developmental phase [20, 22, 23].

In most cases, the few published works related to this topic have described the use of supercritical fluids for the extraction of metal ions or for the degradation of the polymer fraction present in WEEE. Recent studies have obtained excellent results in the direct extraction of metals from the solid phase, with substantial reductions in the time required for the extraction, while in some cases it was also possible to decrease the consumption of reagents [24–26].

2. Processes for recycling of WEEE

There are essentially four methods used for recycling of WEEE. Mechanical processing is used as a pretreatment to concentrate the metal fraction, which is then sent to a subsequent process [27–29]. Pyrometallurgical techniques employ high temperatures that are usually associated with hazardous emissions to the atmosphere, due to the generation of organochlorines, chlorides, and mercury during the process [30]. Hydrometallurgical processes involve the extraction of metals by means of dissolution and purification reactions, with leaching in acid or alkaline media and subsequent operations for concentration and purification [31–33]. In addition, electrometallurgical processes employ electricity to refine or purify metals [34–36].

2.1 Leaching of metals

Leaching constitutes one of the main steps in the recovery of metals from WEEE. Acid leaching is often employed as the first stage of extraction of base metals, especially in the case of copper, for which mineral acids such as HCl, H_2SO_4, or HNO_3 are used in combination with H_2O_2, a strong oxidant that improves the leaching performance of the acid [37]. The combination of H_2SO_4 as a leaching agent and H_2O_2 as an oxidizing agent is commonly used for the recovery of copper from PCBs [38, 39].

The leaching agents most commonly used for precious metals are cyanide, thiocyanate, and halogens [40, 41]. Cyanide provides excellent results in the recovery of precious metals from both minerals and secondary sources. This reagent has been used in mining for over

100 years, offering the main advantages of high efficiency and relatively low cost. However, a series of environmental incidents in gold mines worldwide has led to widespread concern about the use of cyanide. As a result, various environmentally acceptable reagents have been investigated for this purpose, although only thiourea and thiosulfate are generally considered as promising candidates to substitute cyanide [42].

Thiourea [$(NH_2)_2CS$] has shown considerable promise as an agent to extract precious metals from primary and secondary sources. In an acid medium, thiourea dissolves gold, forming a cationic complex. The reaction is rapid and the reagent is highly stable in acidic and neutral media, but decomposes rapidly in basic solutions. The extraction is usually performed at pH 1–2, with success of the operation depending on careful optimization and control of the pH, redox potential, thiourea concentration, and leaching time [42].

The use of thiosulfate in an alkaline medium results in slow dissolution of gold at a rate that is determined by the process temperature and the concentrations of thiosulfate and dissolved oxygen. Copper ions are also added to increase the gold dissolution rate. Although the complex formed between thiosulfate and gold is fairly stable, it is necessary to maintain an alkaline pH, in order to avoid decomposition of the thiosulfate. Despite potential environmental benefits, an important drawback of using thiosulfate for leaching is the high consumption of reagents during the extraction. In addition, the process is generally slow, although acceptable leaching rates can be obtained in the presence of ammonia, with copper (II) as an oxidizing agent [42].

A recent and more environmentally friendly technique for the leaching and recovery of base metals from WEEE involves the use of organic acids. Wu et al. [43, 44] investigated the recovery of Co and Li from Li-ion batteries by leaching with citric and malic acids, in the presence of H_2O_2, in both cases recoveries obtained were 90% for Co and approximately 100% for Li. The use of aspartic acid and H_2O_2 for the leaching and recovery of these metals was also studied, but the recoveries achieved were below 70% [45].

Malic acid ($C_4H_6O_5$) and citric acid ($C_6H_8O_7$) are organic carboxylic acids that dissolve easily in water. Malic acid is widely used in the food and pharmaceutical industries, as well as in the synthesis of various chemicals [46], while citric acid is also employed in many industrial sectors. Both acids degrade much more readily under aerobic and anaerobic conditions, compared to mineral acids (HCl, HNO_3, and H_2SO_4), and can therefore be easily treated when present in waste solutions [43].

The use of citric acid was proposed by Sonmez and Kumar [47] as a new method for the recovery of Pb and PbO from scrap batteries. Leaching with citric acid and H_2O_2 resulted in 99% recoveries for both Pb and PbO. Another organic acid that can be employed for this purpose is acetic acid ($C_2H_4O_2$). Yang et al. [48] used this acid together with sodium citrate for the leaching of PbO, PbO_2, and PbO_4, with recoveries of around 97% achieved. Leaching with citric acid has also been used to extract Co from mineral ores, with 84.5% recovery of Co achieved from an ore containing only 0.13% of the metal [49].

Another important point is that these acids can be produced from organic wastes, such as agroindustrial residues, consequently adding value to materials normally considered wastes, while at the same time reducing the costs of the acids [50–52].

It is evident that there have been many efforts to identify less aggressive methods for the recovery of metals that are both effective and environmentally friendly. In addition to leaching using alternative reagents such as organic acids, thiourea, and thiosulfate, other processes such as extraction using supercritical fluids can be used to substitute conventional solvents, minimize waste generation and reactant consumption, and provide faster extraction of the metal.

2.2 Supercritical fluids

The use of supercritical fluids for the recovery of metals from solid and liquid matrices has become a promising alternative to conventional techniques, due to the higher solvent properties of supercritical fluids, compared to other types of solvents [53].

Supercritical fluids are substances present at pressures and temperatures above their critical points (Fig. 1). They offer higher solvent power for nonpolar or weakly polar substances and can be easily removed from solutes by simple expansion to ambient

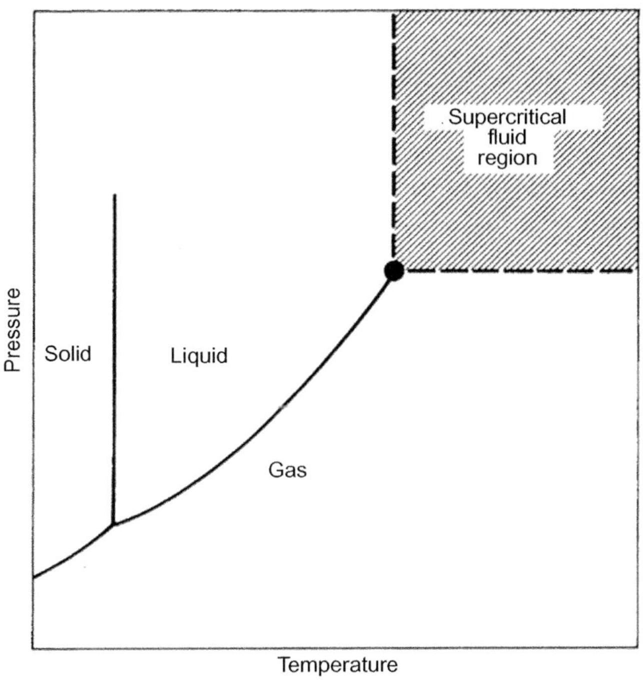

Fig. 1 Phase diagram for a pure component. *(Source: M. McHugh, V. Krukonis, Supercritical Fluid Extraction Principles and Practice, second ed., Butterworth-Heinemann, 1993.)*

Table 1 Properties of solvents used in supercritical extraction

Solvents	Critical temperature (°C)	Critical pressure (bar)
Carbon dioxide	31.1	73.8
Ethane	32.2	48.8
Etanol	240.8	61.4
Ethylene	9.3	50.4
Propane	96.7	42.5
Nitrous oxide	36.4	72.8
Propylene	91.9	46.2
Benzene	289.00	48.9
Water	374.2	220.5

Source: Adapted from F. Lin, D. Liu, S.M. Das, N. Prempeh, Y. Hua, J. Lu, Recent progress in heavy metal extraction by supercritical CO_2 fluids, Ind. Eng. Chem. Res. 53 (5) (2014) 1866–1877; M. McHugh, V. Krukonis, Supercritical Fluid Extraction Principles and Practice, second ed., Butterworth-Heinemann, 1993.

pressure. In the supercritical state, the density of the fluid is similar to the density of liquids, while the viscosity is close to that of normal gases and the diffusivity is about two orders of magnitude higher than typical liquid diffusivities [54]. The low viscosity and high diffusivity of supercritical fluids result in substantial density gradients at the solute-solvent interface, resulting in mass transfer rates superior to those achieved with conventional solvents [55].

In recent years, various fluids have been investigated for use as solvents under supercritical conditions. Some substances, such as ethane and ethylene, require fairly mild operating conditions, but are reactive and leave residues after the process. CO_2 does not have these disadvantages, hence making it an excellent option [56, 57]. The critical pressure and temperature values for CO_2 and other solvents are provided in Table 1.

In cases when the compounds of interest present low solubility in a pure supercritical solvent, small amounts of modifiers can be added to the extractant, altering the polarity of the fluid and increasing the solubility of substances with greater polarity. Supercritical CO_2 is a particularly nonpolar solvent, and the addition of small amounts of substances such as alcohol, or even water, changes the polarity to a much greater degree than obtained by increasing the pressure to high levels [56]. Table 2 provides some examples of cosolvents used in the processing of metals from electronic scrap.

2.2.1 Experimental systems for the extraction of metals using supercritical fluids

For better understanding of the methodology used for solubilization of different metals using supercritical fluids, Fig. 2 shows a schematic diagram of a supercritical extraction process used for the extraction of metals from WEEE.

Firstly, the waste material must be broken up and homogenized. The required amount of material is then loaded into an extractor (4), together with the cosolvent, forming a bed of particulates. The extractor illustrated here is fitted with an outer jacket

Table 2 Examples of cosolvents used in the processing of electronic scrap

Substance extracted	Waste	Cosolvent	Amount extracted (wt %)	Extraction time (min)	Ref.
In	LCD screens	Citric acid +5% H_2O_2	94.6	30	[26]
Co	Li-ion batteries	Sulfuric acid +4% H_2O_2	95.5	5	[24]
Polymers	Printed circuit boards	Ethanol	69.53	30	[23]
Cu	Printed circuit boards	Sulfuric acid +20% H_2O_2	88.79	20	[25]
Pd/Ag	Printed circuit boards	Acetone and KI-I_2	93.7% (Pd) 96.4% (Ag)	10	[58]

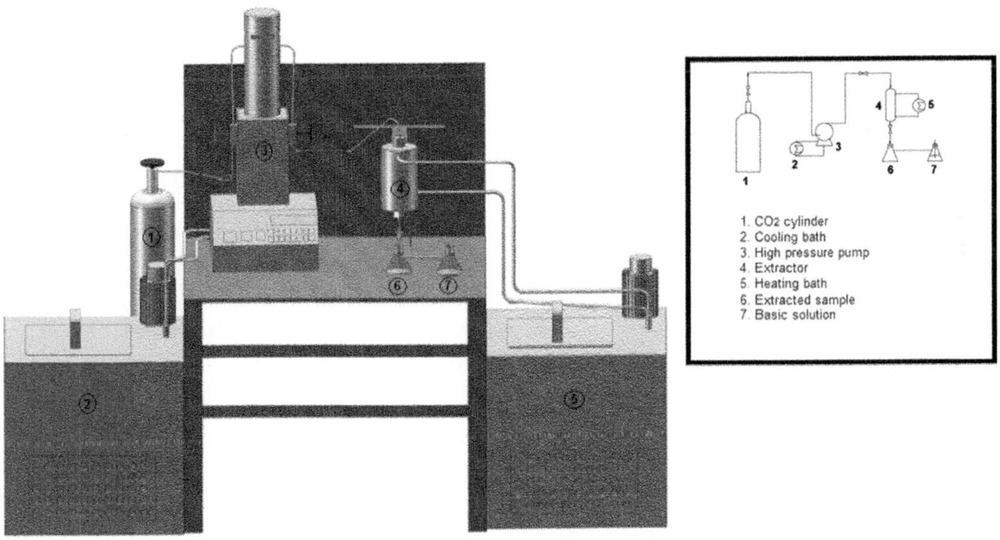

Fig. 2 Schematic representation of a supercritical fluid extraction system [25].

to enable control of the temperature (5). The pressurized solvent (carbon dioxide) is supplied to the system from a cylinder (1). The operating pressure is achieved using a syringe-type high-pressure pump (3). The bed of particulate material in the extractor is leached by the solvent and the solvent-solute mixture is then decompressed using a pressure reduction valve located below the reactor. The mixture becomes gas-solute and the extract is received in the collection flask (6) for subsequent analysis. Downstream of the collection flask, another flask containing a solution of sodium hydroxide is used for scrubbing the CO_2 separated from the extracted solution.

2.2.2 Recycling of WEEE using supercritical CO_2 ($ScCO_2$)

The recovery of valuable metals present in WEEE such as PCBs is hindered by the heterogeneity of the composition of this type of waste, due to the presence of various polymers, different types of metals, and fiberglass. Calgaro et al. [23] investigated the use of supercritical CO_2 in the presence of ethanol (cosolvent), applied to waste PCBs, with the aim of degrading the polymers and concentrating the metals. The results demonstrated that the treatment of this scrap using supercritical fluids was effective and environmentally friendly. The polymeric fraction of the PCBs was converted into oils after treatment with methanol and supercritical CO_2. The oils obtained in this process were mainly phenol and phenol derivatives, which could be used as raw materials in other processes.

Therefore, the use of supercritical CO_2, together with ethanol as a cosolvent, has proved to be highly efficient for the extraction of the organic fraction of PCBs under moderate temperature conditions. Furthermore, the removal of the polymers is essential for the recovery of metals from PCBs, because the metal surface becomes exposed and the metals can be easily leached [23].

Xiu and Zhang [59] obtained excellent results using supercritical water for the extraction of lead and copper from PCBs. It was also found that the technique employed was effective as a pretreatment for the decomposition of the organic compounds of the PCBs. In this process, the copper and lead were oxidized to CuO, Cu_2O, and β-PbO_2, and losses of the metals were insignificant. After the extraction, the solution obtained was treated using an electrochemical process for the recovery of the metals [59]. However, the critical temperature and pressure of water ($Tc=374\,°C$, $Pc=22.1\,MPa$) are very high, resulting in high energy consumption and risk of corrosion of the equipment employed in the process. Hence, it is essential to study the use of other fluids that have milder supercritical points but are still capable of providing efficient extraction [23]. At the present stage of development, it is clear that the use of supercritical fluid technology for the recovery of metals from WEEE is promising, while further studies are needed for better analysis and comparison with conventional methods.

In other work concerning the recycling of copper present in PCBs, comparison was made between conventional leaching at atmospheric pressure and the use of supercritical CO_2 in the presence of H_2O_2 and sulfuric acid. It was found that the supercritical extraction was nine times faster than the extraction at atmospheric pressure. Around 90% of the Cu present in the PCBs was extracted in 20 min of supercritical leaching, while the process at atmospheric pressure required 180 min to achieve the same extraction percentage [25].

Argenta et al. [26] evaluated the use of supercritical CO_2 modified with organic acids and H_2O_2 for the recovery of indium from the LCD screens of discarded cell phones.

Extraction performed for 3h at 90°C and atmospheric pressure, using 1M citric acid, solid:liquid ratio of 1:20, and 5% (*v*/v) H_2O_2, resulted in the extraction of 76.5% of the indium. When supercritical CO_2 is used for the extraction of indium, the process time could be decreased to 30min and the extraction increased to 94.6%, using a temperature of 100°C, pressure of 150bar, solid:liquid ratio of 1:20, and 5% (v/v) H_2O_2 [26].

These results demonstrated the excellent efficiency of the process, indicating the suitability of supercritical CO_2 as an environmentally acceptable and reusable substance for use in the recycling of WEEE. Although application of this technique involves a higher initial cost, the investment made can be rapidly recovered, since the process can provide considerable economy of reagents, in addition to allowing the recycling of CO_2 and providing reaction rates that are much faster compared to those achieved in conventional processes. Such reductions in reaction time and reagent consumption were observed in a study involving the recovery of cobalt from lithium-ion batteries. It was found that the consumption of H_2O_2 was halved and the reaction time was 12 times shorter in comparison with a conventional process performed at atmospheric pressure [24].

3. Extraction of metal ions from aqueous solutions using $ScCO_2$ in the presence of complexing agents

Another important application of $ScCO_2$ is its use in conjunction with complexing agents for the selective extraction of metals present in aqueous solutions. The processes involving liquid–liquid extraction in the presence of different complexing agents are widely employed in hydrometallurgy [60]. In these procedures, chelants are dissolved in an organic medium (such as kerosene). During the extraction, the metal ions present in the aqueous solution are selectively transferred to the organic solution, by controlling the pH, and are then recovered by means of a re-extraction (stripping) from the organic phase to the aqueous phase [57, 61].

The use of supercritical carbon dioxide as a solvent, instead of an organic solvent such as kerosene, results in reduced generation of liquid effluents, shorter extraction times, and easy separation of the phases, since a simple change in pressure causes the carbon dioxide to return to the gaseous state, hence becoming separated from the solute [62].

The complexing agents used in conventional liquid–liquid extraction can also be used in extraction using supercritical CO_2, if they are soluble in $ScCO_2$. The solubility of metal ions in $ScCO_2$ can vary greatly, depending on the chemical nature of the complexes formed. Many complexing agents have been used for the supercritical extraction of heavy metals existing in ionic forms, including Cyanex 302, Aliquat 336, and DEHPA, among others [20, 63].

The use of chelants with supercritical CO_2 is required because this medium is nonpolar, which would make it impossible to solubilize metal ions. When chelants are used as cosolvents, the complexing agents bind to the metal ions, hence enabling extraction of the resulting complexes into the supercritical medium. The structures of the complexing molecules alter the polarity of the fluid medium, hence increasing the extraction of metal ions [64].

The selection of a suitable complexing agent needs to consider characteristics such as satisfactory solubility of the agent and its metal complexes, fast kinetics of reaction between the agent and metal ions, and selectivity toward specific groups of these ions [65]. It has been found that the alkyl groups of chelating agents increase their solubility in $ScCO_2$, with increase in the alkyl groups the polarity of the complexes and complexants reduces [66].

Table 3 presents the structures of some of the complexing agents used industrially, which have been studied as candidates for use in the metallic ions extraction in the presence of $ScCO_2$. As the commercial use of chelating agents for hydrometallurgical and other processes has grown, the number of chelating agents available on the market has also increased.

Laintz et al. [65] reported one of the first uses of supercritical CO_2 in the presence of a chelating agent for metal extraction, using LiFDDC (lithium bis(trifluoroethyl) dithiocarbamate) as a chelating agent for the extraction of Cu^{2+} from an aqueous solution and a silica surface.

Table 4 illustrates the ideal conditions and efficiencies for the extraction of metals from various matrices. Wang and Chiu [21] developed a method to remove chromated copper arsenate from wood, prior to its final disposal. The technique employed supercritical CO_2 in the presence of Cyanex 302, achieving extraction efficiencies of 95% (Cu), 66% (As), and 50% (Cr).

Joo and Shin [67] investigated the use of the commercial chelating agents Cyanex 272, Alamine 304–1, diethylamine (DEA), D2EHPA, and PC884, in different combinations and under defined conditions of pressure and temperature, for the extraction of Co and Mn from a cobalt manganese bromide (Br_4CoMn) leachate solution. It was found that the acidic extractants (Cyanex 272, PC88A, and D2EHPA) gave poor results when used alone (because the reported pH of supercritical CO_2 is around pH 2.8–2.9), while all of these extractants showed considerably improved performance when mixed with DEA. DEA acts as a strong base, deprotonating the extractant and activating it. This effect greatly improved the efficiency and selectivity, with over 90% recoveries achieved for both Co and Mn, with generally greater selectivity toward these ions than toward Mg, Ca, and Na (although Cyanex 272, PC88A, and D2EHPA were also selective for Mg). The efficiencies and operating conditions are presented in Table 4.

Sinclair and Thomson [68] investigated the extraction and separation of neodymium and holmium nitrates using tributyl phosphate in supercritical carbon dioxide. The results demonstrate that maximum Nd and Ho separation was achieved with a TBP:rare earth elements ratio of 1–2. The highest separation efficiency was also reached when the water content was at a minimum. Pressure between 17 and 28 MPa was found to have a minimal effect on the extraction and separation.

Table 3 Structures of some commercial complexing agents

Cyanex 272 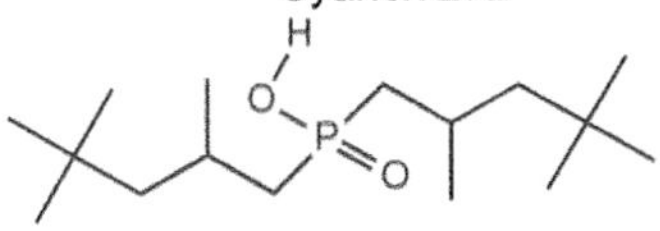 Bis(2,4,4-trimethylpentyl)phosphinic acid	Cyanex 301 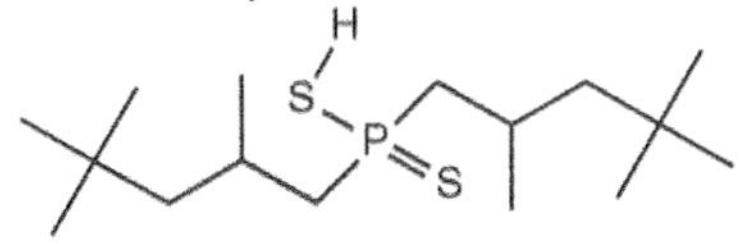 Bis(2,4,4-trimethylpentyl)dithiophosphonic acid
D2EHPA 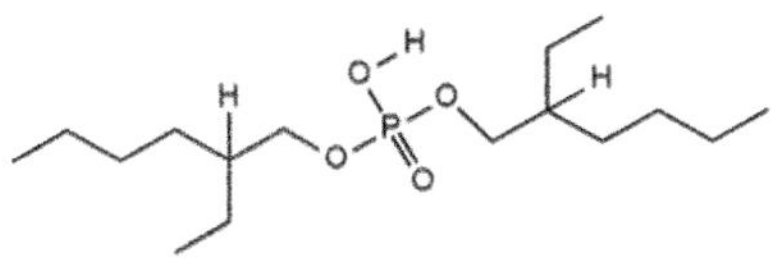 Di-(2-ethylhexyl)phosphoric acid	PC88A 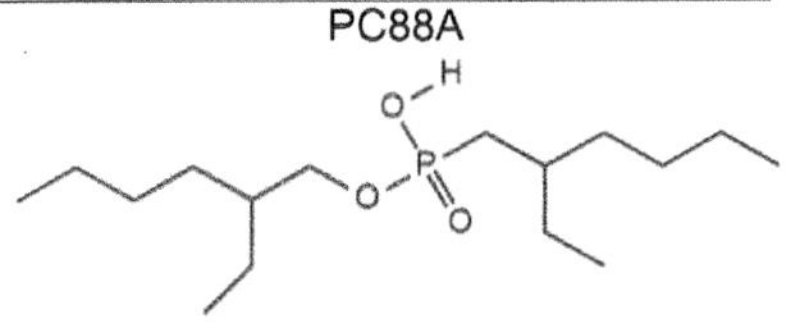 2-Ethylhexyl hydrogen 2-ethylhexylphosphonate
TBP 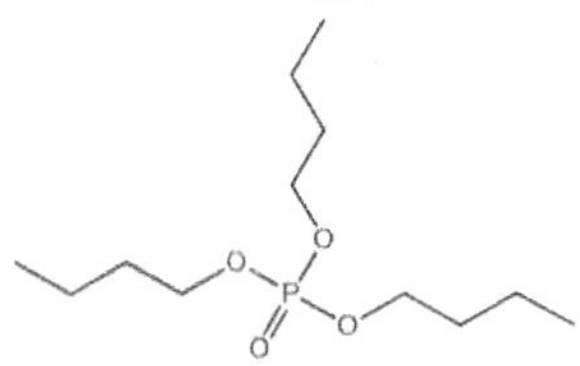 Tributyl phosphate	LiFDDC 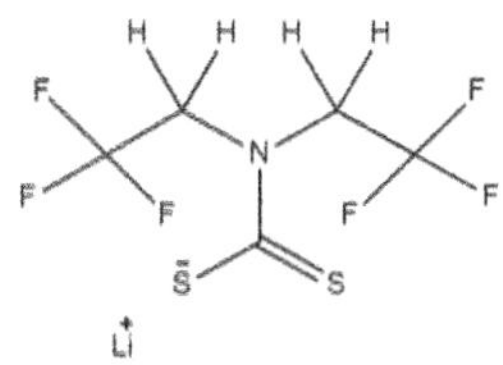 Lithium bis(2,2,2-trifluoroethyl)carbamodithioate
Alamine 304 -1 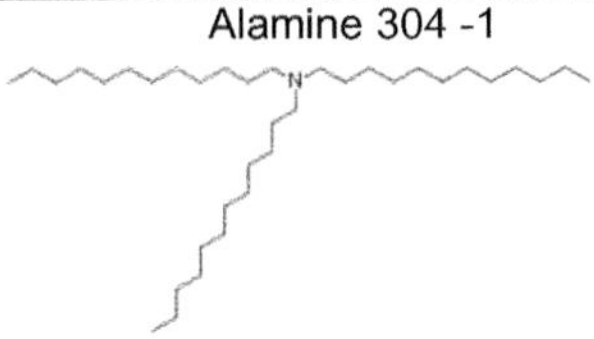 Tridodecylamine	TFA 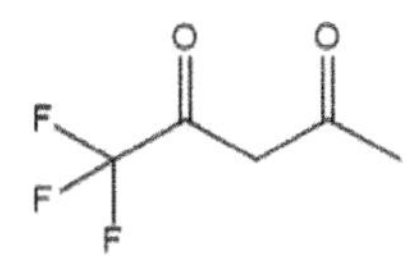 1,1,1-Trifluoropentane-2,4-dione
HFA 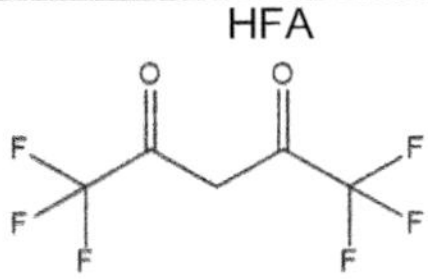 Hexafluoroacetylacetone	TTA 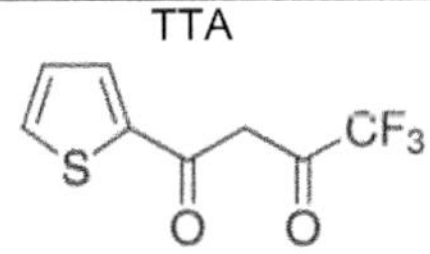 2-Thenoyltrifluoroacetone

$ScCO_2$ can also be used in the process step involving stripping of a complex-loaded solvent. Mekki et al. [69] investigated the use of $ScCO_2$ for the stripping of ionic liquids loaded with complexes of Eu and La. For the media studied, recoveries of over 90% were obtained when TBP was added to the other chelants [HFA (hexafluoroacetylacetone)

Table 4 Reagents, conditions, and efficiencies for the extraction of metal ions from liquid matrices using supercritical CO_2

Matrix	Metal ion	Chelant	Modifier	Pressure (MPa)	Temperature (°C)	Time (min)	Efficiency	Ref.
Cobalt manganese bromide leachate	Co, Mn	Cyanex 272, Alamine 304–1, DEA, D2EHPA, PC884		20	60		99.54% (Co), 99.99% (Mn)	[67]
Synthetic neodymium and holmium nitrate solution	Ho, Nd	TBP	Water	17–28	60	30	87% (Nd), 95% (Ho)	[68]
Ionic liquid containing Eu and La	Eu, La	TBP, HFA, TTA		15	50	60	Higher than 90% for both ions	[69]
Waste water	U, Hf, Zr	Cyanex 301, DEA, Alamine	Methanol	25.54	78	145	98.1% (U), 28.9% (Hf), 65.2% (Zr)	[70]
Wood treated with CCA	As, Cr, Cu	Cyanex 302		30.4	60	40	95% (Cu), 66% (As), 50% (Cr)	[21]

and TTA (thenoyltrifluoroacetylacetone)]. An interesting aspect of these techniques was that all the solvents could be recycled: the extracted ionic liquids, the stripped ligands (for use in another extraction in a fresh ionic liquid), and the dioxide carbon [69].

4. Extraction of metals from solid and particulate matrices using $ScCO_2$

Another field being explored in supercritical fluid extraction is the direct extraction of metal ions from solid matrices (Table 5). Du et al. [71] employed $ScCO_2$, with different types of chelating agents, for the extraction of hazardous metals from urban total suspended particles (TSP). It was found that the use of $ScCO_2$ modified with trifluoroacetylacetone (TFA) and methanol enabled extraction of all the target metal ions (Cu, Pb, Zn, Cd, Cr, and Ni) [71]. This chelation-SFE technique provides a new method for extracting trace metals from urban TSP. The sample preparation procedure avoids the use of chemical digestion followed by extraction with organic solvents, which is necessary for the separation and concentration of trace metals when conventional methods are employed. The authors explored the use of this chelation-SFE technique for improving and possibly substituting the standard preconcentration methods employed for metals present in solid environmental matrices.

A similar approach was used for the extraction of cobalt from fireproof perlite boards. A combination of D2EPHA and n-hexane was employed to extract cobalt ions from the fireproof boards in an environment of $ScCO_2$. A cylindrical perlite matrix spiked with an aqueous solution of cobalt nitrate was treated using $ScCO_2$ modified with D2EPHA and n-hexane, resulting in a complete recovery of the metal [73].

A wide variety of chelating agents are available for the extraction of metal ions, and in some cases, specific molecules have been synthesized for this purpose. Li et al. [72] synthesized different chelating agents, which were successfully applied in the extraction of palladium from commercial Pd/Al_2O_3 catalysts. The extractions with $ScCO_2$ were performed with four fluorinated chelants (since the presence of C—F bonds increases the solubility of a compound in $ScCO_2$), with 42% extraction of palladium achieved under moderate conditions of pressure and temperature [72]. The experimental system consisted of a batch reactor operated under the conditions illustrated in Table 5.

Chirat et al. [74] synthesized the copolymer poly(1,1,2,2-tetrahydroperfluorodecyl acrylate-*co*-vinylbenzylphosphonic diacid) for subsequent use in the extraction of cobalt ions from a contaminated fabric. Determination was made of the amount of water present, and a synergistic effect was observed for the addition of water and copolymer. The efficiency obtained using the reagent mixture (66%) was higher than the sum of the efficiencies for the components used individually (~23% for the copolymer, ~0% for water, and ~11% for CO_2 alone) [74].

Table 5 Reagents, conditions, and efficiencies described in the literature for the extraction of metal ions from solid matrices

Matrix	Metal ions	Chelant	Modifier	Pressure (MPa)	Temp (°C)	Time (min)	Recovery efficiency	Ref.
Urban TSP (particles with diameter <100 μm)	Cu, Pb, Zn, Cd	TFA	Methanol and Triton X-100	20	60	30	99.4% (Cu), 92.6% (Pb), 96.7% (Zn), 92.1% (Cd), 99.8% (Ni), 101.2% (Cr)	[71]
Alumina-supported catalyst with Pd (Pd/Al_2O_3)	Pd	Fluorinated polymers		25	40	15	40% (Pd)	[72]
Fireproof board	Co	D2EHPA	n-Hexane	35	60	120	99% (Co)	[73]
Fabric contaminated with cobalt nitrate	Co	poly(FDA-*co*-VBPDA)	Water	16	40	90	60% (Co)	[74]
Solid matrices: impregnated with ions	Cu, Eu	Perfluoro-polyether surfactants		20	45			[75]
–Filter paper	Cu and Eu	Perfluoro-polyether surfactants		20	45	120	>98%	
–Activated charcoal	Cu	Perfluoro-polyether surfactants		20	45	360	57%	
–Wood	Eu	Perfluoro-polyether surfactants		20	45	120	84%	
–Cement	Eu	Phosphonic acid surfactant		20	45	1440	20%	

Iwao et al. [76] studied the extraction of palladium from a Pd-based solid catalyst, using $ScCO_2$ in the presence of a chelating agent. Almost 100% extraction of palladium was achieved when Cyanex 302 was used as the chelant. The other conditions employed in this study are provided in Table 5.

Campbell et al. [64] investigated the efficiency of $ScCO_2$ extraction of copper and europium from four different types of solid matrices: contaminated filter paper, activated charcoal, wood, and cement. A microemulsion technique was developed, in which the volume of water used was similar to that of the extracted metal, enabling grams of waste to be extracted with microliters of water. Evaluation was made of the effect of the ratio between water and surfactant on the extraction capacity of the process. Following optimization of the ratio, efficiencies higher than 98% were obtained for extraction from the filter paper, using the compound PFPE-COO^-NH4^+ as the surfactant. However, the efficiency decreased with increasing complexity of the medium (see Table 5). In the experiments with cement, the extraction of aluminum and iron was also reported, in addition to europium. All the experiments were performed using perfluoropolyether ammonium carboxylate surfactants, with the exception of the final experiment, in which a phosphonic acid was used as the surfactant [64].

5. Conclusions

The search for processes that are more efficient and environmentally acceptable is a continuing challenge. Increasing efforts are being made to make the alternatives to conventional processes more efficient, from both environmental and economic perspectives. This chapter addressed the various applications of $ScCO_2$ in the extraction of metals from different matrices. The main chelants and cosolvents used in these processes were summarized, together with the best operational conditions and the efficiencies obtained for different applications.

The first part of this chapter provided a description of the main operations involved in the recycling of WEEE, as well as discussion of different studies that employed $ScCO_2$ in the recovery of metals from WEEE, the main cosolvents used, and a schematic illustration of the main components of an experimental system. The next part addressed the selective extraction of metal ions from aqueous solutions using $ScCO_2$ in the presence of complexing agents, indicating the structures of different types of complexing agents employed for this purpose. Finally, a summary was provided of different studies involving the use of $ScCO_2$ for the extraction of metals from solid and particulate matrices.

It can be concluded from the above discussion that the metal extraction process using $ScCO_2$ is effective, versatile, and can be used in applications involving a wide range of matrices. In the specific case of extraction of metals present in WEEE, there have been a number of studies focused on this application. However, this is a relatively new area and further research and development is needed in order to enable the technology to be implemented widely on an industrial scale.

References

[1] Ž. Knez, E. Markočič, M. Leitgeb, M. Primožič, M. Knez Hrnčič, M. Škerget, Industrial applications of supercritical fluids: a review, Energy 77 (2014) 235–243.

[2] Ž. Knez, M. Škerget, M. KnezHrnčič, Principles of supercritical fluid extraction and applications in the food, beverage and nutraceutical industries, in: Separation, Extraction and Concentration Processes in the Food, Beverage and Nutraceutical Industries, 2013.

[3] J. McHardy, S.P. Haber, Supercritical Fluid Cleaning: Fundamentals, Technology and Applications, 1998.

[4] M. Perrut, Sterilization and virus inactivation by supercritical fluids (a review), J. Supercrit. Fluids 66 (2012) 359–371, https://doi.org/10.1016/j.supflu.2011.07.007.

[5] Z. Shen, H. Wang, G. Li, Numerical simulation of the cutting-carrying ability of supercritical carbon dioxide drilling at horizontal section, Pet. Explor. Dev. 38 (2) (2011) 233–236, https://doi.org/10.1016/S1876-3804(11)60028-1.

[6] Y. Jianzhong, Supercritical fluids deposition techniques for the formation of nanocomposites, Prog. Chem. 21 (4) (2009) 606–614.

[7] C.E. Fabre, J.S. Condoret, A. Marty, Extractive fermentation of aroma with supercritical CO_2, Biotechnol. Bioeng. (1999).

[8] V.I. Sokolov, L.N. Nikitin, L.A. Bulygina, V.N. Khrustalev, Z.A. Starikova, A.R. Khokhlov, Supercritical carbon dioxide in organometallic synthesis: combination of Sc-CO_2 with Nafion film as a novel reagent in the synthesis of ethers from hydroxymethylmetallocenes, J. Organomet. Chem. 695 (6) (2010) 799–803, https://doi.org/10.1016/j.jorganchem.2009.12.017.

[9] G. Brunner, Applications of supercritical fluids, Annu. Rev. Chem. Biomol. Eng. 1 (2010) 321–342, https://doi.org/10.1146/annurev-chembioeng-073009-101311.

[10] S. Sthiannopkao, M.H. Wong, Handling e-waste in developed and developing countries: initiatives, practices, and consequences, Sci. Total Environ 463–464 (2013) 1147–1153, https://doi.org/10.1016/j.scitotenv.2012.06.088.

[11] F.O. Ongondo, I.D. Williams, T.J. Cherrett, How are WEEE doing? A global review of the management of electrical and electronic wastes, Waste Manag. (2011).

[12] S.M. Abdelbasir, S.S.M. Hassan, A.H. Kamel, R.S. El-Nasr, Status of electronic waste recycling techniques: a review, Environ. Sci. Pollut. Res. 25 (17) (Jun. 2018) 16533–16547.

[13] C. P. Authored By Baldé, V. Forti, V. Gray, R. Kuehr, and P. Stegmann, Quantities, Flows, and Resources The Global E-waste. 2017.

[14] Associação Brasileira da Industria Elétrica e Eletrônica, "Desempenho Setorial—DECON—Abinee," Desempenho Setorial, 2018. [Online]. Available: http://www.abinee.org.br/abinee/decon/decon15.htm. [Accessed: 07-Nov-2018].

[15] P.P. Sheng, R.D. Rebinsky, T.H. Etsell, Recovery of silver from computer circuit board scrap, J. Solid Waste Technol. Manag. 31 (2) (2005) 78–83.

[16] S. Zhang, E. Forssberg, Intelligent liberation and classification of electronic scrap, in: Powder Technology, 1999.

[17] M. Schlögl, Recycling von Elektro- und Elektronikschrott, Vogel, 1995.

[18] G. Angerer, P. Bars, K. Blätcher, Verwertung von Elektronikschrott. Stand der Technik, Forschungs- und Technologiebedarf. Abschlußbericht, Abfallwirtsch. Forschung Praxis 59 (1993).

[19] A.V.M. Silveira, M.S. Fuchs, D.K. Pinheiro, E.H. Tanabe, D.A. Bertuol, Recovery of indium from LCD screens of discarded cell phones, Waste Manag. (2015) 334–342.

[20] M. Herrero, J.A. Mendiola, A. Cifuentes, E. Ibáñez, Supercritical fluid extraction: recent advances and applications, J. Chromatogr. A 1217 (16) (2010) 2495–2511.

[21] J.S. Wang, K. Chiu, Extraction of chromated copper arsenate from wood wastes using green solvent supercritical carbon dioxide, J. Hazard. Mater. 158 (2–3) (2008) 384–391.

[22] F.-R. Xiu, F.-S. Zhang, Materials recovery from waste printed circuit boards by supercritical methanol, J. Hazard. Mater. 178 (1–3) (2010) 628–634.

[23] C.O. Calgaro, D.F. Schlemmer, M.M. Bassaco, G.L. Dotto, E.H. Tanabe, D.A. Bertuol, Supercritical extraction of polymers from printed circuit boards using CO_2 and ethanol, J. CO_2 Util. (2017).

[24] D.A. Bertuol, C.M. Machado, M.L. Silva, C.O. Calgaro, G.L. Dotto, E.H. Tanabe, Recovery of cobalt from spent lithium-ion batteries using supercritical carbon dioxide extraction, Waste Manag. 51 (2016) 245–251.

[25] C.O. Calgaro, D.F. Schlemmer, M.D.C.R. da Silva, E.V. Maziero, E.H. Tanabe, D.A. Bertuol, Fast copper extraction from printed circuit boards using supercritical carbon dioxide, Waste Manag. 45 (2015) 289–297.

[26] A.B. Argenta, C.M. Reis, G.P. Mello, G.L. Dotto, E.H. Tanabe, D.A. Bertuol, Supercritical CO_2 extraction of indium present in liquid crystal displays from discarded cell phones using organic acids, J. Supercrit. Fluids 120 (2017) 95–101.

[27] D.A. Bertuol, et al., Application of spouted bed elutriation in the recycling of lithium ion batteries, J. Power Sources, 2015.

[28] S. Zhang, E. Forssberg, Mechanical separation-oriented characterization of electronic scrap, Resour. Conserv. Recycl. 21 (4) (1997) 247–269.

[29] C.C.B. Martha De Souza, D. Corrêa De Oliveira, J.A.S. Tenório, Characterization of used alkaline batteries powder and analysis of zinc recovery by acid leaching, J. Power Sources (2001).

[30] D.C.R. Espinosa, V.T. Moraes, J.A.S. Tenório, Pyrometallurgical processing, in: H. Veit, A. Moura Bernardes (Eds.), Electronic Waste, Topics in Mining, Metallurgy and Materials Engineering, Springer, Cham, 2015.

[31] J.S. Cadore, D.A. Bertuol, E.H. Tanabe, Recovery of indium from LCD screens using solid-phase extraction onto nanofibers modified with di-(2-ethylhexyl) phosphoric acid (DEHPA), Process Saf. Environ. Prot. 127 (2019) 141–150.

[32] A. Tuncuk, V. Stazi, A. Akcil, E.Y. Yazici, H. Deveci, Aqueous metal recovery techniques from e-scrap: hydrometallurgy in recycling, Miner. Eng. (2012).

[33] E.B. Pereira, A.L. Suliman, E.H. Tanabe, D.A. Bertuol, Recovery of indium from liquid crystal displays of discarded mobile phones using solvent extraction, Miner. Eng. 119 (2018) 67–72.

[34] D.A. Bertuol, A.M. Bernardes, H. Holeczek, H.J. Fetzer, Spent NiMH batteries-rare earths recovery and leach liquor purification through selective precipitation Spotrebované NiMH Batérie-Znovuzískavanie Kovov Vzácnych Zemín a Čistenie Lúhovacieho Roztoku Cestou Selektívneho Zrážania, 2006.

[35] D.A. Bertuol, A.M. Bernardes, J.A.S. Tenório, Spent NiMH batteries-the role of selective precipitation in the recovery of valuable metals, J. Power Sources 193 (2) (2009) 914–923.

[36] D.A. Bertuol, F.D.R. Amado, H. Veit, J.Z. Ferreira, A.M. Bernardes, Recovery of nickel and cobalt from spent NiMH batteries by electrowinning, Chem. Eng. Technol. 35 (12) (2012) 2084–2092.

[37] D.A. Bertuol, E.H. Tanabe, L. Meili, H.M. Veit, Hydrometallurgical processing, in: H. Veit, A. Moura Bernardes (Eds.), Electronic Waste, Topics in Mining, Metallurgy and Materials Engineering, Springer, Cham, 2015.

[38] I. Birloaga, I. De Michelis, F. Ferella, M. Buzatu, F. Vegliò, Study on the influence of various factors in the hydrometallurgical processing of waste printed circuit boards for copper and gold recovery, Waste Manag. 33 (4) (2013) 935–941.

[39] H. Yang, J. Liu, J. Yang, Leaching copper from shredded particles of waste printed circuit boards, J. Hazard. Mater. 187 (1–3) (2011) 393–400.

[40] J. Li, et al., Thiocyanate hydrometallurgy for the recovery of gold. Part I: chemical and thermodynamic considerations, Hydrometallurgy 113–114 (2012) 1–9.

[41] J. Cui, L. Zhang, Metallurgical recovery of metals from electronic waste: a review, J. Hazard. Mater. 158 (2–3) (2008) 228–256.

[42] S. Syed, Recovery of gold from secondary sources—a review, Hydrometallurgy 115–116 (2012) 30–51.

[43] L. Li, J. Ge, F. Wu, R. Chen, S. Chen, B. Wu, Recovery of cobalt and lithium from spent lithium ion batteries using organic citric acid as leachant, J. Hazard. Mater. 176 (1–3) (2010) 288–293.

[44] L. Li, J. Ge, R. Chen, F. Wu, S. Chen, X. Zhang, Environmental friendly leaching reagent for cobalt and lithium recovery from spent lithium-ion batteries, Waste Manag. (2010).

[45] L. Li, et al., Recovery of metals from spent lithium-ion batteries with organic acids as leaching reagents and environmental assessment, J. Power Sources (2013).

[46] S.Y. Moon, S.H. Hong, T.Y. Kim, S.Y. Lee, Metabolic engineering of Escherichia coli for the production of malic acid, Biochem. Eng. J. 40 (2) (2008) 312–320.

[47] M.S. Sonmez, R.V. Kumar, Leaching of waste battery paste components. Part 1: Lead citrate synthesis from PbO and PbO2, Hydrometallurgy (2009).
[48] X. Zhu, et al., Leaching of spent lead acid battery paste components by sodium citrate and acetic acid, J. Hazard. Mater. (2013).
[49] G.H. Li, M.J. Rao, Q. Li, Z.W. Peng, T. Jiang, Extraction of cobalt from laterite ores by citric acid in presence of ammonium bifluoride, Trans. Nonferrous Met. Soc. China (English Ed.) (2010).
[50] S.R. Ali, Z. Anwar, M. Irshad, S. Mukhtar, N.T. Warraich, Bio-synthesis of citric acid from single and co-culture-based fermentation technology using agro-wastes, J. Radiat. Res. Appl. Sci. (2016).
[51] S.K. Panda, S.S. Mishra, E. Kayitesi, R.C. Ray, Microbial-processing of fruit and vegetable wastes for production of vital enzymes and organic acids: Biotechnology and scopes, Environ. Res. (2016).
[52] A.O. Adeoye, A. Lateef, E.B. Gueguim-Kana, Optimization of citric acid production using a mutant strain of Aspergillus niger on cassava peel substrate, Biocatal. Agric. Biotechnol. (2015).
[53] F.R. Xiu, F.S. Zhang, Materials recovery from waste printed circuit boards by supercritical methanol, J. Hazard. Mater. (2010).
[54] G. Brunner, Supercritical fluids: technology and application to food processing, J. Food Eng. (2005).
[55] C. Erkey, Supercritical carbon dioxide extraction of metals from aqueous solutions: a review, J. Supercrit. Fluids (2000).
[56] A. Bertucco, G. Vetter, High Pressure Process Technology: Fundamentals and Applications, Elsevier, 2001.
[57] R. Banda, H. Jeon, M. Lee, Solvent extraction separation of Pr and Nd from chloride solution containing la using Cyanex 272 and its mixture with other extractants, Sep. Purif. Technol. 98 (2012) 481–487.
[58] K. Liu, Z. Zhang, F.S. Zhang, Direct extraction of palladium and silver from waste printed circuit boards powder by supercritical fluids oxidation-extraction process, J. Hazard. Mater. 318 (2016) 216–223.
[59] F.-R. Xiu, F.-S. Zhang, Recovery of copper and lead from waste printed circuit boards by supercritical water oxidation combined with electrokinetic process, J. Hazard. Mater. 165 (2009) 1002–1007.
[60] F. Nunes da Silva, M.M. Bassaco, D.A. Bertuol, E.H. Tanabe, An eco-friendly approach for metals extraction using polymeric nanofibers modified with di-(2-ethylhexyl) phosphoric acid (DEHPA), J. Clean. Prod. (2018).
[61] B. Gupta, N. Mudhar, I. Singh, Separations and recovery of indium and gallium using bis(2,4,4-trimethylpentyl)phosphinic acid (Cyanex 272), Sep. Purif. Technol. 57 (2) (2007) 294–303.
[62] S. Das, G. Gaustad, A. Sekar, E. Williams, Techno-economic analysis of supercritical extraction of rare earth elements from coal ash, J. Clean. Prod. 189 (2018) 539–551.
[63] X. Ding, Q. Liu, X. Hou, T. Fang, Supercritical fluid extraction of metal chelate: a review, Crit. Rev. Anal. Chem. 47 (2) (Mar. 2017) 99–118.
[64] M.L. Campbell, D.L. Apodaca, M.Z. Yates, J.B. McClain, E.R. Birnbaum, Metal extraction from heterogeneous surfaces using carbon dioxide microemulsions, Langmuir (2001).
[65] Y. Chen, E.R.M. Mariba, L. Van Dyk, J.H. Potgieter, A review of non-conventional metals extracting technologies from ore and waste, Int. J. Miner. Process. 98 (1–2) (2011) 1–7.
[66] F. Lin, D. Liu, S.M. Das, N. Prempeh, Y. Hua, J. Lu, Recent progress in heavy metal extraction by supercritical CO_2 fluids, Ind. Eng. Chem. Res. 53 (5) (2014) 1866–1877.
[67] S.H. Joo, S.M. Shin, Extraction of cobalt from spent cmb catalyst using supercritical CO_2, Arch. Metall. Mater. 60 (2) (2015) 1535–1537.
[68] L.K. Sinclair, J.W. Tester, J.F.H. Thompson, R.V. Fox, The role of water in extraction and separation of rare earth elements in supercritical carbon dioxide, J. Supercrit. Fluids 136 (2018) 180–188.
[69] S. Mekki, et al., Extraction of lanthanides from aqueous solution by using room-temperature ionic liquid and supercritical carbon dioxide in conjunction, Chem. A Eur. J. 12 (6) (Feb. 2006) 1760–1766.
[70] S.M. Ghoreishi, K. Ansari, H.S. Ghaziaskar, Supercritical extraction of toxic heavy metals from aqueous waste via Cyanex 301 as chelating agent, J. Supercrit. Fluids 72 (2012) 288–297.
[71] S. Du, G. Zhang, Z. Cui, Supercritical fluid extraction of hazardous metals from urban total suspended particles, J. Liq. Chromatogr. Relat. Technol. 28 (10) (2005) 1487–1495.
[72] W. S. J. Li et al., "Extraction of palladium from alumina-supported catalyst in supercritical CO_2 using functional fluorinated polymers," J. Supercrit. Fluids, vol. 138, no. April, pp. 207–214, 2018.

[73] M.T. Liang, et al., Metal extraction of a spiked solid with supercritical carbon dioxide, J. Supercrit. Fluids 79 (3) (2013) 324–329.
[74] M. Chirat, T. Ribaut, S. Clerc, F. Charton, B. Fournel, P. Lacroix-Desmazes, Extraction of cobalt ion from textile using a complexing macromolecular surfactant in supercritical carbon dioxide, Ind. Eng. Chem. Res. 52 (1) (2013) 538–542.
[75] M. McHugh, V. Krukonis, Supercritical Fluid Extraction Principles and Practice, second ed., Butterworth-Heinemann, 1993.
[76] S. Iwao, S. Abd El-Fatah, K. Furukawa, T. Seki, M. Sasaki, M. Goto, Recovery of palladium from spent catalyst with supercritical CO_2 and chelating agent, J. Supercrit. Fluids (2007).

Further reading

[77] N. Sanders, Food legislation and the scope for increased use of near-critical fluid extraction operations in the food, flavouring and pharmaceutical industries, in: Extraction of Natural Products Using near-Critical Solvents, 1993.

CHAPTER 6

Use of supercritical carbon dioxide in alkylation reactions

Dhanya Sunil, Nanjangud V Anil Kumar
Department of Chemistry, Manipal Institute of Technology, Manipal Academy of Higher Education, Manipal, India

Contents

1. Introduction

The choice of suitable solvents is a strategic parameter in the initial stage of developing a precise synthetic chemical process, as they not only play a critical part in stabilizing reactive intermediates but also have a pivotal impact on both the reaction rate and product selectivity [1]. Most of the conventional organic solvents are often flammable, toxic, and hazardous to the environment. Though recycling of these solvents is possible, eventually the use of these traditional solvents and catalysts in chemical, pharmaceutical, and polymer industries often contribute to the formation of hazardous waste, the disposal of which is of increasing environmental concern [2]. Besides, the sequestration of the clean product(s), recovery of catalysts or reagents, and finally disposal of the solvent also contribute significantly to the cost of downstream processing procedures in solution-phase synthesis. Therefore, these organic synthetic processes demand to be not only fundamentally selective but also environmentally benign and economical. Hence the need of the hour is the advancement of intrinsically clean reaction approaches that can render new

Green Sustainable Process for Chemical and Environmental Engineering and Science
https://doi.org/10.1016/B978-0-12-817388-6.00006-4

prospects to cost-effective and eco-friendly synthetic processes. Moreover, since most of the currently used solvents are acquired either directly or indirectly from fossil fuels, there are substantial drivers to replace them using greener solvents with reduced environmental impact [3–5]. Research strategies focused toward the design of more sustainable and greener processes for potential replacements of conventional solvents, and reaction media in chemical synthesis are of utmost importance [6] and researchers in both academia as well as in industry are working toward achieving this goal.

2. Supercritical fluids as sustainable solvents

There are several candidates as greener alternatives to traditional volatile organic reaction media including water [7–9], "renewable" solvents obtained from biomass [10, 11] and the "advanced" solvents like ionic liquids [12, 13] and supercritical fluids (SCFs) [14–16] in environmentally benign chemical processes. Each of these has its associated advantages and limitations. The usage of SCFs as a reaction media for catalysis processes has been a dynamic area of exploration over the past few years, attributed to prospective numeral benefits offered by these solvents over traditional liquid media as well as gaseous diluents [17]. SCFs are fluids with pressure and temperature greater than their critical pressure (P_c) and critical temperature (T_c), enabling them to take up the features of a liquid as well as a gas. Fine-tuning of density-dependent features enables their viscosity, solvating strength, diffusivity, and heat capacity to approach liquid like levels and makes them extremely efficient for extraction in porous matrixes. Moreover, separation on post reaction is easy as the liquid/gas phases isolate on cooling the reaction media. Besides, the SCFs can be realized as an excellent solvent medium for the higher boiling reactants which are not reachable in adequately elevated concentration in a gas-phase reaction.

The optimization of the solvent environment of highly compressible SFCs can be exploited in specific reactions, mainly catalytic applications by different strategies such as:

- **(i)** providing a single reaction media to significantly and continuously alter the fluid density and mass transport characteristics,
- **(ii)** altering temperature and pressure to influence both kinetic rates and to stabilize equilibrium constants to yield preferred products [18],
- **(iii)** enhancing the solubility of component [19] and catalyst [16],
- **(iv)** controlling the viscosity or solvent dielectric constant to improve the product selectivity and yields [20],
- **(v)** decreasing the mass transfer restrictions in diffusion-limited reactions [21],
- **(vi)** regulating the temperature in extreme exothermic processes by fine-tuning of the heat capacity of the solvent [22],
- **(vii)** curtailing deactivation of the heterogeneous catalyst via inhibition of coking,
- **(viii)** removal of fouling products [23], and
- **(ix)** enabling the separation and recovery of reaction products [24].

There is immense scope for effectively exploiting these potential advantages in economic commercial processes. Although Jessop and Leitner have compiled a list of 26 SCFs [25], only a few among them have attractive mild critical properties with reasonable cost and could be deliberated as "green" solvents.

3. Supercritical carbon dioxide as unique green solvent

Carbon dioxide (CO_2) is often regarded a perfect choice as a viable green substitute for conventional organic solvents allowing environmentally benign processes.

3.1 Unique properties of supercritical CO_2

Carbon dioxide displays mild critical properties; T_c of 31°C and P_c of 74 bar and distinctive deviations in density and transport characteristics with pressure. The physicochemical characteristics of CO_2-based solvent media can be significantly tweaked by isothermally changing the pressure around the P_c, to obtain unique behavior such as liquid-like solvent property, gas-like transport features, and heat capacities. Consequently, if the reaction conditions are maintained above its T_c, the looked-for combinations of pore transport rates and product desorption, which decides heterogeneous catalyst activity as well as product selectivity can be attained by pressure-tuning the fluid characteristics of CO_2 about its P_c.

3.2 Advantages of $scCO_2$

Highly pure supercritical carbon dioxide ($scCO_2$) can be easily acquired as a by-product from the fermentation of biomass, which is a potential large-scale source of CO_2. Among SCFs, $scCO_2$ is particularly attractive as it is relatively inexpensive, inoffensive, nontoxic, totally noninflammable, unregulated, and easy to recover and reuse. It displays a combination of properties generally related to liquids or gases; solvent power/density of a light liquid alkane while diffusivity/viscosity approaching that of a gas [26]. The nonpolar $scCO_2$ is a potential solvent as it offers striking advantages for chemical reactions such as:

- **(i)** chemically inert reaction medium,
- **(ii)** improvement of heat- and mass-transfer rates compared to organic solvents,
- **(iii)** augmentation of the reaction rate and selectivity,
- **(iv)** increased throughput,
- **(v)** a prolonged catalyst lifespan due to the high solubility of the coke precursors enabling their easy extraction,
- **(vi)** the possibility of its usage in heterogeneous catalysis and regeneration of zeolite catalysts,
- **(vii)** the tunability of extraction power by altering pressure and temperature, and
- **(viii)** simpler product isolation from the media, post reaction [15, 25, 27–32].

The $scCO_2$ is thus a unique green medium allowing environmentally benign processing attributed to its benefits such as replacement of conventional solvents, improved product selectivity that minimalizes waste as well as intrinsically safe reactor operation [32]. Until recently, CO_2 is the most commonly used fluid, which is recycled into the atmosphere, thereby avoiding disposal problems [25, 33].

At moderate pressures and temperatures, $scCO_2$ exhibits solvating features analogous to that of aromatic solvents and is also immiscible with water. Interestingly, it can thus act as an apt solvent media for phase transfer chemical processes from aqueous solutions. The low viscosity, as well as the lack of capillary effects, permits it to penetrate into the porous matrix in reactions involving solid catalysts. Moreover, $scCO_2$ can be safely used in reactors owing to its pressure-adjustable heat capacity avoiding reactor runaway, nonflammable nature, and amplified reaction rates to minimalize holdup. Besides, scaling-out in contrast to scaling-up (parallel processing) can enable the use of smaller reaction containers which can withstand these pressures easily and reduce the safety threats and further concerns of mechanical failures.

3.3 Disadvantages of $scCO_2$

Use of $scCO_2$ demands working with high pressures, when compared to ionic liquids which can be used as suitable liquid-phase replacements for traditional solvents. Moreover, batch reactions in $scCO_2$ have been most often carried out in sealed autoclaves. Hence, the possibility of commercial scale-up is less due to the high cost of large pressure reaction vessels. Though rapid variations in physical characteristics of CO_2 nearer to the critical point can be exploited; it would be challenging to replicate the processes on a commercial scale. These properties are generally experienced in dilute solutions and demand accurate control on both pressure and temperature. Besides, due to the nonpolar nature of CO_2 under accessible conditions of temperature and pressure, reactants or catalysts might be sparingly soluble in $scCO_2$ leading to lower chemical reactivity and product selectivity in many cases. Consequently, $scCO_2$ exhibits limited potential in the extraction of polar substances, and either temperature or pressure has to be increased or 10% polar organic modifiers has to be incorporated in the $scCO_2$ [34].

4. Reaction with $scCO_2$

A consistent understanding of phase behavior is vital for designing chemical processes and product separations using $scCO_2$. A schematic representation of the continuous flow set up for alkylation reaction involving $scCO_2$ is presented in Fig. 1. A chilled pump which is highly compressible at room temperature to be pumped without much difficulty is employed to deliver CO_2 from the cylinder. The substrate current generally carries pure reactants that are mixed with a cosolvent and a catalyst as required. Gaseous reactants are passed into the system discretely using a mass-flow controller or dosing unit. The static mixers confirm a thorough mixing of the reactants before they pass into the reactor.

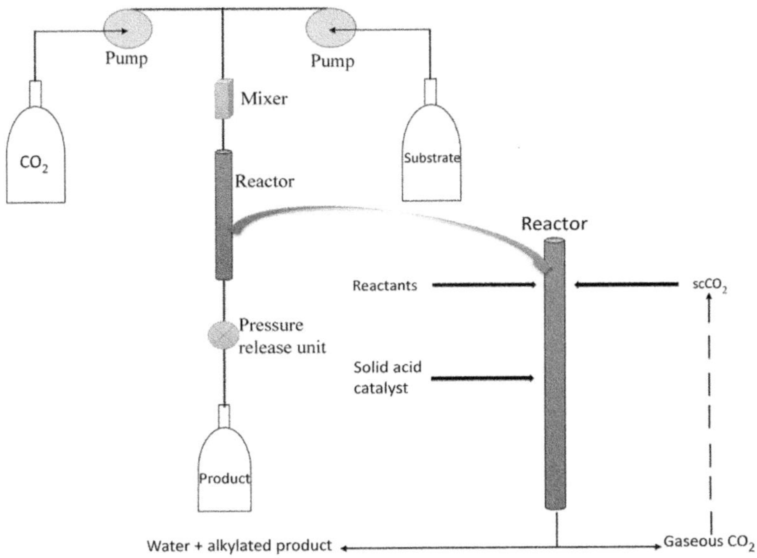

Fig. 1 Schematic representation of a conventional continuous flow device for synthetic processes involving $scCO_2$ [34].

Though tubular reactors are frequently used due to their easy construction, the shape and size of the reactor can be altered to suit a specific synthetic reaction. Many custom designed reactors are also available for particular experiments. The catalyst is packed to arrange in a fixed bed in the heterogeneous catalytic processes, and the process temperature can be controlled by attaching a heating block to the catalyst bed. The mixture is then delivered into the reactor, and the product stream is collected by releasing the system pressure, using a back pressure controller, that also regulates the pressure in the system. The reactions can also be run under pressure without CO_2 if essential, by merely switching off the CO_2 pump [35].

5. ScCO$_2$ in alkylation reactions

There is a mounting requisite for alternatives of strong acid alkylation catalysts such as HF, BF_3, and $AlCl_3$ used in petroleum and fine chemicals industrial processes with replacements that are more environmentally friendly. Though there are many reports on catalysts for solid-acid-based conventional alkylation and isobutane-olefin alkylation reactions [36–38], none of these alkylation catalysts has achieved approval for use in the industry till date. The drawbacks of these solid catalysts which hamper the usage of these new processes from the rapid industrial application are:

(i) rapid deactivation due to coke formation,
(ii) thermal degradation during the regeneration step and
(iii) unacceptable product quality or low alkylate fraction.

The main requirements that need to be satisfied while developing a solid catalyst process which can potentially replace liquid acids as catalysts are enhanced catalyst lifetimes, improved product quality, and better product yields. Catalyst activity and product quality can be improved by either suppressing the production or increasing the removal of coke and also by improving the hydride-transfer rate-limiting step.

It is well known that $scCO_2$ has been viewed as an environment-friendly solvent and employed comprehensively as a carrier media in alkylation reactions. Owing to the unique features and physicochemical characteristics of $scCO_2$, an increase in solid-acid catalyst life can be achieved by the extraction of coke precursor [25]. The liquid-like solvent properties, gas-like transport features, and heat capacities of $scCO_2$ can be incessantly pressure-tuned in the critical region, which could benefit the alkylation in several ways such as (i) improvement of mass and heat transfer rates [25], (ii) augmentation of reaction rate and selectivity, (iii) the option of performing continuous fixed-bed processes, (iv) easier product isolation from the reaction mixture, and (v) enhancement in solid-acid catalyst lifetime [25, 39, 40]. Besides, potential benefits also include replacement of traditional solvents, minimization of waste, and inherently safe reactor operation [25].

The fundamental parameters that influence the activity and product selectivity during the conversion of alkylaromatics in porous catalysts are adsorption/desorption rates and pore-transport features of reactants/products. The unique properties of $scCO_2$ have been exploited for solid acid-catalyzed alkylations in numerous means such as:

(i) the in situ extraction of heavy coke precursors from the catalyst surface before they are transformed into consolidated coke,

(ii) complete miscibility of reactants in the reaction mixture and enhanced pore-transport of these reactants to the catalyst surface, thereby promoting desired reaction pathways, and

(iii) control of temperature in exothermic reactions [31].

The benefits of $scCO_2$-based alkylation media for both optimizing the activity of the solid-acid catalyst and attaining the selectivity of the desired product are highlighted for various processes. Catalytic processes employing microporous solids restricted by diffusion in the liquid phase are favored under SCF conditions, as reactant solubility turn out to be equivalent to that in the liquid phase, whereas molecular diffusivity in $scCO_2$ lies in between a gas and a liquid. In the following sections, a comprehensive, up-to-date review of the use of $scCO_2$ in various types of alkylation reactions is illustrated.

5.1 Friedel Craft's alkylation reaction

The Friedel-Crafts (FC) alkylation is one of the best convenient and beneficial procedures for the formation of carbon-carbon bond in aromatic systems [41] and is of high interest to synthetic chemists. Conventional alkylations are generally performed batchwise with a large surplus substrate concentration in an organic solvent media and require long reaction duration at a low temperature. Generally, the FC reaction is carried out employing an alkyl halide in the presence of environmentally unfriendly Lewis acid catalysts, like $TiCl_4$, $AlCl_3$ or BF_3 [42], or protic acids like concentrated H_2SO_4 or HF [43]. On completion of alkylation, the solvent and the catalyst have to be isolated from the reaction mixture and has to be either purified for recycling or disposed-off. Recently, due to the significant usage of FC reactions in industry, environmental-friendly homogenous catalysts like water or trifluoromethanesulfonic acid (TFSA) modified with trifluoroacetic acid (TFA) [44], *p*-toluenesulfonic acid monohydrate (TsOH) [45], rare earth metal trifluoromethanesulfonates [46], organic resins [47], mesoporous molecular sieves [48], sulfated metal oxides [49], zeolites [50], clays [51], and even graphite [52] are used for heterogeneously catalyzed FC alkylation processes. However, moderately high catalyst concentrations are required; often nearing stoichiometric quantities, mainly when H_2O is produced. Though FC reactions are extensively realized in the production of fine chemicals, these are inherently unfavorable and inefficient in the context of atom efficiency, by-product generation, and catalyst use [53, 54]. Therefore, environmentally acceptable, economically feasible, and greener FC alkylation reactions are highly looked-for. In addition, FC-type alkylation is of continuing research interest, as several syntheses reactions both on laboratory and industrial scales are involved [55–59].

5.1.1 Challenges in FC reaction

The solid alkylation catalysts have not been accepted in industrial scale because of the following limitations: (i) undesirable product quality or low alkylate fraction, (ii) swift catalyst deactivation owing to coke generation, and (iii) thermal degradation of the catalyst during the regeneration step. The primary chemical challenge in these processes is to accomplish selective monoalkylation of the substrate [60] and hence the precise control of the reaction conditions are essential to minimize the generation of polyalkylated by-products, often with high dilutions, extended reaction durations and low temperatures [61].

5.1.2 Use of $scCO_2$ in FC reaction

$ScCO_2$ affords continuous, selective, and clean FC reaction [60]. Though, solid catalysts enable cleaner FC reactions, till lately their implementation in a commercial process has been hindered due to speedy catalyst deactivation [62]. This challenge could be overcome by the extraction of coke precursors using $scCO_2$ to increase the catalyst lifetime. Thus, a very efficient FC process can be achieved with solid acid catalysts, in combination with $scCO_2$. Moreover, the use of solid catalysts enables easy separation from liquid products. Also, it is possible to carry out FC processes in small continuous flow reactors, which are safe, economical, and offer comparable production capacity, rather than batch reactors which are comparatively challenging to scale-up due to the associated high cost of the high-pressure vessels [43, 63, 64]. Interestingly, Thomas Swan & Co. Ltd has already scaled up continuous fixed-bed catalytic flow reactors, with $scCO_2$-based reaction medium, to commercial scale [40], which have been used in FC type alkylation of simple aromatics [65].

Aromatic substrates can be subjected to a sustainable and continuous heterogeneous FC reaction in a flow reactor over solid acid catalysts in SCF media with high selectivity of the desired product. The usage of the $scCO_2$-based medium in the FC reactions can offer certain benefits including better mass and heat transfer at the catalyst surface [25], enhanced catalyst lifespan owing to the removal of coke precursors [32] and elevation in the reaction rate [66]. However, it is necessary to examine and optimize several parameters to realize an extremely selective FC alkylation in $scCO_2$ such as temperature, pressure, $scCO_2$ flow rates, an alkylating agent, the substrate, catalyst selection, and the influence of $scCO_2$ on the catalytic performance. It is important to note that slight variation of the reaction settings brings about intense deviations to the reaction outcome, primarily regarding conversion and selectivity. Therefore, in-depth knowledge and appropriate regulation of the reaction is immensely anticipated from an academic as well as industrial point of view.

5.1.3 Application of $scCO_2$ in FC reactions of aromatic substrates

FC alkylation of mesitylene **1**, generated mono-, di- and tri-alkylated products **3**, **4** and **5** when propene **2** was employed both as a supercritical solvent [$P_c = 46.0$ bar, $T_c = 91.9°C$] and as an alkylating agent as depicted in Fig. 2. However, Hitzler et al. had reported the continuous FC alkylation of mesitylene **1**, and anisole **7**, using propan-2-ol **6** in $scCO_2$ using a polysiloxane-supported Deloxan solid acid heterogeneous catalyst in a small fixed-bed reactor. Mono-alkylated products **3** and **8** with 100% selectivity and 50% conversion were obtained by replacing the organic solvent by $scCO_2$ and adjusting the factors that influence the reaction such as flow rates, pressure, and temperature. Also, the easy separation of the crude product was achieved without further workup by expanding the fluid to atmospheric pressure. Though the process temperatures were high with short residence times, the solid acid Deloxan catalyst having a longer lifetime selectively generated **3** during the alkylation of mesitylene (in excess) with propan-2-ol **6** in $scCO_2$. The water formed during the reaction was easily separated from the organic layer after the $scCO_2$ has been vented. The $scCO_2$ reduced the mass transport limitations at the catalyst surface, compared to reactions in the liquid phase. Besides, the density of the reaction medium increased in comparison to a gas-phase reaction, and thereby the residence time of the substrate. This allowed continuous alkylation performance with high product selectivity on a realistic scale in the small reactor. Besides, the use of $scCO_2$ reduced catalyst coking, thereby avoiding early deactivation of the catalytic sites [60].

Amandi et al. [67] in 1998 investigated the influence of $scCO_2$ on the catalytic performance of an assortment of commercially accessible Bronsted-type solid acids such as:

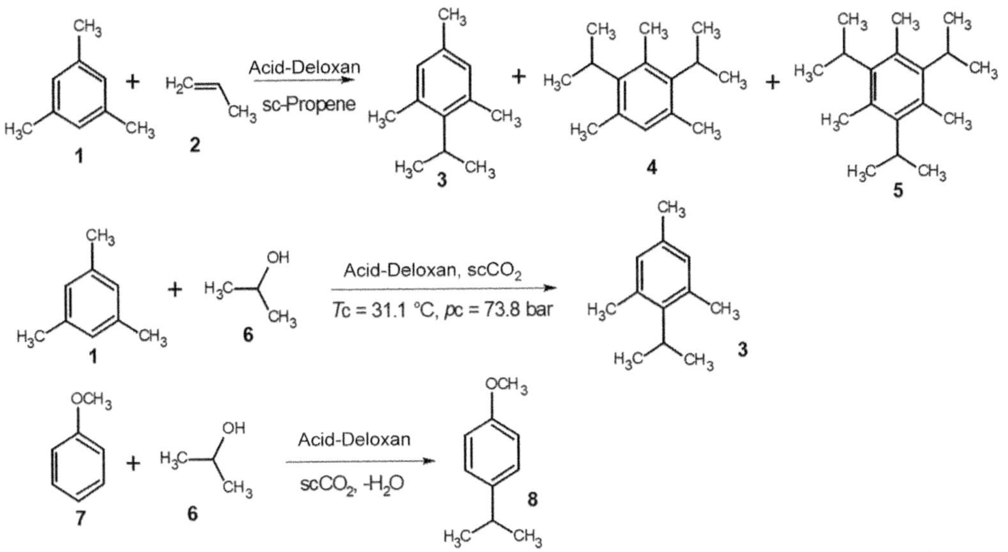

Fig. 2 FC alkylation reaction of mesitylene and anisole using propan-2-ol with high product selectivity [60].

Purolite CT-175 and Amberlyst 15 (organic acidic ion-exchange resins), Deloxan ASP I/7 and Nafion SAC-13 (inorganic supported catalysts), and Zeolyst CBV 600 (a zeolite) for the alkylation of anisole **7** using n-propanol **9**, by varying the reaction pressure and temperature. FC alkylation was successfully performed over these different Bronsted-type solid acid catalytic supports in $scCO_2$, resulting in the very high conversion of **7** in combination with high desired product(s) selectivity as depicted in Fig. 3. Purolite CT-175 and Amberlyst 15, displayed the best catalytic activities within the temperature range spanning from 100°C to 150°C. However, desulfonation of the catalyst above 150°C occurred, decreasing the conversion of **7** with time. The use of Purolite CT-175 and Amberlyst 15 in the two-phase region, led to a pseudo-concentration effect, which modified the catalyst bed behavior, thereby enhancing the reaction rates, enabling

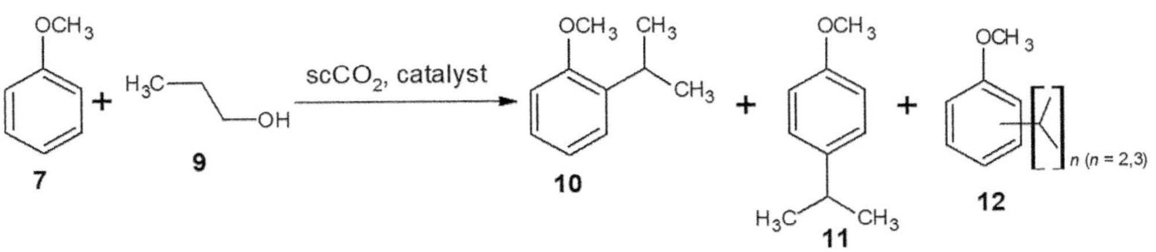

Fig. 3 FC alkylation of anisole in the presence of organic acidic ion-exchange resins, inorganic supported catalysts and zeolite with varying temperature and pressure [67].

a higher conversion of starting material **7** and reducing the selectivity for monoalkylated products. Anisole **7** was dissolved and diluted with $scCO_2$ in the single-phase region, and the diffusion of the single reaction phase into the active catalytic sites determined the overall reaction rate. The phase behavior of the reaction was the decisive factor for the influence of pressure on the reaction rate. In contrast, Deloxan ASP I/7, Nafion SAC-13, and Zeolyst CBV 600 exhibited higher optimum temperature range with increasing conversion of the starting reactants with the rise in temperature, but with lesser selectivity toward the monoalkylated products. However, the increase in the operational pressure while using inorganic supported catalysts led to corresponding escalations in the starting material conversion, attributed to a rise in the residence time through the catalyst bed. Moreover, the acidic terminal -CF_2CF_2-SO_3H group in Nafion SAC-13 being $scCO_2$-philic, [68] enhanced the diffusion of the single phase into the catalytic sites. Regardless of its high surface area, the catalytic behavior of Deloxan ASP I/7 with varying pressure was similar to that of organic supported catalysts. The low conversion of **7** across the studied pressures and temperatures was due to the low catalyst loading. The ortho/para ratio was independent of pressure, temperature, and type of catalyst. The usage of $scCO_2$ enhanced the heat and mass transfer rates and thereby the reaction rate and selectivity, in addition to the catalyst lifespan by extracting the coke precursors and also enabling more comfortable isolation of the products from the solution post reaction.

Hydroxyalkylation of anisole **7** using paraformaldehyde **13** was conveniently performed with $scCO_2$ as the reaction medium, in the presence of microporous zeolites as presented in Fig. 4 [69]. Interestingly, the activity of mordenite (a zeolite) for the hydroxyalkylation process in $scCO_2$ at 60°C was found to be approximately three-fold greater than that detected for the same catalyst in toluene at 110°C. The pore size of mordenite, which is higher than the reactants/products, permits the mass-transfer within the particle, but its performance in the traditional organic solvents is strictly restricted by the limited diffusion owing to the unidirectional pore geometry and the $scCO_2$ was suitable to overcome this significant drawback of zeolites. The usage of $scCO_2$ drastically boosts the diffusion, thereby increasing the catalytic activity in the mentioned FC hydroxyalkylation reaction.

Later, Chateauneuf and Nie [70] in 2000 investigated the FC alkylation process of anisole **7** with triphenylmethanol **15** in $scCO_2$, under two-phase gas/liquid reaction

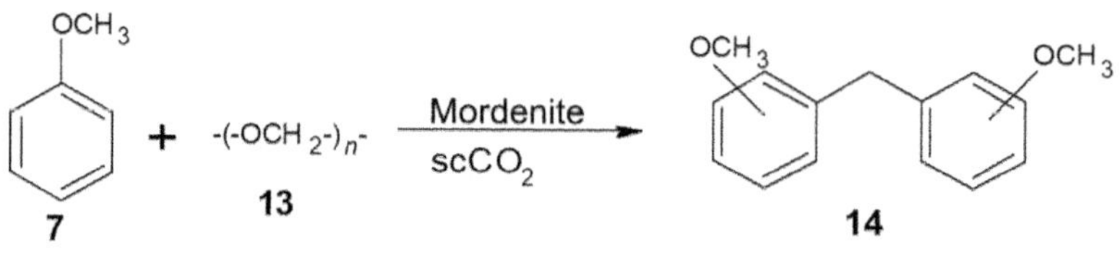

Fig. 4 Friedel-Craft hydroxyalkylation of anisole using paraformaldehyde in $scCO_2$ [69].

environments at moderate temperatures and pressures as presented in Fig. 5. The process was instigated using trifluoroacetic acid to generate triphenylmethyl carbocation as the reaction intermediate and was very competently carried out using neat reaction conditions using $scCO_2$. However, though the FC reaction happened under homogeneous circumstances at the highest CO_2 fluid density of 1.480 g/cm^3, disappointingly the yield of the product was only 3%–4%.

Abbott et al. [71] alkylated anisole **7** with t-BuCl **17** using *p*-toluenesulfonic acid as a catalyst in the $scCO_2$ medium as shown in Fig. 6. Kinetics of the reaction revealed that the solutes behave as polar modifiers and increase the dielectric constant of the reaction mixture. The dielectric constant of the solvent did not influence the reaction and resulted in a higher yield of the desired product.

The alkylated products of naphthalene hold prominent importance in the chemical industry such as; (i) 2-2,6-dialkylated naphthalene which can be oxidized to 2,6-naphthalenedicarboxylic acid, functions as a building block for the preparation of heat-resistant polyester fibers and liquid-crystalline polymers, and (ii) methylnaphthalene is an intermediate obtained in the vitamins K1 and K3 synthesis. The FC alkylation reactions of aromatic molecules using alcohols and acid zeolite catalysts have been efficaciously performed in $scCO_2$ medium. Glaser and Weitkamp [72] have successfully conducted the alkylation of naphthalene **21** with 2-propanol **6** and methanol **22** with mole fractions <0.05, over Bronsted acidic zeolite catalyst LaNaY-73 with $scCO_2$, at 250°C and 1–400 bar pressure range. The pressure tuning was particularly beneficial for zeolite-catalyzed conversion wherein shape-selectivity effect has no

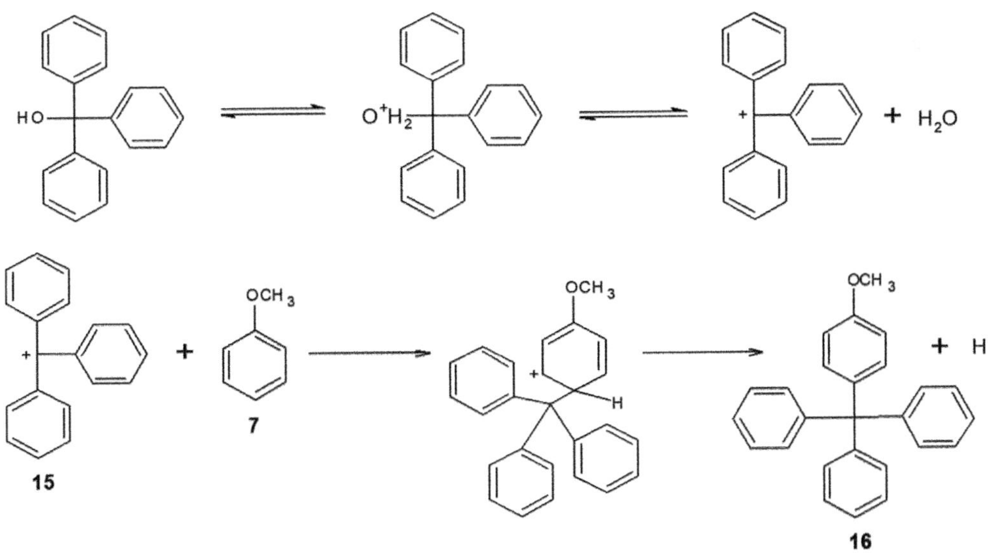

Fig. 5 Friedel-Craft reaction of triphenylmethanol with anisole in $scCO_2$ [70].

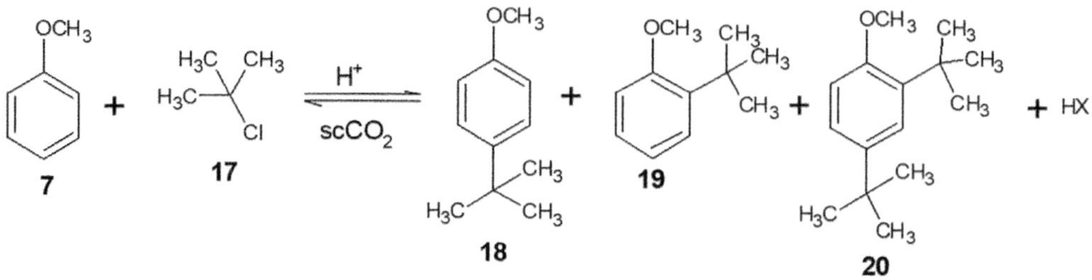

Fig. 6 Acid-catalyzed FC reaction of anisole with t-Butyl chloride [71].

control over the product distribution. As the pore structure of the faujasite-type zeolite enables sufficient space for the reaction mixture for in situ dissolution and subsequent extraction of the higher molecular weight intercrystalline coke precursors generated within the zeolitic pores [73], high-density $scCO_2$ media extended the catalyst lifespan. Also, the deactivation of the catalyst could virtually be inhibited by decreasing the T_c of the reaction mixture by dilution with excess of CO_2. The physical characteristics of the $scCO_2$ medium largely affected the processes occurring on the external surface of the zeolite crystallites, rather than those happening within the zeolite pores. Thus, as the pore entrances were not obstructed by the coke deposits while employing $scCO_2$, shape-selectivity effects, which signify an important property in zeolite catalysis, were wisely exploited to obtain the distribution of the product. The ratio of the yields of the major reaction products: mono- **23**/**25** and dialkylnaphthalenes **24**/**26** (Fig. 7), was mostly governed by the alcohol/naphthalene ratio in the feed and was independent of the degree of catalyst deactivation as well as the reaction pressure. Moreover, the isomer distribution within the products **23**/**25** and **24**/**26** changed with time on stream, based on a transition from thermodynamically to kinetically controlled product generation, the magnitude being reliant on the state of the fluid and the reaction pressure. More highly tri- and tetra alkylated naphthalenes were detected only in a collective yield of less than 0.1%, whereas isopropyl and ethyl naphthalene were also obtained in trace quantities in the product condensate.

Kamalkar et al. carried out selective tert-butylation of phenol **27** to give 65% yield of 2,4-di-tert-butylphenol (2,4-DTBP) **31** over H-Y (SiO_2/Al_2O_3) zeolites and 40% yield of 2,4,6-tri-tert-butylphenol (2,4,6-TTBP) **32** over $Sc(OTf)_3$ supported on MCM-41 [$Sc(OTf)_3$/MCM-41] in $scCO_2$ [74] as depicted in Fig. 8. The catalyst performances in $scCO_2$ were striking compared to those in solvent-free vapor-phase processes under nitrogen atmosphere and liquid-phase reactions with hexane as the solvent. Also, there was minimal coke formation in $scCO_2$ compared to those in other media. These excellent reaction presentations can be attributed to the enhanced solubility of both starting materials and products as well as the removal of heavy aromatic coke precursors, from phenol and/or isobutene oligomers, which retains the catalytic spots clean during catalysis. Moreover, the recycled catalysts can be reused without substantial loss of their

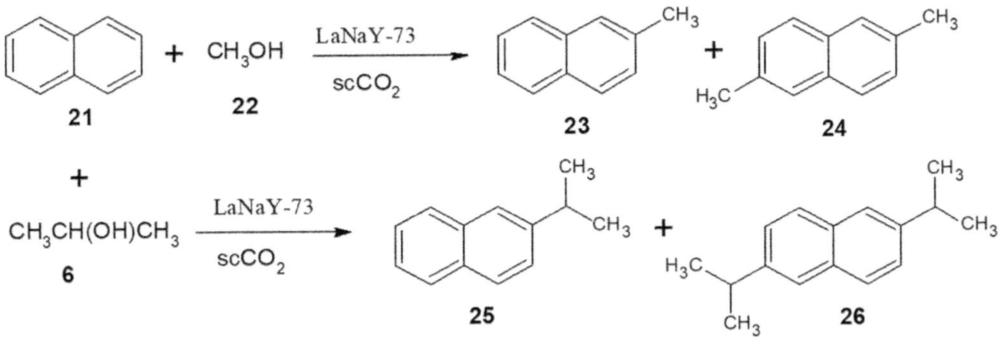

Fig. 7 Zeolite catalyzed FC alkylation of naphthalene in $scCO_2$ [72].

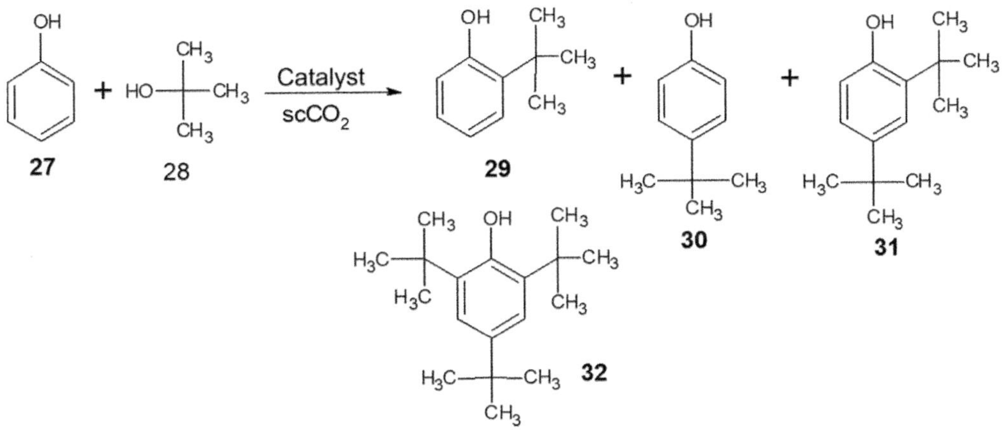

Fig. 8 Tert-butylation of phenol in $scCO_2$ for efficient synthesis of 2,4-di-tert-butylphenol and 2,4,6-tri-tert-butylphenol [74].

activity even after three cycles. The spatial requirement for active sites of the catalysts is a key parameter for the selective synthesis of **31** and **32**. Though both H-Y zeolites and $Sc(OTf)_3$/MCM-41 can offer a suitable environment for the formation of **31** and its diffusion from the channels, H-Y zeolites do not allow the formation of **32**. The $scCO_2$ pressure is also another significant factor that influenced the yields of **31** and **32**. The catalytic performances were maximum at an optimum CO_2 pressure of 10 MPa at 130°C, and in addition, a low viscosity of $scCO_2$ facilitated the better solubility and mass transfer of reactants and products and retained the catalytic active acid sites clean for the catalysis. However, highly dense CO_2 obtained by further escalation in pressure slows down the accessibility of the reactants to the active catalytic sites by its preferred filling of the catalyst channels and quickly reduced the product formation. [74]

Wimmer et al. [75] synthesized thymol **36** starting from m-cresol **33** and isopropanol/propene using activated alumina as a catalyst, which is also the industrial method as presented in Fig. 9. The by-products, 2-isopropyl-3-methylphenol **35** and 4-isopropyl-3-methylphenol **37** being isomers, show boiling points close to **36**. Thus the formation of undesirable products and excess usage of **33** to reduce the dialkylated and polyalkylated derivatives formed have been the main limitations of this conventional alkylation. Amandi et al. [76] attempted selectivity of the reaction through a green approach by using the catalysts such as; γ-Al_2O_3, a solid Lewis acid, and Nafion1 SAC-13, a solid Brønsted acid in the $scCO_2$ medium. γ-Al_2O_3 minimized the by-product formation and the $scCO_2$ used in the reaction facilitated the removal of a trace quantity of water formed and suppressed the formation of minor products **35** and **37**.

The successful para alkylation of phenol **27** with 2-chloro-2,4,4-trimethylpentane in $scCO_2$ was performed by Deak and his team [77], as depicted in Fig. 10. The two-step synthesis involving $scCO_2$ involves (i) the preparation of ter-Cl ended PIB prepolymer (PIB-Cl) **38** and (ii) the quantitative para alkylation of phenol by the prepolymer in the presence of BF_3OEt_2 to yield **39**. Phenol was highly soluble in $scCO_2$. However, the researcher's effort to alkylate 2-bromoethylbenzene with PIB in the $scCO_2$ medium was futile.

Alkylation of toluene **40** with different catalysts like $AlCl_3$, BF_3, H_2SO_4, Y or β zeolites results in the formation of *p*-cymene **41**. But, the major problem with these processes is the selectivity of *p*-cymene, which is less than 60% due to the larger pore size of the zeolites or the formation of side products such as *o*- and *m*-cymenes at the acid sites near the pore mouth [78, 79]. The HZSM-5 pellets modified by chemical liquid deposition (CLD) of $SiCl_4$ with $scCO_2$ by Kuo and Tan [65] resulted in high yield with 98% selectivity of *p*-cymene with an improved catalyst lifetime under optimum conditions of 523 K temperature and 11.72 MPa pressure, and a molar ratio of toluene to propylene

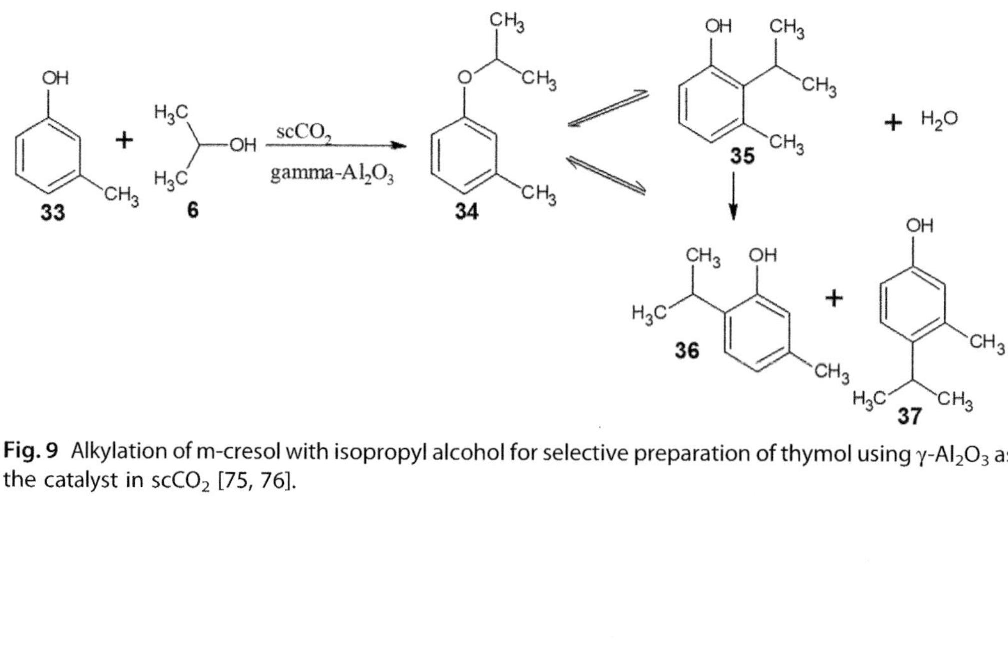

Fig. 9 Alkylation of m-cresol with isopropyl alcohol for selective preparation of thymol using γ-Al_2O_3 as the catalyst in $scCO_2$ [75, 76].

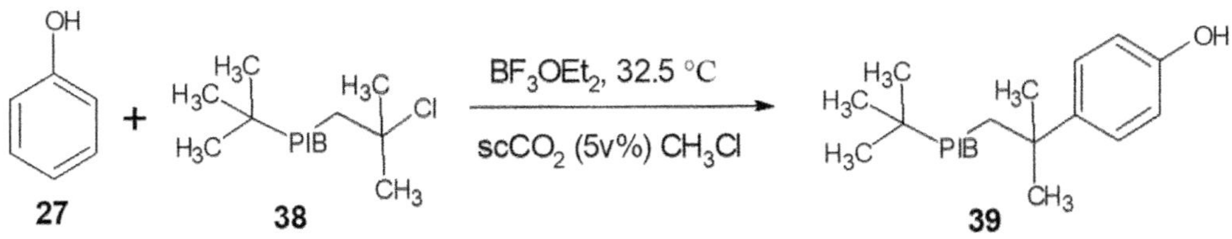

Fig. 10 Synthesis of phenol-terminated polyisobutylene by carbocationic polymerization in $scCO_2$ [77].

2 of 7.7 as represented by Fig. 11. The better results were attributed to the extended retention time under supercritical operation conditions and the solubilizing ability of $scCO_2$ toward coke precursors and reduced cracking of propylene during the alkylation process. Chiang et al. [80] also, reported the alkylation of toluene with isopropyl alcohol **6** in $scCO_2$ where toluene played the role of both as reactant and carrier as shown in Fig. 11. Here also, the pore size reduction helped in achieving the selectivity and yield of *p*-cymene. But the pressure necessary for a steady alkylation process at 543 K was about 5.6 MPa that is, higher than required for the alkylation of toluene with propylene in $scCO_2$, owing to the requirement to eliminate the generated water from the catalyst pellets. Isopropyl alcohol was acknowledged to be a better alkylating agent as propylene cracked to a significant extent even under $scCO_2$ operations.

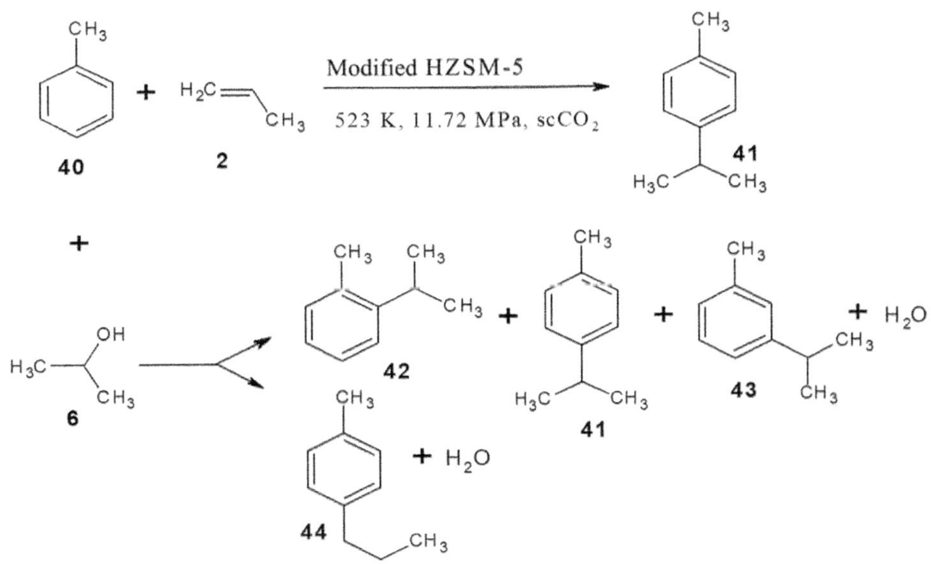

Fig. 11 Formation of *p*-cymene from toluene in $scCO_2$ using propene and isopropyl alcohol as alkylating agents [65, 80].

Yu and Tan synthesized para-ethyltoluene **46** by alkylating toluene **40** with ethylene **45** in $scCO_2$ by employing the solid acid catalysts HZSM-5 crystals obtained through CLD with tetraethyl orthosilicate (TEOS) as presented in Fig. 12 [81, 82]. As reported the selectivity was increased by reducing the pore size of the catalyst at a temperature of 623 K and a pressure of 13.1 MPa [65].

Harmer et al. [68] studied the alkylation of benzene to get linear alkylbenzenes (LABs), which are used as the building blocks in detergents industries. Inspired by the alkylation in SCFs using polysiloxane supported solid acid catalyst, they alkylated *p*-xylene **49** with dodec-1-ene **50** using $CF_3(CF_2)_7SO_3H$ as a catalyst as depicted in Fig. 13. Despite the increased solubility of perfluorosulfonic acid in $scCO_2$ and initiated activity, the availability of limited catalyst sites resulted in a poor conversion of LABs **51** with 68% yield.

The selectivity and reactivity of reactants or catalysts in many cases are limited due to their sparingly solubility in $scCO_2$. As reactants that bear perfluorinated side chains display increased solubility in $scCO_2$, Francio et al. [83] designed an experiment, where the reactant has both CO_2-philic and a lipophilic unit. This motivated Manabe et al. [84] to successfully alkylate indoles **52** using perfluoroalkyl benzene to obtain high yields of the alkylated product. Accordingly, Komoto and Kobayashi [85] used dodecyloxy-4-perfluoroalkyl benzene **54**, which accelerated FC alkylation using alkylating agents carrying electron withdrawing groups to give the alkylated product with 92% yield as evident in Fig. 14.

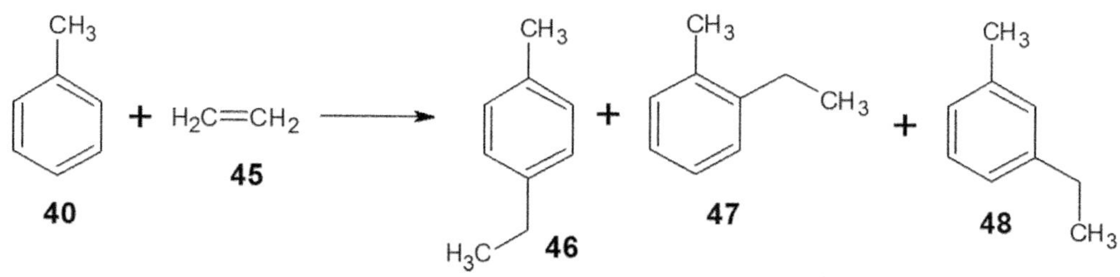

Fig. 12 Selective ethylation of toluene in $scCO_2$ using CLD modified HZSM-5 pellets [81].

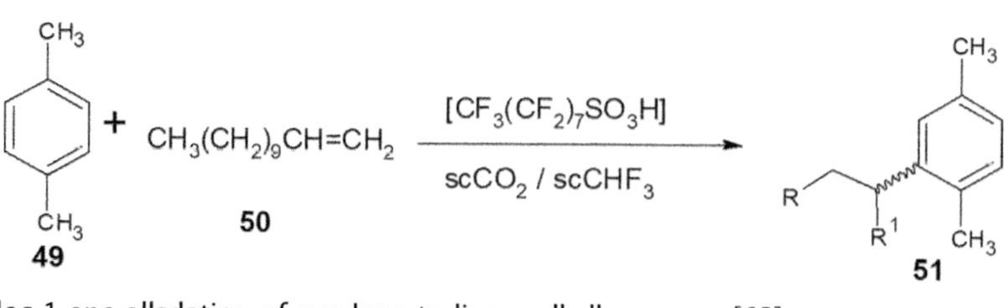

Fig. 13 Dodec-1-ene alkylation of *p*-xylene to linear alkylbenzenes [68].

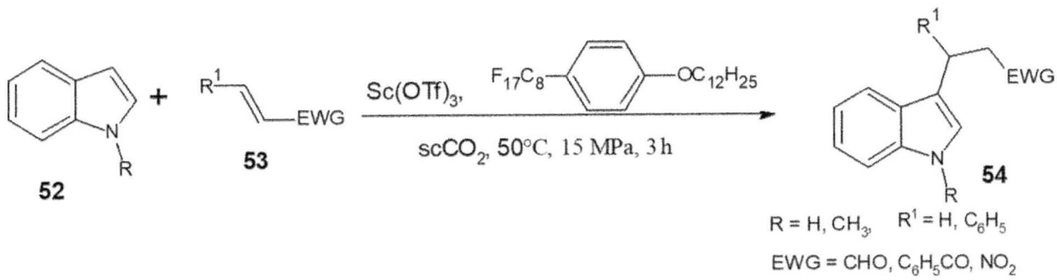

Fig. 14 Friedel-Crafts alkylation of indole derivatives using 1-dodecyloxy-4-perfluoroalkylbenzene as an efficient additive in $scCO_2$ [85].

5.2 Alkylation of olefins

Alkylate is a low-sulfur, ultraclean gasoline blending stream produced in a refinery and is an essential requisite for generating environmentally sound reformulated gasoline fuel component. Trimethylpentane (TMP) isomers are the desired products in the alkylation of isobutane. However, the less branched C_8 products, including dimethylhexanes (DMHs), as well as heavy alkylation products (C_{9+}) and lighter cracked products (C_5–C_7) are also formed apart from TMPs. The total liquid product (C_{5+}) during the process is referred to as alkylate.

Olefins interact with strong acids which catalyze alkylation reaction, resulting in conjugate oligomerization reaction and rapid deactivation of the catalyst [86]. The deactivation of acid sites of catalysts in alkylation of isobutane **55** with olefins eliminates the possibility of implementing the functions of the hydride transfer from isobutane and disrupts the course of the target alkylation reactions. The selectivity was increased when the temperature and pressure were increased, due to high solubility in $scCO_2$. Shiriyazdanov [87] regenerated the catalyst in alkylation of isobutene with butane-butene **56–57** fraction (Fig. 15) in $scCO_2$. The alkylate yield increased with the rise in pressure and temperature owing to the rising solubility of the deposit in $scCO_2$ with increasing distance from the critical point. Moreover, alkylation in $scCO_2$ led to complete regeneration of the zeolite-containing catalysts with its performance characteristics at par with a fresh catalyst with no morphological

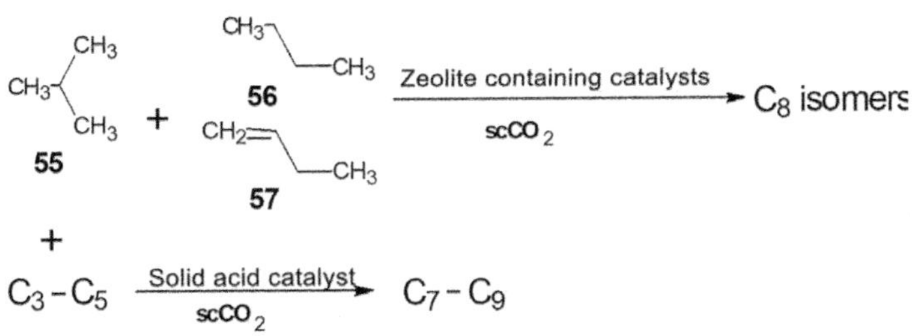

Fig. 15 Alkylation of isobutane in $scCO_2$.

and structural characteristic variations. Clark and Subramaniam [50] reported the alkylation of isobutene/1-butene by using a molar excess of $scCO_2$ at P_c of 71.8 bar and T_c of 31.1°C, lower than the critical temperature of isobutane (<135°C). Steady alkylate (trimethylpentanes and dimethylhexanes) production was obtained with both microporous zeolite and mesoporous sulfated zirconia catalysts. Solid-acid-catalyzed fixed-bed alkylation process using $scCO_2$ aided in coke mitigation and provided better catalyst pore accessibilities leading to better yield. Moreover, reduction in pressure enabled easy product separation from reactants and CO_2, thereby avoiding distillation.

Ginosar et al. [49] studied the effect of six solid-acid catalysts and SCFs in the alkylation of isobutane **55** with trans 2-butene **58** to get the desired trimethylpentane **59** as presented in Fig. 16. Apart from **59**, the by-product of this reaction has carbon atoms of five or more (C_5–C_9). The six solid acids studied belonged to zeolites, nafion and sulfated metal oxide catalysts. The media used were liquid, near-critical liquid, and SCFs. The use of $scCO_2$ as a cosolvent in the alkylation reaction did not exhibit any positive impact on catalyst lifespans, product yields, and product quality.

Salinas et al. reported the alkylation of isobutene **55** with 1-butene **57** using a solid-acid catalyst and supercritical 2-methylpropane [88]. The undesirable side reactions such as oligomerization and cracking that occurred at a supercritical temperature of 410 K for 2-methylpropane prompted Li et al. [89] to tune the critical temperature of the same reaction using binary (2-methylpropane + CO_2) and ternary (2-methylpropane + 1-butene + CO_2, and 2-methylpropane + 2,2,4-trimethylpentane + CO_2) mixtures. The presence of CO_2 in binary or ternary mixture reduced the critical temperature significantly, preventing undesirable side reactions. Lyon et al. [90] alkylated isobutene **55** with 1-butene **57** using silica-supported catalyst giving rise to isooctane. They observed that mild supercritical pressure facilitated the desorption of C_8 products. But, at the same pressure, the selectivity of C_8 decreased by more than 50%, with an increase in heavier (C_{12} and higher) products formation. Besides oligomerization predominated at high temperature.

The conventional reaction between isobutane and n-butene that involves liquid HF or H_2SO_4 or solid-acids forms highly branched isoparaffins, which is an important process to produce high-octane and clean-burning gasoline. As observed in the majority of the alkylation reactions, oligomerization and cracking are observed with high reaction temperature (>408 K) to reach the supercritical conditions. Santana and Akgerman [91] employed the SCFs having a low critical temperature, and the alkylated product

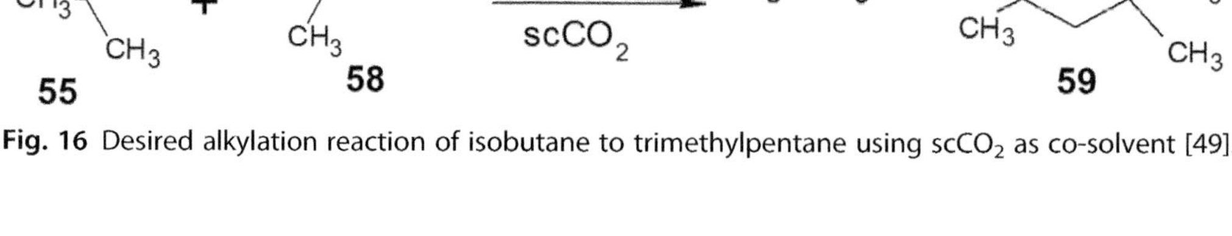

Fig. 16 Desired alkylation reaction of isobutane to trimethylpentane using $scCO_2$ as co-solvent [49].

selectivity increased. Isononanes and isononenes, the heavier saturated/unsaturated products were observed, by adding CO_2 as a diluent which facilitates a reduction in critical temperature, alkylate selectivity and coke precursor removal. Also, trimethylpentanes were formed in high concentration over dimethylhexanes and methylheptanes.

5.3 Allylic alkylation

The substitution reaction catalyzed by palladium, where the allylic position has a leaving group is known as Tsuji-Trost or Trost allylic alkylation reaction. The nucleophile attacks the metal—allyl group complex, forming the carbon-carbon and carbon-heteroatom bonds in the substituted product. Mild conditions with phosphine ligands yield high enantioselectivity and high diastereoselectivity [92]. The allylic alkylation has high utility in medicinal and natural product synthesis.

Lyubimov et al. [93] alkylated 1,3-diphenylallyl acetate **61** with dimethyl malonate **60** using Pd-catalyst at 170 atm. The first report on allylic alkylation resulted in enantioselectivity of 81% with 75% conversion. A year later Lyubimov and his group [94] carried out Tsuji-Trost in $scCO_2$ as shown in Fig. 17, which resulted in decreased pressure (70 atm) and temperature (60°C), with increased product **62** selectivity, avoiding the formation of the diallylated by-product. Later, in 2011, Lyubimov et al. [95] used phosphite-type ligand along with $scCO_2$, resulting in regioselectivity of 99% for linear products.

5.4 Transalkylation

Transfer of an alkyl group from an organic compound to another is known as transalkylation. The petrochemical industry employs this reaction [96] to manufacture aromatic compounds.

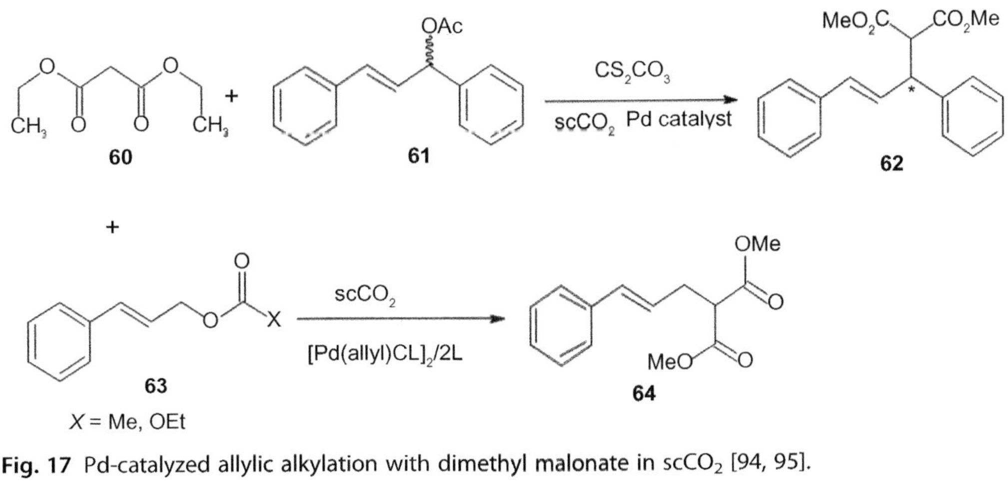

Fig. 17 Pd-catalyzed allylic alkylation with dimethyl malonate in $scCO_2$ [94, 95].

Sotelo et al. [97] carried out transalkylation of *p*-diisopropyl benzene **66** to get cumene **67** (Fig. 18) employing different acidic commercial zeolite materials as catalysts with very high conversion in combination with high selectivity for the desired product. In comparison with the liquid-phase transalkylation conditions, the addition of $scCO_2$ leads to augmented activity and improved product quality only with the beta and mordenite structures and not with large-pore Y-zeolite.

Though the catalyst activity in both alkylation and transalkylation reactions is directly related to its acidic properties, transalkylation requires higher acidic strength than the alkylation reaction due to the low stability of the carbenium ion intermediate. Cavani et al. [98] alkylated benzene and transalkylated diethyl benzene using zeolite catalyst to get ethylbenzene**69**. The bulky carbonium ion with low stability deactivates transalkylation [99]. Uguina et al. [100] used $scCO_2$ along with large pore zeolites, in solvent-free systems, which gave better catalyst performance. Another group Rodriguez et al. [105] from the same lab varied the experimental parameters to optimize the conditions. The supercritical condition, free-solvent system, and high temperature favored the transalkylation reaction.

5.5 N-alkylation

Amine alkylation is a nucleophilic substitution reaction that takes place between alkyl halide and ammonia or an amine to generate substituted amine [101]. Alkylation of amines is an essential reaction in the synthesis of drugs [5].

Amino alcohols undergo intramolecular and intermolecular alkylation reactions in the presence of heterogeneous catalyst, $\gamma\text{-}Al_2O_3$ to get N-alkylated heterocycles **72**. The reactions as presented in Fig. 19 were beneficial in $scCO_2$, as cyclic amines, cyclic ureas, and urethanes were formed in higher yields [102].

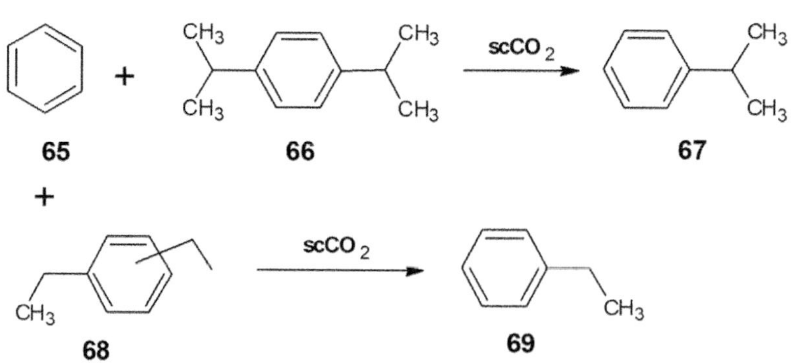

Fig. 18 Transalkylation of benzene with *p*-diisopropylbenzene and diethyl benzene [97, 98].

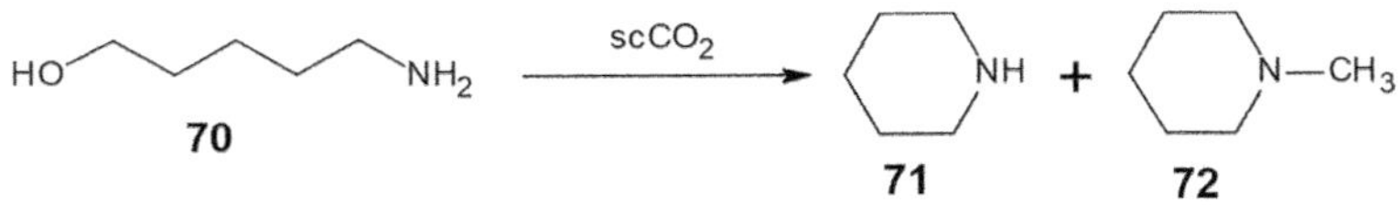

Fig. 19 N-alkylation of amino alcohols using γ-Al_2O_3 [102].

5.6 Alkylation of alcohols and phenols

Licence et al. [103] reported methylation of long-chain alcohols and diols at high temperature using excess of methanol. The associated problem of high-temperature reaction was that secondary or tertiary alcohols can give alkenes. Selva and Perosa [104] evaluated the methylating agents and reported dimethyl carbonate as environmentally benign. Accordingly, Gooden et al. [105] carried out the alkylation of 1-octanol **73** using dimethyl carbonate (DMC) to get ethers as shown in Fig. 20. The usage of acidic heterogeneous catalysts in $scCO_2$ resulted in the desirable ethers in high selectivity and yield.

o-Cyclohexylphenol and *p*-cyclohexylphenol are the major constituents of dyes, resins, and biocides [106]. Alkylation of phenol **27** with cyclohexene or cyclohexanol using γ-Al_2O_3 and $scCO_2$ produces cyclohexylphenol (Fig. 21). As water is formed with cyclohexanol, cyclohexene is preferred over it. Amandi et al. [107] prepared *o*-cyclohexylphenol **76** from cyclohexanol using γ-Al_2O_3 and $scCO_2$ in a two-step process with high selectivity and yield. The first step involved the formation of cyclohexene from cyclohexanol through dehydration, followed by alkylation with phenol.

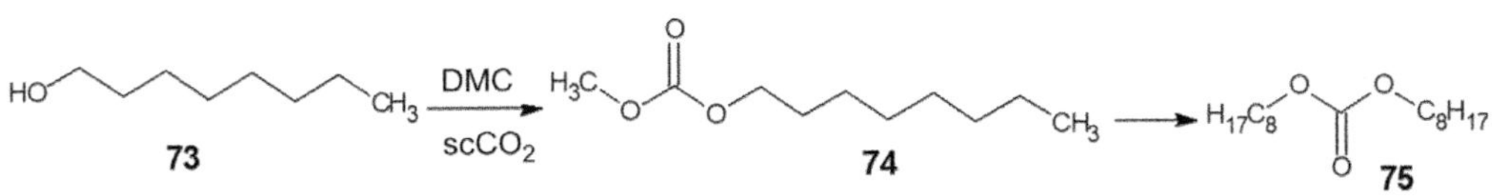

Fig. 20 Methylation of 1-octanol using DMC in $scCO_2$ [105].

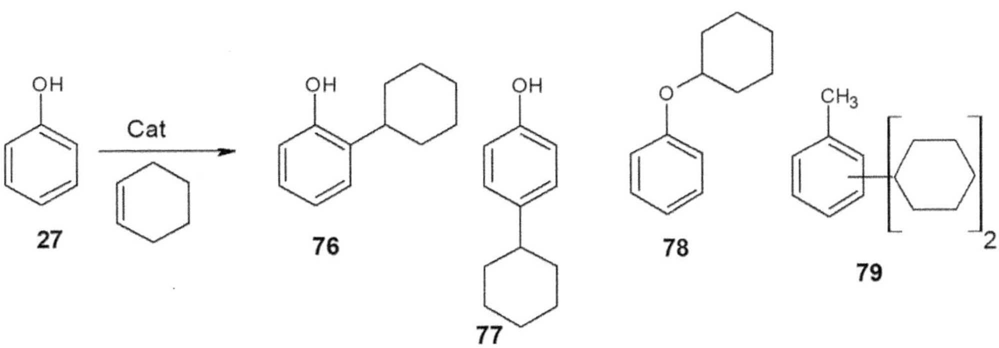

Fig. 21 Preparation of *o*-cyclohexylphenol from cyclohexene or cyclohexanol [107].

6. Conclusion

The processes in supercritical carbon dioxide ($scCO_2$) are one step nearer to environmentally benign Friedel-Crafts and other alkylation reactions and have potential important features essential for the manufacture of fine chemicals. The $scCO_2$ facilitates reaction conditions to be tuned to obtain high product selectivity with an added advantage of easy solvent removal and product separation. The use of $scCO_2$ in conjunction with a heterogeneous catalyst and a continuous flow reactor can circumvent many environmental and practical issues of alkylation processes. Recently established solid acid catalyzed alkylation reactions are undeniably commercially sustainable in conjugation with the use of $scCO_2$ reaction media. The heterogeneous Bronsted acidic zeolite catalyst LaNaY-73, solid acid catalysts H-Y zeolite and $Sc(OTf)_3$ supported on MCM-41 in $scCO_2$ can be used to provide environmentally sentient syntheses of bulk chemicals. The clean technology in $scCO_2$ mediated alkylation processes enables recycling of catalyst and fewer by-products. Rational design of catalyst as well as tailoring various factors such as pore structure and acidity, further enhance the possibility of better product selectivity in $scCO_2$ mediated alkylation.

The perspective of employing dense phase $scCO_2$ in environmentally green solid-acid catalysis for pharmaceutical and fine chemical syntheses is quite promising. More research can be steered to spread out the usability of $scCO_2$ as reaction media in the transformations of other higher molecular weight aromatic molecules and to explore the role of mass transfer inside the pores of zeolite catalysts compared to the mass transfer from the $scCO_2$ to the exterior surface of the catalyst particles. The present economic and environmental impact studies illustrate the benefits and limitations of the $scCO_2$ assisted alkylation processes and provide rational guidance for upcoming research.

References

[1] W. Leitner, Supercritical carbon dioxide as a green reaction medium for catalysis, Acc. Chem. Res. 35 (2002) 746–756.
[2] S. Mayadevi, Reactions in supercritical carbon dioxide, Indian J. Chem. A 51A (2012) 1298–1305.
[3] R.K. Henderson, C. Jiménez-González, D.J.C. Constable, S.R. Alston, G.G.A. Inglis, G. Fisher, J. Sherwood, S.P. Binks, A.D. Curzons, Expanding GSK's solvent selection guide – embedding sustainability into solvent selection starting at medicinal chemistry, Green Chem. 13 (2011) 854–862.
[4] M. Poliakoff, P. Licence, Green chemistry, Nature 450 (2007) 810–812.
[5] D.J.C. Constable, P.J. Dunn, J.D. Hayler, G.R. Humphrey, J.J.L. Leazer, R.J. Linderman, K. Lorenz, J. Manley, B.A. Pearlman, A. Wells, A. Zaks, T.Y. Zhang, Key green chemistry research areas—a perspective from pharmaceutical manufacturers, Green Chem. 9 (2007) 411–420.
[6] D. Adam, Industrial solvents: Clean and green...but are they mean? Nature 407 (2000) 938–940.
[7] M. Raj, V.K. Singh, Organocatalytic reactions in water, Chem. Commun. 44 (2009) 6687–6703.
[8] C.-J. Li, L. Chen, Organic chemistry in water, Chem. Soc. Rev. 35 (2006) 68–82.
[9] C.-J. Li, Organic reactions in aqueous media with a focus on carbon–carbon bond formations: a decade update, Chem. Rev. 105 (2005) 3095–3166.

[10] R.A. Sheldon, Utilisation of biomass for sustainable fuels and chemicals: molecules, methods and metrics, Catal. Today 167 (2011) 3–13.

[11] U. Biermann, U. Bornscheuer, M.A.R. Meier, J.O. Metzger, H.J. Schäfer, Oils and fats as renewable raw materials in chemistry, Angew. Chem. Int. Ed. 50 (2011) 3854–3871.

[12] K.R. Seddon, A taste of the future, Nat. Mater. 2 (2003) 363–365.

[13] M.J. Earle, K.R. Seddon, Ionic liquids: green solvents for the future, Pure Appl. Chem. 72 (2000) 1391–1398.

[14] J.R. Hyde, P. Licence, D. Carter, M. Poliakoff, Continuous catalytic reactions in supercritical fluids, Appl. Catal. A 222 (2001) 119–131.

[15] A. Baiker, Supercritical fluids in heterogeneous catalysis, Chem. Rev. 99 (1999) 453–474.

[16] P.G. Jessop, T. Ikariya, R. Noyori, Homogeneous catalysis in supercritical fluids, Chem. Rev. 99 (1999) 475–494.

[17] R.S. Oakes, A.A. Clifford, C.M. Rayner, The use of supercritical fluids in synthetic organic chemistry, J. Chem. Soc. Perkin Trans. 1 (2001) 917–941.

[18] K.P. Johnston, Safer solutions for chemists, Nature 368 (1994) 187–188.

[19] Y. Sun, R.N. Landau, J. Wang, C. LeBlond, D.G. Blackmond, A re-examination of pressure effects on enantioselectivity in asymmetric catalytic hydrogenation, J. Am. Chem. Soc. 118 (1996) 1348–1353.

[20] D.C. Wynne, P.G. Jessop, Cyclopropanation enantioselectivity is pressure dependent in supercritical fluoroform, Angew. Chem. Int. Ed. 38 (1999) 1143–1144.

[21] P.G. Jessop, T. Ikariya, R. Noyori, Homogeneous catalytic hydrogenation of supercritical carbon dioxide, Nature 368 (1994) 231–233.

[22] D.J. Bochniak, B. Subramaniam, Fischer-tropsch synthesis in near-critical n-hexane: Pressure-tuning effects, AICHE J. 44 (1998) 1889–1896.

[23] B. Subramaniam, A. Jooma, In situ mitigation of coke buildup in porous catalysts with supercritical reaction media: effect of feed peroxide, in: K.W. Hutchenson, N.R. Foster (Eds.), Innovations in Supercritical Fluids: Science and Technology, American Chemical Society, 1995, , pp. 246–256.

[24] C.A. Eckert, K. Chandler, Tuning fluid solvents for chemical reactions, J. Supercrit. Fluids 13 (1998) 187–195.

[25] P.G. Jessop, W. Leitner, Chemical Synthesis Using Supercritical Fluids, Wiley, 2008.

[26] M. Škerget, Z.E. Knez, M.A. Knez-Hrnčič, Solubility of solids in sub- and supercritical fluids: a review, J. Chem. Eng. Data 56 (2011) 694–719.

[27] M. McHugh, V. Krukonis, H. Brenner, Supercritical Fluid Extraction: Principles and Practice, Elsevier Science, 2013.

[28] C. White, J. Burnett, Integration of supercritical fluid chromatography into drug discovery as a routine support tool, J. Chromatogr. A 1074 (2005) 175–185.

[29] J. Durand, E. Teuma, M. Gómez, Ionic liquids as a medium for enantioselective catalysis, C. R. Chim. 10 (2007) 152–177.

[30] P.G. Jessop, Homogeneous catalysis using supercritical fluids: Recent trends and systems studied, J. Supercrit. Fluids 38 (2006) 211–231.

[31] B. Subramaniam, Enhancing the stability of porous catalysts with supercritical reaction media, Appl. Catal. A 212 (2001) 199–213.

[32] B. Subramaniam, C.J. Lyon, V. Arunajatesan, Environmentally benign multiphase catalysis with dense phase carbon dioxide, Appl. Catal. B 37 (2002) 279–292.

[33] S.B. Hawthorne, D.J. Miller, D.E. Nivens, D.C. White, Supercritical fluid extraction of polar analytes using in situ chemical derivatization, Anal. Chem. 64 (2002) 405–412.

[34] X. Han, M. Poliakoff, Continuous reactions in supercritical carbon dioxide: problems, solutions and possible ways forward, Chem. Soc. Rev. 41 (2012) 1428–1436.

[35] Y. Zhang, C. Erkey, Preparation of supported metallic nanoparticles using supercritical fluids: a review, J. Supercrit. Fluids 38 (2006) 252–267.

[36] O. Pascu, V. Liautard, M. Vaultier, M. Pucheault, C. Aymonier, Catalysed stereodivergent hydrosilylation with Onium Salts stabilised M(0) nanocatalysts prepared in $scCO_2$, RSC Adv. 4 (2014) 59953–59960.
[37] A. Corma, A. Martínez, Chemistry, catalysts, and processes for isoparaffin–olefin alkylation: actual situation and future trends, Catal. Rev. 35 (1993) 483–570.
[38] P. Botella, A. Corma, J.M. López-Nieto, The influence of textural and compositional characteristics of nafion/silica composites on isobutane/2-butene alkylation, J. Catal. 185 (1999) 371–377.
[39] A. de Angelis, P. Ingallina, D. Berti, L. Montanari, M.G. Clerici, Solid acid catalysts for alkylation of hydrocarbons, Catal. Lett. 61 (1999) 45–49.
[40] P. Licence, J. Ke, M. Sokolova, S.K. Ross, M. Poliakoff, Chemical reactions in supercritical carbon dioxide: from laboratory to commercial plant, Green Chem. 5 (2003) 99–104.
[41] K. Tanabe, Industrial application of solid acid–base catalysts, Appl. Catal. A 181 (1999) 399–434.
[42] R.P. Singh, R.M. Kamble, K.L. Chandra, P. Saravanan, V.K. Singh, An efficient method for aromatic Friedel–Crafts alkylation, acylation, benzoylation, and sulfonylation reactions, Tetrahedron 57 (2001) 241–247.
[43] M.A. Harmer, Q. Sun, A.J. Vega, W.E. Farneth, A. Heidekum, W.F. Hoelderich, Nafion resin–silica nanocomposite solid acid catalysts. Microstructure–processing–property correlations, Green Chem. 2 (2000) 7–14.
[44] G.A. Olah, P. Batamack, D. Deffieux, B. Török, Q. Wang, Á. Molnár, G.K. Surya Prakash, Acidity dependence of the trifluoromethanesulfonic acid catalyzed isobutane-isobutylene alkylation modified with trifluoroacetic acid or water, Appl. Catal. A 146 (1996) 107–117.
[45] M.P.D. Mahindaratne, K. Wimalasena, Detailed characterization of p-toluenesulfonic acid monohydrate as a convenient, recoverable, safe, and selective catalyst for alkylation of the aromatic nucleus, J. Organomet. Chem. 63 (1998) 2858–2866.
[46] S. Kobayashi, M. Sugiura, H. Kitagawa, W.W.L. Lam, Rare-earth metal triflates in organic synthesis, Chem. Rev. 102 (2002) 2227–2302.
[47] M. Hart, G. Fuller, D.R. Brown, C. Park, M.A. Keane, J.A. Dale, C.M. Fougret, R.W. Cockman, Acidities and catalytic activities of persulfonated poly(styrene-co-divinylbenzene) ion-exchange resins, Catal. Lett. 72 (2001) 135–139.
[48] M. Selvaraj, A. Pandurangan, K.S. Seshadri, P.K. Sinha, V. Krishnasamy, K.B. Lal, Comparison of mesoporous Al-MCM-41 molecular sieves in the production of p-cymene for isopropylation of toluene, Mol. Catal. A 186 (2002) 173–186.
[49] D.M. Ginosar, D.N. Thompson, K. Coates, D.J. Zalewski, The effect of supercritical fluids on solid acid catalyst alkylation, Ind. Eng. Chem. Res. 41 (2002) 2864–2873.
[50] M.C. Clark, B. Subramaniam, Extended alkylate production activity during fixed-bed supercritical 1-butene/isobutane alkylation on solid acid catalysts using carbon dioxide as a diluent, Ind. Eng. Chem. Res. 37 (1998) 1243–1250.
[51] J.M. Miller, M. Goodchild, L.J. Lakshmi, D. Wails, J.S. Hartman, Friedel–Crafts catalysis using supported reagents. Synthesis, characterization, and catalytic application of $ZnCl_2$ supported on sol–gel-derived alumina, Mater. Lett. 44 (2000) 164–169.
[52] M. Kodomari, Y. Suzuki, K. Yoshida, Graphite as an effective catalyst for Friedel–Crafts acylation, Chem. Commun. (1997) 1567–1568.
[53] D.J.C. Constable, A.D. Curzons, V.L. Cunningham, Metrics to 'green' chemistry—which are the best? Green Chem. 4 (2002) 521–527.
[54] R.A. Sheldon, Consider the environmental quotient, ChemTech 25 (1994) 38–46.
[55] G.A. Olah, Friedel-Crafts Chemistry, Wiley, 1980.
[56] G.A. Olah, R. Krishnamurti, G.K.S. Prakash, Friedel–Crafts Alkylations, in: B.M. Trost, I. Fleming (Eds.), Comprehensive Organic Synthesis, Oxford, Pergamon Press, 1991, , pp. 293–339.
[57] J.H. Clark, D.J. Macquarrie, Heterogeneous catalysis in liquid phase transformations of importance in the industrial preparation of fine chemicals, Org. Process. Res. Dev. 1 (1997) 149–162.
[58] R. Sreekumar, R. Padmakumar, Friedel-Crafts acylation of aromatic hydrocarbons using zeolites, Synth. Commun. 27 (1997) 777–780.

[59] M.Y. Croft, E.J. Murby, R.J. Wells, Simultaneous extraction and methylation of chlorophenoxyacetic acids from aqueous solution using supercritical carbon dioxide as a phase transfer solvent, Anal. Chem. 66 (2002) 4459–4465.
[60] M.G. Hitzler, F.R. Smail, M. Poliakoff, M.G. Hitzler, S.K. Ross, Friedel–Crafts alkylation in supercritical fluids: continuous, selective and clean, Chem. Commun. 133 (1998) 359–360.
[61] L. Fan, I. Nakamura, S. Ishida, K. Fujimoto, Supercritical-phase alkylation reaction on solid acid catalysts: mechanistic study and catalyst development, Ind. Eng. Chem. Res. 36 (1997) 1458–1463.
[62] K.P.C. Vollhardt, N.E. Schore, Organic Chemistry: Structure and Function, seventh ed., W. H. Freeman and Company, England, 2014.
[63] G. Ertl, Handbook of heterogeneous catalysis, eighth ed., 5, Wiley-VCH, Weinheim, 2008.
[64] P. Tundo, Continuous Flow Methods in Organic Synthesis, Ellis Horwood, New York, 1991.
[65] T.-W. Kuo, C.-S. Tan, Alkylation of toluene with propylene in supercritical carbon dioxide over chemical liquid deposition HZSM-5 Pellets, Ind. Eng. Chem. Res. 40 (2001) 4724–4730.
[66] E.J. Beckman, Supercritical and near-critical CO_2 in green chemical synthesis and processing, J. Supercrit. Fluids 28 (2004) 121–191.
[67] R. Amandi, P. Licence, S.K. Ross, O. Aaltonen, M. Poliakoff, Friedel – Crafts alkylation of anisole in supercritical carbon dioxide: a comparative study of catalysts, Org. Process. Res. Dev. 9 (2005) 451–456.
[68] M.A. Harmer, K.W. Hutchenson, Super acid catalysis in supercritical fluid reaction media for the formation of linear alkyl benzenes, Chem. Commun. 1 (2002) 18–19.
[69] M. Álvaro, Friedel–Crafts hydroxyalkylation: reaction of anisole with paraformaldehyde catalyzed by zeolites in supercritical CO_2, J. Catal. 219 (2003) 464–468.
[70] J.E. Chateauneuf, K. Nie, An investigation of a Friedel-Crafts alkylation reaction in homogeneous supercritical CO_2 and under subcritical and splitphase reaction conditions, Adv. Environ. Res. 4 (2000) 307–312.
[71] A.P. Abbott, S. Corr, N.E. Durling, E.G. Hope, Pressure effects on Friedel-Crafts alkylation reactions in supercritical difluoromethane, Chem. Phys. Chem. 6 (2005) 466–472.
[72] R. Gläser, J. Weitkamp, Supercritical carbon dioxide as a reaction medium for the zeolite-catalyzed alkylation of naphthalene, Ind. Eng. Chem. Res. 42 (2003) 6294–6302.
[73] G. Manos, H. Hofmann, Coke removal from a zeolite catalyst by supercritical fluids, Chem. Eng. Technol. 14 (1991) 73–78.
[74] G. Kamalakar, K. Komura, Y. Sugi, tert-Butylation of phenol over ordered solid acid catalysts in supercritical carbon dioxide: efficient synthesis of 2,4-di-tert-butylphenol and 2,4,6-tri-tert-butylphenol, Ind. Eng. Chem. Res. 45 (2006) 6118–6126.
[75] P. Wimmer; H. Buysch, L. Puppe, Burscheid, B. Aktiengesellschaft, Germany, US5,030,770, 1991.
[76] R. Amandi, J.R. Hyde, S.K. Ross, T.J. Lotz, M. Poliakoff, Continuous reactions in supercritical fluids; a cleaner, more selective synthesis of thymol in supercritical CO_2, Green Chem. 7 (2005) 288–293.
[77] G. Deák, T. Pernecker, J.P. Kennedy, Carbocationic polymerization in Supercritical CO_2. Synthesis of phenol-terminated polyisobutylene, J. Macromol. Sci. Part A 32 (1995) 979–984.
[78] C. Perego, S. Amarilli, A. Carati, C. Flego, G. Pazzuconi, C. Rizzo, G. Bellussi, Mesoporous silica-aluminas as catalysts for the alkylation of aromatic hydrocarbons with olefins, Microporous Mesoporous Mater. 27 (1999) 345–354.
[79] P.A. Parikh, N. Subrahmanyam, Y.S. Bhat, A.B. Halgeri, Toluene isopropylation over zeolite β and metallosilicates of MFI structure, Appl. Catal. A 90 (1992) 1–10.
[80] T.-C. Chiang, J.-C. Chan, C.-S. Tan, Alkylation of toluene with isopropyl alcohol over chemical liquid deposition modified HZSM-5 under atmospheric and supercritical operations, Ind. Eng. Chem. Res. 42 (2003) 1334–1340.
[81] C.-C. Yu, C.-S. Tan, Selective toluene ethylation in supercritical CO_2 using chemical liquid deposition modified HZSM-5 pellets, J. Supercrit. Fluids 44 (2008) 341–346.
[82] D.M. Ginosar, K. Coates, D.N. Thompson, The effects of supercritical propane on the alkylation of toluene with ethylene over USY and sulfated zirconia catalysts, Ind. Eng. Chem. Res. 41 (2002) 6537–6545.

[83] G. Franciò, K. Wittmann, W. Leitner, Highly efficient enantioselective catalysis in supercritical carbon dioxide using the perfluoroalkyl-substituted ligand (R,S)-3-H2F6-BINAPHOS, J. Organomet. Chem. 621 (2001) 130–142.

[84] K. Manabe, N. Aoyama, S. Kobayashi, Friedel–Crafts-type conjugate addition of indoles using a lewis acid–surfactant-combined catalyst in water, Adv. Synth. Catal. 343 (2001) 174–176.

[85] I. Komoto, S. Kobayashi, 1-Dodecyloxy-4-perfluoroalkylbenzene as a novel efficient additive in aldol reactions and Friedel – Crafts alkylation in supercritical carbon dioxide, Org. Lett. 4 (2002) 1115–1118.

[86] W. Zhang, M. Sun, R. Prins, Multinuclear MAS NMR identification of fluorine species on the surface of fluorinated γ-Alumina, J. Phys. Chem. B 106 (2002) 11805–11809.

[87] R.R. Shiriyazdanov, Regeneration of zeolite-containing catalysts of alkylation of isobutane with the butane-butene faction in supercritical carbon dioxide, Russ. J. Phys. Chem. B 5 (2012) 1080–1083.

[88] A.L.M. Salinas, D. Kong, Y.B. Tâarit, N. Essayem, Continuous iC4/C4 = alkylation under iC4supercritical conditions over K2.5H0.5PW12O40 and H – beta solid acids, Ind. Eng. Chem. Res. 43 (2004) 6355–6362.

[89] J. Li, Z. Qin, G. Wang, M. Dong, J. Wang, Critical temperatures and pressures of several binary and ternary mixtures concerning the alkylation of 2-methylpropane with 1-butene in the presence of methane or carbon dioxide, J. Chem. Eng. Data 52 (2007) 1736–1740.

[90] C. Lyon, B. Subramaniam, C. Pereira, Enhanced isooctane yields for 1-butene/Isobutane alkylation on SiO_2-supported Nafion®®® in Supercritical Carbon Dioxide, Stud. Surf. Sci. Catal. 139 (2001) 221–228.

[91] G.M. Santana, A. Akgerman, Alkylation of isobutane with 1-butene on a solid acid catalyst in supercritical reaction media, Ind. Eng. Chem. Res. 40 (2001) 3879–3882.

[92] I.G. Rios, A. Rosas-Hernandez, E. Martin, Recent advances in the application of chiral phosphine ligands in Pd-catalysed asymmetric allylic alkylation, Molecules 16 (2011) 970–1010.

[93] S.E. Lyubimov, I.V. Kuchurov, A.A. Vasil'ev, A.A. Tyutyunov, V.N. Kalinin, V.A. Davankov, S.G. Zlotin, The use of a new carboranylamidophosphite ligand in the asymmetric Pd-catalysed allylic alkylation in organic solvents and supercritical carbon dioxide, J. Organomet. Chem. 694 (2009) 3047–3049.

[94] S.E. Lyubimov, I.V. Kuchurov, A.A. Vasil'ev, S.G. Zlotin, V.A. Davankov, Asymmetric allylic alkylation in supercritical carbon dioxide using P*-chiral diamidophosphite ligands, Mendeleev Commun. 20 (2010) 143–144.

[95] S.E. Lyubimov, T.A. Verbitskaya, E.A. Rastorguev, P.V. Petrovskii, V.N. Kalinin, V.A. Davankov, Allyl substitution in a supercritical carbon dioxide medium catalyzed by complexes of palladium with phosphite-type ligands, Russ. J. Phys. Chem. B 5 (2012) 1130–1134.

[96] K. Griesbaum, A. Behr, D. Biedenkapp, H.-W. Voges, D. Garbe, C. Paetz, G. Collin, D. Mayer, H. Höke, Hydrocarbons, Wiley Online Library(2000).

[97] J. Sotelo, L. Calvo, A. Velazquez, D. Capilla, F. Cavani, M. Bolognini, A comparative study on the transalkylation of diisopropylbenzene with benzene over several zeolitic materials in supercritical CO_2 and liquid phase, Appl. Catal. A 312 (2006) 194–201.

[98] F. Cavani, Liquid-phase transalkylation of diethylbenzenes with benzene over β-zeolite: effect of operating parameters on the distribution of the products, Appl. Catal. A 226 (2002) 31–40.

[99] R. Roldan, Transformation of mixtures of benzene and xylenes into toluene by transalkylation on zeolites, Appl. Catal. A 266 (2004) 203–210.

[100] M.A. Uguina, A. Rodríguez, A. Pérez-Velazquez, D. Capilla, F. Cavani, Effect of supercritical conditions on the transalkylation of diethylbenzene with benzene, J. Supercrit. Fluids 158 (2005) 1969–1976.

[101] J. March, Advanced organic chemistry: reactions, mechanisms, and structure, John Wiley & Sons, USA, 1992.

[102] E.S. Streng, D.S. Lee, M.W. George, M. Poliakoff, Continuous N-alkylation reactions of amino alcohols using γ-Al_2O_3 and supercritical CO_2: unexpected formation of cyclic ureas and urethanes by reaction with CO_2, Beilstein J. Org. Chem. 13 (2017) 329–337.
[103] P. Licence, W.K. Gray, M. Sokolova, M. Poliakoff, Selective monoprotection of 1,n-terminal diols in supercritical carbon dioxide: a striking example of solvent tunable desymmetrization, J. Am. Chem. Soc. 127 (2005) 293–298.
[104] M. Selva, A. Perosa, Green chemistry metrics: a comparative evaluation of dimethyl carbonate, methyl iodide, dimethyl sulfate and methanol as methylating agents, Green Chem. 10 (2008) 457–464.
[105] P.N. Gooden, R.A. Bourne, A.J. Parrott, H.S. Bevinakatti, D.J. Irvine, M. Poliakoff, Continuous acid-catalyzed methylations in supercritical carbon dioxide: Comparison of methanol, dimethyl ether and dimethyl carbonate as methylating agents, Org. Process. Res. Dev. 14 (2010) 411–416.
[106] G.D. Yadav, P. Kumar, Alkylation of phenol with cyclohexene over solid acids: Insight in selectivity of O- versus C-alkylation, Appl. Catal. A 286 (2005) 61–70.
[107] R. Amandi, K. Scovell, P. Licence, T.J. Lotz, M. Poliakoff, The synthesis of o-cyclohexylphenol in supercritical carbon dioxide: towards a continuous two-step reaction, Green Chem. 9 (2007) 797–801.

CHAPTER 7

Extraction of phytochemicals from saffron by supercritical carbon dioxide

Emma Suali
Faculty of Engineering, Jalan UMS, Universiti Malaysia Sabah, Kota Kinabalu, Sabah, Malaysia

Contents

Nomenclature

R_s recovery of safranal
R_c recovery of crocin
T temperature (°C)
p pressure (MPa)
f flow rate (cm^3/min)
Y oil yield (%)
t extraction time

1. Introduction

Phytochemicals are chemical compounds that exist naturally in saffron and others plants. Phytochemical function as constitutive metabolites for growth and reproduction while enabling plants to overcome threats from the environment. According to Molyneux et al. [1], phytochemical roles are advantageous for the plant itself but might have an adverse effect on other organisms. Nevertheless, phytochemicals have proven useful for human health when used in moderation. Phytochemicals can be classified into polyphenols and carotenoids, and further grouped into many types based on the chemical structure.

Green Sustainable Process for Chemical and Environmental Engineering and Science
https://doi.org/10.1016/B978-0-12-817388-6.00007-6

Saffron, which is scientifically known as *Crocus sativus*, is a well-known spice that has many benefits to humans [2–5]. It is regarded as the most expensive spice due to its high demand [6], and the fact that the component of the spice is localized in the saffron flower, which is harvested manually for further processes, thus incurring high labor costs. Approximately 200 man-days per hectare are required to harvest large-scale production of saffron [7]. This contributes to the cost of saffron production. Furthermore, saffron as a commodity is expensive enough that it causes adulteration in some of the local markets of its producers [8, 9]. According to Caiola and Canini [10], saffron production is an expensive process. The process includes cultivation, harvesting, and extraction of the specific compounds for various purposes. Early studies have shown a few attempts to cultivate saffron in vitro to reduce the saffron production cost [11–13]. However, conventional cultivation is more practical and preferred by the major saffron producers. The routes of saffron production are summarized in Fig. 1.

The extraction of valuable compounds of saffron has been studied for decades, mainly for coloring, flavoring, and medicinal purposes [14–16]. Scientific analysis has shown more than 150 chemical compounds in saffron stigma, mainly crocin (80%), safranal (70%), and a small amount of crocetin and picrocrocin [17]. An earlier study on the comparison of saffron compounds from the main producers such as Azerbaijan, India, Iran, and Spain has shown similar trends in the chemical compositions [18]. This trend results in high demand for saffron for its medicinal properties. According to a review conducted by Khazdair et al. [19], the chemical compounds in saffron have been used for the treatment of a wide range of disorders, which include learning and memory impairment, hypertension, coronary artery diseases, and so on. Furthermore, antioxidants in saffron are effective to relieve symptoms of premenstrual syndrome (PMS) [20] and also as an

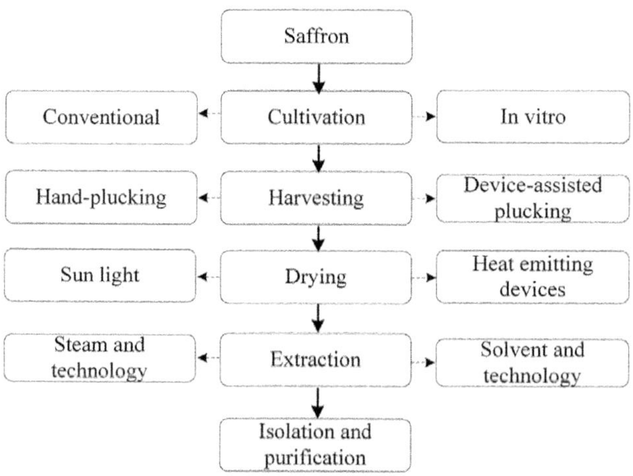

Fig. 1 Summary of saffron compounds production processes.

antiobesity drug [21]. When consumed at the permitted dosage, crocin from saffron has no adverse effect on human organs [22]. However, the challenge is in the method of obtaining the compounds through extractions processes. Like others edible plants, saffron extraction is temperature sensitive and could affect the quality of the chemical compounds. Thus, in this study, extraction method with a focus on supercritical carbon dioxide (CO_2) is discussed in detail.

2. Extraction of saffron

Extraction is a crucial step in obtaining high quality of compounds from saffron. There are a number of available methods that can be used, depending on the end product target. Saffron compounds are usually extracted for medicinal properties, therefore, the safety of the end product is the main concern. Early studies on the extraction of volatile compounds of saffron show that it is mostly conducted in a conventional way using chemicals or water as a solvent. In Zharghami and Heinz [23], monoterpene aldehydes and isophorone-related compounds were extracted using diethyl ether in a batch process. A similar method using methanol as a solvent was reported by Saito and Miyakawa [24] to extract quinoidal chalcone—a dyer pigment of the saffron flower. Sometimes, solvent extraction is combined with steam distillation and often conducted in a simple set-up apparatus such as beakers, soxhlets, and so on. Such extraction methods can be operated easily. However, such methods are impractical due to high chemicals requirement as a solvent. In addition, the solvents involved in the processing are often hazardous and could pose a risk to the quality of the compound.

The separation of the residual chemical from the extracts is also challenging when using the conventional methods. Many studies show that extraction of saffron is more preferred in a conventional way for lab scale extraction. However, recently, technology-assisted extraction has received attention due to its advantage in prevention of cell disruption of the essential compound in saffron. Table 1 shows some extracted compounds with respect to the method of extraction. However, comparison of the method is challenging because the result was reported in various units. The comparison in Table 1 indicates that crocin, picrocrocin, safranal, antioxidant compounds, and phenolic content are the most extracted compounds in saffron. It also indicates that a simple conventional extraction is able to extract the phytochemical of saffron with the aid of chromatography analysis, either on its own or in combination with GC-MS, HPLC, UV-Vis, and other analysis technologies. Nevertheless, technology-assisted extraction, such as ultrasound and supercritical CO_2, has resulted in a higher yield of crocin compounds compared to conventional methods. This shows that technology-assisted extraction is better compared to the conventional method in terms of final yield, time consumption, and purity when operated at optimum operating conditions [42]. Both ultrasound and supercritical extraction are the most

Table 1 Summary of methods used for various saffron compound extractions

Compound	Method	Results	References
Crocin	Ultrasound-assisted (USAE)	• 35.0%–75.7%	[25]
		• 16.9%–31.1%	[26]
		• 498.7[a] $E_{440}^{1\%}$	[27]
	Supercritical CO_2 with water and methanol	• 4.466–21.92 area C/area IS	[28]
	Thermal-assisted high hydrostatic pressure	• More than 250 mg/g	[29]
	Conventional method	• 7.42–25.18 mg/100 mg IS	[30]
Safranal	Thermal-assisted high hydrostatic pressure	• 0%–71%[b] $E_{330}^{1\%}$ higher than 36.4	[29]
	Supercritical CO_2	• 0.012–0.279 area C/area IS	[28]
	Conventional method with ethanol	• 0.012–12.96 mg/100 mg IS	[30]
	Ultrasound-assisted	• 151 $E_{330}^{1\%}$	[27]
	Conventional methods with diethyl ether, hexane, chloroform	• 0.03–0.07 mg/mL	[31]
Picrocrocin	Thermal-assisted high hydrostatic pressure	• 0%–68%[c] $E_{257}^{1\%}$ higher than 56.3	[29]
	Supercritical CO_2	• 0.012–0.279 area C/area IS	[28]
	Conventional method with ethanol	• 4.63–12.96 mg/100 mg IS	[30]
	Solid-phase extraction with cyclohexane	• 7.1%–27%	[32]
	Ultrasound assisted (USAE)	• 9.1%–12.1%	[26]
		• 49.71 $E_{440}^{1\%}$	[27]
Phenolic content	Conventional way with water and ethanol	• Sigma:8.28 mg/g • Corm:7.07 mg/g • Leaf:5.62 mg/g	[33]
	Supercritical CO_2	• 1423 mg/100 mg sample	[34]
	Subcritical water extraction	• 1616 mg/100 mg sample	[35]

Antioxidant	Ultrasonic water bath with ether	• 14.17 to 203.88 IC_{50}μg/mL	[36]
	Ultrasound Assisted Extraction (UAE)	• ABTS:0.042 mmol trolox/100 g • DPPH:14.692 mmol trolox/100 g • FRAP:8.556 mmol trolox/100 g	[37]
	Rapid Solid-Liquid Dynamic Extraction	• ABTS:0.858 mmol trolox/100 g • DPPH:16.673 mmol trolox/100 g • FRAP:260.723 mmol trolox/100 g	
Others	• Acetate-Urea buffer • SDS Buffer • NP-40 Buffer	• 5–250 kDa (total Protein)[d]	[38]
	Conventional way with water and ethanol	• 180 mg/100 mg sample (Flavonoid contents)	[34]
	Supercritical CO_2	• 103.4 mg/100 mg (anthocyanin)	[35]
	Subcritical water extraction	• 86.05% (DPPH)	[35]
	Solvent-assisted flavor evaporation (SAFE)	• 8699 μg/g (aroma compounds)	[39]
	Enzymatic water (EW)	• 6.7 mg/g (anthocyanin)	[40]
	Microwave-assisted extraction (MAE)	Metabolic content, ppm in crocins • β-d-gentiobiosyl: 5.54 • β-d-glucosyl:5.56 • all-trans-crocetin:7.44 • 13-cis-crocetin:7.51	[41]
	Solid-phase with cyclohexane	• 17.2%–29.4% (crocetin glycosides)	[35]

[a] $E_{440}^{1\%}$ indicates absorbance at 440 nm with 1% dilute solution.
[b] $E_{330}^{1\%}$ indicates absorbance at 330 nm with 1% dilute solution.
[c] $E_{257}^{1\%}$ indicates absorbance at 257 nm with 1% dilute solution.
[d] Result shown as overall for all three types of extractions.

preferred technology to extract compounds of saffron [25, 43, 44]. According to Sereshti et al. [44], the ultrasound method is capable of producing high yield but resulted in the destruction of the cell membrane.

The emerging of green technology as early as the 1990s has improved the yield of chemical compounds, not only from saffron, but from other edible plants. This technology is known as green as it requires less energy and chemicals. Since the technology for extraction is very much affected by the operating conditions, further discussion on the topic is presented with emphasis on the supercritical CO_2 extraction method. The extraction of phytochemicals by supercritical CO_2 from plants is possible and resulted in purer final yield compared to conventional method [45].

3. Supercritical CO_2 system for saffron extraction

Supercritical CO_2 has gained more interest recently due to its role in chemical extraction. It has proven safer and environmental friendlier compared to conventional extraction using chemicals as a solvent. The CO_2 employed behaves like a liquid solvent, but has diffusion properties like a gas, making it suitable for extraction of compounds from a complex matrix [46]. The extraction is achieved above the critical pressure of 7.4 MPa and critical temperature of 31°C [47]. Fig. 2 shows the general working principle of supercritical CO_2.

In general, most supercritical CO_2 systems consist of three main components: liquid CO_2 supplier, extraction vessel, and separator. The sample in the form of powder is supplied as a batch process into the extraction vessel. The cold CO_2 is supplied at a controlled flow rate to the extraction vessel until it reaches the desired pressure and temperature. The condition in which the CO_2 reaches supercritical causes reaction within the extraction vessel and releases the compound within the sample cell. It is then pumped to the low-pressure separator causing separation of the compounds due to differences in the compound volatility. Nowadays, supercritical CO_2 extractors are widely available commercially. However, each compound of saffron requires specific temperature, pressure, and extraction time for optimum yield. This is due to the location of the compound reside within the sample. The compounds that reside at the surface of the sample required less effort to release compared to the compound that lies deep within the cell.

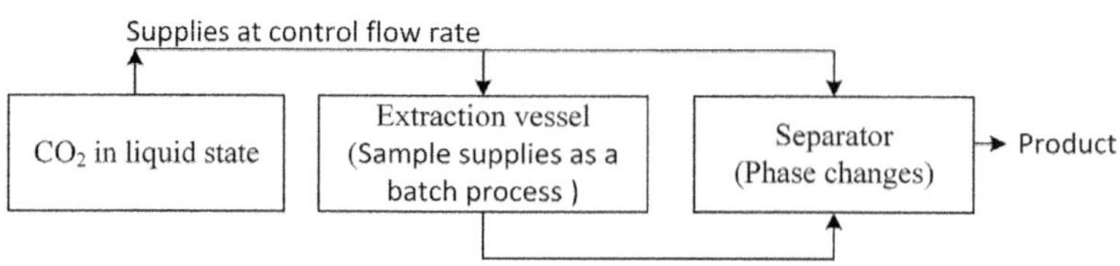

Fig. 2 General system of a supercritical CO_2 extractor.

3.1 Effect of sample condition on extraction performance

The particle size of a sample is reported to affect the extraction using supercritical CO_2 [48]. There is not much study conducted on the effect of particle size and moisture of saffron as feed to supercritical extraction so far. However, there are many studies that have reported on a similar case for other types of samples such as soybean, turmeric, papaya, and so on. Snyder et al. [49] found that less than 0.01-in. sample resulted in a higher yield compared to unground sample, in which they have tested soybeans, peanuts, and cottonseed. A recent study by Jahongir et al. [50] indicates that the smaller the particle size of a sample, the higher the yield of oil that can be extracted from the sample. There are also many studies that confirm this. The main reason for the high yield is best explained in terms of surface area contact between the sample and solvents: smaller particles lead to better surface contact compared to larger particles. This is probably the reason for saffron extraction using supercritical CO_2 mostly being conducted in the form of powder to increase the mass transfer of CO_2, thus increasing the contact time with the cell. Besides particle size, the moisture content of a sample is reported to affect the extraction yield.

Water could affect the extraction as moisture in a sample, or used as an entrainer for supercritical CO_2 extraction. It is reported that water as an entrainer can elevate the yield of a certain compound in saffron. Nerome et al. [28] show that α-crocin is better extracted with water compared to methanol, where approximately 21.92 (area comp./area IS) was obtained at 80°C and 30 MPa. However, other compounds do not show better results when applying similar entrainer and conditions. The moisture in a saffron sample for extraction using supercritical CO_2 is not as much of a concern as the temperature and pressure effect on the extraction yield. This is probably the reason for less study on the moist effect of the saffron sample when using supercritical extraction. However, the effect of sample moisture in another sample that is not saffron, such as soybean and so on, has been widely studied since the 1980s. According to Synder et al. [49] less than 12% moisture does not have an apparent effect on the yield of oil from soybean. A similar finding was found by Nagy and Simandi [51] for moisture content of 7% and 18% of paprika seeds. Another finding by Jinap et al. [52] shows that 9.79% of moisture in the sample of cocoa butter resulted in a higher yield compared to 1.95% and 17.64% of moisture. These are general reports on the effect of moisture in a sample to the efficiency of supercritical CO_2. Thus, it can be concluded that the effect on extraction of a certain type of compound in a sample affected by the presence of water varied for each compound. The moisture content in a sample could interfere with the final yield calculation of a sample if included as a raw weight. Thus, a mean value of moisture in a sample should be known for correct calculation of the final yield.

3.2 Acceptable range of extraction conditions

The most important variable for saffron extraction using supercritical CO_2 is the temperature and pressure inside the extraction vessel. It is basically used to control the contact time between CO_2 at supercritical conditions and the sample. Table 2 shows a comparison of operating conditions of supercritical CO_2 and the product yield of saffron compounds from various authors. As stated, only a few works were conducted so far in saffron extraction using supercritical CO_2 indicating that more study needs to be done in the field. As shown in Table 2, each author claims various optimum conditions for extraction of the same compound. For example, Lazano et al. [43] claim that 100°C operated at 20 MPa for 30 min are the best conditions to extract safranal. They state that extraction time exceeding 30 min is not practical as it does not increase the extraction yield and will result in high CO_2 consumption. Meanwhile, Nerome et al. [28] claim that the optimum conditions for extraction of the same compounds are at 80°C, 30 MPa for 240 min with the addition of methanol as entrainer. This resulted in a slightly higher yield compared to extraction without entrainer as described by Lazano et al. [43]. Other entrainers can be used for supercritical CO_2 including water, ethanol, diethyl ether, and so on. However, the use of an entrainer could contaminate the product yield and this is another challenge for using entrainer for supercritical CO_2 extraction. The list in Table 2 shows that the optimum conditions for saffron extraction using supercritical CO_2 was in the range of 44.9–120°C, 16.4–34.9 MPa, and 30–240 min for temperature, pressure, and extraction time, respectively.

Table 2 Practical ranges of supercritical CO_2 extraction of saffron

T (°C)[a]	*P* (MPa)[b]	QCO_2 (mL/min)[c]	*t* (min)[d]	Yield (%)	Reference
100	20	0.7	30	Safranal: 0.045–0.149	[43]
120	22.1	1	45[e]	Safranal: 0.05–1.5[f]	[53]
44.9	34.9	168	150.2	Volatile: 1.09	[42]
80	30	3	240	Picrocrocin: 2.419	[28]
80	30	3	240	Crocins: 34.94	[28]
80	30	3	240	Safranal: 0.09	[28]
92	21.3	0.9	122	Safranal: 33.05[g]	[54]
44	19.3	1	110	Crocin: 90.03[g]	[54]
62	16.4	Up to 100	47	Phenolic: 1.423[h]	[34]
62	16.4	Up to 100	47	flavonoid: 1.80[h]	[34]
50	30	N.d	100	Oil recipe: 4.8	[55]

N.d: no data.
[a]Extraction temperature.
[b]Extraction pressure.
[c]CO_2 flow rate.
[d]Extraction time.
[e]Total of 30 min statistic and 15 min dynamic.
[f]Unit in mg/mL of sample.
[g]Result as recovery percentage by Goleroudbary and Ghoreishi [54].
[h]Reported by Ahmadian-Kouchaksaraie and Niazmand [34].

The optimum range of the supercritical CO_2 operation was in agreement with extraction conditions for others edible plants. For example, a recent study conducted by Abdullah et al. [56] found that optimum conditions for extraction of oil from red pitaya seed were at 47°C and 32.75 MPa. This resulted in 6.88% yield. Other work conducted by Gustinelli et al. [57] found that the optimum conditions for oil extraction from bilberry were at 20 MPa and 60°C. This shows that pressure and temperature of extraction have a significant effect on the product yield. The key to lower energy consumption lies in the ability to control these variables. Nowadays, the optimization of the variables gets easier with the application of Response Surface Methodology (RSM).

3.3 Optimization using RSM

In general, RSM is a method to optimize process conditions. It is a technique to interpret experimental data and convert it into a polynomial equation. It is the most useful technique when dealing with the effect of various factors during operation, and makes experiment design much easier. Theory and principle application of RSM is described in Bezzera et al. [58]. The application of RSM in saffron extraction by supercritical CO_2 shows that a second-order polynomial model is able to provide the optimum conditions for the extraction of oil yield of saffron in Shao et al. [42] They have developed a model to represent four major variables, namely temperature (T), pressure (P), extraction time (t), and flow rate (f), supported with 7.22 F-test value for $P=0.0002$, which indicates that the model was significant. The model is represented in Eq. (1). In 2016, Goleroudbary and Ghoreishi [54] applied a similar approach to develop two models to estimate optimum conditions for recovery of safranal (Eq. (2)) and crocin (Eq. (3)) during supercritical extraction. The RSM technique also proved useful for extraction optimization when using other extraction technology such as ultrasound extraction [59] and the conventional method [60]. It is shown that the application of RSM would benefit the extraction process. From the model, a high probability estimation of the pressure and temperature conditions can be achieved; thus unnecessary trial and error can be avoided and this could save energy consumption which leads to high cost.

$$\begin{aligned} Y = {} & 0.9668 + 0.0078T + 0.0584P + 0.0187t + 0.0267f \\ & - 0.0168TP - 0.0064Tt + 0.0152Tf + 0.0096Pt \\ & - 0.0213Pf - 0.0254tf - 0.031T^2 - 0.0346P^2 + 0.0048t^2 - 0.0073f^2 \end{aligned} \quad (1)$$

$$\begin{aligned} R_s = {} & 84.936 + 7.138T + 2.065P + 6.492dt - 6.299T^2 - 6.9548p^2 \\ & - 8.252f^2 - 4.383dt^2 + 3.132T*dt - 2.285p*f \end{aligned} \quad (2)$$

$$\begin{aligned} R_c = {} & 31.434 - 1.071T - 0.595P + 1.534f + 2.476dt - 2.523T^2 \\ & - 3.198p^2 - 1.852f^2 - 1.586dt^2 + 0.679T*P - 0.799T*f - 0.871f*dt \end{aligned} \quad (3)$$

The application of a model to estimate optimum conditions for supercritical extraction is proven could reduce the time and energy consumption of extraction. As mentioned in a few studies, RSM could reduce the amount of laboring and the number of the experimental trials [61, 62].

3.4 Purification of saffron extract

The use of supercritical CO_2 extraction has resulted in high selectivity compared to other extraction methods [47, 63]. Despite the fact that supercritical resulted in high compound selectivity, it is notable that none of the final results yield a single compound, as summarized in Table 2. Thus, a further process is required to separate the extraction yield. A patent by Ziyu [64] proposed a step-by-step extraction method of essential oil from saffron by performing steam distillation followed by supercritical CO_2 extraction. The method improved yield up to 20%. Furthermore, the purification of insoluble and soluble components in a saffron solution can be done through a few methods. An insoluble component often requires an easier process compared to soluble. In general, the purification of a compound in the final extract of sa0ffron is quite challenging due to high solubility. Fig. 3 shows general purification steps to purify compounds of saffron.

The purification of saffron extracts can be grouped into two, based on the solubility of components. Usually, the conventional method of extraction using methanol involves other analytical methods to discard the physical component of saffron residue from the solution. The analytical method includes filtration and thin layer chromatography (TLC). The TLC was mentioned as a useful purification tool for picrocrocin, HTCC, and crocin [65]. The extract solution of saffron using supercritical CO_2 usually contains only soluble components as a liquid-liquid mixture. The compounds in saffron extract have specific boiling, melting, and freezing points, which differ from each other, thus making the compound separable under set up temperature conditions.

There are many possible ways of saffron extracts purification. This includes crystallization, column chromatography, and fractional distillation. Crystallization occurs in two major steps: nucleation and crystal growth. This usually occurs by tuning at a very low temperature and supersaturated solution. According to Hadizadeh et al. [66], crocin has high purity under the crystallization at $-5°C$. Safranal, on the other hand, was reported to be isolated by applying cold trap at $-78°C$ [67].

The use of flash or column chromatography for saffron extract isolation and purification was reported few times. Purification based on chromatography depends on the principle where molecules in the mixture are applied onto the surface and fluid stationary phase. There are three components of chromatography: stationary phase, mobile phase, and molecules separated from the earlier phases [68]. The use of flash or column chromatography for purification of antioxidants crocin and kaempferol of saffron petal was achieved up to 98% [69]. According to Weber et al. [70], the efficiency of flash or column

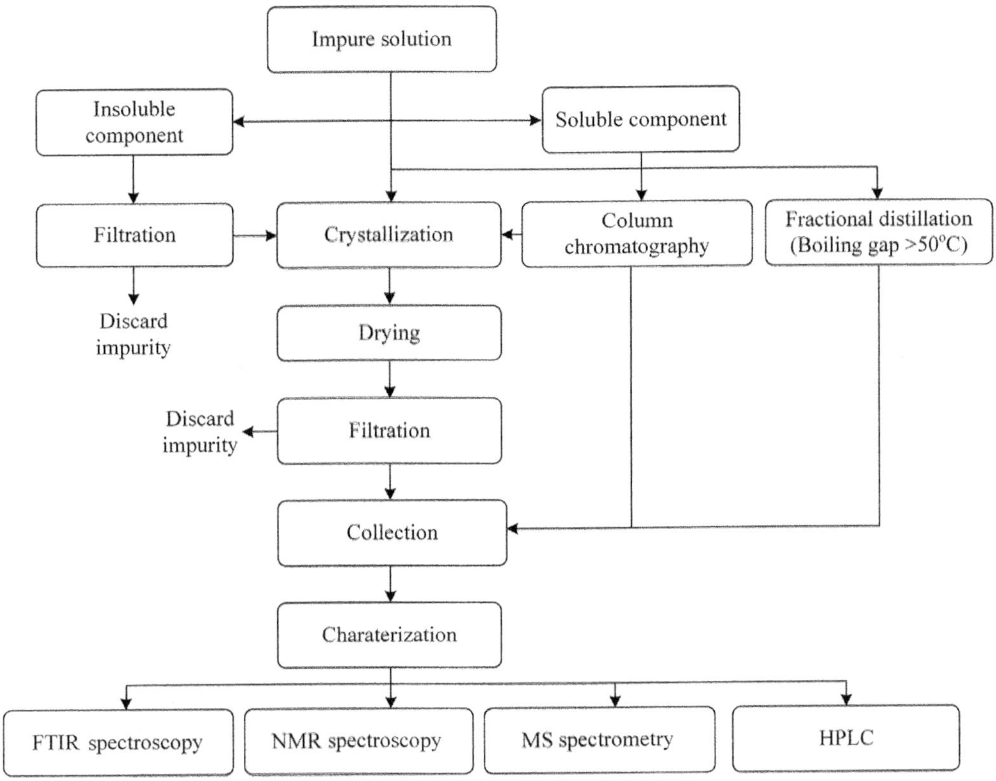

Fig. 3 Purification of saffron extract.

chromatography as separator depends on the suitability of cartridge, whereas *Curcuma xanthorrhiza*, *Piper nigrum*, and *Salvia milthiorrhiza* were proven have suitable properties for medicinal plants separation.

Fractional distillation is useful by controlling the boiling temperature of each compound. However, at least a 50°C gap in boiling temperature would make it useable, otherwise the compound's purity would not be achieved. The main compounds of saffron safranal, crocin and picrocrocin have 70, 1169, and 520°C boiling points, respectively. Based on the boiling point, separation of these compounds is possible. Although there is not much data of saffron separation using fractional distillation, Farah et al. [71] have successful separated essential oil compounds of *Myrtus communis* L. However, further purification of the separated compound is still required to get the purer component.

4. Economic assessment on supercritical CO_2 extraction of saffron

The cost of production of saffron, which includes farming and labor for a 5-year cycle is estimated by Manzo et al. [72]. A gram of dried saffron stigma was estimated to be 27.45

US$ in 2015. Recently, Iran, the biggest producer of saffron, has increased the export value of saffron by 46% in the first 8 months of 2018, which resulted in the production of US$240 million per 152.6 tons of saffron (~U$1.57/g) [73]. However, this price can be considered as the lowest among the market as it is exported in bulk. Thus, further processing includes the extraction of the raw material expected to increase the value of the compound extract. The cost of electricity has a major effect on the cost of manufacture of supercritical CO_2 extraction techniques. It is estimated that the CO_2 cost is US$0.3/kg of a sample [74]. In addition, approximately 4% total flow for CO_2 loss. To date, there is no official report on the cost of supercritical CO_2 for saffron extraction, however, it is estimated that €88.89 (US$ 101.76) would be required to produce 1 kg of wax from maize stover [75] whereas the price of maize in 2015 was estimated to be 170 US$ per metric ton (~US$0.17/g); about 11% lower compared to saffron produced by Iran. Thus, the extraction of saffron is estimated to be higher than US$101.76/ton. The cost of supercritical fluid extraction is considered higher compared to solvent extraction, steam distillation, ultrasonic extraction, membrane processes, and solid phase extraction [76].

Many studies have reported that the setup of supercritical CO_2 itself is more expensive compared to the setup for conventional extraction. The handmade supercritical CO_2 by Horvat et al. [77] was €12,500 (~US$14332) whereas the commercial supercritical is more expensive. A supercritical CO_2 operation would require electricity to generate heat and for the mechanical function such as pumps. It was estimated that US0.0703/Mcal is required for electric power to operate supercritical CO_2 [78]. This covers approximately 30% to 70% of the total production expenses. In order to estimate the cost of extraction using supercritical CO_2, the formula for the labor cost of operating supercritical, utility costs including electricity, and raw materials can be found in Shariaty-Niassar et al. [74].

5. Conclusion

Supercritical CO_2 extraction of saffron compounds is a technique that has gained interest recently in saffron extraction. However, not many studies in the field have been reported, which indicates that more study is required especially on the effect of entrainer to aid the supercritical extraction. Most studies have reported that the use of supercritical is expensive, although it was proven capable of producing high purity of the chemical compound. The major constituents of saffron, which are safranal, crocin, and picrocrocin, can be further purified through various methods include crystallization, column chromatography, and fractional distillation. It was found that supercritical extraction would require less chemical as a solvent and this would result in the production of a safer product. The safety of the final yield is important as saffron constituents are mostly incorporated into the medicinal field.

Acknowledgment

This work was financially supported by UMS grant no. SGI0082-2018.

References

[1] R.J. Molyneux, S.T. Lee, D.R. Gardner, K.E. Panter, L.F. James, Phytochemicals: the good, the bad and the ugly? Photochemistry 68 (2007) 2973–2985.
[2] B. Dov, M. Negbi, Uses of saffron, Econ. Bot. 37 (1983) 228–236.
[3] M.S. Zeinab, S.Z. Bathaie, Historical uses of saffron: identifying potential new avenues for modern research, Avicenna J. Phytomed. 1 (2011) 57–66.
[4] I. Hasan, A.H. Ansari, A.M.K. Sherwani, M. Zulkifle, The incredible health benefits of saffron: a review, Int. J. Pharm. Res. 4 (2011) 2156–2158.
[5] M. Rameshrad, B.M. Razavi, H. Hosseinzadeh, Saffron and its derivatives, crocin, crocetin, and safranal: a patent review, Expert. Opin. Ther. Pat. 28 (2018) 147–165.
[6] J.P. Melnyk, S. Wang, M.F. Marcone, Chemical and biological properties of the world's most expensive spice: saffron, Food Res. Int. 43 (2010) 1981–1989.
[7] M. Ghorbani, The economics of saffron in Iran, in: International Symposium on Saffron Biology and Technology, vol. 739, 2006, , pp. 321–331.
[8] K. Heidarbeigi, S.S. Mohtasebi, A. Foroughirad, M. Ghasemi-Varnamkhasti, S. Rafiee, K. Rezaei, Detection of adulteration in saffron samples using electronic nose, Int. J. Food Prop. 18 (2015) 1391–1401.
[9] V. Khilare, T. Anita, P. Bharathi, U. Balasaheb, K. Ganesh, N. Dinesh, A. Nadeem, K. Chandraprakash, K. Gulab, Multiple tests on saffron find new adulterant materials and reveal that Ist grade saffron is rare in the market, Food Chem. 272 (2019) 635–642.
[10] M.G. Caiola, A. Canini, Looking for saffron's (Crocus sativus L.) parents, Funct, Plant. Sci. Biotechnol. 4 (2010) 1–14.
[11] H. Himeno, K. Sano, Synthesis of crocin, picrocrocin and safranal by saffron stigma-like structures proliferated in vitro, Agric. Biol. Chem. 51 (1987) 2395–2400.
[12] O. Plessner, M. Ziv, M. Negbi, In vitro corm production in the saffron crocus (Crocus sativus L.), Plant Cell Tissue Organ Cult. 20 (1990) 89–94.
[13] J.A. Parray, A.N. Kamili, R. Hamid, A.M. Husaini, In vitro cormlet production of saffron (Crocus sativus L. Kashmirianus) and their flowering response under greenhouse, GM Crops Food 3 (2012) 289–295.
[14] K. Saito, A. Fukushima, A new affinity-trapping method for isolation of quinoidal chalcone pigments from aqueous extracts of dyer's saffron flowers, Biochem. Physiol. Pflanz. 184 (1989) 145–153.
[15] W. Peter, M. Straubinger, Saffron—renewed interest in an ancient spice, Food Rev. Int. 16 (2000) 39–59.
[16] A.R. Gohari, S. Saeidnia, M.K. Mahmoodabadi, An overview on saffron, phytochemicals, and medicinal properties, Pharmacogn. Rev. 7 (2013) 61–66.
[17] T. Shahi, E. Assadpour, S.M. Jafari, Main chemical compounds and pharmacological activities of stigmas and tepals of 'red gold'; saffron, Trends Food Sci. Technol. 58 (2016) 69–78.
[18] H. Caballero-Ortega, L. Riveron-Negrete, Chemical composition of saffron (Crocus sativus L.) from four countries, Acta Hort (ISHS) (2004).
[19] M.R. Khazdair, M.H. Boskabady, M. Hosseini, R. Rezaee, A.M. Tsatsakis, The effects of Crocus sativus (saffron) and its constituents on nervous system: a review, Avicenna J. Phytomed. 5 (2015) 376.
[20] M. Agha-Hosseini, L. Kashani, A. Aleyaseen, A. Ghoreishi, H. Rahmanpour, A.R. Zarrinara, S. Akhondzadeh, Crocus sativus L. (Saffron) in the treatment of premenstrual syndrome: a double-blind, randomized and placebo-controlled trial, BJOG. Int. J. Obstet. Gy. 115 (2008) 515–519.
[21] M. Mashmoul, A. Azlan, H. Khaza'ai, B.N.M. Yusof, S.M. Noor, Saffron: a natural potent antioxidant as a promising anti-obesity drug, Antioxidants 2 (2013) 293–308.

[22] S.H. Alavizadeh, H. Hosseinzadeh, Bioactivity assessment and toxicity of crocin: a comprehensive review, Food Chem. Toxicol. 64 (2014) 65–80.

[23] N.S. Zarghami, D.E. Heinz, Monoterpene aldehydes and isophorone-related compounds of saffron, Phytochemistry 10 (1971) 2755–2761.

[24] K. Saito, K.I. Miyakawa, Sono-oscillatory mobilization of bound precarthamine from flower florets of dyer's saffron, Food Chem. 54 (1995) 269–271.

[25] R. Rocchi, M. Mascini, M. Sergi, D. Compagnone, D. Mastrocola, P. Pittia, Crocins pattern in saffron detected by UHPLC-MS/MS as marker of quality, process, and traceability, Food Chem. 264 (2018) 241–249.

[26] A. Kyriakoudi, A. Chrysanthou, F. Mantzouridou, M.Z. Tsimidou, Revisiting extraction of bioactive apocarotenoids from Crocus sativus L. dry stigmas (saffron), Anal. Chim. Acta 755 (2012) 77–85.

[27] S. Gazeran, F. Tajalli, A.M. Sani, Effect of ultrasonic extraction on qualitative parameters of saffron edible extract, J. Essent. Oil Bear. Plants 19 (2016) 1286–1291.

[28] H. Nerome, M. Ito, S. Machmudah, H. Kanda, M. Goto, Extraction of phytochemicals from saffron by supercritical carbon dioxide with water and methanol as entrainer, J. Supercrit. Fluids 107 (2016) 377–383.

[29] K.J. Shinwari, P.S. Rao, Thermal-assisted high hydrostatic pressure extraction of nutraceuticals from saffron (Crocus sativus): process optimization and cytotoxicity evaluation against cancer cells, Innov. Food Sci. Emerg. Technol. 48 (2018) 296–303.

[30] M. Vahedi, M. Kabiri, S.A. Salami, H. Rezadoost, M. Mirzaie, M.R. Kanani, Quantitative HPLC-based metabolomics of some Iranian saffron (Crocus sativus L.) accessions, Ind. Crop. Prod. 118 (2018) 26–29.

[31] L. Maggi, A.M. Sánchez, M. Carmona, C.D. Kanakis, E. Anastasaki, P.A. Tarantilis, M.G. Polissiou, G.L. Alonso, Rapid determination of safranal in the quality control of saffron spice (Crocus sativus L.), Food Chem. 127 (2011) 369–373.

[32] A.M. Sanchez, M. Carmona, C.P. del Campo, G.L. Alonso, Solid-phase extraction for picrocrocin determination in the quality control of saffron spice (Crocus sativus L.), Food Chem. 116 (2009) 792–798.

[33] S.A. Baba, A.H. Malik, Z.A. Wani, T. Mohiuddin, Z. Shah, N. Abbas, N. Ashraf, Phytochemical analysis and antioxidant activity of different tissue types of Crocus sativus and oxidative stress alleviating potential of saffron extract in plants, bacteria, and yeast, S. Afr. J. Bot. 99 (2015) 80–87.

[34] Z. Ahmadian-Kouchaksaraie, R. Niazmand, Supercritical carbon dioxide extraction of antioxidants from Crocus sativus petals of saffron industry residues: optimization using response surface methodology, J. Supercrit. Fluids 121 (2017) 19–31.

[35] Z. Ahmadian-Kouchaksaraie, R. Niazmand, M.N. Najafi, Optimization of the subcritical water extraction of phenolic antioxidants from Crocus sativus petals of saffron industry residues: Box-Behnken design and principal component analysis, Innov. Food Sci. Emerg. Technol. 36 (2016) 234–244.

[36] E. Kakouri, D. Daferera, S. Paramithiotis, K. Astraka, E.H. Drosinos, M.G. Polissiou, Crocus sativus L. tepals: the natural source of antioxidant and antimicrobial factors, J. Appl. Res. Med. Aromat. Plants. 4 (2017) 66–74.

[37] L. Ferrara, D. Naviglio, M. Gallo, Extraction of bioactive compounds of saffron (Crocus sativus L.) by ultrasound-assisted extraction (UAE) and by rapid solid-liquid dynamic extraction (RSLDE), Eur. Sci. J. 10 (2014).

[38] S.S. Mehraj, A.N. Kamili, R. Nazir, E. Haq, H.M. Balkhi, Comparative evaluation of extraction methods for total proteins from Crocus sativus L. (Saffron), Saudi J. Biol. Sci. 25 (2018) 1603–1608.

[39] A. Amanpour, A.S. Sonmezdag, H. Kelebek, S. Selli, GC–MS–olfactometric characterization of the most aroma-active components in a representative aromatic extract from Iranian saffron (Crocus sativus L.), Food Chem. 182 (2015) 251–256.

[40] L. Lotfi, A. Kalbasi-Ashtari, M. Hamedi, F. Ghorbani, Effects of enzymatic extraction on anthocyanins yield of saffron tepals (Crocus sativus) along with its color properties and structural stability, J. Food Drug Anal. 23 (2015) 210–218.

[41] A.P. Sobolev, S. Carradori, D. Capitani, S. Vista, A. Trella, F. Marini, L. Mannina, Saffron samples of different origin: an NMR study of microwave-assisted extracts, Foods 3 (2014) 403–419.
[42] Q. Shao, Y. Huang, A. Zhou, H. Guo, A. Zhang, Y. Wang, Application of response surface methodology to optimize supercritical carbon dioxide extraction of volatile compounds from Crocus sativus, J. Sci. Food Agric. 94 (2014) 1430–1436.
[43] P. Lozano, D. Delgado, D. Gomez, M. Rubio, J.L. Iborra, A non-destructive method to determine the safranal content of saffron (Crocus sativus L.) by supercritical carbon dioxide extraction combined with high-performance liquid chromatography and gas chromatography, J. Biochem. Bioph. Meth. 43 (2000) 367–378.
[44] H. Sereshti, R. Heidari, S. Samadi, Determination of volatile components of saffron by optimized ultrasound-assisted extraction in tandem with dispersive liquid-liquid microextraction followed by gas chromatography-mass spectrometry, Food Chem. 143 (2014) 499–505.
[45] M. Oman, M. Škerget, Z. Knez, Application of supercritical fluid extraction for separation of nutraceuticals and other phytochemicals from plant material, Maced. J. Chem. Chem. Eng. 32 (2013) 183–226.
[46] S. Junco, T. Casimiro, N. Ribeiro, M.N. Da Ponte, H.M.C. Marques, Optimisation of supercritical carbon dioxide systems for complexation of naproxen: beta-cyclodextrin, J. Incl. Phenom. Macrocycl. Chem. 44 (2002) 69–73.
[47] G.N. Sapkale, S.M. Patil, U.S. Surwase, P.K. Bhatbhage, Supercritical fluid extraction, Int. J. Chem. Sci. 8 (2010) 729–743.
[48] J.M. del Valle, E.L. Uquiche, Particle size effects on supercritical CO_2 extraction of oil-containing seeds, J. Am. Oil Chem. Soc. 79 (2002) 1261–1266.
[49] J.M. Snyder, J.P. Friedrich, D.D. Christianson, Effect of moisture and particle size on the extractability of oils from seeds with supercritical CO_2, J. Am. Oil Chem. Soc. 61 (1984) 1851–1856.
[50] H. Jahongir, Z. Miansong, I. Amankeldi, Z. Yu, L. Changheng, The influence of particle size on supercritical extraction of dog rose (Rosa canina) seed oil, King Saud Univ. Eng. Sci. (2018) (In press).
[51] B. Nagy, B. Simandi, Effects of particle size distribution, moisture content, and initial oil content on the supercritical fluid extraction of paprika, J. Supercrit. Fluids 46 (2008) 293–298.
[52] S. Jinap, A.E. Kusnadi, J.H. Md, R.A. Akanda, S. Rahman, M.Z.I.S. Ferdosh, Effects of moisture and pH on supercritical fluid extraction of cocoa butter, Food Bioprocess Technol. 6 (2013) 2455–2465.
[53] M. Zougagh, A. Ríos, M. Valcarcel, Determination of total safranal by in situ acid hydrolysis in supercritical fluid media: application to the quality control of commercial saffron, Anal. Chim. Acta 578 (2006) 117–121.
[54] M.G. Ghorbandoust, S.M. Ghoreishi, Response surface optimization of Safranal and Crocin extraction from Crocus sativus L. via supercritical fluid technology, J. Supercrit. Fluids 108 (2016) 136–144.
[55] J. Li, E. Guan, L. Chen, X. Zhang, L. Yin, L. Dong, Q. Pan, X. Fu, L. Zhang, Optimization for extraction of an oil recipe consisting of white pepper, long pepper, cinnamon, saffron, and myrrh by supercritical carbon dioxide and the protective effects against oxygen-glucose deprivation in PC12 cells, Rev. Bras. Farm. (2018).
[56] A. Abdullah, S.S.A. Gani, N.F.M. Mokhtar, T.Y.Y. Hin, Z.A. Haiyee, S. Mustafa, Supercritical carbon dioxide extraction of red pitaya (Hylocereus polyrhizus) seeds: response surface optimization, fatty acid composition and physicochemical properties, Malays. Appl. Biol. 47 (2018) 39–46.
[57] G. Gustinelli, L. Eliasson, C. Svelander, M. Alminger, L. Ahrne, Supercritical CO_2 extraction of bilberry (Vaccinium myrtillus L.) seed oil: fatty acid composition and antioxidant activity, J. Supercrit. Fluids 135 (2018) 91–97.
[58] M.A. Bezerra, R.E. Santelli, E.P. Oliveira, L.S. Villar, L.A. Escaleira, Response surface methodology (RSM) as a tool for optimization in analytical chemistry, Talanta 76 (2008) 965–977.
[59] K.N. Khazaei, S.M. Jafari, M. Ghorbani, A. Hemmati Kakhki, M. Sarfarazi, Optimization of anthocyanin extraction from saffron petals with response surface methodology, Food Anal. Methods 9 (2016) 1993–2001.
[60] M. Sarfarazi, S.M. Jafari, G. Rajabzadeh, Extraction optimization of saffron nutraceuticals through response surface methodology, Food Anal. Methods 8 (2015) 2273–2285.

[61] X. Tang, L. Yan, J. Gao, H. Ge, H. Yang, N. Lin, Optimization of extraction process and investigation of antioxidant effect of polysaccharides from the root of Limonium sinense Kuntze, Pharmacogn. Mag. 7 (2011) 186.

[62] K.A.M. Said, M. Afizal, M. Amin, Overview on the response surface methodology (RSM) in extraction processes, J. Appl. Sci Pro. Eng. 2 (2016).

[63] A.P. Sánchez-Camargo, J.A. Mendiola, E. Ibáñez, M. Herrero, Supercritical fluid extraction, in: Reference Module in Chemistry, Molecular Sciences, and Chemical Engineering, Elsevier, 2014.

[64] Z. Ziyu, Method for extracting saffron crocus essential oil, Patent No. CN103805340A. 2014, https://patents.google.com/patent/CN103805340A/en (Accessed on 28 December 2018).

[65] J.O.L. Iborra, M.R. Castellar, M.A. Canovas, A.R. Manjón, TLC preparative purification of picrocrocin, HTCC and crocin from saffron, J. Food Sci. 57 (1992) 714–716.

[66] F. Hadizadeh, S.A. Mohajeri, M. Seifi, Extraction, and purification of crocin from saffron stigmas employing a simple and efficient crystallization method, Pak. J. Biol. Sci. 13 (2010) 691–698.

[67] A. Bolhasani, S.Z. Bathaie, I. Yavari, A.A. Moosavi-Movahedi, M. Ghaffari, Separation and purification of some components of Iranian saffron, Asian J. Chem. 17 (2005) 725–729.

[68] O. Coskun, Separation techniques: Chromatography, North. Clin. Istanb. 3 (2016) 156–160.

[69] K. Zeka, C.R. Ketan, A.C. Maria, D. Stagos, F. Veglio, R.R.J. Arroo, Petals of Crocus sativus L. as a potential source of the antioxidants crocin and kaempferol, Fitoterapia 107 (2015) 128–134.

[70] P. Weber, M. Hamburger, N. Schafroth, O. Potterat, Flash chromatography on cartridges for the separation of plant extracts: rules for the selection of chromatographic conditions and comparison with medium pressure liquid chromatography, Fitoterapia 82 (2011) 155–161.

[71] A. Farah, A. Afifi, M. Fechtal, A. Chhen, B. Satrani, M. Talbi, A. Chaouch, Fractional distillation effect on the chemical composition of Moroccan myrtle (Myrtus communis L.) essential oils, Flavour Fragr. J. 21 (2006) 351–354.

[72] A. Manzo, S. Panseri, D. Bertoni, A. Giorgi, Economic and qualitative traits of Italian Alps saffron, J. Mt. Sci. 12 (2015) 1542–1550.

[73] MEHR Newsagency, Saffron Exports Value Up by 46% in First 8 Months, https://en.mehrnews.com/news/140525/Saffron-exports-value-up-by-46-in-first-8-months. Accessed 28 December 2018.

[74] M. Shariaty-Niassar, B. Aminzadeh, P. Azadi, S. Soltanali, Economic evaluation of herb extraction using supercritical fluid, Chem. Ind. Chem. Eng. Q. 15 (2009) 143–148.

[75] T.M. Attard, C.R. McElroy, A.J. Hunt, Economic assessment of supercritical co_2 extraction of waxes as part of a maize stover biorefinery, Int. J. Mol. Sci. 16 (2015) 17546–17564.

[76] S. Heydari, G.H. Haghayegh, Extraction and microextraction techniques for the determination of compounds from saffron, Can. Chem. Trans. 2 (2014) 221–247.

[77] G. Horvat, K. Aladic, S. Jokic, Supercritical CO_2 extraction pilot plant design-towards IoT integration, Teh. Vjesn. 24 (2017) 925–934.

[78] P.T.V. Rosa, M.A.A. Meireles, Rapid estimation of the manufacturing cost of extracts obtained by supercritical fluid extraction, J. Food Eng. 67 (2005) 235–240.

CHAPTER 8

Extraction of bioactive compounds

F.W.F. Bezerra[a], M.S. de Oliveira[a], P.N. Bezerra[a], V.M.B. Cunha[a], M.P. Silva[a], W.A. da Costa[b], R.H.H. Pinto[a], R.M. Cordeiro[b], J.N. da Cruz[c], A.M.J. Chaves Neto[b,c], R.N. Carvalho Junior[a,b]

[a]Technology Institute, Program of Post-Graduation in Food Science, Technology and Engineering, Federal University of Pará, Belém, Brazil

[b]Technology Institute, Program of Post-Graduation in Amazon Natural Resources Engineering, Federal University of Pará, Belém, Brazil

[c]Laboratory of Preparation and Computation of Nanomaterials, Federal University of Pará, Belém, Brazil

Contents

1. Introduction

Secondary metabolites, such as antioxidants, vitamins, and minerals, are bioactive compounds produced by different plants species, and are also used as medicines and flavorings. Their antioxidant properties and possible uses in food industry have been investigated in the recent years [1].

Among the various methods of obtaining natural substances, such as maceration, solvent extraction, ultrasound-assisted extraction, and microwave-assisted extraction, the supercritical fluid extraction (SFE) has become an important alternative for food, pharmaceutical, and cosmetic industries. The preparation of solvent-free products is one of the most important advantages of this process. Carbon dioxide (CO_2) is the most used solvent in SFE due to its inertness, nontoxicity, high solubility, and perfect conditions for thermosensitive compounds extraction [1, 2].

Green Sustainable Process for Chemical and Environmental Engineering and Science
https://doi.org/10.1016/B978-0-12-817388-6.00008-8

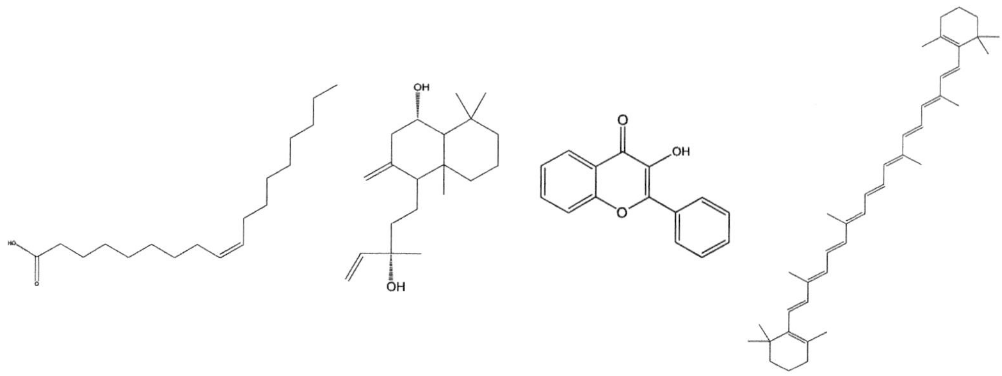

Fig. 1 Representation of bioactive compounds of the class of fatty acids (oleic acid), essential oils (larixol), phenolic compounds (flavonol), and carotenoids (β-carotene).

A fluid is said to be supercritical when it reaches a certain point above its critical temperature and pressure (T_c and P_c). Upon reaching the supercritical state, some properties of the fluid such as viscosity, density, and diffusivity present changes with intermediate values to the gaseous and liquid states [3, 4]. Such characteristics are important to extract high added-value compounds with biological activities, such as fatty acids, essential oils, phenolic compounds, and carotenoids, etc. (Fig. 1).

2. CO_2 as supercritical fluid

Over the years, several extraction methods have been developed with the purpose of obtaining bioactive compounds from natural raw materials. However, such methods (maceration, hydrodistillation, solvent extraction, etc.) are associated with some disadvantages such as the need for high extraction time, use of toxic solvents which are harmful to health and the environment, and they need of high temperature, which can promote the degradation of thermosensitive compounds [5].

Some solvents used in industries bring impacts on economic, ecological, and health issues, in this way, the concept of "green solvents" involves the reduction of environmental impacts that are caused by organic solvents. Thus, some characteristics are important for a given solvent to be considered "green," such as being biodegradable, having low volatility, low potential of ozone layer degradation, being produced with renewable resources, and having a low impact on human health and in the environment [6–9].

The extraction with supercritical fluids (SFs), in particular with CO_2, provides products free of toxic waste, having a higher quality when compared to the products obtained by conventional methods, becoming a process of great interest for the food, pharmaceutical, and cosmetic industries [10, 11]. The CO_2 offers several additional advantages such as:

residual toxicity elimination of organic solvents, low critical temperature (31.1°C), and critical pressure (73.7 bar), low thermosensitive compounds deterioration, selectivity for desired compounds, and sensory properties preservation of the material in food and pharmaceutical applications [12, 13]. Fig. 2 shows the pressure *versus* temperature (*P*–*T*) CO_2 phase diagram, marking its different aggregation regions (solid, liquid, and gaseous), phase change lines (sublimation, melting, and boiling, or vapor pressure). It is also possible to observe the triple point that occurs in 5.2 bar pressure and −56.6°C temperature, at this point it starts the vapor pressure, which ends at the critical point at 73.7 bar and 31.1°C and from this point it starts the supercritical region.

CO_2 is considered to extract apolar and moderately polar compounds. However, when smaller volumes of more polar solvents (modifiers or cosolvents) such as water, ethyl acetate, ethanol, or methanol are added the polarity is modified and more polar compounds can be extracted due to the CO_2 increased solvency and the polar compounds solubility. According to the type of sample and the analyte-sample affinity, the cosolvents may influence the extraction of polar compounds in three ways: (i) the cosolvent may produce an analyte-cosolvent interaction, increasing the analytes solubility during extraction; (ii) the polar cosolvents can interact with the sample facilitating the analytes desorption; and (iii) the sample swelling that occurs in the cosolvent presence may favor SF penetration, increasing the polar compounds extraction [4, 5, 13, 14].

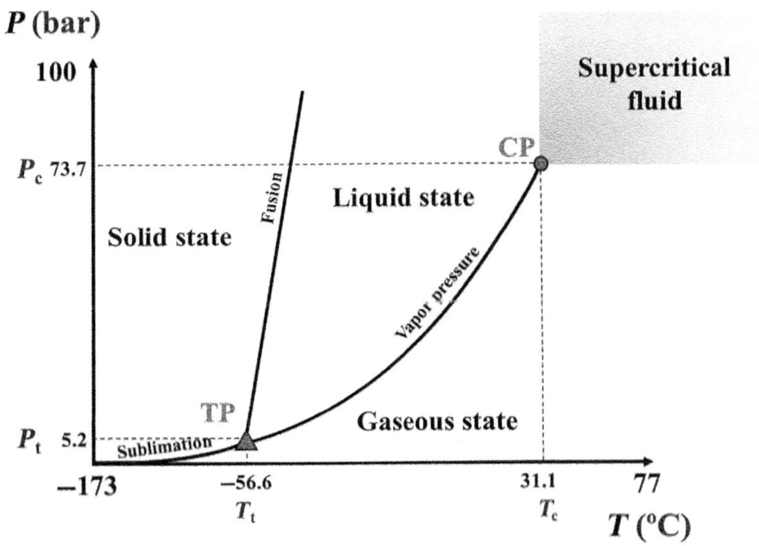

Fig. 2 *P-T* phase diagram for CO_2. *P* = pressure, P_c = critical pressure, *CP* = critical point, P_t = triple point pressure, *T* (°C) = temperature (°Celsius), T_P = triple point, T_t = triple point temperature, and T_c = critical temperature.

3. Supercritical carbon dioxide properties

3.1 Thermodynamics proprieties

3.1.1 Density

Density is defined as the inverse of the specific volume ($\rho = 1/V$) and can be obtained from thermodynamic tables or calculated by volumetric properties using an equation of state (EoS). When there is addition of cosolvents in the extraction process, the calculation can be performed by applying a cubic EoS such as Peng-Robinson or other EoS combined with mixing rules with appropriate interaction parameters [15, 16].

This thermodynamic property is extremely important in the SFE process. The supercritical carbon dioxide (SC-CO_2) density is highly dependent on temperature and pressure variation. The increase in pressure favors the increase in density at constant temperature, while the increase in temperature decreases the density at constant pressure, this behavior is due to the state of molecular approximation that occurs when the pressure increases, in this case the SC-CO_2 density is similar to a liquid, presenting high intermolecular interaction of its respective electron clouds (hydrogen bonds). The increase of density causes increase in its power solvation [15, 17, 18].

3.1.2 Solubility

Solubility is defined as the amount of a solute (mass fraction or mole fraction) that is dissolved by the supercritical solvent in thermodynamic equilibrium [19]. In this way, this equilibrium determines the mass transfer limit between the system phases, providing information about the phase compositions, the distribution of the individual components, and their quantities in equilibrium, as well as their variations with temperature and pressure, and the various components concentration. The parameters that influence solubility are temperature, pressure, and solute properties (molecular mass, polarity, and vapor pressure). Among these, the temperature deserves more attention due to its detrimental effect on solubility of SF [20, 21].

3.2 Transport properties

CO_2 is well established for use as a processing solvent in applications in the food and pharmaceutical industries and is a good alternative to replace organic solvents in conventional processes [22, 23]. Studies have reported that SC-CO_2 offers many advantages, for example in improving the selectivity of the compounds of interest present in a plant matrix in a binary system where mass and heat transfer occur between different phases equilibrium involved in separation processes [24]. Beyond temperature and pressure, other factors must be considered, such as the physicochemical properties of transport that depend on these two variables which are responsible for their changes. Although there are several studies on phase equilibrium thermodynamics, experimental data on transport properties (viscosity and diffusivity) are scarce in the literature [25].

In the supercritical state, CO_2 has diffusion coefficient close to that of a gas, which facilitates the penetration of the fluid in the solid matrix, being of great interest not only to study the structure and the thermophysical behavior of the fluid, but also to perform practical engineering applications and, depending on the process conditions, viscosity reduction is especially important as it determines not only the mass transfer but also the convective heat transfer [26, 27].

4. SC-CO_2 extraction

SFE is a unitary contact operation based on the equilibrium and the physicochemical properties of the SF, influenced by operational conditions as: temperature, pressure, solvent flow, material morphology, bed porosity, and particle size [20]. The SFE process can be performed through two stages of operation: extraction and separation. The extraction step involves the relation of the solvent solubility capacity by manipulating the thermodynamic properties (pressure and temperature) and/or modifying the chemical nature of the fluid upon addition of a cosolvent. Separation is achieved by depressurization, heating, or gradually cooling the extract, thereby providing a controlled fractionation. The separation can also be obtained by combining of the extraction process to purification/enrichment methods of extracts, for example the adsorption [28].

Fig. 3 shows an SC-CO_2 extraction system formed by a solvent feed system CO_2, a high-pressure pump, an extraction cell and a separation system. The extraction process is relatively simple, in which the preprocessed solid sample (dried and comminuted) is inserted into the extractor vessel which is heated. The system is subsequently pressurized to predetermined values by the action of a high-pressure pump, which allows the CO_2 to flow through the system and favor the solubilization of the compounds of interest present in the sample. Finally, the SC-CO_2 and extract obtained pass through an expansion valve and the CO_2 has its pressure reduced until the ambient pressure. The extract then precipitates in the collection flask and the CO_2 in the gas phase passes through a flow meter for the control and recording of its flow and subsequent release to the environment or recycle for a new process.

Pressure is the preferred variable for adjusting density values in SC-CO_2 extraction behavior, since it offers higher manipulation margins compared to other properties [15]. The temperature effect on the extraction can be observed through inflection points (intersection) between isotherms. The increase of the temperature causes a greater intermolecular distance, and consequently, the reduction of the CO_2 solubility power, by decreasing the density. The presence of inflection points will depend on the interaction between SF and the extract to be obtained [29]. From the date listed in Table 1, it is possible to observe the influence of temperature, pressure, flow rate, and density of the supercritical CO_2 in the bioactive compounds extraction process from vegetable raw materials.

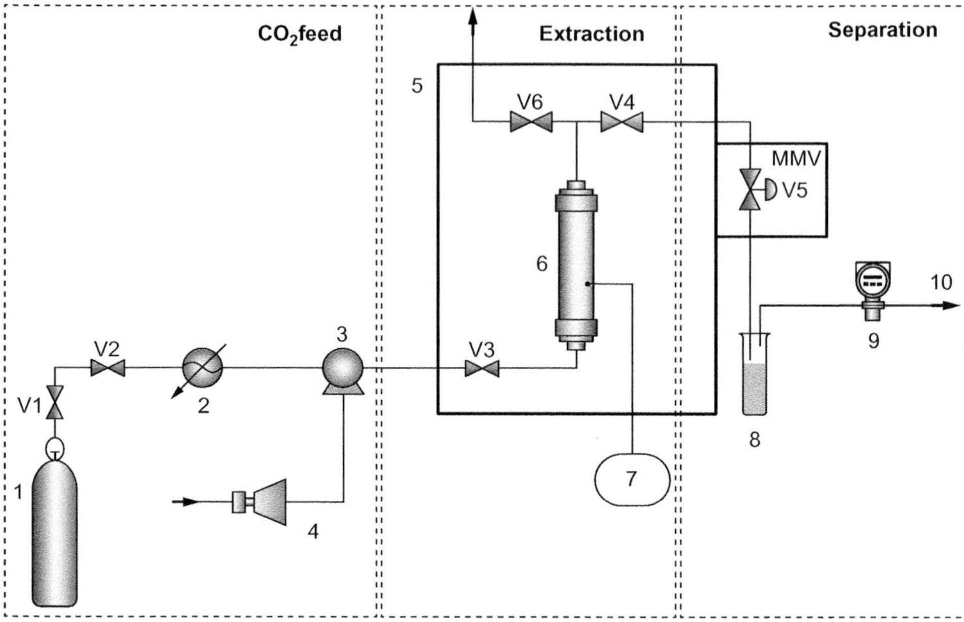

Fig. 3 Flow diagram of SC-CO_2 extraction. (1) CO_2 cylinder; (2) cooling bath; (3) pump; (4) compressor; (5) oven; (6) extractor vessel; (7) monitor; (8) collecting bottle; (9) flowmeter; (10) CO_2 outlet; V1–V6 flow control valves. *(Adapted from F.W.F. Bezerra, W.A. da Costa, M.S. de Oliveira, E.H. de Aguiar Andrade, R.N. de Carvalho Jr., Transesterification of palm pressed-fibers (Elaeis guineensis Jacq.) oil by supercritical fluid carbon dioxide with entrainer ethanol, J. Supercrit. Fluids 136 (2018) 136–143. doi:10.1016/j.supflu.2018.02.020.)*

5. Bioactive compounds extraction by SC-CO_2

5.1 Fatty acids

Fatty acids are a class of lipid compounds that have from 2 to 80 carbon atoms in their chain, and may or may not have double bonds and as a radical function may have hydroxyl groups, epoxy, and halogen atoms. Fatty acids maintain the energy balance of human organism, and when ingested properly can reduce the risk of some diseases like diabetes, hypertension, and inflammation [18]. These compounds can be obtained by some methods like solvent accelerated extraction, microwave-assisted extraction, ultrasonic-assisted extraction, and Bligh and Dyer [39–41].

However, some conventional techniques may imply losses of compounds by thermal degradation, which may lead to fixed oil hydrolytic rancidity [40, 42, 43]. Ranjith Kumar et al. [44] have reported that solvent free methods may be promising to enable large-scale commercialization of high-quality oils. With this, the supercritical technology as an alternative has presented good results for lipid compounds extraction from several vegetal matrices, as evident from Table 2.

Table 1 Extraction parameters of bioactive compounds from vegetable raw materials

Raw material	Compounds	*T* (°C)	*P* (bar)	CO_2 density ($kg\,m^{-3}$)	Flow rate CO_2 ($kg\,s^{-1}$)	Yield (%)	References
Euterpe oleracea	Unsaturated fatty acids (C18:1 ω-9 and C18:2 ω-6)	50–70	150–490	700–900	8.85×10^{-5}	9.07–45.40	[30]
Oenocarpus bacaba	Unsaturated fatty acids (C18:1 ω-9, C18:2 ω-6, C18:3 ω-3)	40–60	120–420	700–900	8.85×10^{-5}	4.32–60.39	[31]
Oenocarpus distichus Mart.	Unsaturated fatty acids (C18:1 ω-9, C16:0, C18:2 ω-6)	50–60	150–350	~700–~900	8.85×10^{-5}	17.51–45.90	[32]
Rosmarinus officinalis	Camphor and 1,8-cineole	30–40	100–300	~650–~950	8.33×10^{-5}	1.00–5.00	[33]
Sesamum indicum L.	Unsaturated fatty acids and phytosterols	40–60	200–400	~700–~1000	5.67×10^{-5}	~32.5–~55.0	[34]
Spilanthes acmella	Spilanthol	50–70	150–490	700–900	7.40×10^{-5}	6.89–8.94	[35]
Syzygium aromaticum	Essential oil (Eugenol, *E*-caryophyllene, eugenol acetate)	40–50	100–300	~400–~900	8.85×10^{-5}	12.70–13.14	[36]
Cissus sicyoides L.	Terpenes, phenolic compounds.	40–60	200–400	~760–953	1.34×10^{-4}	1.27–3.21	[37]
Virola surinamensis	Saturated fatty acids (C14:0, C12:0)	40–80	350	~790–~936	7.90×10^{-5}	59.21–64.39	[38]
Piper divaricatum	Essential oil (methyl eugenol, eugenol, β-elemene)	35–55	100–500	325–1005	8.85×10^{-5}	4.68–7.40	[39]

Table 2 Fatty acids obtained by SC-CO_2 from vegetable raw materials

Raw material	Condition	Yield (%)	Major compounds	References
Watermelon seeds	T=60°C; P=400 bar	51.83	Linoleic acid	[45]
Scenedesmus sp.	T=53°C; P=500 bar	7.41	Trans-9-elaidic methyl ester	[46]
Tetraselmis sp.	T=40°C; P=150 bar	10.88	Palmitic acid	[47]
Scenedesmus obliquus	T=60°C;	8.00	Palmitic acid	[48]
Chlorella protothecoides	P=300 bar;	30.40		
Nannochloropsis salina	5% ethanol	13.00		
Italian Walnuts	T=50°C; P=270 bar	62.30	Linoleic acid and sitosterol	[49]
Euterpe oleracea	T=70°C; P=490 bar;	45.40	Oleic acid	[30]
Nannochloropsis sp.	T=50°C; P=400 bar	6.90	Oleic acid	[50]
Elaeis guineensis Jacq.	T=40°C; P=450 bar	6.09	Hexadecanoic acid	[29]

5.2 Essential oils

Essential oils are secondary metabolites produced by plants for protection against diseases and herbivorous insects. Several industrial segments use essential oils for the most varied applications due to their beneficial biological properties for the human health maintenance [50, 51].

The isolation of essential oils from aromatic plants usually occurs by conventional methods, the most widespread being hydrodistillation [51, 52]. However, this technique has limitations, as losses of compounds that may occur during extraction due to hydrolysis [52]. In this sense, it is recommended to use techniques that can be selective and do not degrade their bioactive compounds. The essential oils extraction with SC-CO_2 can be a viable alternative, mainly because it is a green technology suitable for lipophilic molecules extraction and capable of providing high-quality extracts and totally organic solvents free [35, 43, 53, 54]. Table 3 lists work reported during the last years using SC-CO_2 as extractant to obtain essential oils.

5.3 Phenolic compounds

Phenolic compounds are a secondary metabolites group characterized by having in their structure an aromatic ring attached to one or more hydroxyl substituents and can be

Table 3 Major compounds of essential oils obtained by SC-CO_2 from vegetable raw materials

Raw material	Condition	Yield (%)	Major compounds	References
Eryngium billardieri	$T=35°C$; $P=300$ bar	0.85	γ-cadinene	[53]
Juniperus communis L.	$T=55°C$; $P=300$ bar	6.55	1-octadecene	[55]
Helichrysum italicum	$T=52°C$; $P=200$ bar	4.09	1-[2-(2-Methyl-2,3-dihydroxypropyl)-2,3-dihydro-1-benzofuran-5-yl] ethanon	[56]
Echinophora platyloba	$T=55°C$; $P=240$ bar	1.18	Linalool	[57]
Cleome coluteoides	$T=35°C$; $P=220$ bar	0.65	α-cadinol	[58]
Mentha spicata	$T=48°C$; $P=151$ bar	1.40	Carvone	[59]
Rosmarinus officinalis	$T=50°C$; $P=103$ bar	1.41	Camphor	[60]
Launaea acanthodes B.	$T=55°C$; $P=240$ bar	1.02	n-dodecanal	[61]
Croton zehntneri	$T=15°C$; $P=66.7$ bar	3.40	*E*-anethole	[62]

classified as individual phenols or polyphenols according to the molecule phenol units number [15]. Their antioxidant properties are important in determining their role as protective agents against processes of free-radical-mediated diseases [62]. When added to foods, antioxidants control the rancidity development, delay the formation of toxic oxidation products, maintain nutritional quality, and extend product shelf life [63].

Depending on the plant species, SC-CO_2 can be used to obtain extracts rich in various phenolic groups such as coumarins, cinnamic acids, quinones, flavonoids, and lignans [15]. However, to modify the CO_2 apolar nature, some cosolvents need to be added, to increase the phenolic compounds extraction yield by the increase of solvation power. Table 4 lists some recent researches performed on phenolic compounds extraction using SC-CO_2 and cosolvents.

5.4 Carotenoids

The term "Carotenoids" is used to denote most of the pigments found naturally in animal and plant kingdoms. This group of fat soluble pigments has biological properties and contains more than 700 compounds responsible for red, orange, and yellow colors. Most carotenoids are hydrocarbons containing 40 carbon atoms and two terminal rings. These

Table 4 Phenolic compounds obtained by SC-CO_2 from vegetable raw materials

Raw material	Condition	Cosolvent	Major compounds	References
Paullinia cupana	t=40 min; T=40°C; P=300 bar	Ethanol, Methanol	Caffeine, catechin epicatechin	[64]
Vaccinium myrtillus L.	T=40°C; P=200 bar	Ethanol, Water	Anthocyanins	[65]
Syzygium cumini	T=50°C; P=162 bar	Ethanol	Anthocyanin, phenolic compounds	[66]
Citrus sinensis	T=40°C; P=350 bar	Ethanol	Phenolic compounds	[67]
Feijoa sellowiana	T=50°C; P=250 bar	Ethanol, Methanol	Gallic acid, catechin, rutin, ferulic acid, apigenin, quercetin	[68]
Opuntia stricta	T=40°C; P=350 bar	–	Polyphenols	[69]

pigments have been used mainly in the food, pharmaceutical, and cosmetic industries due to the potential for health benefits [69]. Nowadays, industrial production is performed through chemical synthesis, but natural alternatives of carotenoid production/obtaining are under development, mainly through the extraction of plants or algae [70].

The carotenoids extraction usually occurs with organic solvents, but the supercritical technology is an alternative technique for these compounds recovery, presenting many advantages when compared to the conventional processes. As a solvent for supercritical extraction, CO_2 has several advantages that make it the ideal solvent for thermosensitive molecules extraction such as carotenoids [71, 72]. Table 5 presents information from recent studies on SC-CO_2 extraction of carotenoids from plant materials.

5.5 Applications of bioactive compounds obtained from SC-CO_2

The production of new drugs from natural products and their analogous molecules obtained from the SC-CO_2 use has gained great importance. In the literature, the pharmacological activities of natural products against pathologies caused by bacteria, fungi, protozoa, viruses, besides cytoprotective activity, anticancer, among others have been documented [80–83]. Table 6 shows some biological activities that have already been related to extracts obtained through SC-CO_2.

SC-CO_2 has also been applied as a solvent and antisolvent in other industrial uses such as obtaining micro and nanoparticles for encapsulation [94, 95]. Particularly in the nanocomposite synthesis process, the use of SC-CO_2 is advantageous due to its high permeability of the structure and can be used in various compounds impregnation [96, 97].

Table 5 Carotenoids obtained by SC-CO_2 from vegetable raw materials

Raw material	Condition	Major compounds	References
Cucurbita sp.	T=47.75°C; P=300 bar	β-carotene and α-tocopherol	[73]
Solanum lycopersicum L.	T=50°C; P=400 bar	Lycopene	[74]
Dunaliella salina	T=55°C; P=400 bar	Total carotenoids	[75]
Lycopersicum esculentum L.	T= 80°C; P= 400 bar	Lycopene and β-carotene	[76]
Diospyros kaki L.	T=60°C; P=300 bar	Xanthophylls, β-carotene	[77]
Spinacia oleracea	T=56°C; P=390 bar	Lutein	[78]
Bactris gasipaes	T=40°C; P=300 bar	Total carotenoids	[79]
Mangifera indica L.	T=40°C; P=300 bar	Total carotenoids, β-carotene	[80]

Table 6 Biological activity related to the plant and its chemical compounds

Plant	Major Compounds	Activity	References
Cinnamomum zeylanicum	Cinnamaldehyde (77.10%), β-caryophyllene (6.00%), and α-terpineo (4.40%)	Antityrosinase	[84]
Piper nigrum L.	B-caryophyllene (25.38%), limonene (15.64%), and sabinene (13.63%)	Antioxidant	[43]
Stellera chamaejasme	Squalene (37.24%), 9,12-octadecadienoic acid *Z,Z*-(26.61%), and n-hexadecanoic acid (8.72%)	Acaricidal	[85]
Origanum vulgare	Carvacrol (39.89%), trans-sabinene hydrate (22.07%), and cis-piperitol (7.77%)	Antimicrobial	[86]
Lavandula angustifolia L.	Linalool (42.82%), linalyl acetate (23.40%), and camphor (8.05%)	Antioxidant and Antimicrobial	[87]
Rosmarinus eriocalyx	β-Amyrin (17.70%), olean-12-ene (8.2%), and epi-α-bisabolol (6.10%)	Antioxidant	[88]
Melaleuca cajuputi	β-caryophyllene (31.50%), Humulene (14.60%), and eugenin (13.60%)	Allelopathic	[89]
Apium nodiflorum	Dillapiol (57.5%), Phytol (12.7%), and *Z*-Falcarinol (6.8%)	Antifungal	[54]
Syzygium aromaticum	Eugenol (62.88%), Eugenol acetate (18.67%), and *E*-caryophyllene (15.60%)	Phytotoxic	[36]

Continued

Table 6 Biological activity related to the plant and its chemical compounds—cont'd

Plant	Major Compounds	Activity	References
Myristica fragrans	Myristicin (32.8%), Sabinene (16.1%), and α-Pinene (9.8%)	Cytotoxic	[90]
Zingiber officinale R.	A-Zingiberene (18.18%), β-Sesquiphellandrene (11%), and α-curcumene	Antibacterial	[91]
Tanacetum parthenium	Camphor (75.54%), trans-chrysanthenyl acetate (28.01%), and *Z*-Spiroether (16.78%)	Insecticidal	[92]
Pelargonium graveolens	Citronellol (14.44%), Citronellylformate (7.74%), and geraniol (7.73%)	Pesticidal and Antifungal	[93]
Chrysanthemum indicum	D-Camphor (8.58%), Caryophyllene oxide (8.46%), and α-curcumene (5.93%)	Antiinflammatory	[94]

The encapsulation technique has been successfully applied for stabilization and controlled release of nutrients such as vitamins and antioxidants [98–100]. The process has been widely used by the food, cosmetic, and pharmaceutical industries in order to stabilize certain compounds which under normal conditions would be unstable so that they can be used in different media [101–103]. Karim et al. [104] used hydroxypropyl methyl cellulose (HPMC) as a carrier material for the fish oil microencapsulation using the supercritical antisolvent process and obtained 81.75% efficiency in the process. Thus, micronized fish oil can be introduced into a variety of food matrices since it is a source rich in omega-3, omega-6, eicosapentaenoic acid, and docosahexaenoic acid that reduce the risks of cardiovascular disease, hypertension, and inflammation.

Another application involving the supercritical technology that can also be cited is the supercritical solvent impregnation (SSI), which is defined as a saturation process of a given material by compression and mixing with the pressurized CO_2 which may be in the sub or supercritical state [105]. The use of this technique in active packaging production brings an innovative alternative for several products conservation, because it maintains its nutritional and sensorial quality, prolonging its shelflife due to the action of an active agent added to the packaging material that can confer antimicrobial and/or antioxidant activity [106–108].

6. Considerations

The SC-CO_2 extraction is an important technique for obtaining biocompounds, since the obtained extracts are of high quality and strongly concentrated. The advantages of this process include the rapid separation of CO_2 from the extracts (due to the simple depressurizing) and the not requirement of the use of organic solvents. In addition, bioactive compounds can be used in other technologies that employ SC-CO_2, obtaining materials of high added value for application in several sectors in an environmentally safe way.

Acknowledgment

This study was financed in part by the Coordenação de Aperfeiçoamento de Pessoal de Nível Superior—Brasil (CAPES)—Finance Code 001.

References

[1] H. Bandar, A. Hijazi, H. Rammal, A. Hachem, Z. Saad, B. Badran, Techniques for the Extraction of Bioactive Compounds from Lebanese *Urtica dioica*, Am. J. Phytomed. Clin. Ther. 1 (2013) 507–513, https://doi.org/10.1080/17524032.2018.1436579.

[2] W.A. da Costa, M.S. de Oliveira, M.P. da Silva, V.M.B. Cunha, R.H.H. Pinto, F.W.F. Bezerra, R.N. de Carvalho Junior, Açaí (*Euterpe oleracea*) and Bacaba (*Oenocarpus bacaba*) as Functional Food, in: Superfood and Functional Food—An Overview of Their Processing and Utilization, 2017, pp. 155–172. https://doi.org/10.5772/65881.

[3] N. Budisa, D. Schulze-Makuch, Supercritical carbon dioxide and its potential as a life-sustaining solvent in a planetary environment, Life 4 (2014) 331–340, https://doi.org/10.3390/life4030331.

[4] M.S. Uddin, M.Z.I. Sarker, S. Ferdosh, M.J.H. Akanda, M.S. Easmin, S.H. Bt Shamsudin, K. Bin Yunus, Phytosterols and their extraction from various plant matrices using supercritical carbon dioxide: a review, J. Sci. Food Agric. 95 (2015) 1385–1394, https://doi.org/10.1002/jsfa.6833.

[5] L. Campone, R. Celano, A.L. Piccinelli, I. Pagano, M. Russo, E. Ibañez, A. Cifuentes, L. Rastrelli, Response surface methodology to optimize supercritical carbon dioxide/co-solvent extraction of brown onion skin by-product as source of nutraceutical compounds, Food Chem. 269 (2018) 495–502, https://doi.org/10.1016/j.foodchem.2018.07.042.

[6] C.J. Clarke, W.-C. Tu, O. Levers, A. Bröhl, J.P. Hallett, Green and sustainable solvents in chemical processes, Chem. Rev. 118 (2018) 747–800, https://doi.org/10.1021/acs.chemrev.7b00571.

[7] K. Häckl, W. Kunz, Some aspects of green solvents, Drill. Fluid Complet. Fluid. 21 (2018) 572–580, https://doi.org/10.1016/j.crci.2018.03.010.

[8] Y. Medina-Gonzalez, S. Camy, J.-S. Condoret, ScCO2/Green solvents: biphasic promising systems for cleaner chemicals manufacturing, ACS Sustain. Chem. Eng. 2 (2014) 2623–2636, https://doi.org/10.1021/sc5004314.

[9] T. Welton, Solvents and Sustainable Chemistry, 471, Royal Society Publishing, 2015, pp. 1–26, https://doi.org/10.1098/rspa.2015.0502.

[10] F. Sahena, I.S.M. Zaidul, S. Jinap, A.A. Karim, K.A. Abbas, N.A.N. Norulaini, A.K.M. Omar, Application of supercritical CO_2 in lipid extraction—a review, J. Food Eng. 95 (2009) 240–253, https://doi.org/10.1016/j.jfoodeng.2009.06.026.

[11] S.G. Özkal, M.E. Yener, Supercritical carbon dioxide extraction of flaxseed oil: effect of extraction parameters and mass transfer modeling, J. Supercrit. Fluids 112 (2016) 76–80, https://doi.org/10.1016/j.supflu.2016.02.013.

[12] R. Scopel, C.F. Silva, A.M. Lucas, J.J. Garcez, A.T. Espirito Santo, R.N. Almeida, E. Cassel, R.M.F. Vargas, Fluid phase equilibria and mass transfer studies applied to supercritical fluid extraction of Illicium verum volatile oil, Fluid Phase Equilib. 417 (2016) 203–211.

[13] M. Herrero, J.A. Mendiola, A. Cifuentes, E. Ibáñez, Supercritical fluid extraction: recent advances and applications, J. Chromatogr. A 1217 (2010) 2495–2511.
[14] L. Wang, C.L. Weller, Recent advances in extraction of nutraceuticals from plants, Trends Food Sci. Technol. 17 (2006) 300–312, https://doi.org/10.1016/j.tifs.2005.12.004.
[15] M.M.R. De Melo, A.J.D. Silvestre, C.M. Silva, Supercritical fluid extraction of vegetable matrices: applications, trends and future perspectives of a convincing green technology, J. Supercrit. Fluids 92 (2014) 115–176, https://doi.org/10.1016/j.supflu.2014.04.007.
[16] D. Peng, D.B. Robinson, A new two-constant equation of state, Ind. Eng. Chem. Fundam. 15 (1976) 59–64. http://pubs.acs.org/doi/abs/10.1021/i160057a011.
[17] R.L. Mendes, B.P. Nobre, M.T. Cardoso, A.P. Pereira, Supercritical carbon dioxide extraction of compounds with pharmaceutical importance from microalgae, Inorg. Chim. Acta 356 (2003) 328–334, https://doi.org/10.1016/S0020-1693(03)00363-3.
[18] V.M.B. Cunha, M.P. da Silva, W.A. da Costa, M.S. de Oliveira, F.W.F. Bezerra, A.C. de Melo, R.H.H. Pinto, N.T. Machado, M.E. Araujo, R.N. de Carvalho Junior, Carbon dioxide use in high-pressure extraction processes, in: Carbon Dioxide Chem. Capture Oil Recover, IntechOpen, 2018, pp. 211–240, https://doi.org/10.5772/intechopen.71151.
[19] R.B. Gupta, J.-J. Shim, Solubility in Supercritical Carbon Dioxide, CRC Press–Taylor & Francis Group, 2007.
[20] G. Brunner, Gas Extraction An Introduction to Fundamentals of Supercritical Fluids and the Application to Separation Processes, Springer, New York, 1994.
[21] C.G. Pereira, M.A.A. Meireles, Supercritical fluid extraction of bioactive compounds: Fundamentals, applications and economic perspectives, Food Bioprocess. Technol. 3 (2010) 340–372, https://doi.org/10.1007/s11947-009-0263-2.
[22] D. Čuček, T. Perko, L. Ilić, B. Znoj, P. Venturini, Ž. Knez, M. Škerget, Phase equlibiria and diffusivity of dense gases in various polyethylenes, J. Supercrit. Fluids 78 (2013) 54–62, https://doi.org/10.1016/j.supflu.2013.03.020.
[23] G. Kravanja, Ž. Knez, M. Knez Hrnčič, Density, interfacial tension, and viscosity of polyethylene glycol 6000 and supercritical CO_2, J. Supercrit. Fluids 139 (2018) 72–79, https://doi.org/10.1016/j.supflu.2018.05.012.
[24] G. Brunner, Supercritical fluids: technology and application to food processing, J. Food Eng. 67 (2005) 21–33.
[25] K. Kian, A.M. Scurto, Viscosity of compressed CO_2-saturated n-alkanes: CO_2/n-hexane, CO_2/n-decane, and CO_2/n-tetradecane, J. Supercrit. Fluids 133 (2018) 411–420.
[26] H.M. Avelino, J.M. Fareleira, D. Gourgouillon, J.M. Igreja, M.N. da Ponte, Viscosity of poly(ethyleneglycol) 200 [PEG 200] saturated with supercritical carbon dioxide, J. Supercrit. Fluids 128 (2017) 300–307.https://en.wikipedia.org/wiki/Supercritical_carbon_dioxide.
[27] J. Leite, A.L. Magalhães, A.A. Valente, C.M. Silva, Measurement and modelling of tracer diffusivities of gallic acid in liquid ethanol and in supercritical CO_2 modified with ethanol, J. Supercrit. Fluids 131 (2017) 130–139, https://doi.org/10.1016/j.supflu.2017.09.004.
[28] R.S. Mohamed, Extração e fracionamento de produtos de ocorrência natural com fluidos supercríticos, Ciênc. Tecnol. Aliment. 17 (1997) 344–353, https://doi.org/10.1590/S0101-20611997000400000.
[29] C. Turner, J.W. King, L. Mathiasson, Supercritical fluid extraction and chromatography for fat-soluble vitamin analysis, J. Chromatogr. A 936 (2001) 215–237, https://doi.org/10.1016/S0021-9673(01)01082-2.
[30] R.H.H. Pinto, C. Sena, O.V. Santos, W.A. Da Costa, A.M.C. Rodrigues, R.N. Carvalho Junior, Extraction of bacaba (*Oenocarpus bacaba*) oil with supercritical CO_2: Global yield isotherms, fatty acid composition, functional quality, oxidative stability, spectroscopic profile and antioxidant activity, Grasas Aceites 69 (2018) 246, https://doi.org/10.3989/gya.0883171.
[31] V.M.B. Cunha, M.P. da Silva, S.H.B. de Sousa, P.d.N. Bezerra, E.G.O. Menezes, N.J.N. da Silva, D. A.D.d.S. Banna, M.E. Araújo, R.N. de Carvalho Junior, Bacaba-de-leque (*Oenocarpus distichus* Mart.) oil extraction using supercritical CO_2 and bioactive compounds determination in the residual pulp, J. Supercrit. Fluids 144 (2019) 81–90, https://doi.org/10.1016/j.supflu.2018.10.010.
[32] R.N. Carvalho, L.S. Moura, P.T.V. Rosa, M.A.A. Meireles, Supercritical fluid extraction from rosemary (*Rosmarinus officinalis*): kinetic data, extract's global yield, composition, and antioxidant activity, J. Supercrit. Fluids 35 (2005) 197–204, https://doi.org/10.1016/j.supflu.2005.01.009.

[33] J.R.S. Botelho, N.G. Medeiros, A.M.C. Rodrigues, M.E. Araújo, N.T. Machado, A. Guimarães Santos, I.R. Santos, W. Gomes-Leal, R.N. Carvalho, Black sesame (*Sesamum indicum* L.) seeds extracts by CO_2 supercritical fluid extraction: Isotherms of global yield, kinetics data, total fatty acids, phytosterols and neuroprotective effects, J. Supercrit. Fluids 93 (2014) 49–55, https://doi.org/10.1016/j.supflu.2014.02.008.

[34] A.M.A. Dias, A.C.S. da Silva, J.R.S. Botelho, R.N.C. Júnior, H.C. de Sousa, M.E.M. Braga, Temperature and density effects of the $scCO_2$ extraction of spilanthol from *Spilanthes acmella* flowers, J. Supercrit. Fluids 121 (2017) 32–40, https://doi.org/10.1016/j.supflu.2016.11.004.

[35] M.S. de Oliveira, W.A. da Costa, D.S. Pereira, J.R.S. Botelho, T.O. de Alencar Menezes, E.H. de Aguiar Andrade, S.H.M. da Silva, A.P. da Silva Sousa Filho, R.N. de Carvalho, Chemical composition and phytotoxic activity of clove (*Syzygium aromaticum*) essential oil obtained with supercritical CO_2, J. Supercrit. Fluids 118 (2016) 185–193, https://doi.org/10.1016/j.supflu.2016.08.010.

[36] M.A.R. Salazar, J.V. Costa, G.R.O. Urbina, V.M.B. Cunha, M.P. Silva, P.d.N. Bezerra, W.B.S. Pinheiro, W. Gomes-Leal, A.S. Lopes, R.N. Carvalho Junior, Chemical composition, antioxidant activity, neuroprotective and anti-inflammatory effects of cipó-pucá (*Cissus sicyoides* L.) extracts obtained from supercritical extraction, J. Supercrit. Fluids 138 (2018) 36–45, https://doi.org/10.1016/j.supflu.2018.03.022.

[37] R.M. Cordeiro, A.P.d.S.e. Silva, R.H.H. Pinto, W.A. da Costa, S.H.M. da Silva, W.B. de Souza Pinheiro, M.S.P. Arruda, R.N. Carvalho Junior, Supercritical CO_2 extraction of ucuúba (*Virola surinamensis*) seed oil: global yield, kinetic data, fatty acid profile, and antimicrobial activities, Chem. Eng. Commun. 0 (2018) 1–12, https://doi.org/10.1080/00986445.2018.1474741.

[38] M.S. de Oliveira, J.N. da Cruz, S. Gomes Silva, W.A. da Costa, S.H.B. de Sousa, F.W.F. Bezerra, E. Teixeira, N.J.N. da Silva, E.H. de Aguiar Andrade, A.M. de Jesus Chaves Neto, R.N. de Carvalho, Phytochemical profile, antioxidant activity, inhibition of acetylcholinesterase and interaction mechanism of the major components of the *Piper divaricatum* essential oil obtained by supercritical CO_2, J. Supercrit. Fluids 145 (2019) 74–84, https://doi.org/10.1016/j.supflu.2018.12.003.

[39] M. Mubarak, A. Shaija, T.V. Suchithra, A review on the extraction of lipid from microalgae for biodiesel production, Algal Res. 7 (2015) 117–123, https://doi.org/10.1016/j.algal.2014.10.008.

[40] F. Al Juhaimi, M.M. Özcan, Effect of cold press and soxhlet extraction systems on fatty acid, tocopherol contents, and phenolic compounds of various grape seed oils, J. Food Process. Preserv. 42 (2018) e13417, 1–8. https://doi.org/10.1111/jfpp.13417.

[41] M. Kozłowska, E. Gruczyńska, I. Ścibisz, M. Rudzińska, Fatty acids and sterols composition, and antioxidant activity of oils extracted from plant seeds, Food Chem. 213 (2016) 450–456, https://doi.org/10.1016/j.foodchem.2016.06.102.

[42] H. Bagheri, M.Y. Bin Abdul Manap, Z. Solati, Antioxidant activity of *Piper nigrum* L. essential oil extracted by supercritical CO_2 extraction and hydro-distillation, Talanta 121 (2014) 220–228, https://doi.org/10.1016/J.TALANTA.2014.01.007.

[43] A. Zermane, O. Larkeche, A.H. Meniai, C. Crampon, E. Badens, Optimization of *Algerian rosemary* essential oil extraction yield by supercritical CO_2 using response surface methodology, C. R. Chim. 19 (2016) 538–543, https://doi.org/10.1016/j.crci.2015.08.011.

[44] R.R. Kumar, P. Hanumantha Rao, M. Arumugam, Lipid extraction methods from microalgae: a comprehensive review, Front. Energy Res. 2 (2015) 1–9, https://doi.org/10.3389/fenrg.2014.00061.

[45] H. Taher, S. Al-Zuhair, A.H. Al-Marzouqi, Y. Haik, M. Farid, Enzymatic biodiesel production of microalgae lipids under supercritical carbon dioxide: process optimization and integration, Biochem. Eng. J. 90 (2014) 103–113, https://doi.org/10.1016/j.bej.2014.05.019.

[46] Y. Li, F. Ghasemi Naghdi, S. Garg, T.C. Adarme-Vega, K.J. Thurecht, W.A. Ghafor, S. Tannock, P.M. Schenk, A comparative study: the impact of different lipid extraction methods on current microalgal lipid research, Microb. Cell Factories 13 (2014) 1–9, https://doi.org/10.1186/1475-2859-13-14.

[47] M. Solana, C.S. Rizza, A. Bertucco, Exploiting microalgae as a source of essential fatty acids by supercritical fluid extraction of lipids: comparison between *Scenedesmus obliquus*, Chlorella protothecoides and *Nannochloropsis salina*, J. Supercrit. Fluids 92 (2014) 311–318, https://doi.org/10.1016/j.supflu.2014.06.013.

[48] N. De Zordi, A. Cortesi, I. Kikic, M. Moneghini, V. Baldan, S. Sut, D. Solinas, S. Dall'Acqua, The Supercritical carbon dioxide extraction of Ω-3, Ω-6 lipids and B-sitosterol from italian walnuts: a

central composite design approach, J. Supercrit. Fluids 127 (2017) 223–228, https://doi.org/10.1016/j.supflu.2017.02.020.

[49] K.-S. Choi, J.-H. Ryu, D.-J. Park, S.-C. Oh, H. Kwak, Lipid extraction from *Nannochloropsis* sp. microalgae for biodiesel production using supercritical carbon dioxide, Korean Chem. Eng. Res. 53 (2015) 205–210, https://doi.org/10.9713/kcer.2015.53.2.205.

[50] S.S. Dahham, Y.M. Tabana, M.A. Iqbal, M.B.K. Ahamed, M.O. Ezzat, A.S.A. Majid, A.M.S. A. Majid, The anticancer, antioxidant and antimicrobial properties of the sesquiterpene β-caryophyllene from the essential oil of *Aquilaria crassna*, Molecules 20 (2015) 11808–11829, https://doi.org/10.3390/molecules200711808.

[51] J. Hafsa, M. ali Smach, M.R. Ben Khedher, B. Charfeddine, K. Limem, H. Majdoub, S. Rouatbi, Physical, antioxidant and antimicrobial properties of chitosan films containing *Eucalyptus globulus* essential oil, LWT Food Sci. Technol. 68 (2016) 356–364, https://doi.org/10.1016/j.lwt.2015.12.050.

[52] E.S.S. Abdel-Hameed, M.S. Salman, M.A. Fadl, A. Elkhateeb, M.M. Hassan, Chemical composition and biological activity of *Mentha longifolia* L. essential oil growing in Taif, KSA extracted by hydrodistillation, solvent free microwave and microwave hydrodistillation, J. Essent. Oil Bear. Plants 21 (2018) 1–14, https://doi.org/10.1080/0972060X.2018.1454343.

[53] A. Maxia, D. Falconieri, A. Piras, S. Porcedda, B. Marongiu, M.A. Frau, M.J. Gonçalves, C. Cabral, C. Cavaleiro, L. Salgueiro, Chemical composition and antifungal activity of essential oils and supercritical CO_2 extracts of *Apium nodiflorum* (L.) lag, Mycopathologia 174 (2012) 61–67, https://doi.org/10.1007/s11046-011-9519-2.

[54] M.-T. Golmakani, K. Rezaei, Comparison of microwave-assisted hydrodistillation withthe traditional hydrodistillation method in the extraction of essential oils from *Thymus vulgaris* L, Food Chem. 109 (2008) 925–930, https://doi.org/10.1016/j.foodchem.2007.12.084.

[55] I. Jerković, M. Rajić, Z. Marijanović, M. Bilić, S. Jokić, Optimization of supercritical CO_2 extraction of dried *Helichrysum italicum* flowers by response surface methodology: GC-MS profiles of the extracts and essential oil, Sep. Sci. Technol. 51 (2016) 2925–2931, https://doi.org/10.1080/01496395.2016.1237967.

[56] G. Sodeifian, S.A. Sajadian, Investigation of essential oil extraction and antioxidant activity of *Echinophora platyloba* DC. using supercritical carbon dioxide, J. Supercrit. Fluids 121 (2017) 52–62, https://doi.org/10.1016/j.supflu.2016.11.014.

[57] G. Sodeifian, S.A. Sajadian, N. Saadati Ardestani, Optimization of essential oil extraction from *Launaea acanthodes* Boiss: utilization of supercritical carbon dioxide and cosolvent, J. Supercrit. Fluids 116 (2016) 46–56, https://doi.org/10.1016/j.supflu.2016.05.015.

[58] N.M. Shrigod, N.R. Swami Hulle, R.V. Prasad, Supercritical fluid extraction of essential oil from mint leaves (*Mentha spicata*): process optimization and its quality evaluation, J. Food Process Eng. 40 (2017) 1–9, https://doi.org/10.1111/jfpe.12488.

[59] L.A. Conde-Hernández, J.R. Espinosa-Victoria, A. Trejo, J. Guerrero-Beltrán, CO_2-supercritical extraction, hydrodistillation and steam distillation of essential oil of rosemary (*Rosmarinus officinalis*), J. Food Eng. 200 (2017) 81–86, https://doi.org/10.1016/j.jfoodeng.2016.12.022.

[60] G. Sodeifian, N. Saadati Ardestani, S.A. Sajadian, S. Ghorbandoost, Application of supercritical carbon dioxide to extract essential oil from *Cleome coluteoides* Boiss: experimental, response surface and grey wolf optimization methodology, J. Supercrit. Fluids 114 (2016) 55–63, https://doi.org/10.1016/j.supflu.2016.04.006.

[61] E.M.B.D. Sousa, J. Martínez, O. Chiavone-Filho, P.T.V. Rosa, T. Domingos, M.A.A. Meireles, Extraction of volatile oil from *Croton zehntneri* Pax et Hoff with pressurized CO_2: solubility, composition and kinetics, J. Food Eng. 69 (2005) 325–333, https://doi.org/10.1016/j.jfoodeng.2004.08.023.

[62] D. De Beer, E. Joubert, W.C.A. Gelderblom, M. Manley, Phenolic compounds: a review of their possible role as in vivo antioxidants of wine, S. Afri. J. Enol. Vitic. 23 (2002) 48–61.

[63] F. Shahidi, P. Ambigaipalan, Phenolics and polyphenolics in foods, beverages and spices: Antioxidant activity and health effects—a review, J. Funct. Foods 18 (2015) 820–897, https://doi.org/10.1016/j.jff.2015.06.018.

[64] J. Paes, R. Dotta, G.F. Barbero, J. Martínez, Extraction of phenolic compounds and anthocyanins from blueberry (*Vaccinium myrtillus* L.) residues using supercritical CO_2 and pressurized liquids, J. Supercrit. Fluids 95 (2014) 8–16, https://doi.org/10.1016/j.supflu.2014.07.025.

[65] J.P. Maran, B. Priya, S. Manikandan, Modeling and optimization of supercritical fluid extraction of anthocyanin and phenolic compounds from *Syzygium cumini* fruit pulp, J. Food Sci. Technol. 51 (2014) 1938–1946, https://doi.org/10.1007/s13197-013-1237-y.

[66] F.A. Espinosa-Pardo, V.M. Nakajima, G.A. Macedo, J.A. Macedo, J. Martínez, Extraction of phenolic compounds from dry and fermented orange pomace using supercritical CO_2 and cosolvents, Food Bioprod. Process. 101 (2017) 1–10, https://doi.org/10.1016/j.fbp.2016.10.002.

[67] M. Mousavi, M. Bimakr, S.M. Ghoreishi, A. Ganjloo, Supercritical Carbon Dioxide Extraction of Bioactive Compounds from Feijoa (*Feijoa sellowiana*) Leaves, 5, 2018, pp. 15–23.

[68] M. Koubaa, H. Mhemdi, F.J. Barba, A. Angelotti, F. Bouaziz, S.E. Chaabouni, E. Vorobiev, Seed oil extraction from red prickly pear using hexane and supercritical CO_2: assessment of phenolic compound composition, antioxidant and antibacterial activities, J. Sci. Food Agric. 97 (2017) 613–620, https://doi.org/10.1002/jsfa.7774.

[69] J. Prado, P. Veggi, M. Meireles, Extraction methods for obtaining carotenoids from vegetables—review, Curr. Anal. Chem. 10 (2013) 29–66, https://doi.org/10.2174/1573411011410010005.

[70] N. Mezzomo, S.R.S. Ferreira, Carotenoids functionality, sources, and processing by supercritical technology: a review, J. Chem. 2016 (2016), https://doi.org/10.1155/2016/3164312.

[71] M. Durante, M.S. Lenucci, G. Mita, Supercritical carbon dioxide extraction of carotenoids from pumpkin (*Cucurbita* spp.): a review, Int. J. Mol. Sci. 15 (2014) 6725–6740, https://doi.org/10.3390/ijms15046725.

[72] J.L. Martínez, S.W. Vance, Supercritical Extraction Plants: Equipment, Process, and Costs, in: Supercrit. Fluid Extr. Nutraceuticals Bioact. Compd, CRC Press–Taylor & Francis Group, New York, 2007, p. 420.

[73] H.D.F.Q. Barros, R. Grimaldi, F.A. Cabral, Lycopene-rich avocado oil obtained by simultaneous supercritical extraction from avocado pulp and tomato pomace, J. Supercrit. Fluids 120 (2017) 1–6, https://doi.org/10.1016/j.supflu.2016.09.021.

[74] S.R.P. Hosseini, O. Tavakoli, M.H. Sarrafzadeh, Experimental optimization of SC-CO_2 extraction of carotenoids from Dunaliella salina, J. Supercrit. Fluids 121 (2017) 89–95, https://doi.org/10.1016/j.supflu.2016.11.006.

[75] M. Kehili, M. Kammlott, S. Choura, A. Zammel, C. Zetzl, I. Smirnova, N. Allouche, S. Sayadi, Supercritical CO_2 extraction and antioxidant activity of lycopene and β-carotene-enriched oleoresin from tomato (*Lycopersicum esculentum* L.) peels by-product of a Tunisian industry, Food Bioprod. Process. 102 (2017) 340–349, https://doi.org/10.1016/j.fbp.2017.02.002.

[76] K. Zaghdoudi, X. Framboisier, C. Frochot, R. Vanderesse, D. Barth, J. Kalthoum-Cherif, F. Blanchard, Y. Guiavarc'h, Response surface methodology applied to Supercritical Fluid Extraction (SFE) of carotenoids from Persimmon (*Diospyros kaki* L.), Food Chem. 208 (2016) 209–219, https://doi.org/10.1016/j.foodchem.2016.03.104.

[77] M. Derrien, M. Aghabararnejad, A. Gosselin, Y. Desjardins, P. Angers, Y. Boumghar, Optimization of supercritical carbon dioxide extraction of lutein and chlorophyll from spinach by-products using response surface methodology, LWT Food Sci. Technol. 93 (2018) 79–87, https://doi.org/10.1016/j.lwt.2018.03.016.

[78] F.A. Espinosa-Pardo, J. Martinez, H.A. Martinez-Correa, Extraction of bioactive compounds from peach palm pulp (*Bactris gasipaes*) using supercritical CO_2, J. Supercrit. Fluids 93 (2014) 2–6, https://doi.org/10.1016/j.supflu.2014.05.010.

[79] M.P. Garcia-Mendoza, J.T. Paula, L.C. Paviani, F.A. Cabral, H.A. Martinez-Correa, Extracts from mango peel by-product obtained by supercritical CO_2 and pressurized solvent processes, LWT Food Sci. Technol. 62 (2015) 131–137, https://doi.org/10.1016/j.lwt.2015.01.026.

[80] N. Garg, T. Luzzatto-Knaan, A.V. Melnik, A.M. Caraballo-Rodríguez, D.J. Floros, D. Petras, R. Gregor, P.C. Dorrestein, V.V. Phelan, Natural products as mediators of disease, Nat. Prod. Rep. 34 (2017) 194–219, https://doi.org/10.1039/c6np00063k.

[81] J. Janata, Z. Kamenik, R. Gazak, S. Kadlcik, L. Najmanova, Biosynthesis and incorporation of an alkylproline-derivative (APD) precursor into complex natural products, Nat. Prod. Rep. 35 (2018) 257–289, https://doi.org/10.1039/c7np00047b.

[82] D.G.I. Kingston, Modern Natural Products Drug Discovery and Its Relevance to Biodiversity Conservation, J. Nat. Prod. 74 (2011) 496–511, https://doi.org/10.1021/np100550t.

[83] D.J. Newman, G.M. Cragg, Natural Products as Sources of New Drugs from 1981 to 2014, J. Nat. Prod. 79 (2016) 629–661, https://doi.org/10.1021/acs.jnatprod.5b01055.

[84] W. Liang, J. Cheng, L.-Q. Ma, Y.-B. Liu, G.-L. Shi, Y.-N. Wang, Componential Analysis and Acaricidal Activities of *Stellera chamaejasme* Extracts by Supercritical Fluid Extraction, in: Advances in Intelligent and Soft Computing, Springer, Berlin, Heidelberg, 2012, pp. 643–652. https://doi.org/10.1007/978-3-642-13214-8{_}21.

[85] S. SANTOYO, S. CAVERO, L. JAIME, E. IBAÑEZ, F.J. SEÑORÁNS, G. REGLERO, Supercritical carbon dioxide extraction of compounds with antimicrobial activity from Origanum vulgare L.: determination of optimal extraction parameters, J. Food Prot. 69 (2006) 369–375, https://doi.org/10.4315/0362-028X-69.2.369.

[86] L.T. Danh, L.N. Han, N.D.A. Triet, J. Zhao, R. Mammucari, N. Foster, Comparison of chemical composition, antioxidant and antimicrobial activity of Lavender (*Lavandula angustifolia* L.) essential oils extracted by supercritical CO_2, hexane and hydrodistillation, Food Bioprocess Technol. 6 (2013) 3481–3489, https://doi.org/10.1007/s11947-012-1026-z.

[87] H. Bendif, M.D. Miara, Z. Kalboussi, D. Grauzdytė, D. Povilaitis, P.R. Venskutonis, F. Maggi, Supercritical CO_2 extraction of *Rosmarinus eriocalyx* growing in Algeria: chemical composition and antioxidant activity of extracts and their solid plant materials, Ind. Crop. Prod. 111 (2018) 768–774, https://doi.org/10.1016/J.INDCROP.2017.11.056.

[88] B.W. Bin Kueh, S. Yusup, N. Osman, Supercritical carbon dioxide extraction of *Melaleuca cajuputi* leaves for herbicides allelopathy: Optimization and kinetics modelling, J. CO_2 Util. 24 (2018) 220–227, https://doi.org/10.1016/J.JCOU.2018.01.005.

[89] A. Piras, A. Rosa, B. Marongiu, A. Atzeri, M.A. Dessì, D. Falconieri, S. Porcedda, Extraction and separation of volatile and fixed oils from seeds of *Myristica fragrans* by supercritical CO_2: chemical composition and cytotoxic activity on Caco-2 cancer cells, J. Food Sci. 77 (2012) C448–C453, https://doi.org/10.1111/j.1750-3841.2012.02618.x.

[90] M.C. Mesomo, M.L. Corazza, P.M. Ndiaye, O.R. Dalla Santa, L. Cardozo, A.D.P. Scheer, Supercritical CO_2 extracts and essential oil of ginger (*Zingiber officinale* R.): Chemical composition and antibacterial activity, J. Supercrit. Fluids 80 (2013) 44–49, https://doi.org/10.1016/j.supflu.2013.03.031.

[91] R. Pavela, M. Sajfrtová, H. Sovová, M. Bárnet, J. Karban, The insecticidal activity of *Tanacetum parthenium* (L.) Schultz Bip. extracts obtained by supercritical fluid extraction and hydrodistillation, Ind. Crop. Prod. 31 (2010) 449–454, https://doi.org/10.1016/J.INDCROP.2010.01.003.

[92] Z. Machalova, M. Sajfrtova, R. Pavela, M. Topiar, Extraction of botanical pesticides from *Pelargonium graveolens* using supercritical carbon dioxide, Ind. Crop. Prod. 67 (2015) 310–317, https://doi.org/10.1016/J.INDCROP.2015.01.070.

[93] X.-L. Wu, C.-W. Li, H.-M. Chen, Z.-Q. Su, X.-N. Zhao, J.-N. Chen, X.-P. Lai, X.-J. Zhang, Z.-R. Su, Anti-inflammatory effect of supercritical-carbon dioxide fluid extract from flowers and buds of *Chrysanthemum indicum* Linnén, Evid. Based Complement Alternat. Med. 2013 (2013) 1–14, https://doi.org/10.1155/2013/413237.

[94] E. Keven Silva, M. Angela, A. Meireles, Encapsulation of food compounds using supercritical technologies: applications of supercritical carbon dioxide as an antisolvent, Food Public Health 4 (2014) 247–258, https://doi.org/10.5923/j.fph.20140405.06.

[95] D. Sanli, S.E. Bozbag, C. Erkey, Synthesis of nanostructured materials using supercritical CO_2: Part I. Physical transformations, J. Mater. Sci. 47 (2012) 2995–3025, https://doi.org/10.1007/s10853-011-6054-y.

[96] Y. Haldorai, J.-J. Shim, K.T. Lim, Synthesis of polymerâ inorganic filler nanocomposites in supercritical (CO_2), J. Supercrit. Fluids 71 (2012) 45–63, https://doi.org/10.1016/j.supflu.2012.07.007.

[97] X. Han, C. Zeng, L.J. Lee, K.W. Koelling, D.L. Tomasko, Extrusion of polystyrene nanocomposite foams with supercritical CO_2, Polym. Eng. Sci. 43 (2003) 1261–1275, https://doi.org/10.1002/pen.10107.

[98] N.W. Clifford, K.S. Iyer, C.L. Raston, Encapsulation and controlled release of nutraceuticals using mesoporous silica capsules, J. Mater. Chem. 18 (2008) 162–165, https://doi.org/10.1039/B715100D.

[99] D.J. McClements, Requirements for food ingredient and nutraceutical delivery systems, in: Encapsulation Technol. Deliv. Syst. Food Ingredients Nutraceuticals, 2012, pp. 3–18, https://doi.org/10.1533/9780857095909.1.3.

[100] C.I. Onwulata, Encapsulation of new active ingredients, Annu. Rev. Food Sci. Technol. 3 (2012) 183–202, https://doi.org/10.1146/annurev-food-022811-101140.

[101] P.J. Chawda, J. Shi, S. Xue, S. Young Quek, Co-encapsulation of bioactives for food applications, Food Qual. Saf. 1 (2017) 302–309, https://doi.org/10.1093/fqsafe/fyx028.
[102] L. Chen, R. Luque, Y. Li, Encapsulation of metal nanostructures into metal-organic frameworks, Dalton Trans. 47 (2018) 3663–3668, https://doi.org/10.1039/c8dt00092a.
[103] M.R. Mozafari, K. Khosravi-Darani, G.G. Borazan, J. Cui, A. Pardakhty, S. Yurdugul, Encapsulation of food ingredients using nanoliposome technology, Int. J. Food Prop. 11 (2008) 833–844, https://doi.org/10.1080/10942910701648115.
[104] F.T. Karim, K. Ghafoor, S. Ferdosh, F. Al-Juhaimi, E. Ali, K. Bin Yunus, M.H. Hamed, A. Islam, M. Asif, I.S.M. Zaidul, Microencapsulation of fish oil using supercritical antisolvent process, J. Food Drug Anal. 25 (2017) 654–666, https://doi.org/10.1016/j.jfda.2016.11.017.
[105] E. Weidner, Impregnation via supercritical CO_2–What we know and what we need to know, J. Supercrit. Fluids 134 (2018) 220–227, https://doi.org/10.1016/j.supflu.2017.12.024.
[106] C.C. Bastante, L.C. Cardoso, C.M. Serrano, E.M. de la Ossa, Supercritical impregnation of food packaging films to provide antioxidant properties, J. Supercrit. Fluids 128 (2017) 200–207, https://doi.org/10.1016/j.supflu.2017.05.034.
[107] M. Belizón, M.T. Fernández-Ponce, L. Casas, C. Mantell, E.J. Martínez De La Ossa-Fernández, Supercritical impregnation of antioxidant mango polyphenols into a multilayer PET/PP food-grade film, J. CO_2 Util. 25 (2018) 56–67, https://doi.org/10.1016/j.jcou.2018.03.005.
[108] C. Villegas, A. Torres, M. Rios, A. Rojas, J. Romero, C.L. de Dicastillo, X. Valenzuela, M.J. Galotto, A. Guarda, Supercritical impregnation of cinnamaldehyde into polylactic acid as a route to develop antibacterial food packaging materials, Food Res. Int. 99 (2017) 650–659, https://doi.org/10.1016/j.foodres.2017.06.031.

Further reading

[109] F.W.F. Bezerra, W.A. da Costa, M.S. de Oliveira, E.H. de Aguiar Andrade, R.N. de Carvalho Jr., Transesterification of palm pressed-fibers (*Elaeis guineensis* Jacq.) oil by supercritical fluid carbon dioxide with entrainer ethanol, J. Supercrit. Fluids 136 (2018) 136–143, https://doi.org/10.1016/j.supflu.2018.02.020.
[110] C.d.C.R. Batista, M.d.S. Oliveira, M.E. Araújo, A.M.C. Rodrigues, J.R.S. Botelho, A.P.S. S. Filho, N.T. Machado, R.N. Carvalho Jr., Supercritical CO_2 extraction of açaí (*Euterpe oleracea*) berry oil: Global yield, fatty acids, allelopathic activities, and determination of phenolic and anthocyanins total compounds in the residual pulp, J. Supercrit. Fluids 107 (2015) 364–369, https://doi.org/10.1016/j.supflu.2015.10.006.
[111] A. Rai, B. Mohanty, R. Bhargava, Modeling and response surface analysis of supercritical extraction of watermelon seed oil using carbon dioxide, Sep. Purif. Technol. 141 (2015) 354–365, https://doi.org/10.1016/j.seppur.2014.12.016.
[112] O. Larkeche, A. Zermane, A.H. Meniai, C. Crampon, E. Badens, Supercritical extraction of essential oil from *Juniperus communis* L. needles: Application of response surface methodology, J. Supercrit. Fluids 99 (2015) 8–14, https://doi.org/10.1016/j.supflu.2015.01.026.
[113] L.L.M. Marques, G.P. Panizzon, B.A.A. Aguiar, A.S. Simionato, L. Cardozo-Filho, G. Andrade, A. G. de Oliveira, T.A. Guedes, J.C.P. de Mello, Guaraná (*Paullinia cupana*) seeds: Selective supercritical extraction of phenolic compounds, Food Chem. 212 (2016) 703–711, https://doi.org/10.1016/j.foodchem.2016.06.028.
[114] X. Wang, C. Wang, X. Zha, Y. Mei, J. Xia, Z. Jiao, Supercritical carbon dioxide extraction of β-carotene and α-tocopherol from pumpkin: A Box-Behnken design for extraction variables, Anal. Methods 9 (2017) 294–303, https://doi.org/10.1039/c6ay02862d.
[115] B. Marongiu, A. Piras, S. Porcedda, E. Tuveri, E. Sanjust, M. Meli, F. Sollai, P. Zucca, A. Rescigno, Supercritical CO_2 extract of *Cinnamomum zeylanicum*: Chemical characterization and anti-tyrosinase activity, J. Agric. Food Chem. 55 (2007) 10022–10027, https://doi.org/10.1021/jf071938f.

CHAPTER 9

Extraction of propolis using supercritical carbon dioxide

Bruna Aparecida Souza Machado[a], João Henrique de Oliveira Reis[b], Ana Lúcia Barbosa de Souza[a], Janice Izabel Druzian[b], Fernando Luiz Pellegrini Pessoa[a]
[a]University Center SENAI CIMATEC, Health Institute of Technologies (CIMATEC ITS), National Service of Industrial Learning—SENAI, Salvador, Bahia, Brazil
[b]Federal University of Bahia (UFBA), Faculty of Pharmacy, Salvador, Bahia, Brazil

Contents

1. Introduction

Propolis is a resinous substance collected by bees from the shoots and leaves of different trees and plants that is mixed with pollen and enzymes secreted by the bees [1–3]. Propolis has varied appearance and, categorized as yellow, green, brown, or red. The color and characteristics of propolis vary according to the plant source, collection season, seasonality, and geographical origin [4, 5]. It is considered a traditional folk remedy and has been the object of several pharmacological and chemical studies in recent years [6–8].

This matrix is widely used in cosmetic formulations, foodstuffs and pharmaceuticals, and thus is considered as one of the most useful products obtained from natural sources [9, 10]. Many compounds were identified in different samples of propolis [11–14], and new compounds are still being identified during the chemical characterization of the new samples. Phenolic compounds, particularly flavonoids, have been extensively related to various properties of this natural product [5]. Many studies on the biological properties of propolis and the differences in propolis compositions have been published [10, 15, 16].

New extraction methods are being investigated to replace the classic methods, such as solvent extraction, maceration, and vacuum distillation, for obtaining extracts from natural products [17]. One of the most promising methods is supercritical fluid extraction (SFE), specifically the method that uses carbon dioxide (CO_2) as the extraction fluid ($scCO_2$) [18–21]. This technology has proven effective for use in different processes,

Green Sustainable Process for Chemical and Environmental Engineering and Science
https://doi.org/10.1016/B978-0-12-817388-6.00009-X

including chemical, petrochemical, pharmaceutical, environmental, and food processes, mainly because it is considered a clean technology. Particularly, $scCO_2$ has a lower critical temperature and pressure (31°C and 74bar) than other supercritical solvents, which favors the extraction of thermosensitive compounds [18]. In addition, CO_2 is considered nonexplosive and nontoxic, is available in high purity, fully recoverable by the process, and has a low cost [22–24]. Another advantage of SFE is the ability to preserve the chemical properties (including antioxidant capacity) of the extracts obtained due to the use of low temperatures [17, 25–27].

It is important to highlight that one of the main aspects that should be considered in the SFE is the choice of operating conditions in the extraction process. Thus, for each sample the process parameters must be optimized with the objective of increasing the extraction of the compounds of interest. Aiming to effectively optimize the conditions that affect the process (temperature, pressure, time, types and percentages of modifiers or cosolvents, and sample particle size, among others), different approaches have been applied in studies on SFE [17, 21, 25]. Thus, the first step of SFE from natural matrices, such as propolis, is to optimize the temperature and pressure conditions to maximize the yield of the compounds of interest and to minimize the coextraction of undesirable compounds [28–32]. In addition, it is possible to extend and modify the selectivity and solubility of the compounds in $scCO_2$ through the addition of polar species as cosolvents. Thus, the extraction yields of propolis components can be increased due to the polar nature of phenolic compounds.

It should be noted that propolis extracts have been preferably obtained using conventional methods, and using low pressures, such as ethanol extraction, hydroalcoholic extraction, or oil extraction [13, 33, 34]. However, the application of SFE using supercritical CO_2 combined with a cosolvent for the extraction of propolis compounds is quite feasible, as already demonstrated in different studies. In this chapter, the supercritical extraction method for obtaining propolis extracts using $scCO_2$ and the biotechnological potential of this natural matrix is described. A survey and evaluation of the main scientific papers and patent documents involving the technology of interest are also presented.

2. Propolis: Geographical origin and biological properties

Propolis has been extensively studied since 1960s, with numerous studies revealing the potential of this matrix due mainly to the presence of several chemical compounds with proven biological activities [3, 35, 36]. The studies on propolis involve different lines of research. These studies mainly involve the chemical characterization and biological activity of the extracts; standardization and comparative analysis of the extracts obtained under different extraction conditions; evaluation of the composition of the extracts relative to the geographical origin, plant source, and season; and application of the extracts for product development.

More than 300 compounds (polyphenols, terpenoids, steroids, sugars, amino acids, and others) have been detected in raw propolis from different sources. The presence and concentration of these compounds are influenced by botanical and geographical factors, as well as by the sampling season [37–39]. Thus, the composition of the propolis varies according to the geographical origin that is influenced by the plant species that inhabit the region. As already observed during some studies, the chemical composition of propolis extracts, and consequently their biological activity, also varies according to the extraction methods employed [15, 33, 40].

The propolis varieties produced in different regions of the world have been categorized according to their botanical origin and predominant chemical composition. The poplar type (*Populus* spp.) originates mainly from Europe, the nontropical regions of Asia, New Zealand, and North America and is rich in esters of caffeic acid [41]. The birch type (*Betula verrucosa*) is rich in flavonols and flavones [34] and found mainly in Russia.

The green type, also called Brazilian propolis, originates from the plant species *Baccharis dracunculifolia* [42, 43]. An important phenolic acid present in samples of Brazilian green propolis that has already been reported in different studies is Artepillin C (3,5-diprenyl-4-hydroxycinnamic acid [4, 44–46]. Chen et al. [47] and Machado et al. [30] investigated the use of different organic solvents and supercritical CO_2 in the extraction processes to recover Artepillin C from Brazilian propolis. Some studies have pointed out important biological activities associated with Artepillin C, mainly its antitumor effects [48–51]. The inhibitory effects of Artepillin C have been reported on a variety of human and murine malignant solid tumor cell lines [52–54]. Thus, green propolis is currently the most valued worldwide and has been extensively studied in recent years due to its chemical composition that is rich in different compounds, such as diterpenes, lignans, and prenylated derivatives of p-coumaric acid, as well as different flavonoids.

Another important type of propolis that has attracted the attention of researchers around the world is the red propolis, which is derived from the plant species *Dalbergia ecastaphyllum* [38, 55, 56]. The main chemical components identified in samples of red propolis have been isoflavones, flavones, flavonoids, phenolic compounds, aurones, chalcones, pterocarpans, and xanthones [57, 58]. The biological activity of red propolis is mainly due to isoflavones, formononetin and kaempferol, which act in synergy with the other compounds. Formononetin is considered one of the biomarkers of red propolis, and several studies have confirmed its presence in extracts obtained from this matrix [38, 59], and kaempferol has a wide application in traditional Chinese medicine due to its antioxidant and free-radical scavenging properties. Red propolis is produced in Brazil, Mexico, and Cuba [55, 60].

Clusia propolis (*Clusia* spp.) is found in Cuba and Venezuela. It contains polyisoprenylated benzophenone [61]; Pacific propolis (*Macaranga tanarius*) originates in Indonesia, Taiwan, and Okinawa [62]; and the most recently identified Mediterranean propolis has its main source plant species of the family as *Cupressaceae*, located in Greece, Sicily, and Malta, and is generally rich in diterpenes [63]. Brazil and China are the main propolis producing markets in the world. However, the propolis produced in these countries differs considerably due to the variety and diversity of the botanical sources. For the application of $scCO_2$ to obtain propolis extracts, Brazilian propolis, specifically the green type, has been the most investigated.

Several experimental and clinical investigations have shown the biological properties of propolis extracts obtained through classical processes (such as liquid-liquid extraction) and alternative processes (such as ultrasound and SFE). Due to the variety of chemical compounds, especially phenolic compounds, many biological activities such as antioxidant [37], antibacterial [60], fungicidal [64, 65], antiparasitic [40, 57], anticancer [66], antiviral [67], anticarious [68, 69], antiinflammatory [70], healing [6], and immunomodulatory [71] activities have been reported in recent years for propolis extracts of different types and origins. In addition to biological research, propolis extracts have been used as active ingredients for the development of food products, active packaging [72], new drugs [73], and vaccine adjuvant formulations [71, 74], among other uses.

Raw propolis cannot be used as a raw material and therefore must be purified. The process should remove the waxy material and preserve the polyphenol fraction. Thus, the industrial extraction method commonly used to obtain biocompounds from propolis is infusion, where the sample is submerged in an extraction solvent for days, weeks, or months, thus requiring a substantial amount of time on an industrial scale and still resulting in a low yield [56, 63, 75].

Due to the chemical characteristics of propolis, ethanol is currently the most used extracting solvent to obtain propolis extracts. Ethanol has interesting chemical characteristics, such as suitable polarity, to obtain the target compounds present in propolis, mainly flavonoids [28, 29]. Different studies indicate that propolis extracts can be obtained using classical or alternative methods, such as the application of ultrasound technology [76] or SFE [40, 47, 77].

3. SFE using CO_2

SFE has very interesting attributes, mainly due to its high flexibility. It is possible to obtain products of excellent quality, principally when compared to classical techniques. Other advantages of SFE are the low use of polluting organic solvents, as well as the complete elimination of the solvent at the end of the process.

Several solvents, such as water [78], ammonia [79], benzene [80], ethanol [81], methanol [82], and toluene [83] have been used in SFE. It is emphasized that, any solvent can be used as a supercritical solvent. However, the temperature and pressure conditions to reach the critical properties, as well as their toxicity, price and their solvation are the main characteristics that define the solvent most suitable for application in SFE. Due to the reasons previously discussed, CO_2 is the first-choice solvent and the solvent most used in SFE in different processes.

scCO_2 is considered a green technology for the extraction of biologically active compounds from different natural matrices [84]. This technique is very efficient in terms of selectivity and separation; however, depending on the polarity of the compounds of interest, it may be necessary to add small amounts of a modifier or cosolvent to the system to improve the extraction yield [37].

As evidenced in different studies, flavonoids are the main compounds present in propolis and are associated with the biological effects of this matrix. Among the classes of flavonoids found in propolis, the most important are: flavones, flavonols, flavanones, flavanonols, chalcones and dihydrochalcones, isoflavones, isodihydroflavones, flavans, isoflavans, and neoflavonoids [34, 85].

Most of these compounds present in propolis samples are not extractable in pure CO_2, due to the chemical characteristics. However, the extraction can be improved when a mixture of CO_2 with ethanol or ethanol and water is used, allowing its separation, and consequently the extraction, based on the molecular weight of each compound and polarity [28, 30, 77].

Given the great relevance to scCO_2 extraction as an alternative extraction process, which mainly aims to replace the use of organic solvents in the extraction of materials at the scientific and industrial scales, and considering the important and promising biological effects of propolis, some studies have been published in the last decades on the application of this technology to obtain propolis extracts and evaluate their biological effects. All the 29 scientific articles identified on the use of SFE as extraction method to obtain propolis extracts performed that CO_2 was used as the extraction fluid, with or without different cosolvents.

The identified studies focus mainly on the determination of the ideal operating conditions for the extraction of propolis extracts rich in specific components [29, 86], with or without other techniques [87], on the biological evaluation of the extracts obtained by scCO_2 [88–90], and on comparative analyses of the extracts obtained by scCO_2 with those obtained by conventional methods [28, 40, 90]. Table 1 presents the main studies on the use of scCO_2 as an extraction medium for obtaining propolis extracts of different types from geographical origins.

One of the first studies describing the application of scCO_2 for obtaining propolis extracts was by Stahl et al. [96]. The cleaning of raw propolis to remove the wax and obtain extracts rich in flavonoids was performed using scCO_2 at 600 bar and a temperature of 39.8°C. Later, Kimoto et al. [97] carried out a study on a commercial propolis extract (Japan) obtained by scCO_2 and identified its potent effect against mammary carcinogenesis in rats.

Table 1 Main studies performed using $scCO_2$ to obtain propolis extracts of different types and geographical origins

Sample type and source	Compounds of interest	Parameters used	References
New Zealand—Propolis tincture	Flavonoids and cinnamic acid derivatives	CO_2, ethanol as cosolvent, 300 bar, 60°C	Catchpole et al. [91]
Italy	Flavonoids	CO_2, 82.3–317.7 bar, 31.6–41.4°C, 30 min	Zordi et al. [29]
Southeast Brazil—green propolis	Artepillin C, p-coumaric acid and kaempferide	CO_2, ethanol + water as cosolvent, 250 bar, 50°C	Monroy et al. [92]
Southern Brazil—green propolis	Phenolic compounds and flavonoids	CO_2, ethanol as cosolvent (2%, 5%, and 7%), 100–250 bar, 30°C, 40°C, and 50°C	Biscaia et al. [28]
Southern Brazil—green	Artepillin C and p-coumaric acid	CO_2, ethanol as cosolvent (1% and 2%), 250, 350 and 400 bar and 40°C and 50°C	Machado et al. [30]
Brazil—green, red, and brown	Phenolic compounds	CO_2, ethanol as cosolvent (1%), 350 bar and 50°C	Silva et al. [40]
Brazil—green	Cinnamic acid derivatives	CO_2 ethanol as cosolvent, 207 bar, 40°C, 50°C, and 60°C	Chen et al. [47]
Brazil—dry green propolis extract	Artepillin C, p-coumaric acid and kaempferide	CO_2, 150 and 350 bar, 60°C	Paviani et al. [93]
Brazil	Phenolic compounds	CO_2, 50, 100, 150, and 200 bar and 50°C	Wang et al. [94]
Brazil—green, red, and brown	Artepillin C, p-coumaric acid and kaempferide	CO_2, ethanol as cosolvent (1%), 350 bar and 50°C	Machado et al. [90]
Brazil—green	Artepillin C	CO_2, ethanol as cosolvent, 50 bar, 50°C	Lee et al. [32]
China	Flavonoids	CO_2, ethanol as cosolvent, 276–345 bar, 45°C	You et al. [95]

The influences of the different parameters on the extraction of $scCO_2$ from Italian propolis were studied by Zordi et al. [29], focusing on the extraction yields and chemical compositions of the fractions obtained. Thus, the authors showed that pressure and time have significant linear effects on the yield and chemical profile of the extract obtained under the conditions employed. A comparative analysis with other methods (ultrasound-assisted extraction and conventional ethanol extraction) indicated that it is possible to use $scCO_2$ as a pretreatment of propolis and thus to enable more efficient extraction of the antioxidant compounds with ethanol. In addition, another option for $scCO_2$ would be to obtain compounds of interest in propolis with polar characteristics.

Some studies on the extraction of Artepillin C (the main compound of interest of green propolis) using organic solvents have been published [44, 71]; however, the selectivity demonstrated by SFE has motivated the study of the process parameters for obtaining and fractionating Artepillin C-rich extracts. Lee et al. [32] used $scCO_2$ modified with ethyl acetate to extract Artepillin C from Brazilian propolis and obtained a highly pure compound using $scCO_2$ extraction. A positive relationship between the purity of the Artepillin C in the extracts obtained by $scCO_2$ and the antioxidant capacity was also demonstrated.

According to Machado et al. [30], who specifically evaluated the process conditions for obtaining p-coumaric acid and Artepillin C from green propolis (Brazil), the presence of ethanol as a cosolvent (1%) increased the extraction capacity of these compounds on using $scCO_2$. In addition, high p-coumaric acid and Artepillin C contents were obtained at a pressure of 350 bar and temperature of 50°C. It was observed that extraction of these compounds was difficult at high temperatures and pressures, as a high degree of solubility required [98]. In the study, a cross of isotherms between 360 and 380 bar pressures was identified. In the crossover region, it was observed that the effect of temperature exerted greater influence on the increase of vapor pressure, in relation to the decrease of the density of the solvent. Paviani et al. [93] evaluated the selectivity of the process in the fractionation of dry green propolis extract to obtain some specific compounds, including Artepillin C. The supercritical process showed that the CO_2 selectivity was higher at lower pressures (150 bar) but with a lower extraction yield.

A comparative evaluation of extracts obtained by $scCO_2$ and by other conventional methods was performed. Green propolis extracts were obtained by different conventional methods. When applied to SFE, CO_2, ethanol, and water (and mixtures) were used as process solvents or as cosolvents of CO_2 [92]. In spite of the advantages of SFE, high overall yields were obtained with conventional ethanol extraction (80%) or with a two-step extraction procedure involving $scCO_2$ extraction followed by conventional ethanol extraction, resulting in higher phenolic and total flavonoid contents. The results indicated that ethanol was more selective to obtain compounds present in green propolis, such as Artepillin C and p-coumaric acid.

Biscaia et al. [28] investigated the yields of propolis extracts obtained under different conditions, SFE in one step, using only CO_2 or CO_2 with ethanol as cosolvent, and SFE in two steps, as well as other low-pressure extraction methods using different liquid organic solvents. The highest yield was obtained by Soxhlet extraction using chloroform, while for $scCO_2$, the maximum yield was obtained using 5% ethanol as a cosolvent. It has already validated in the literature that the use of ethanol as a cosolvent in the process activated the extraction yield of propolis compounds of interest [31, 86, 93].

As previously discussed, the chemical composition of propolis depends on various factors including the botanical and geographical origins, as well as the sampling time; in addition, the extraction method used directly influences the compounds of interest to be obtained [13, 33, 45]. Thus, propolis extracts of different types and origins may have different biological effects. Thus, the problem of the standardization of the chemical composition and concentrations of the compounds limits the uses of propolis.

Machado et al. [90], evaluated samples of green, red, and brown varieties of propolis of different geographical origins and the extracts obtained by $scCO_2$ were compared with those obtained by liquid ethanol extraction at atmospheric pressure, it was observed that the concentrations of specific phenolic compounds (Artepillin C, p-coumaric acid, kaempferol, and formononetin) in the extracts obtained by $scCO_2$ were much higher. However, the higher amount of extracted components of interest, the extraction yield was lower, which may indicate a greater selectivity of the extraction process [18]. Similar results were obtained by Wang et al. [94] for propolis extracts obtained by $scCO_2$, and these extracts exhibited biological effects superior to those of extracts obtained with organic solvents.

Silva et al. [40] investigated the antioxidant, antimicrobial, antiparasitic, and cytotoxic activities of several Brazilian propolis extracts (green, red, and brown) obtained by $scCO_2$ (using ethanol as a cosolvent) and by conventional ethanol extraction. It was evidenced that the propolis extracts obtained by the two techniques studied showed high antioxidant activity that varied according to the concentration of the extracted compounds; however, the propolis extracts obtained by conventional ethanolic extraction showed highest antioxidant activities and biological activities. In the most recent study involving the application of SFE, Devequi-Nunes et al. [99] investigated propolis extracts (green, red and brown) obtained by $scCO_2$ and ethanol extraction. Extracts with high antioxidant capacity were obtained; however, the extracts obtained with $scCO_2$ showed higher IC_{50} values in relation to extracts obtained with conventional ethanol extraction (infusion), which showed lower antioxidant activity. It is noteworthy that Silva et al. [40] and Devequi-Nunes et al. [99] used process parameters adopted in previous studies [28, 30] (the same pressure conditions, temperature, and type and percentage of cosolvent) to obtain extracts by $scCO_2$, which may lead to lower process efficiencies since other types of samples were used.

To obtain extracts rich in specific components, some studies have evaluated the application of different concentrations and types of cosolvents in the system [28, 30, 100] or the application of CO_2 as an antisolvent [100–102]. For example, Monroy et al. [92] obtained extracts of green propolis rich in phenolic compounds using $scCO_2$ as antisolvent. In this process, a total of four separators were used in series to selectively fractionate the ethanolic and hydroalcoholic extracts of the green propolis by precipitation.

Catchpole et al. [91] developed a process using $scCO_2$ as an antisolvent for the fractionation of a propolis tincture. The proposed method was efficient in obtaining flavonoids as well as essential oil fractions and removing high-molecular-weight components by precipitation.

$scCO_2$ has also been used in some studies to obtain propolis microparticles for application in different pharmaceutical and food products. Wu et al. [101] and Yang et al. [103] used an antisolvent process to obtain micronization and encapsulation of propolis using polyethylene glycol as the encapsulating agent. More recently, Di-Capua et al. [87] successfully encapsulated propolis through $scCO_2$-assisted atomization. Two carriers were used in the process (hydroxypropyl-β-cyclodextrin and polyvinylpyrrolidone) with the aim of increasing the shelf life of the compounds extracted from propolis, and consequently improving their bioavailability.

In general, when evaluating only the total content of phenolic compounds and flavonoids (by spectrophotometric methods), the ethanolic extracts obtained by conventional techniques were more efficient than those obtained with SFE. However, the application of $scCO_2$ becomes extremely viable for the extraction of target compounds from propolis samples, as confirmed by the chromatographic analysis of the extracts. In this case, studies of the extraction process for determining the best conditions of pressure, temperature, and type and concentration of cosolvent, among other factors, should be carried out to maximize the extraction of the compound(s) of interest. Better biological activities of propolis extracts obtained by $scCO_2$ have also been demonstrated in some studies because of the greater concentrations of the active compounds of interest. Additional advantages include the reduced use of organic solvents and preservation of the bioactivity of some compounds due to the low temperature of the extraction.

4. Patents

A survey of patent documents was conducted at Espacenet (European Patent Office), which includes applications for patents filed in more than 90 countries worldwide. A total of 27 patent documents referring to the application of $scCO_2$ for obtaining specific propolis compounds as well as extracts or applications in food, cosmetic, or pharmaceutical products were identified.

The first patent document identified was from 1993 (JPH06276966), filed in Japan, and referred to a process of cleaning the waxy and lipophilic material from raw propolis using $scCO_2$, with a flow rate of 5 mL/min at 40°C and 250 bar for 6 h. The patent documents identified had their countries of origin as China (17), Japan (4), Taiwan (4), Korea (1), and Brazil (1).

All identified documents applying SFE in propolis used CO_2 as the extraction fluid. The use of a cosolvent was also reported in some documents, with the goal of the extraction and purification of target compounds from propolis. For example, propolis extracts obtained by $scCO_2$ and ethanol as cosolvent of the process, improved the extraction efficiency of biologically active compounds present in the raw sample, including Artepillin C and p-coumaric acid.

5. Conclusion

From the results found during numerous studies involving SFE, extracts obtained by this technology showed a differentiated profile when compared to extracts obtained by classical techniques, such as liquid-liquid extraction. For propolis, the majority of the studies showed that extracts obtained by $scCO_2$ (using ethanol as cosolvent) have higher biological potential. In view of the different types and the diverse chemical composition of propolis, studying the process conditions to obtain compounds of interest is extremely feasible. For example, optimized parameters for the extraction of specific compounds present in green propolis may not be efficient for obtaining red propolis markers due to the differentiated characteristics of these compounds. Thus, $scCO_2$ proved to be very efficient in obtaining specific components of propolis samples of different types. As regard the application of $scCO_2$ to obtain propolis extracts, Brazilian propolis, specifically green propolis, has been the most investigated. Studies applying SFE with optimized conditions to obtain compounds from red propolis are necessary, considering the biotechnological potential of this new type of propolis.

References

[1] A.S. Freitas, A. Cunha, S.M. Cardoso, R. Oliveira, C. Almeida-Aguiar, Constancy of the bioactivities of propolis samples collected on the same apiary over four years, Food Res. Int. (2018), https://doi.org/10.1016/J.FOODRES.2018.10.040.
[2] K. Pobiega, K. Kraśniewska, M. Gniewosz, Application of propolis in antimicrobial and antioxidative protection of food quality—a review, Trends Food Sci. Technol. (2018), https://doi.org/10.1016/J.TIFS.2018.11.007.
[3] G.A. Burdock, Review of the biological properties and toxicity of bee propolis (propolis), Food Chem. Toxicol. 36 (1998) 347–363.
[4] J.K.S. Andrade, et al., Development and characterization of microencapsules containing spray dried powder obtained from Brazilian brown, green and red propolis, Food Res. Int. 109 (2018) 278–287.
[5] J.K.S. Andrade, M. Denadai, C.S. de Oliveira, M.L. Nunes, N. Narain, Evaluation of bioactive compounds potential and antioxidant activity of brown, green and red propolis from Brazilian northeast region, Food Res. Int. 101 (2017) 129–138.
[6] A. Oryan, E. Alemzadeh, A. Moshiri, Potential role of propolis in wound healing: biological properties and therapeutic activities, Biomed. Pharmacother. 98 (2018) 469–483.
[7] A. Kurek-Górecka, et al., Structure and antioxidant activity of polyphenols derived from propolis, Molecules 19 (2013) 78–101.
[8] M.F. Najafi, F. Vahedy, M. Seyyedin, H.R. Jomehzadeh, K. Bozary, Effect of the water extracts of propolis on stimulation and inhibition of different cells, Cytotechnology 54 (2007) 49–56.
[9] S.R. Gregory, N. Piccolo, M.T. Piccolo, M.S. Piccolo, J.P. Heggers, Comparison of propolis skin cream to silver sulfadiazine: a naturopathic alternative to antibiotics in treatment of minor burns, J. Altern. Complement. Med. 8 (2002) 77–83.

[10] H. Shokri, F. Katiraee, M. Fatahinia, M.H. Minooeianhaghighi, Chemical composition and antifungal potential of Iranian propolis against *Candida krusei* strains, J. Apic. Res. 56 (2017) 581–587.
[11] K. Wang, et al., Polyphenol-rich propolis extracts from China and Brazil exert anti-inflammatory effects by modulating ubiquitination of TRAF6 during the activation of NF-κB, J. Funct. Foods 19 (2015) 464–478.
[12] S.I. Falcão, et al., Phenolic characterization of Northeast Portuguese propolis: usual and unusual compounds, Anal. Bioanal. Chem. 396 (2010) 887–897.
[13] J. Zhang, et al., Comparisons of ethanol extracts of Chinese Propolis (Poplar Type) and poplar gums based on the antioxidant activities and molecular mechanism, Evid. Based Complement. Alternat. Med. 2015 (2015) 1–15.
[14] J.M. Sforcin, V. Bankova, Propolis: Is there a potential for the development of new drugs? J. Ethnopharmacol. 133 (2011) 253–260.
[15] S.M. Cottica, et al., Antioxidant activity and composition of propolis obtained by different methods of extraction, J. Braz. Chem. Soc. 22 (2011) 929–935.
[16] N. Kalogeropoulos, S.J. Konteles, E. Troullidou, I. Mourtzinos, V.T. Karathanos, Chemical composition, antioxidant activity and antimicrobial properties of propolis extracts from Greece and Cyprus, Food Chem. 116 (2009) 452–461.
[17] C.G. Pereira, M.A.A. Meireles, Supercritical fluid extraction of bioactive compounds: fundamentals, applications and economic perspectives, Food Bioprocess Technol. 3 (2010) 340–372.
[18] B.A.S. Machado, C.G. Pereira, S.B. Nunes, F.F. Padilha, M.A. Umsza-Guez, Supercritical fluid extraction using CO2: main applications and future perspectives, Sep. Sci. Technol. 48 (2013) 2741–2760.
[19] W.A. da Costa, et al., Supercritical CO2 extraction and transesterification of the residual oil from industrial palm kernel cake with supercritical methanol, J. Supercrit. Fluids (2018), https://doi.org/10.1016/j.supflu.2018.10.012.
[20] C.F. Silva, F.C. Moura, M.F. Mendes, F.L.P. Pessoa, Extraction of citronella (Cymbopogon nardus) essential oil using supercritical co2: experimental data and mathematical modeling, Braz. J. Chem. Eng. 28 (2011) 343–350.
[21] C. Grosso, et al., Mathematical modelling of supercritical CO2 extraction of volatile oils from aromatic plants, Chem. Eng. Sci. 65 (2010) 3579–3590.
[22] X.-L. Jin, et al., Antioxidant and anti-inflammatory effects of Chinese propolis during palmitic acid-induced lipotoxicity in cultured hepatocytes, J. Funct. Foods 34 (2017) 216–223.
[23] G. Sodeifian, S. Sajadian, A. Investigation of essential oil extraction and antioxidant activity of Echinophora platyloba DC. using supercritical carbon dioxide, J. Supercrit. Fluids 121 (2017) 52–62.
[24] G. Sodeifian, S.A. Sajadian, N. Saadati Ardestani, Optimization of essential oil extraction from Launaea acanthodes Boiss: utilization of supercritical carbon dioxide and cosolvent, J. Supercrit. Fluids 116 (2016) 46–56.
[25] P.C. Frohlich, et al., Evaluation of the effects of temperature and pressure on the extraction of eugenol from clove (Syzygium aromaticum) leaves using supercritical CO2, J. Supercrit. Fluids 143 (2019) 313–320.
[26] T. Baysal, S. Ersus, D.A.J. Starmans, Supercritical CO_2 extraction of β-carotene and lycopene from tomato paste waste, J. Agric. Food Chem. 48 (2000) 5507–5511.
[27] M.M. Barth, C. Zhou, K.M. Kute, G.A. Rosenthal, Determination of optimum conditions for supercritical fluid extraction of carotenoids from carrot (Daucus carota L.) tissue, J. Agric. Food Chem. 43 (1995) 2876–2878.
[28] D. Biscaia, S.R.S. Ferreira, Propolis extracts obtained by low pressure methods and supercritical fluid extraction, J. Supercrit. Fluids 51 (2009) 17–23.
[29] N. De Zordi, et al., The supercritical carbon dioxide extraction of polyphenols from propolis: a central composite design approach, J. Supercrit. Fluids 95 (2014) 491–498.
[30] B.A.S. Machado, et al., Determination of parameters for the supercritical extraction of antioxidant compounds from green propolis using carbon dioxide and ethanol as co-solvent, PLoS ONE 10 (2015). e0134489.
[31] L. Wang, B. Yang, X. Du, C. Yi, Optimisation of supercritical fluid extraction of flavonoids from Pueraria lobata, Food Chem. 108 (2008) 737–741.

[32] Y.-N. Lee, C.-R. Chen, H.-L. Yang, C.-C. Lin, C.-M.J. Chang, Isolation and purification of 3,5-diprenyl-4-hydroxycinnamic acid (artepillin C) in Brazilian propolis by supercritical fluid extractions, Sep. Purif. Technol. 54 (2007) 130–138.
[33] F. Hu, et al., Effects of ethanol and water extracts of propolis (bee glue) on acute inflammatory animal models, J. Ethnopharmacol. 100 (2005) 276–283.
[34] N. Zabaiou, et al., Biological properties of propolis extracts: something new from an ancient product, Chem. Phys. Lipids 207 (2017) 214–222.
[35] J. Starzyk, S. Scheller, J. Szaflarski, M. Moskwa, A. Stojko, Biological properties and clinical application of propolis. II. Studies on the antiprotozoan activity of ethanol extract of propolis, Arzneimittelforschung 27 (1977) 1198–1199.
[36] Y.K. Park, S.M. Alencar, C.L. Aguiar, Botanical origin and chemical composition of brazilian propolis, J. Agric. Food Chem. 50 (2002) 2502–2506.
[37] M.A. Salazar, et al., Chemical composition, antioxidant activity, neuroprotective and anti-inflammatory effects of cipó-pucá (Cissus sicyoides L.) extracts obtained from supercritical extraction, J. Supercrit. Fluids 138 (2018) 36–45.
[38] B.B. Silva, et al., Chemical composition and botanical origin of red propolis, a new type of brazilian propolis, Evid. Based Complement. Alternat. Med. 5 (2008) 313–316.
[39] S.M. Alencar, et al., Chemical composition and biological activity of a new type of Brazilian propolis: red propolis, J. Ethnopharmacol. 113 (2007) 278–283.
[40] R.P. Dantas Silva, et al., Antioxidant, antimicrobial, antiparasitic, and cytotoxic properties of various Brazilian propolis extracts, PLoS ONE 12 (2017). e0172585.
[41] B. Anđelković, et al., Metabolomics study of Populus type propolis, J. Pharm. Biomed. Anal. 135 (2017) 217–226.
[42] J.T. Paula, I.M.O. Sousa, M.A. Foglio, F.A. Cabral, Selective fractionation of supercritical extracts from leaves of Baccharis dracunculifolia, J. Supercrit. Fluids 127 (2017) 62–70.
[43] J.T. Paula, et al., Scale-up study of supercritical fluid extraction process for Baccharis dracunculifolia, J. Supercrit. Fluids 107 (2016) 219–225.
[44] N. Paulino, et al., Anti-inflammatory effects of a bioavailable compound, Artepillin C, in Brazilian propolis, Eur. J. Pharmacol. 587 (2008) 296–301.
[45] B.C.B.S. Mello, M.D. Hubinger, Antioxidant activity and polyphenol contents in Brazilian green propolis extracts prepared with the use of ethanol and water as solvents in different pH values, Int. J. Food Sci. Technol. 47 (2012) 2510–2518.
[46] R. Lima Cavendish, et al., Antinociceptive and anti-inflammatory effects of Brazilian red propolis extract and formononetin in rodents, J. Ethnopharmacol. 173 (2015) 127–133.
[47] C.-R. Chen, Y.-N. Lee, M.-R. Lee, C.-M.J. Chang, Supercritical fluids extraction of cinnamic acid derivatives from Brazilian propolis and the effect on growth inhibition of colon cancer cells, J. Taiwan Inst. Chem. Eng. 40 (2009) 130–135.
[48] Y. Uto, et al., Artepillin C isoprenomics: Design and synthesis of artepillin C isoprene analogues as lipid peroxidation inhibitor having low mitochondrial toxicity, Bioorg. Med. Chem. 14 (2006) 5721–5728.
[49] M.-R. Ahn, et al., Suppression of tumor-induced angiogenesis by Brazilian propolis: major component artepillin C inhibits in vitro tube formation and endothelial cell proliferation, Cancer Lett. 252 (2007) 235–243.
[50] I.J. Camuri, A.B. Costa, A.S. Ito, W.M. Pazin, Optical absorption and fluorescence spectroscopy studies of Artepillin C, the major component of green propolis, Spectrochim. Acta Part A Mol. Biomol. Spectrosc. 198 (2018) 71–77.
[51] S. Endo, et al., Autophagy inhibition enhances anticancer efficacy of artepillin C, a cinnamic acid derivative in Brazilian green propolis, Biochem. Biophys. Res. Commun. 497 (2018) 437–443.
[52] T. Kimoto, et al., Apoptosis and suppression of tumor growth by artepillin C extracted from Brazilian propolis, Cancer Detect. Prev. 22 (1998) 506–515.
[53] T. Kimoto, et al., Pulmonary carcinogenesis induced by ferric nitrilotriacetate in mice and protection from it by Brazilian propolis and artepillin C, Virchows Arch. 438 (2001) 259–270.
[54] Y. Konishi, Transepithelial transport of artepillin C in intestinal Caco-2 cell monolayers, Biochim. Biophys. Acta Biomembr. 1713 (2005) 138–144.

[55] A.L. Piccinelli, et al., Cuban and Brazilian red propolis: botanical origin and comparative analysis by high-performance liquid chromatography-photodiode array detection/electrospray ionization tandem mass spectrometry, J. Agric. Food Chem. 59 (2011) 6484–6491.
[56] B.G.-C. López, E.M. Schmidt, M.N. Eberlin, A.C.H.F. Sawaya, Phytochemical markers of different types of red propolis, Food Chem. 146 (2014) 174–180.
[57] Á. Sena-Lopes, et al., Chemical composition, immunostimulatory, cytotoxic and antiparasitic activities of the essential oil from Brazilian red propolis, PLoS ONE 13 (2018). e0191797.
[58] B. Bueno-Silva, et al., Brazilian red propolis attenuates inflammatory signaling cascade in LPS-activated macrophages, PLoS ONE 10 (2015). e0144954.
[59] B. Trusheva, et al., Bioactive constituents of brazilian red propolis, Evid. Based Complement. Alternat. Med. 3 (2006) 249–254.
[60] M.S. Regueira, et al., Seasonal variation of Brazilian red propolis: antibacterial activity, synergistic effect and phytochemical screening, Food Chem. Toxicol. 107 (2017) 572–580.
[61] O. Cuesta-Rubio, et al., Chemical characterization of Cuban propolis by HPLC – PDA, HPLC – MS, and NMR: the *Brown*, *Red*, and *Yellow* Cuban varieties of propolis, J. Agric. Food Chem. 55 (2007) 7502–7509.
[62] S. Kumazawa, M. Murase, N. Momose, S. Fukumoto, Analysis of antioxidant prenylflavonoids in different parts of Macaranga tanarius, the plant origin of Okinawan propolis, Asian Pac J Trop Med 7 (2014) 16–20.
[63] K. Graikou, M. Popova, O. Gortzi, V. Bankova, I. Chinou, Characterization and biological evaluation of selected Mediterranean propolis samples. Is it a new type? LWT Food Sci. Technol. 65 (2016) 261–267.
[64] B. Pippi, et al., *In vitro* evaluation of the acquisition of resistance, antifungal activity and synergism of Brazilian red propolis with antifungal drugs on *Candida* spp, J. Appl. Microbiol. 118 (2015) 839–850.
[65] M.V.M. das Neves, T.M.S. da Silva, E. de O. Lima, E.V.L. da Cunha, E. de J. Oliveira, Isoflavone formononetin from red propolis acts as a fungicide against Candida sp., Braz. J. Microbiol. 47 (2016) 159–166.
[66] K.R. Begnini, et al., Brazilian red propolis induces apoptosis-like cell death and decreases migration potential in bladder cancer cells, Evid. Based Complement. Alternat. Med. 2014 (2014) 639856.
[67] V. Bankova, A.S. Galabov, D. Antonova, N. Vilhelmova, B. Di Perri, Chemical composition of Propolis Extract ACF® and activity against herpes simplex virus, Phytomedicine 21 (2014) 1432–1438.
[68] B. Bueno-Silva, et al., Effect of neovestitol-vestitol containing Brazilian red propolis on accumulation of biofilm in vitro and development of dental caries in vivo, Biofouling 29 (2013) 1233–1242.
[69] A. Sabir, A. Sumidarti, Interleukin-6 expression on inflamed rat dental pulp tissue after capped with Trigona sp. propolis from south Sulawesi, Indonesia, Saudi J. Biol. Sci. 24 (2017) 1034–1037.
[70] J.L. Machado, et al., Brazilian Green propolis: anti-inflammatory property by an immunomodulatory activity, Evid. Based Complement. Alternat. Med. 2012 (2012) 1–10.
[71] G. Fischer, et al., Green propolis phenolic compounds act as vaccine adjuvants, improving humoral and cellular responses in mice inoculated with inactivated vaccines, Mem. Inst. Oswaldo Cruz 105 (2010) 908–913.
[72] S.S. Costa, J.I. Druzian, B.A.S. Machado, C.O. De Souza, A.G. Guimaraes, Bi-functional biobased packing of the cassava starch, glycerol, licuri nanocellulose and red propolis, PLoS ONE 9 (2014).
[73] M. Franchin, et al., The use of Brazilian propolis for discovery and development of novel anti-inflammatory drugs, Eur. J. Med. Chem. 153 (2018) 49–55.
[74] E.S.H. El Ashry, T.A. Ahmad, The use of propolis as vaccine's adjuvant, Vaccine 31 (2012) 31–39.
[75] K. Doi, et al., Chemopreventive action by ethanol-extracted Brazilian Green Propolis on post-initiation phase of inflammation-associated rat colon tumorigenesis, In Vivo (Brooklyn) 31 (187–198) (2017).
[76] V.A. Taddeo, F. Epifano, S. Fiorito, S. Genovese, Comparison of different extraction methods and HPLC quantification of prenylated and unprenylated phenylpropanoids in raw Italian propolis, J. Pharm. Biomed. Anal. 129 (2016) 219–223.
[77] L.C. Paviani, et al., Supercritical CO2 extraction of raw propolis and its dry ethanolic extract, Braz. J. Chem. Eng. 29 (2012) 243–251.

[78] T. Anekpankul, M. Goto, M. Sasaki, P. Pavasant, A. Shotipruk, Extraction of anti-cancer damnacanthal from roots of Morinda citrifolia by subcritical water, Sep. Purif. Technol. 55 (2007) 343–349.

[79] J. Bludworth, F. Carl Knopf, Reactive extraction of lignin from wood using supercritical ammonia-water mixtures, J. Supercrit. Fluids 6 (1993) 249–254.

[80] J.-W. Shen, K.B. Domański, O. Kitao, K. Nakanishi, Computer simulation on supercritical carbon dioxide fluid a potential model for the benzene-carbon dioxide system from ab initio calculations, Fluid Phase Equilib. 104 (1995) 375–390.

[81] Y. Marcus, Total and partial solubility parameters of sub- and supercritical ethanol, J. Chem. Thermodyn. 126 (2018) 187–189.

[82] N. Lamba, S. Adhikari, J.M. Modak, G. Madras, Catalytic synthesis of fatty acid methyl esters from Madhuca indica oil in supercritical methanol, Energy Convers. Manag. 173 (2018) 412–425.

[83] J.R. Kershaw, P.J. Smart, Extraction of coal-tar pitch with supercritical toluene, J. Supercrit. Fluids 6 (1993) 155–163.

[84] E. Reverchon, I. De Marco, Supercritical fluid extraction and fractionation of natural matter, J. Supercrit. Fluids 38 (2006) 146–166.

[85] V. Bankova, Recent trends and important developments in propolis research, Evid. Based Complement. Alternat. Med. 2 (2005) 29–32.

[86] P.C. Veggi, R.N. Cavalcanti, M.A.A. Meireles, Production of phenolic-rich extracts from Brazilian plants using supercritical and subcritical fluid extraction: Experimental data and economic evaluation, J. Food Eng. 131 (2014) 96–109.

[87] A. Di Capua, A. Bejarano, R. Adami, E. Reverchon, Preparation and characterization of Chilean propolis coprecipitates using supercritical assisted atomization, Chem. Eng. Res. Des. 136 (2018) 776–785.

[88] P. Bhargava, et al., Anticancer activity of the supercritical extract of Brazilian green propolis and its active component, artepillin C: bioinformatics and experimental analyses of its mechanisms of action, Int. J. Oncol. 52 (2018) 925–932.

[89] C.-R. Chen, Y.-N. Lee, C.-M.J. Chang, M.-R. Lee, I.-C. Wei, Hot-pressurized fluid extraction of flavonoids and phenolic acids from Brazilian propolis and their cytotoxic assay in vitro, J. Chin. Inst. Chem. Eng. 38 (2007) 191–196.

[90] B.A.S. Machado, et al., Chemical composition and biological activity of extracts obtained by supercritical extraction and ethanolic extraction of brown, green and red propolis derived from different geographic regions in Brazil, PLoS ONE 11 (2016). e0145954.

[91] O. Catchpole, J. Grey, K. Mitchell, J. Lan, Supercritical antisolvent fractionation of propolis tincture, J. Supercrit. Fluids 29 (2004) 97–106.

[92] Y.M. Monroy, et al., Brazilian green propolis extracts obtained by conventional processes and by processes at high pressure with supercritical carbon dioxide, ethanol and water, J. Supercrit. Fluids 130 (2017) 189–197.

[93] L.C. Paviani, C. Dariva, M.C. Marcucci, F.A. Cabral, Supercritical carbon dioxide selectivity to fractionate phenolic compounds from the dry ethanolic extract of propolis, J. Food Process Eng. 33 (2010) 15–27.

[94] B.-J. Wang, Y.-H. Lien, Z.-R. Yu, Supercritical fluid extractive fractionation—study of the antioxidant activities of propolis, Food Chem. 86 (2004) 237–243.

[95] G.-S. You, S.-C. Lin, C.-R. Chen, W.-C. Tsai, C.J. Chang, W.-W. Huang, Supercritical carbon dioxide extraction enhances flavonoids in water-soluble propolis, J. Chin. Inst. Chem. Eng. 33 (2002) 233–242.

[96] E.(.E.). Stahl, K.-W. Quirin, D. Gerard, Dense Gases for Extraction and Refining, Springer, Berlin Heidelberg, 1988.

[97] N. Kimoto, et al., Post-initiation effects of a super critical extract of propolis in a rat two-stage carcinogenesis model in female F344 rats, Cancer Lett. 147 (1999) 221–227.

[98] C.-H. Lin, et al., Extraction of lignans from the seed of Schisandra chinensis by supercritical fluid extraction and subsequent separation by supercritical fluid simulated moving bed, J. Supercrit. Fluids 98 (2015) 17–24.

[99] D. Devequi-Nunes, et al., Chemical characterization and biological activity of six different extracts of propolis through conventional methods and supercritical extraction, PLoS ONE 13 (2019). e0207676.

[100] Y.M. Monroy, R.A.F. Rodrigues, M.V.N. Rodrigues, F.A. Cabral, Fractionation of ethanolic and hydroalcoholic extracts of green propolis using supercritical carbon dioxide as an anti-solvent to obtain artepillin rich-extract, J. Supercrit. Fluids 138 (2018) 167–173.
[101] J.-J. Wu, et al., Supercritical carbon dioxide anti-solvent process for purification of micronized propolis particulates and associated anti-cancer activity, Sep. Purif. Technol. 70 (2009) 190–198.
[102] C.-R. Chen, et al., Precipitation of sub-micron particles of 3,5-diprenyl-4-hydroxycinnamic acid in Brazilian propolis from supercritical carbon dioxide anti-solvent solutions, J. Supercrit. Fluids 50 (2009) 176–182.
[103] S.-J. Yang, et al., Encapsulation of propolis flavonoids in a water soluble polymer using pressurized carbon dioxide anti-solvent crystallization, J. Supercrit. Fluids 94 (2014) 138–146.

CHAPTER 10

Solubility of pharmaceutical compounds in supercritical carbon dioxide: Application, experimental, and mathematical modeling

Mojhdeh Baghbanbashi[a], Naghmeh Hadidi[b], Gholamreza Pazuki[a]
[a]Department of Chemical Engineering, Amirkabir University of Technology (Tehran Polytechnic), Tehran, Iran
[b]Department of Clinical Research and Electronic Microscope, Pasteur Institute of Iran, Tehran, Iran

Contents

1. Introduction

Supercritical fluid (SCF) technique was reported as an applicable method in extraction, separation, and crystallization procedure that are main steps in active pharmaceutical ingredients (APIs) synthesis and particle engineering. Particle design of APIs is very critical in making solid and liquid dosage forms. It might be used to modify and control various particle properties including various sizes, shapes, crystallinity structures, polymorphism, processability, and solubilities. For example, in pharmaceutics and medical chemistry, supercritical carbon dioxide (SC-CO_2) is applied for preparation of solid dispersion to enhance the solubility of drugs in water. This would be achieved by

Green Sustainable Process for Chemical and Environmental Engineering and Science
https://doi.org/10.1016/B978-0-12-817388-6.00010-6

polymorphic control and complex formation behavior of APIs and excipients by carbon dioxide (CO_2) processing in SCF technique.

It is said that the main target of pharmaceutical industry is to formulate and manufacture safe and stable drug products with required content uniformity and labeled amount along with specified qualities and acceptance criteria stated in the pharmacopeia. The quality depends on organoleptic properties of the substances and reagents as well as production procedure. So that it would be inconceivable to gain the specification needed for consistency in products without considering the role of particle engineering [1].

APIs may exist in a number of solid states such as polymorphs, solvents, desolvates (e.g., hydrous/anhydrous), and amorphous solutes. These solid states are different in physical appearance, chemical properties, melting temperature, solubility, and bioavailability. Solid dispersions, conjugation, and micronization are technological approaches for enhancing dissolution and bioavailability of drugs, which are practically insoluble or slightly soluble in water [1]. In this chapter, we would focus first on current literatures about supercritical technology applications in particle engineering including crystal and polymorph modification, complex formation, solid dispersion, and other methods, which has been used for enhancement of API solubility, bioavailability, and processability. Next, the solubility measurement methods of drugs in SC-CO_2 and the application of SC-CO_2 as an antisolvent or cosolvent in pharmaceutical processes have been explained. Finally, the mathematical models such as equations of state (EOSs), solution model, conductor-like screening model (COSMO), artificial neural network (ANN) system, and molecular dynamic simulation have been studied for prediction of pharmaceutical compounds solubility in SC-CO_2.

1.1 Crystal modification

Particle design is a critical step in controlling physicochemical properties of drug substances. These characteristics have key roles in formulation and industrial scale-ups. For example, polymorphic form of a drug substance would directly influence solubility, dissolution profile and stability of pharmaceuticals as well as absorption and bioavailability. So that the polymorphic control is vital to reach required therapeutic plasma levels of APIs. Among newly developed technologies, SCF technique has been found as more accessible and relatively newer methods to produce particles for medicinal applications. SC-CO_2 is most widely used among other methods. In this method, the operating temperature and pressure is lower than other methods. Nonflammability, nontoxicity, and inexpensiveness have been reported as other advantages of this technique. SC-CO_2 is a micronization method which fine drug particle would be formed either by the rapid expansion of supercritical solution (RESS) or though gas-saturated solution (PGSS) in SC-CO_2 [1]. Particle size reduction would be achieved by super saturation of drug substances in SC-CO_2 during RESS process. Alternatively, gas or supercritical antisolvent

(GAS or SAS) are methods in which SC-CO_2 is used as an antisolvent. Solution-enhanced dispersion and aerosol solvent extraction systems (ASES) are other by SCF methods, which have been run by antisolvent approach [1].

1.1.1 Application of SC-CO_2 in polymorphism and polymorphic transformation

Drug substances and pharmaceutical excipients can be categorized as polymorphs and amorphous. Polymorphs are similar in chemical formula. However, they are different in molecular arrangement and crystallinity. If a drug substance is in pesudopolymorph state, it means that there are crystalline adducts containing water or other solvent molecules within the crystal structure. On the other hands, amorphous state has defective molecular arrangements and higher surface energy which resulted in less stability and greater dissolution rate. That is why polymorphs have various thermodynamic properties and in consequences the less thermodynamically stable, the more solubility and dissolution rate would be seen in amorphous state. The polymorphic transformation from less to more stable state would occur by alteration of physicochemical properties. Conventional crystallization techniques are multistep process with mixture of different polymorphs as final products but SCF technique is a single-step process and would be preferable in controlling the crystallization of a single polymorph state as final product [1].

1.1.2 Application of SC-CO_2 in crystal polymorphs preparation by nonsolvent method

SCF systems are different in crystallization segment. In the RESS technique, SC-CO_2 is applied as a solvent. During rapid expansion of sprayed solution, sudden decrease in solvent density would cause super saturation and resultant precipitation of drug. The RESS method is used for generation of phenyl butazone particles. Phenyl butazone, a nonsteroidal antiinflammatory drug (NSAID), has five polymorphs and two pseudopolymorphs with different stabilities and dissolution profile. It has been reported that RESS method is successfully applied for the production of fine particles of phenyl butazone with improved wettability but less stability. Tolbutamide, an antidiabetic medicine, has more than four polymorphs. Researches revealed that the crystalline structure of tolbutamide depends on the conditions of the RESS procedure such as pressure, temperature, and the supersaturation level. The purity and single polymorph formation specially depend on temperature and pressure during application of the SC-CO_2 in particle engineering.

1.1.3 Application of SC-CO_2 as antisolvent in crystal polymorphs preparation (SEDS)

Solution-enhanced dispersion by supercritical fluids (SEDS) is an antisolvent technique, similar to GAS and ASES. The advantages of this method are that the product would be crystallizing micronized particles with pure polymorphs state. SEDS is a one-step procedure in which the drug solution and the SCF are entered at the same time into the particle production chamber under controlled condition of temperature and pressure. Fine particle formation would be facilitated in one-step process. Sulfathiazole which has four polymorphs was used as a model drug in one study. The SEDS process with low temperature would lead to the production of forms II, III, and IV while at high-temperature form I was the output. Three polymorphs were produced by using methanol as solvent . The polymorph state was also dependent on running temperature [1].

1.1.4 Application of SC-CO_2 in drug particle design

According to the biopharmaceutical classification system (BCS), APIs for oral administration would be classified into four groups of I, II, III, and IV depending on the solubility and permeability. Studies showed nearly 40% of the new molecular entities (NMEs) belong to class II and class IV of BCS which are known to have low intrinsic solubility. This means that water solubility is a big problem with direct negative effect on oral bioavailability [2].

It has been reported that low solubility would affect drug absorption and would lead to nonlinear pharmacokinetics along with intra- and interindividual absorption variability. Therefore, improving drug solubility with pharmaceutical techniques would be desirable. Numerous approaches were applied to enhance solubility of hydrophobic drugs. Table 1 summarizes some of the technologies that their products were launched in the market [2].

Solubility would be normally improved by three methods. Micronization is the first method, in which, dissolution rate is related to drug surface area. Micronization would become a method of choice for high drug loading and is more expensive than other technologies. In the second method of solubility enhancement, the main target is to alter crystal structure to more entropy and enthalpy than the steady-state crystalline form. The second approach would be focus on composite particle formation of drug substances by dispersion technologies. The basis of this approach would be alteration of crystal structures to obtain isoforms with higher entropy and enthalpy to improve wettability and dissolution rate of nonsoluble drug substances in water. But this method has some disadvantages such as less chemical and physical stability which would make more difficulties during production, industrial scale-up, sales programming, and market launch. The third method is practical in the formulation of liquid dosages forms. The liquid formulation is the dosages form of choice to launch pharmaceuticals with low solubility especially in parenteral form because of ease of processing. The SCF technology has been applied to enhance both solubility and processability of products. Table 1 summarizes marketed products using SCF technology.

It should be noted that, SCF technology has found its way in various industries. However, its outstanding advantages, which are producing drug substances with appropriate aerodynamic diameters with low polydispersity index (PDI). Resulted in gaining popularity in pharmaceutical and nutraceutical especially in pulmonary drug delivery of dry powders. CO_2 is widely used as SCF due to its low critical point and nontoxic nature and it suitable for various biological productions (marketed approaches to solubility

Table 1 Marketed approaches to solubility improvement of hydrophobic active pharmaceutical ingredients (APIs) [2]

Categories	Description	Representative	Owners	Marketed products
Micronization (nanoparticle) Downsizing	Milling	NanoCrystal	Elan	Rapamune, Emen
	Homogenization	IDD	SkyePharma	Triglide
		Nanopure	PharmaSol	Under development
		Nanoedge	Baxter	Under development
	SCF	SCF milling	DuPont	Under development
Build-up	SCF	SFEE	Ferro Corp	Under development
		SAS-EM	Thar Tech	Under development
	Percipitation	BioAqueous^{SM}	Dow	Under development
Composite particle (solid dispersion, solid solution, etc)	HMT	Meltrex	SOLIQS	Verapamil SR
	Modeling	Lipid base	Camurus AB	Elyzol
		Microtrol	Supernus Pharm	Carbatrol
	Spray freezing	BioAqueous^{SM}	Dow	Under development
	SCF	Expansion	Ferro Corp	Under development
		Infuse-X	Lavipharma	Under development
Liquid-like formulation	Complexation	Captisol	Cydex	Geodon, Vfend
	Soft gelatin	RPScherersol	Cardinal Health	Lots of OTC
	Lipid emulsion	TOCOSOL	SONUS	Under development
		SolEmuls	PharmaSol	Under development
	Self-emulsions	SMEDDS	Gattefosse	Under development
	Micelles	NanoCap	NanoCarrier	Under development
	SCF	Nanosomes	Aphios Corporation	Under development

Table 2 Classification of SCF methods in pharmaceutical industry [2]

Rule of SCF	Available process candidate	Critical factors of SCF for particle production	Typical types of produced particles
Solvent	RESS	Solubility of materials, preexpansion conditions, spray device	Micronized particles, composite particles
	Impregnation	Solubility of materials, plasticization, diffusion property	composite particles
	RESS-coating	Solubility of materials, preexpansion conditions, spray device	Microcapsules
Antisolvent	Antisolvent processes	Solubility of materials, solvent extraction, mass transfer	Micronized particles, composite particles
Others	Expansion (PGSS)	Preexpansion, diffusion property, mass transfer	Porous materials, composite particles
	Milling	Solubility of materials, diffusion property, viscosity	Micronized particles

improvement of hydrophobic APIs Table 2). SC-CO_2 is applied (1) for APIs production, (2) as antisolvent for the salting out of materials solved in organic solvents, and (3) as a medium [2].

1.2 Application of SC-CO_2 in separation and reaction processes

SCFs were used for the purification of pharmaceutical APIs from mixtures. SCF technology is effective in controlling the individual and total impurities. This would result in improving physical and chemical stability of produced drug substances Fig. 1 summarizes SCF application in separation, purification, and reaction process of chemicals [3].

In the 1980s, SCF application was explored in various industrial features including extraction processes. The most investigated field was food industry with more focus on decaffeination. Nowadays, SCF is extensively used in the extraction of essential oils and flavors to be used in natural products, herbal medicines, nutraceuticals, cosmecuticals, chemical reactions, and analytical applications. Safety considerations and low costs have made SC-CO_2 more preferable in comparison to ethylene, methane, nitrogen, and xenon [3]. SC-CO_2, with preferable critical properties of 3041.25 K and 73.8 bar is suitable for heat-sensitive materials such as proteins and peptides [4].

SCFs are also achieving more importance as alternative to organic solvents in chemical synthesis. SCF technique showed more advantages in replacing traditional hazardous organic solvents and yield reaction optimization by controlling temperature, pressure, and fluid density [3].

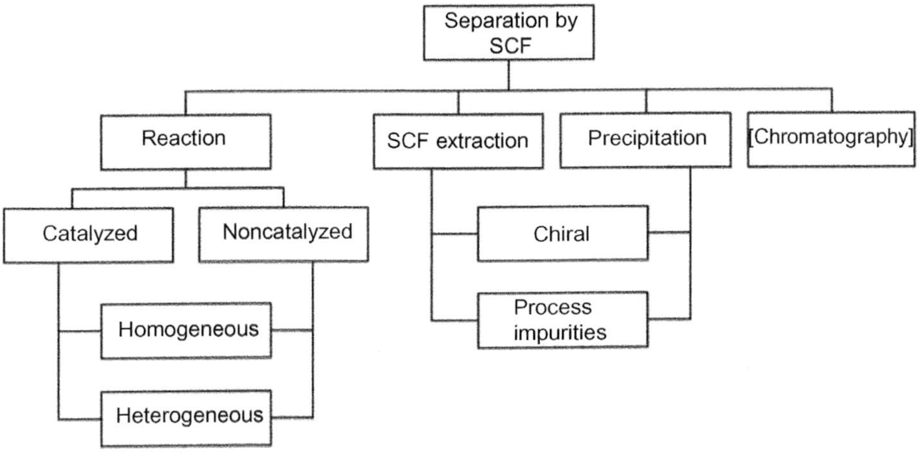

Fig. 1 Overview of SCF separation techniques [3].

Literature review demonstrated that SCF reaction-mediated synthesis were more rapid than traditional ones with higher alteration rates and more simple downstream processing. Using of stereo-specific biocatalysts is very interesting and demonstrated great future prospects. Crystallization is one of the most common purification methods in chemical synthesis (Fig. 2). The implementation of the GAS crystallization to purification aims was investigated in numerous scholars. These include the purification of carotenoids, proteins, anthraquinone, lecithin from egg yolk, racemic mixtures, and so on [3].

Several investigators also reported that the residual solvent levels are low after processing with SC-CO_2. It seems that the affinity and/or solubility of a solute in a particular solvent and extraction time affect the final of residual solvent level. Several surveys showed that supercritical processing may be more efficient in isolation of small quantities of metastable forms than conventional techniques. Besides, by SCF method, we can

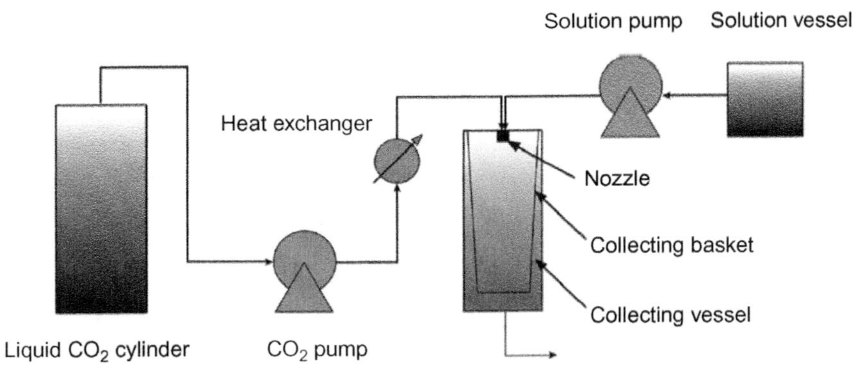

Fig. 2 Antisolvents type in SCF process [3].

produce pure polymorphic forms. Separation is often dependent on the phase behavior and so that are predictable by using thermodynamic principles.

Currently, more than half of the marketed pharmaceuticals are chiral and administered as racemic mixtures. Increasing information regarding the benefits of enantio pure chiral medicines together with their different pharmacological profiles causes more interest in marketing of these products. Investigations showed that SCFs technique has been applied successfully to produce highly pure enantiomers. However, antisolvent SCFs show lower yields in comparison with conventional organic solvents. Other interesting field in the research is implementation of the GAS process for chiral resolution of diastereomeric salts. The GAS process is thermodynamically driven and it is proposed to be a more commercially viable process for the separation of conglomerates together with simultaneous production of particles of desired physical properties in comparison to the RESS process [3, 5].

2. Solubility of pharmaceutics in SC-CO_2

Solubility data are very important and useful thermodynamic property to design and increase the scale of the extraction system in all processes that use SCF technology. Efficiency, product rate, design, and process economics can be optimized by knowledge of solubility. In addition, solubility affects the size, shape, surface structure, and morphology of the products. Therefore, the measurement and prediction of the solubility of a particular solute in the supercritical solvent is necessary before performing any process using SCFs.

2.1 Solubility measurement methods of solid solutes in supercritical solvent

In the supercritical conditions, the solute can create a metastable equilibrium with the solvent. Adding cosolvent or crystalline grains also increases the solubility. In general, the static and dynamic methods are applied for solubility measurement of a solid in the supercritical solvent, which will be discussed in the following.

2.1.1 Static method

In static method, sampling is performed to check solubility after achieving the equilibrium between solute and solvent. In this method, the equilibrating procedure is achieved using many ways, such as recirculation, agitation made by a magnetic stirrer, or by solvent settling in equilibrium cell for a specified period. Depending on the type of vessel, the static method can be divided into three categories: analytic, synthetic, and gravimetric [6]. Fig. 3 shows a schematic of the static method [7].

Static-analytic method: in this method, the volume of the vessel is constant and the solute is exposed to a certain volume of SCF. The resulting fluid is sampled and the solubility of the solute is measured. The main part of static analytic method is the

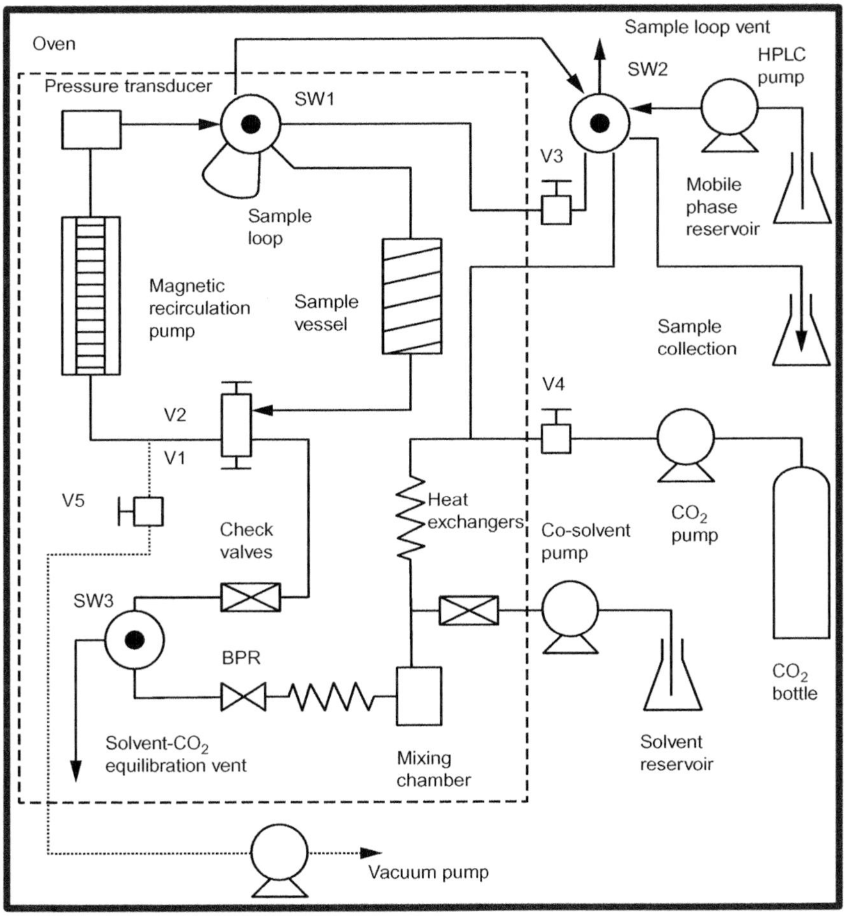

Fig. 3 Experimental system for static method for determining the solubility in the SCF [7].

sampling procedure in which a small volume of equilibrium solution is removed from the cell and transferred quantitatively to the sample flask. In this way, the cell should always remain in equilibrium state, which is very difficult in reality. Therefore, different methods are used to keep the system as close to ideal state as possible. For instance, few samples are removed and the sampling is carried out in such a way to have a small effect on the pressure drop and also the valves are opened carefully for the sampling procedure and only very small amount of the solution is taken for sampling to keep the sample lines air tight. A narrow capillary performs sampling in static analytic method. The specimen is then expanded into a flask and separated into gaseous and condensed phases. The amount of gas phase is determined by measuring its volume or obtaining its pressure in a known expansion volume. The amount of condensed phase is obtained by weighing or by more accurate analysis such as gas chromatography (GC) for multicomponent systems.

Static-synthetic method: this is used to examine the equilibrium of multicomponent systems. In this method, the temperature of the vessel is controlled and a liquid mercury or a movable piston in the cell adjusts the pressure of the system. In static-synthetic method, as well as other methods, the solute is well dissolved in the supercritical solvent. The vessel has a window for monitoring the cloud point. Typically changing the pressure is preferred to temperature variation since it is easier to control volume and pressure variations.

Static-gravimetric method: in static-gravimetric method, the solute is placed in a small vial in a pressure vessel containing SCF. The vial is designed in such a way that only the dissolved solute can penetrate from the lid to the pressure vessel. First, a certain amount of solute is inserted into the vial and the pressure vessel is filled with SCF. By dissolving the solute in the vial within the SCF, the dissolved part is eliminated from the vial by SCF and finally reaches the equilibrium. The pressure of the vessel is then decreased and the residual solute in the vial is weighed.

2.1.2 Dynamic method

In this method, the condensed solute remains in the cell, while the SCF streams continuously. The SCF is in contact with solute and achieve equilibrium during the residence of the SCF in the equilibrium cell. Then, the solution is removed and the concentrations are measured. Although the SCF residence time is not adequate for full contact with the solute and is not completely mixed with the solute, but it is assumed that the output flow has reached equilibrium. With the precise selection of operating conditions, the assumption will be acceptable. This is a disadvantage of the dynamic method and can be improved by recirculation of the solvent flow [6]. The dynamic method for determining the solubility in the SCF is shown in Fig. 4 [7].

2.1.3 Validity of the experimental values

Before measuring the drug solubility in SC-CO_2, it is necessary to evaluate the accuracy of the used equipment and the values measured by these devices. Therefore, the solubility of a common composition in SC-CO_2, reported in literature, is measured using the device applied and its results are compared with the existing experimental values.

2.2 Solubility of pharmaceutical compounds in SC-CO_2

One of the needs in the pharmaceutical industry is the supply of products with a high purity. On the other hand, SCFs have been used in various processes because of their great permeability and small amount of viscosity. One of the applications of SCF is drug treatment and extraction without contamination with other compounds, in which it is necessary to know the drug solubility in SCF. So far, many studies have been conducted to measure the solubility of various drug compounds in SC-CO_2. As an

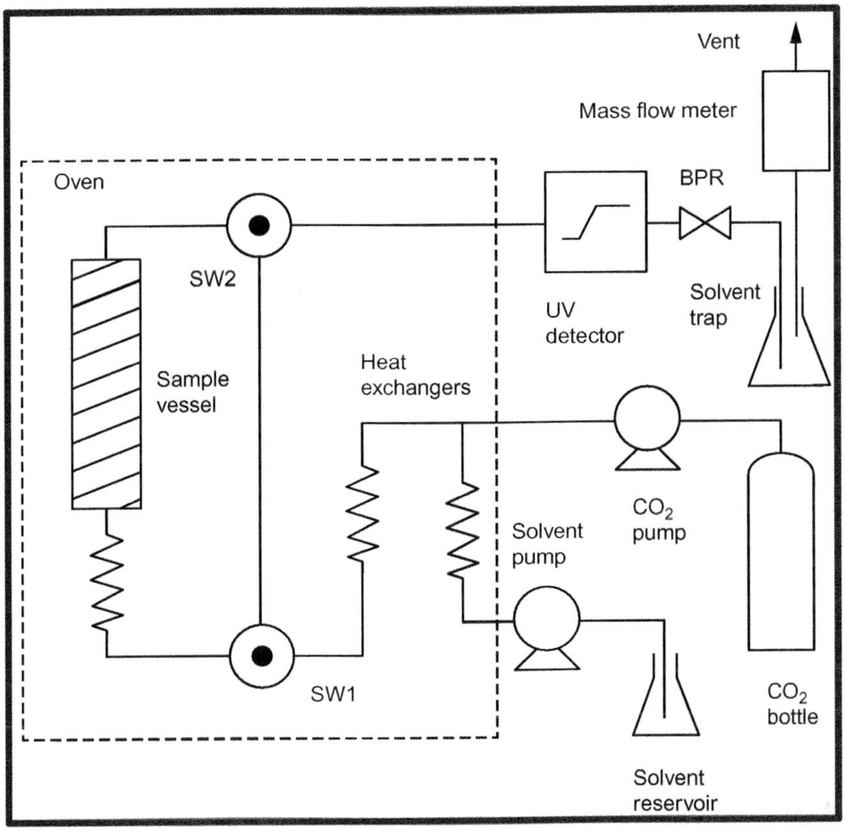

Fig. 4 Experimental system for dynamic method for determining the solubility in the SCF [7].

example, the solubility of cetirizine [8] in SC-CO_2 was measured by the static method. This drug's solubility was measured at 308.15–338.15 K and 16–40 MPa. The solubility of cetirizine in terms of the mole fraction, presented in Fig. 5, has a maximum standard error of 7.1%.

The density of SC-CO_2 increases by increasing the pressure, which enhances the dissolution ability of SC-CO_2, reduces the intermolecular space between the CO_2 molecules, and consequently increases interaction between the CO_2 and cetirizine molecules. Therefore, by increasing the operating pressure, at all temperatures, the drug solubility has increased. Temperature is another effective factor on the solubility of cetirizine. Temperature affects drug's vapor pressure and density of solvent in fluid state. Two factors affecting the solubility of cetirizine are sublimation pressure of drug and solvent density. These two temperature-dependent factors, which are function of temperature result in a crossover, pressure about 20 MPa for cetirizine. When the pressure is less than crossover, the density effect is predominant and the solute is more soluble at lesser

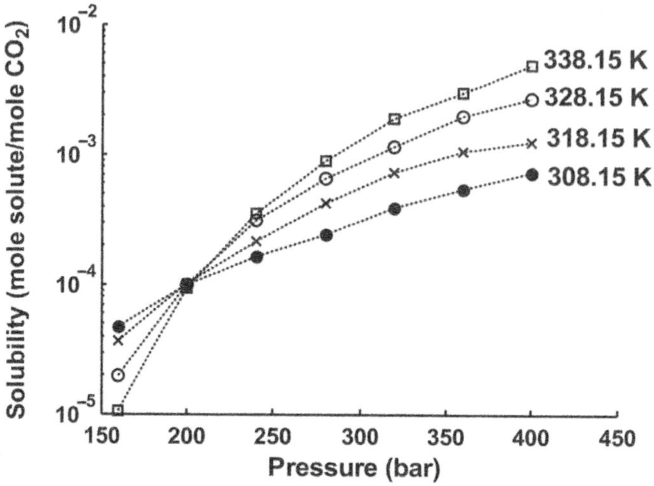

Fig. 5 Cetirizine solubility at different pressures and temperatures [8].

temperature. However, the effect of drug sublimation pressure is more effective at pressures above crossover pressure, thus the solubility increases at high temperatures.

In another study, the solubility of loratadine in SC-CO_2 is measured by static method [9]. The experiments were carried out at 308.15–338.15 K and 12–27 MPa. The solubility of loratadine increases with pressure at constant temperature. The effect of temperature on drug's sublimation pressure and solvent density was also investigated. By increasing the temperature, density of the solvent decreases and therefore, its dissolution ability decreases. At the same time, the solute's sublimation pressure increases with temperature, which increases the solubility in the SCF. Therefore, the increase in the temperature changes the solubility of solute in SC-CO_2 based on whether the effective factor is density of solvent or the drug's sublimation pressure. The crossover zone for loratadine is about 18–21 MPa [9].

The solubility of macrolide antibiotics including azithromycin, erythromycin, clindamycin, and clarithromycin has been measured in SC-CO_2 at the range of 308–348 K and 12.2–35.5 MPa by static method. Results showed azithromycin solubility at high temperatures and pressures was lower than other drugs, which is due to its polarity and its solubility in SC-CO_2 as a nonpolar solvent is low. The SC-CO_2 does not have dipole moment and it just has a small quadrapole moment and a very low polarizability and permeability. The solubility data of drugs at the constant temperature in SC-CO_2 would show a similar trend if drawn in terms of pressure. The crossover pressure regions for azithromycin, clarithromycin, clindamycin, and erythromycin were 13.8–14, 15–15.2, 14.8–15.2, and 16.8–17.1 MPa, respectively. The crossover phenomenon, as mentioned, is due to the competition of solute's sublimation pressure and density of solvent, which have opposite effects on drug solubility [10].

Applying a static measurement system along with a gravimetric system, the solubility of gabapentin has been obtained [11]. The solubility of this drug was obtained 8.97×10^{-5} to 7.36×10^{-3} at 16–40 MPa and 308–338 K based on mole fraction. The crossover pressure for gabapentin is about 20 MPa. As shown in Fig. 6, it can be seen that the effect of crossover pressure is due to the density variation and in this case, the solubility isotherms have not interrupted each other. The solubility of many drug compounds in the SC-CO_2 is measured, which are presented in classification of SCF methods in pharmaceutical industry (Table 3).

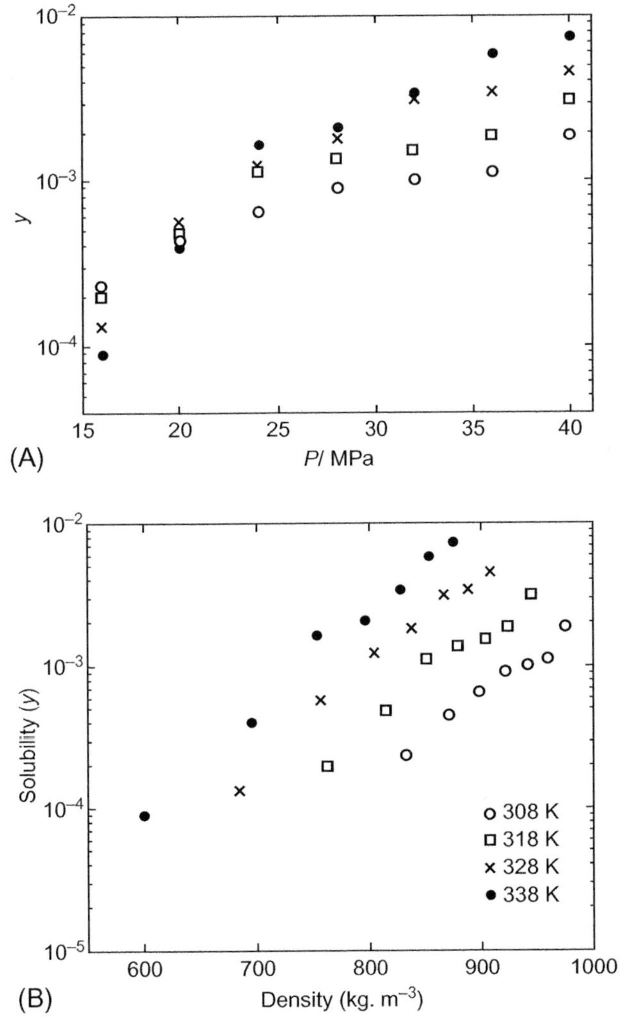

Fig. 6 Solubility of gabapentin (A) at various temperatures and pressures (B) as a function of density [11].

Table 3 Solubility of drug compounds in the SC-CO_2

Drug name	Temperature [K]	Pressure [MPa]	Solubility (mole fraction)	Crossover pressure [MPa]	Measurement method	Ref
Loratadine	308.15–338.15	12–27	0.45×10^{-5}–1.30×10^{-3}	18–21	Static method	[9]
Cetirizine	308.15–338.15	16–40	1.05×10^{-5}–4.92×10^{-3}	20	Static method	[8]
Azithromycin	308–348	12.2–35.5	0.69×10^{-4}–2.73×10^{-4}	13.8–14.0	Static method	[10]
Erythromycin	308–348	12.2–35.5	0.43×10^{-4}–3.12×10^{-4}	16.8–17.1	Static method	[10]
Clindamycin	308–348	12.2–35.5	1.77×10^{-4}–11.46×10^{-4}	14.8–15.2	Static method	[10]
Clarithromycin	308–348	12.2–35.5	1.31×10^{-4}–3.28×10^{-4}	15.0–15.2	Static method	[10]
Gabapentin	308–328	16–40	8.97×10^{-5}–7.36×10^{-3}	20	Static method coupled with gravimetric	[11]
Niflumic acid	313.2–353.2	Up to 31	0.80×10^{-5}–2.09×10^{-5}	19–25	Semiflow type phase equilibrium apparatus	[66]
Celecoxib	313.2–353.2	Up to 31	0.26×10^{-5}–1.52×10^{-5}	15–18	Semiflow type phase equilibrium apparatus	[66]
Spironolactone	308–338	16–40 MPa	6.25×10^{-5}–5.01×10^{-5}	–	Static method coupled with a gravimetric method	[26]
Methimazole	308–348	12.2–35.5 MPa	2.2×10^{-5}–19.0×10^{-5}	18	Static method	[67]
Methylparaben	308–348	12.2–35.5 MPa	11.3×10^{-5}–121.3×10^{-5}	15	Static method	[67]
Bisacodyl	308–348	12.2–35.5 MPa	3.6×10^{-5}–58.3×10^{-5}	18	Static method	[67]
Ketoprofen	312.5 and 331.5	10–22	1.33×10^{-5}–15.5×10^{-5}	16	Dynamic technique	[68]
Ketoprofen	308.15–338.15	16–40	2.21×10^{-5}–7.12×10^{-4}	–	Gravimetric method	[69]
Piroxicam	312.5 and 331.5	10–22	0.45×10^{-5}–3.89×10^{-5}	22	Dynamic technique	[68]
Nimesulide	312.5 and 331.5	10–22	1.89×10^{-5} to 9.85×10^{-5}	16	Dynamic technique	[68]
Sulfamethazine	313.15	10–40	2.23×10^{-6}–2.11×10^{-6}	–	Static technique	[70]
Sulfadimethoxine	313.15	10–40	2.69×10^{-6}–2.26×10^{-6}	–	Static technique	[70]
Tolbutamide	313.15–353.15	10–30	1.66×10^{-5}–40.5×10^{-5}	17–20	Semi flow apparatus	[71]

Chlorpropamide	313.15–353.15	10–30	2.29×10^{-6}–72.2×10^{-6}	17–20	Semi flow apparatus	[71]
Isoniazid	308–313	13–18.5	0.53×10^{-6}–6.05×10^{-6}	–	Dynamic analytical method	[72]
Naproxen	313–353	15–40	1.54×10^{-5}–1.84×10^{-4}	15	Static method	[73]
Naproxen	313.15	12.12–27.98	1.0×10^{-5}–4.2×10^{-5}	15–20	Chromatographic technique and dynamic solubility	[74]
Ibuprofen	313.15	12.12–23.10	2.1×10^{-3}–7.7×10^{-3}	12–13	Chromatographic technique and dynamic solubility	[74]
Paclitaxel	318.15 and 328.15	10–27.5	1.2×10^{-6}–5.9×10^{-6}	24–26	Chromatographic technique and dynamic solubility	[74]
5-Fluorouracil	318.15 and 328.15	12.5–25	3.6×10^{-6}–14.6×10^{-6}	22–23	Chromatographic technique and dynamic solubility	[74]
Thymidine	318.15 and 328.15	10–30	1.2×10^{-6}–8.0×10^{-6}	17–18	Chromatographic technique and dynamic solubility	[74]
Azodicarbonamide	308.15 and 328.15	10–30	0.9×10^{-5}–2.6×10^{-5}	15–25	Chromatographic technique and dynamic solubility	[74]
2-Phenyl-4H-1,3-Benzoxazin-4-one	308.15 and 328.15	10–30	0.8×10^{-4}–4.5×10^{-4}	25	Chromatographic technique and dynamic solubility	[74]
Benzocaine	298–318	7–28	1.75×10^{-5}–20.9×10^{-5}	8–10	Variable-volume stirred vessel	[75]
Lidocaine	298–318	7–28	1.55×10^{-5}–27.5×10^{-5}	8–15	Variable-volume stirred vessel	[75]
Procaine	298–318	7–28	5.82×10^{-5}–16.5×10^{-5}	8–10	Variable-volume stirred vessel	[75]
Carbamazepine	308–343	12.2–35.5	0.3×10^{-5}–9.4×10^{-5}	18	Static method	[76]

Continued

Table 3 Solubility of drug compounds in the SC-CO_2—cont'd

Drug name	Temperature [K]	Pressure [MPa]	Solubility (mole fraction)	Crossover pressure [MPa]	Measurement method	Ref
Diazepam	308–348	12.2–35.5	1.6×10^{-4}–11.1×10^{-4}	15	Static method	[76]
Codeine	308–348	12.2–35.5	0.4×10^{-4}–12.3×10^{-4}	15	Static method	[76]
Atropine	308–348	12.2–35.5	0.6×10^{-4}–16.7×10^{-4}	15	Static method	[76]
5-Fluorouracil[a]	308–328	10–22	3.72×10^{-7}–2.86×10^{-6}	–	Dynamic technique	[77]
β-Estradiol[a]	308–328	10–22	1.05×10^{-5}–3.05×10^{-4}	–	Dynamic technique	[77]
Sulfamerazine	313.15–333.15	13.1–48.8	2.0×10^{-7}–4.83×10^{-6}	25	Sample preparation accessory (SPA)	[78]
Sulfamethazine	313.15–333.15	13.1–48.8	8.82×10^{-7}–3.14×10^{-6}	–	Sample preparation accessory (SPA)	[78]
Sulfadimethoxine	313.15–333.15	13.1–48.8	2.58×10^{-7}–3.08×10^{-6}	20–30	Sample preparation accessory (SPA)	[78]
Penicillin V	314.85–334.85	7.99–28.02	6.2×10^{-5}–5.7×10^{-4}	10	Dynamic technique	[79]
Nabumetone	308.2–328.2	10–22	3.93×10^{-5}–2.68×10^{-3}	15	Semiflow apparatus	[80]
Phenylbutazone	308.2–328.2	10–22	1.99×10^{-5}–2.65×10^{-3}	17	Semiflow apparatus	[80]
Salicylamide	308.2–328.2	10–22	8.45×10^{-6}–2.10×10^{-4}	13	Semiflow apparatus	[80]
Meloxicam sodium	303–323	14.9–25.5	4.41×10^{-6}–12.76×10^{-6}	15–17	Static analytical method	[81]
Cefixime trihydrate	308–328	18.3–33.5	1.6×10^{-7}–3.02×10^{-7}	–	Static method	[82]
Oxymetholone	308–328	12.1–30.5	1.6×10^{-5}–1.49×10^{-4}	18	Static method	[82]
Fluoxetine hydrochloride	308.15–338.15	16–40	2.65×10^{-5}–8.12×10^{-4}	20	Static method coupled with gravimetric	[83]
Phenylephrine hydrochloride	308.15–338.15	16–40	1.01×10^{-4}–2.89×10^{-3}	–	Static method coupled with gravimetric	[84]
Cyproheptadine	308–338	16–40	3.35×10^{-5}–3.09×10^{-3}	20	Static method	[85]
Chlorpheniramine maleate	308–338	20–40	1.54×10^{-5}–4.26×10^{-4}	–	Static method	[86]
Chlormezanone	308.2–328.2	12–24	0.05×10^{-4}–5.45×10^{-4}	16.8–18.5	Semiflow equipment	[87]

Metaxalone	308.2–328.2	12–24	0.58×10^{-6}–148.0×10^{-6}	15.1–16.9	Semiflow equipment	[87]
Methocarbamol	308.2–328.2	12–24	0.28×10^{-4}–4.87×10^{-4}	12.8–14.5	Semiflow equipment	[87]
All-trans retinoic acid (ATRA)	308.2–328.2	10–30	0.19×10^{-6}–18.41×10^{-6}	12–13	Static analytical method	[88]
10-Undecenoic acid	308–323	10–18	0.4×10^{-3}–17.4×10^{-3}	–	Dynamic method	[89]
Geraniol	308–323	10–18	2.7×10^{-3}–25×10^{-3}	–	Dynamic method	[89]
Clobetasol propionate	315–345	15.2–30.5	1×10^{-6}–2.1×10^{-5}	24.5	static method	[90]
Desoxycorticosterone acetate	315–345	15.2–30.5	5×10^{-6}–1.03×10^{-4}	24.5	Static method	[90]
Artemisinin	313–333	11–31	0.50×10^{-3}–2.92×10^{-3}	20–23	Static method	[91]
Juglone	308.2–328.2	9.2–24.4	2.0×10^{-5}–1.6×10^{-3}	21–22.	Static analytical method	[92]
Mefenamic acid	308.15–328.15	16–40	8.31×10^{-5}–5.98×10^{-3}	20	Static method	[24]
Ketotifen fumarate	308.2–338.2	12–30	0.21×10^{-4}–10.77×10^{-4}	19–20	Static method	[43]
Amiodarone hydrochloride	313.2–343.2	12–30	2.51×10^{-5}–1.01×10^{-3}	18	Static method	[29]
Norfloxacin	308.2–328.2	10–30	1.4×10^{-6}–24.4×10^{-6}	–	Static analytical method	[93]
Ofloxacin	308.2–328.2	10–30	0.4×10^{-6}–1.3×10^{-6}	–	Static analytical method	[93]

[a]Solubility based on mass fraction.

2.3 Application of cosolvent in drug extraction using SC-CO_2

CO_2 is not suitable for drugs with polarity characteristic due to the lack of dipole moments and the inability to create specific interactions between the solvent and the solute. Adding a cosolvent with polarity to SC-CO_2 would improve the dissolution ability [12, 13].

Adding a cosolvent to the system alters local and bulk compositions. It changes the composition of the cybotactic region (region around solute) and makes conditions more complicated in a supercritical system modified by cosolvent. Adding a cosolvent increases the solute's solubility due to additional interactions among solute and cosolvent. The effect of cosolvent is because of different mechanisms. Adding cosolvent increases mixture density enhancing the solubility and the physical interactions such as dipole-dipole, dipole-induced dipole, and induced dipole-induced dipole interactions. Although, using polar cosolvents for polar drugs can lead to a significant increase in solubility due to special chemical interactions such as chemical bonds or charge transfer complex formation [14].

So far, many studies have been done on polar compounds in SC-CO_2 in the presence of polar cosolvent. It has been shown that small amounts of cosolvents can enhance the solubility of the compound in SC-CO_2. Some of the drugs that their solubility has been increased by adding cosolvent have been brought in Table 4.

2.4 SC-CO_2 as antisolvent

Improvement of the thermophysical properties of the API is a key issue in the pharmaceutical industry [15]. The crystallinity and chemical activity of a drug should be preserved during production process. It should be noted that the drug size and size distribution obtained by conventional methods are not suitable [16]. Several methods are used to micronize the drug compounds. Each of these methods has disadvantages [17]. In order to overcome these disadvantages, many studies have been done to utilize new technologies using SCFs. The main methods are the RESS and GAS. In RESS method, SCF is the solvent and drug that is supposed to be micronized is solved by an extractor in a SCF. Using the RESS method, low PDI particles can be made at low temperatures. However, one limitation of this method is the low solubility of some drug compounds in SCFs.

In the GAS process, SCF plays an antisolvent role. Accordingly, the drug is dissolved in an organic solvent, and then SCF is inserted in the solution, which causes the expansion. Decreasing the density reduces solvation strength and dissolves the solute in precipitator. The GAS process is batch [17]. The use of SCF as antisolvent has several advantages. For example, antisolvent can be eliminated by reducing the pressure to the gas phase pressure and no further treatment procedures are required. In addition, the fluid diffusivity of the SCF is twice that of liquid solvents. Therefore, it penetrates

Table 4 Solubility of drug compounds in presence of cosolvent

Drug name	Temperature [K]	Pressure [MPa]	Solubility (mole fraction)	Crossover pressure [MPa]	Measurement method	Cosolvent	Solubility after addition of cosolvent	Ref
Aspirin	308.15–328.15	10–25	0.72×10^{-4}–3.47×10^{-4}	12.5	Dynamic method	Acetone	2.18×10^{-4}–7.57×10^{-4}	[13]
Benzamide	308–328	11–21	1.91×10^{-5}–16.05×10^{-5}	11	Dynamic flow method	Ethanol Acetone Ethylene glycol	5.79×10^{-5}–74.83×10^{-5} 4.93×10^{-5}–33.99×10^{-5} 5.07×10^{-5}–24.03×10^{-5}	[94]
Dexamethasone	313.20–323.20	10–25	1.19×10^{-6}–1.52×10^{-6}	14–16	Static analytical method	Ethanol	1.97×10^{-6}–2.98×10^{-6}	[12]
Eflucimibe	308.15 and 318.15	8–30	0.14×10^{-7}–15.67×10^{-7}	10	Analytical open circuit method	Ethanol Dimethyl sulphoxide	11.09×10^{-7}–90.82×10^{-7} 32.84×10^{-7}–365.37×10^{-7}	[95]
Flurbiprofen	303–323	8–25	1.67×10^{-5}–19.68×10^{-6}	12	Static analytical method	Ethanol	36.75×10^{-4}	[96]
Naproxen	313.1–333.1	8.96–19.31	0.19×10^{-5}–3.18×10^{-5}	–	Flow technique coupled with gravimetric analysis	Ethyl acetate Acetone Methanol Ethanol 2-Propanol 1-Propanol	0.64×10^{-5}–14.33×10^{-5} 0.67×10^{-5}–16.97×10^{-5} 0.87×10^{-5}–28.54×10^{-5} 1.26×10^{-5}–47.78×10^{-5} 1.37×10^{-5}–56.60×10^{-5} 2.23×10^{-5}–61.82×10^{-5}	[14]

faster into the liquid solvent, providing the solute's super saturation and precipitation of particles with smaller dimensions compared with liquid solvents. In addition, size distribution controlling is possible with this method [18].

CO_2 as an antisolvent has been used [16] to enhance the bioavailability of atenolol by modifying the morphology of the drug's particles. Hence, three organic solvents including methanol, ethanol, and isopropanol were chosen and the saturated conditions of these three solvents were investigated in three ternary systems containing organic solvent, CO_2, and atenolol. The atenolol precipitation process was performed with SAS solution containing three different organic solvents in countercurrent configuration. The precipitated powder was subjected to various analyses to determine the characteristics of the drug. The particles precipitated from various organic solvents, compared with untreated drug, were spherical but agglomerated, smaller and larger crystals. To prevent agglomeration, the process was performed in equicurrent configuration and continuously. Various experiments with different currents and solvents to achieve optimal particles showed that using isopropanol as organic solvent produced smaller particles and size distribution with spherical morphology with a diameter of 5–7 μm, connected to each other by a bridge. As the flow rate increased, the resulting network was denser and the particles had shorter bond length.

In another study, SC-CO_2 has been used as an antisolvent (SAS method) to produce a cocrystal to improve physicochemical properties of an API [15]. Therefore, the resulting cocrystal has two antiinflammatory drugs, diflunisal and nicotinamide, prepared with SAS method. The results showed that the effect of parameters of SAS method including temperature, pressure, drug concentration, and solvent type (acetone and ethanol) on the produced cocrystal is weak. The resulting crystals were needle shaped with uniform width and variable lengths and compared to the cocrystals obtained by liquid assisted ball mill grinding and solution crystallization methods, they had similar crystalline structure, melting point and Fourier-transform infrared (FTIR) spectroscopy, and the resulting cocrystals were produced in a one-step procedure and free of solvent [15].

In another study, an antibiotic, cefonicid, has been micronized from dimethylsulfoxide using SC-CO_2 [18]. They achieved dimensions around 0.2 up to 50 μm by changing the concentration of the drug in the solvent or the change in the temperature of the process. The scanning electron microscopy (SEM) analysis of the precipitated powder proved that particles with dimensions of 0.2 μm and concentrations up to 50 mg/mL were obtained. While at concentrations in the range of 75 mg/mL microparticles and particulates in the form of balloons or empty shells were produced.

3. Solid solubility in SC-CO_2 using EOSs

For avoid experimental measurements and high costs, the prediction drug compounds solubility in SC-CO_2 seems to be necessary. One of the most common solubility prediction methods is based on applying EOSs. However, in order to avoid computer calculations, empirical relations have been applied. Since EOSs are divided into two categories of cubic and noncubic EOSs including EOSs based on the statistical association fluid theory (SAFT) model. In this chapter, two mentioned categories of EOSs and empirical relations are explained.

This approach (the use of EOSs and empirical relations) is a common method for estimation the thermodynamic properties of fluids. Accordingly, the solubility modeling of pharmaceuticals compounds in SC-CO_2 is divided into four parts.

In the first section, the prediction of the solubility of pharmaceutical compounds will be investigated by means of cubic EOSs. In the second section, using noncubic EOSs, pharmaceuticals solubility in SC-CO_2 is assessed. In third section, solubility parameter-based model has been used to predict solubility of drugs in SC-CO_2. In fourth part, pharmaceuticals solubility in SC-CO_2 is studied using mathematical methods such as ANN, COSMO as structural models, and molecular simulation model. In the last section, drug solubility calculations by means of empirical correlations in SC-CO_2 will be discussed.

3.1 Calculation of solids solubility in SCFs

In this section, the solubility of a solid such as drug component would be calculated from phase equilibrium calculation. Fugacity is the basis of phase equilibrium calculations in thermodynamics. Accordingly, the fugacity of species i in the two phases at equilibrium state must be equal. Therefore,

$$\hat{f}_i^{I} = \hat{f}_i^{II} \tag{1}$$

In the above equation, the superscripts (I) and (II) stand for the two phases. For calculate drug compound solubility in SC-CO_2, we assume that the two phases, solid and the supercritical phase, are in equilibrium. The subscript (1) is considered for the gas component in a supercritical state and the subscript (2) is considered for solid drug compound. The solid phase can be considered as pure, thus, the fugacity for species (2) in two solid and supercritical phases, is obtained as follows:

$$\hat{f}_2^{V} = y_2 \hat{\varphi}_2^{V} P \tag{2}$$

$$f_2^{S} = \varphi_2^{sub} P_2^{sub} \exp\left(\int_{P_2^{sub}}^{P} \frac{v_2^{S}}{RT} dP\right) \tag{3}$$

In the above equation, v_2^s is molar volume of species (2) in solid phase and the superscript (sub) refers to saturation state. If v_2^S is constant, solid-phase fugacity can be derived as

$$f_2^{S} = \varphi_2^{sub} P_2^{sub} \exp\left[\frac{v_2^{S}}{RT}\left(P - P_2^{sub}\right)\right] \tag{4}$$

Therefore, at equilibrium state for two solid and supercritical phases we have

$$y_2\hat{\varphi}_2^V P = \varphi_2^{sub} P_2^{sub} \exp\left[\frac{v_2^S}{RT}\left(P - P_2^{sub}\right)\right] \tag{5}$$

As a result, solid pharmaceutical component solubility in SC-phase is obtained from the following equation:

$$y_2 = \frac{P_2^{sub}}{P}\frac{\varphi_2^{sub}}{\hat{\varphi}_2^V} \exp\left[\frac{v_2^S}{RT}\left(P - P_2^{sub}\right)\right] \tag{6}$$

The saturation pressure of solid so-called sublimation pressure of the solid component (P_2^{sub}) is very low, the quantity of φ_2^{sub} approaches unity. Therefore, Eq. (6) can be simplified as follows [19]:

$$y_2 \cong \frac{P_2^{sub}}{\hat{\varphi}_2^v P} \exp\left[\frac{v_2^s}{RT}\left(P - P_2^{sub}\right)\right] \tag{7}$$

To calculate solid solubility in SC-phase, it is necessary to calculate $\hat{\varphi}_2^v$. Fugacity coefficient is derived by using a suitable EOS from the following equation [20]:

$$\ln \hat{\varphi}_i = \int_V^{\infty} \left[\frac{1}{RT}\left(\frac{\partial P}{\partial n_i}\right)_{T,v,n_{j\neq i}} - \frac{1}{V}\right] - \ln Z \tag{8}$$

In order to calculate this coefficient, an EOS is being used which will be discussed in the following section.

3.2 Solubility prediction of drug components using cubic EOS

Cubic EOSs are the most commonly used methods that used to calculate thermodynamics functions of fluids. Parameters of EOS are calculated by critical point properties (T_c, P_c, or v_c) as well as acentric factor (ω). Since these properties are hardly found in literature for pharmaceutical compounds, applying the EOSs to calculate pharmaceutical components solubility in SC-CO_2 becomes difficult.

In order to use these equations, mixing rules should be used for multicomponent mixtures. The mixing rules make the EOS's parameters applicable for multicomponent systems. In drug-CO_2 binary mixtures, there are differences between size and shape of drug and CO_2. In addition, drug is a polar component and CO_2 is a nonpolar component. Therefore, the simple and usual mixing rules cannot estimate precise solubility in SC-CO_2. In this section, with a review of papers for modeling drug solubility in SC-CO_2, the accuracy of cubic EOSs along with the mixing rules will be evaluated to estimate pharmaceutical components solubility.

3.2.1 Peng-Robinson, Soave-Redlich-Kwong, and Petal-Taja-Valderrama as two and three parameters cubic EOSs

In this section, modeling of solubility of ibuprofen, ketoprofen, naproxen, and flurbiprofen in SC-CO_2 will be discussed [21]. Peng-Robinson (PR), Soave-Redlich-Kwong (SRK), and Petal-Taja-Valderrama (PTV) and also mixing rules of van der Waals-1 parameter (vdW1), van der Waals-2 parameter (vdW2), Panagiotopoulos-Reid (mrPR), and Mukhopadhyay-Rao (MR) are applied for this reason. The general form of the EOSs that has been used is given in Eq. (9):

$$P = \frac{RT}{v-b} - \frac{a_c\alpha(T_r,\omega)}{v(v+b)+c(v-b)} \quad (9)$$

Parameters of EOSs (m, n), as well as the coefficients a_c, α, b, and c that has been used for the three EOSs has been represented in Table 5.

Also, the mixing rules that are utilized in the current research can be found in Table 6.

Table 5 Parameters of Eq. (1) for EOSs of PR, SRK and PTV [21]

EOS	a_c	α	b	c
PR	$0.45724\frac{R^2T_c^2}{P_c}$	$[1+K(1-\sqrt{T_r})]^2$ $K=0.37464+1.54226\omega-0.26992\omega^2$	$0.07780\frac{RT_c}{P_c}$	0
SRK	$0.42747\frac{R^2T_c^2}{P_c}$	$[1+m(1-\sqrt{T_r})]^2$ $m=0.480+1.574\omega-0.176\omega^2$	$0.08664\frac{RT_c}{P_c}$	0
PTV	$\left(0.66121-0.761057Z_c\right)\frac{R^2T_c^2}{P_c}$	$[1+F(1-\sqrt{T_r})]^2$ $F=0.46283+3.58230\omega Z_c+8.19417(\omega Z_c)^2$	$(0.02207+0.20868Z_c)\frac{RT_c}{P_c}$	$\left(0.57765-1.87080Z_c\right)\frac{RT_c}{P_c}$

Table 6 Mixing rules and combining rules used for cubic EOSs [21]

Mixing Rule	Parameters	a	b	c
vdW1	k_{ij}	$\sum_i\sum_j y_iy_ja_{ij}$ $a_{ij}=\sqrt{a_ia_j}(1-k_{ij})$	$\sum_i\sum_j y_iy_jb_{ij}$ $b_{ij}=\frac{b_i+b_j}{2}$	$\sum_i\sum_j y_iy_jc_{ij}$ $c_{ij}=\frac{c_i+c_j}{2}$
vdW2	k_{ij} and l_{ij}	$\sum_i\sum_j y_iy_ja_{ij}$ $a_{ij}=\sqrt{a_ia_j}(1-k_{ij})$	$\sum_i\sum_j y_iy_jb_{ij}$ $b_{ij}=\frac{b_i+b_j}{2}(1-l_{ij})$	$\sum_i\sum_j y_iy_jc_{ij}$ $c_{ij}=\frac{c_i+c_j}{2}$
mrPR	k_{ij} and k_{ji}	$\sum_i\sum_j y_iy_ja_{ij}$ $a_{ij}=\sqrt{a_ia_j}\left[1-k_{ij}+(k_{ij}-k_{ji})y_i\right]$	$\sum_i\sum_j y_iy_jb_{ij}$ $b_{ij}=\frac{b_i+b_j}{2}$	$\sum_i\sum_j y_iy_jc_{ij}$ $c_{ij}=\frac{c_i+c_j}{2}$
MR	m_{ij}	$\sum_i\sum_j y_iy_ja_{ij}\left(\frac{b}{b_{ij}}\right)^{m_{ij}}$ $m_{ii}=m_{jj}=1$ $a_{ij}=\sqrt{a_ia_j}$	$\sum_i y_ib_i$	$\sum_i y_ic_i$

The parameters of the mixing rules are obtained from optimization of average absolute relative deviation (AARD%):

$$AARD(\%) = \frac{100}{N}\sum_{i}^{N}\frac{|y_{calc} - y_{exp}|}{y_{exp}} \tag{10}$$

It should be noted that (T_c, P_c) and ω for various drugs are obtained from methods such as Fedors (set1) [22], Ambrose-Walton (set2) [23], and Joback (set3) [23].

The sublimation pressure plays an important role in correlation of the experimental data related to solubility of drug components with the EOSs. In the present study, the sublimation pressure is obtained from the experimental relationships respect to the critical properties of the drug.

The results obtained for flurbiprofen solubility in SC-CO_2 using the intended EOSs and four mixing rules various series of solid thermophysical as well as critical properties is shown in Table 7.

As the error values reported in Table 7 show, the estimated properties of set2 give the best results for PR and SRK EOSs, but this result does not always exist for the PTV EOS. However, the mentioned result valid for the vdw2 mixing rules in all temperatures. In addition, the obtained results related to the correlation of naproxen, ibuprofen as well as, ketoprofen solubility in SC-CO_2 at 313.15 K with the above equations have been shown in Table 8.

Table 7 The solubility correlation results of fluriprofen in SC-CO_2 at 303.15, 313.15, and 323.15 K using PR, SRK, and PTV EOSs and four mixing rules including vdW1, vdW2, mrPR, and MR [21]

EOS	Mixing rule	T = 303.15 K			T = 313.15 K			T = 323.15 K		
		Set1	Set2	Set3	Set1	Set2	Set3	Set1	Set2	Set3
PR	vdW1	24.67	10.83	39.05	32.63	12.22	35.97	38.13	8.49	30.11
	vdW2	6.61	5.97	5.61	15.01	9.43	5.16	13.70	8.14	5.21
	mrPR	24.71	10.15	39.05	36.69	12.19	35.97	38.19	8.49	30.09
	MR	23.47	10.92	39.06	32.21	12.32	36.01	37.78	8.49	30.24
SRK	vdW1	23.70	13.83	41.99	30.20	13.93	32.77	36.79	8.69	31.80
	vdW2	7.41	6.03	5.49	15.31	9.64	5.48	13.92	8.54	5.24
	mrPR	22.37	13.86	41.99	30.40	13.90	37.77	36.76	8.70	31.79
	MR	22.11	13.93	41.98	29.74	14.02	37.80	36.39	8.69	31.92
PTV	vdW1	72.45	52.34	41.57	66.21	45.42	23.64	59.71	33.93	11.58
	vdW2	18.69	9.35	6.03	16.57	9.77	5.39	12.38	6.46	3.25
	mrPR	48.34	30.32	24.62	29.13	6.83	5.99	12.24	10.45	24.28
	MR	71.98	52.13	41.36	65.90	44.77	23.25	59.18	33.45	11.24

Table 8 The AARD% obtained for solubility correlation of ibuprofen, ketoprofen and naproxen in SC-CO_2 using PR, SRK and PTV EOSs and four mixing rules including vdW1, vdW2, mrPR and MR [21]

EOS	Mixing rule	Ibuprofen	Ketoprofen	Naproxen
PR	vdW1	6.42	27.85	14.69
	vdW2	4.21	9.02	3.92
	mrPR	5.22	27.90	15.03
	MR	12.50	27.07	15.06
SRK	vdW1	8.63	29.58	16.25
	vdW2	4.62	8.30	4.42
	mrPR	4.52	30.01	16.28
	MR	14.61	29.74	16.27
PTV	vdW1	54.81	75.24	30.61
	vdW2	7.96	25.31	7.88
	mrPR	19.23	41.67	26.44
	MR	42.17	75.09	30.56

Table 9 Solubility correlation results of fluriprofen at 303.15, 313.15 and 323.15 K using PR EOS and the vdW1mixing rule [21]

T(K)	Set1	Set2	Set3
303.15	6.67	6.06	6.06
313.15	11.95	11.08	11.09
323.15	8.47	7.66	7.64

The results reported in Table 8 also confirm that the PR and SRK EOSs along with the vdW2 mixing rule provide the best results. In the following, sublimation pressure is considered as an adjustable parameter. The result of fluriprofen solubility in SC-CO_2 at various temperatures using PR EOS with vdW1 mixing rule with different sets is reported in Table 9. The results reported in Table 9 show that the model has a very good accuracy for each temperature in three sets.

Based on the obtained results, authors suggested that collection of PR-vdW1, SRK-vdW1 has a good accuracy for estimation the flurbiprofen solubility data in SC-CO_2. From molecular point of view, it can be said that the drugs solubility in SC-CO_2 is related to molecular interactions between drug and CO_2, also molecular weight of drug. Solid pharmaceutical compounds that have similar molecules (identical structural groups), have the same polarization and similar molecular weight and the same solubility in SC-CO_2 are expected at equal thermodynamics conditions(T, P). As a result, drug compounds with polarization and high molecular weight hardly dissolve in the SCF of nonpolar CO_2.

3.2.2 Esmaeilzadeh and Roshanfekr EOS

In this section, correlation of solubility of mefenamic acid in SC-CO_2 using Esmaeilzadeh and Roshanfekr (ER) EOS is studied [24] and the obtained results are compared with PR EOS. The ER EOS is written as follows [25]:

$$P = \frac{RT}{v-b} - \frac{a(T)}{v(v+c)+c(v-c)} \tag{11}$$

$$a(T) = \Omega_a \left(\frac{R^2 T_c^2}{P_c}\right) \alpha(T_r) \tag{12}$$

$$b = \Omega_b \left(\frac{RT_c}{P_c}\right) \tag{13}$$

$$c = \Omega_c \left(\frac{RT_c}{P_c}\right) \tag{14}$$

$$\Omega_b = 2\Omega_c - 1 + 3\zeta_c \tag{15}$$

$$\Omega_a = 3\zeta_c^2 + \Omega_c^2 + 2\Omega_b\Omega_c + 2\Omega_c \tag{16}$$

$$\Omega_c^3 + \left(3\zeta_c - \frac{5}{8}\right)\Omega_c^2 + \left(\zeta_c^2 - \frac{3}{4}\zeta_c\right)\Omega_c + \left(\zeta_c^3 - \frac{3}{8}\zeta_c^2\right) = 0 \tag{17}$$

$$\alpha(T_r) = \left[m_1 + m_2\left(1 - T_r^{0.5}\right)\right]^2 \tag{18}$$

$$\zeta_c = 0.3284438 - 0.0690264\omega + 0.0078711\omega^2 \tag{19}$$

$$m_1 = 0.999035 - 0.01061842\omega - 0.0081174\omega^2 \tag{20}$$

$$m_2 = 0.4400108 + 1.5297151\omega - 0.4710752\omega^2 \tag{21}$$

The ER EOS was coupled with the vdW2 mixing rule for drug solubility correlation in SC-CO_2. $k_{ij}$and l_{ij} adjustable parameters are tuned by optimization of AARD%. AARD% values for correlating mefenamic acid solubility data in SC-CO_2 are represented in Table 10.

Table 10 AARD% values for correlating solubility data of mefenamic acid with vdW2 mixing rule [24]

Temperature (K)	ER-EOS	PR-EOS
308.15	21.25	30.56
318.15	12.10	23.29
328.15	2.27	9.86
338.15	3.54	6.81

As the results of AARD% indicate, the ER EOS has a better accuracy than PR EOS. However, mentioned model is much better on high temperatures than low temperatures. The authors of the article have claimed that this behavior is related to SCF density (transient behavior), which is not too far from critical point. Also, ER EOS has a better accuracy than PR EOS since it contains the parameter of ζ_c. The value of the parameter ζ_c for PR EOS is constant. Of course, ER EOS yields different values for ζ_c which increase the predictive ability of the EOS, especially in critical and supercritical regions.

In addition, spironolactone solubility in SC-CO_2 has been modeled [26]. The spironolactone solubility in SC-CO_2 is calculated using PR EOS-vdW2 set. However, thermodynamic properties such as (T_c, P_c, ω) as well as $P^{sub.}$ are obtained by group contribution method. The values of AARD% at various temperatures have been represented in Table 11.

The AARD% results that are reported in Table 11 show that PR EOS is not able to estimate spironolactone solubility in SC-CO_2. The best error is for temperature of 318 K and the worst error is for 308 K. This inaccuracy on the behavior of transient CO_2 is related to supercritical at this temperature.

3.2.3 Kwak-Mansoori-PR EOS

The solubility of antibiotics including penicillin G, penicillin V, and antiinflammatory drugs including aspirin , naproxen, ibuprofen, ketoprofen, flurbiprofen, and diflunisal in SC-CO_2 is correlated by using the modified version of PR EOS along the modified mixing rule proposed by Kwak and Mansoori (mKM), which is temperature independent [27]. The PR EOS is modified based on the mKM mixing rule as follows [28]:

$$Z = \frac{v}{v-b} - \frac{(a/RT) + c - 2\sqrt{ac/RT}}{(v+b) + (b/v)(v-b)} \tag{22}$$

where constants can be written as

$$a = a(T_c)(1+\kappa)^2 \tag{23}$$

$$b = 0.0778\frac{RT_c}{P_c} \tag{24}$$

Table 11 AARD% values for correlating spironolactone solubility with vdW2 mixing rule [26]

Temperature (K)	AARD%
308	48.38
318	34.23
328	34.36
338	38.12

$$c = a(T_c)\frac{\kappa^2}{RT_c} \quad (25)$$

$$a(T_c) = 0.45724\frac{R^2T_c^2}{P_c} \quad (26)$$

$$\kappa = 0.37464 + 1.54226\omega - 0.26992\omega^2 \quad (27)$$

a and b parameters are calculated from vdW2 mixing rule and parameter c can be calculated as below:

$$c = \sum_i^n \sum_j^n x_i x_j c_{ij} \quad (28)$$

$$c_{ij} = \frac{(c_{ii} + c_{jj})}{2}(1 - m_{ij}) \quad (29)$$

The adjustable parameters [m_{ij} in Eq. (29) along with k_{ij} and l_{ij} in vdW2 mixing rule] are the adjustable binary interaction parameters that are temperature independent. The critical properties of drug can be obtained from the group contribution method. As mention in previous works, sublimation pressure of drug is important for prediction drug solubility in SC-CO_2. In this research, sublimation pressure along with binary interaction parameters (k_{ij}, l_{ij}, and m_{ij}) are estimated from optimization of AARD% as objective function. The relationship between sublimation pressure and temperature is considered as

$$R\ln(P^{sub}) = A + \frac{B}{T} + \Delta_{sub}C_P \ln\left(\frac{T}{T_0}\right) \quad (30)$$

where T_0 is an arbitrary temperature and it is assumed that $T_0 = 298.15$ K. According to the sublimation pressure relation, adjustable parameters A/R,B/R, $\Delta_{sub}C_P/R$ and binary interaction parameters can be obtained from drug solubility in SC-CO_2. AARD% values for the PR EOS along with mMK mixing rule for antibiotics and antiinflammatory drugs are shown in Table 12.

In addition, results for the PR EOS along with vdW common mixing rule are represented in Table 13.

The results reported in Tables 12 and 13 show that the AARD% values for mKM mixing rule are low. However, PR-mKM set is more accurate than PR-VdW set.

3.2.4 Pazuki et al. 1 (PAZ1) EOS

In other study, amiodarone hydrochloride solubility in SC-CO_2, as a low solubility drug component in water was correlated using cubic EOSs [29]. The EOSs used in this research were the EOS proposed equation by Pazuki et al. (PAZ1 EOS) [30] and the

Table 12 The AARD% values of PR-mMK EOS prediction of various pharmaceutics solubility [27]

Drug	AARD% (PR-mMK)
Penicillin G	23.95
Penicillin V	11.03
Ibuprofen	10.76
Ketoprofen	10.93
Naproxen	2.97
Diflunisal	15.82
Aspirin	3.27
Flurbiprofen	4.90
Average	10.45

results of PAZ EOS were compared with those obtained from the PR EOS. The EOS suggested by Pazuki et al. is written by as follows:

$$P = \frac{RT}{v - b(T)} \frac{v + 2b(T)}{v} - \frac{a(T)}{v(v + b(T))} \tag{31}$$

Table 13 The AARD% values of PR EOS correlating of various pharmaceutics solubility using vdW mixing rule [27]

Drug	*T* (K)	AARD% (PR-vdW)
Penicillin V	314.85	36.23
	324.85	40.25
	334.85	41.30
Ibuprofen	308.15	27.30
	313.15	19.20
	318.15	43.60
Ketoprofen	313	9.50
	318	10.40
Naproxen	313.1	13.90
	323.1	13.70
	333.1	9.40
Diflunsial	308.2	17.30
	318.2	20.60
	328.2	23.30
Aspirin	308.15	2.24
	318.15	8.19
	328.15	7.45
Flurbiprofen	303	6.06
	313	11.08
	323	7.66
Average		20.43

$$a = 0.51301\frac{R^2T_c^2}{P_c}\alpha(T_r, \omega) \quad (32)$$

$$\alpha(T_r, \omega) = \left[1 + m_1\left(1 - T_r^{0.5}\right) + m_2\left(1 - T_r^{0.5}\right)^2 + m_3\left(1 - T_r^{0.5}\right)^3\right]^2 \quad (33)$$

$$m_1 = 0.4690 + 0.709\omega - 0.266\omega^2 \quad (34)$$

$$m_2 = -0.6548 - 1.2625\omega - 1.9727\omega^2 \quad (35)$$

$$m_3 = 0.9553 + 5.0064\omega + 0.4159\omega^2 \quad (36)$$

$$b(T) = 0.058743\frac{RT_c}{P_c}\beta(T_r, \omega) \quad (37)$$

$$\beta(T_r, \omega) = [1 + n(1 - T_r)]^2 \quad (38)$$

$$n = 0.1723 - 0.3858\omega - 0.1683\omega^2 \quad (39)$$

Critical properties of the drug component are calculated from the Marrero-Gani thermodynamic model [31]. The acentric factor as well as the sublimation pressure of the drug component is also derived from the Ambrose-Walton corresponding states method [23]. In order to calculate the parameters of the utilized EOS, vdW2 mixing rule has been used. The results of AARD% for correlating solubility data of antiarrhythmic drug component in the SC-CO_2 for both PAZ1 and PR EOSs are reported in Table 14.

The average value of AARD% error of the PR EOS was equal to 6.03%, which indicates that PR EOS results and experimental solubility are in accordance. In addition, the average value of AARD% for PAZ1 EOS is equal to 8.88% which shows that the results obtained from this EOS are also suitable for prediction of solubility. Likewise, the value of interaction parameters for the both EOSs demonstrates a linear correlation between the interaction parameters and temperature of the system.

3.2.5 Modified PR and Pazuki et al. 2 (PAZ2)

The solubility of three drug components containing methimazole, phenazopyridine, and propranolol, in SC-CO_2 has been correlated [32]. By using three EOSs including PR, Pazuki et al. (PAZ2), and the modified Peng-Robinson (MPR), along with the vdW1, vdW2, and Adachi-Sugie mixing rules [32] were applied. The PAZ2 EOS is written as follows [33]:

Table 14 AARD% values for correlating solubility of amiodorane hydrochloride with vdW2 mixing rule [29]

EOS	$T = 313.2$ K	$T = 323.2$ K	$T = 333.2$ K	$T = 343.2$ K
PAZ1	7.59	5.56	11.20	11.19
PR	8.52	2.59	6.30	6.72

$$P = \frac{RT}{v}\left(\frac{v+b}{v-b}\right) - \frac{a}{v(v+b)} \tag{40}$$

In which

$$a = a_c\alpha(T_r) \tag{41}$$

$$a_c = 0.47448\frac{R^2T_c^2}{P_c} \tag{42}$$

$$\alpha(T_r) = \left[1 + m_1(1-T_r) + m_2(1-T_r)^2 + m_3(1-T_r)^3\right]^2 \tag{43}$$

$$m_1 = 0.2513 + 0.4178\omega - 0.0207\omega^2 \tag{44}$$

$$m_2 = -0.1382 - 0.5923\omega - 1.0493\omega^2 \tag{45}$$

$$m_3 = 0.1931 + 1.3955\omega + 0.1961\omega^2 \tag{46}$$

and

$$b = b_c\beta(T_r) \tag{47}$$

$$b_c = 0.06824\frac{RT_c}{P_c} \tag{48}$$

$$\beta(T_r) = [1 + n(1-T_r)] \tag{49}$$

$$n = 0.2155 - 0.9069\omega - 0.206\omega^2 \tag{50}$$

The general form of MPR EOS is similar to the original PR EOS. Only difference with the PR EOS is the $\alpha(T_r)$ parameter. The parameter of MPR EOS is obtained as follows [34]:

$$\alpha(T_r) = \exp\left[(2 + 0.836T_r)\left(1 - T_r^{0.134+0.508\omega-0.0467\omega^2}\right)\right] \tag{51}$$

The Adachi-Sugie mixing rule is also defined as [35]

$$a = \sum_i\sum_j y_i y_j a_{ij} \tag{52}$$

$$a_{ij} = \sqrt{a_i a_j}\left(1 - k_{ij} - \lambda_{ij}(y_i - y_j)\right) \tag{53}$$

$$b = \sum_i\sum_j y_i y_j b_{ij} \tag{54}$$

$$b_{ij} = \left(\frac{b_i + b_j}{2}\right)(1 - l_{ij}) \tag{55}$$

In this equation, λ_{ij} and l_{ij} are interaction parameters of the mixing rules which are obtained from optimization of the AARD%. The critical properties of T_c and P_c are obtained from the Joback group contribution method. The acentric factor is also calculated from the Lee-Kesler equation [23].

Sublimation pressure was calculated utilizing Neau et al. [36] method which was proposed for heavy pure compounds. This method does not require experimental data. In this method, the sublimation pressure is obtained from the Clapyron equation. The Clapyron integral will be calculated from triple-point temperature and pressure to sublimation temperature and pressure, assuming that $\Delta h^{sub.}$ is constant.

The results of the AARD% error for correlating the solubility data of phenazopyridine, propranolol, and methimazole at 308, 318, 328, 338, and 348 K temperatures have been reported in Table 15.

The AARD% error values indicated that Pazuki EOS has less error than PR and MPR EOSs. Parameter *b* in Pazuki EOS is temperature dependent, while in PR and MPR EOSs parameter *b* is temperature independent.

Since the parameter *b* shows volume of molecules, it is important for calculation of dense phase densities. In the investigated systems, size of the components is asymmetric, so the parameter *b* plays a critical role in these systems. The reason why Pazuki EOS has a better accuracy in compare to the PR and MPR EOSs can be related to the role of this parameter.

Table 15 The AARD% values for solubility of phenazoprifine, propranolol and methimazole in SC-CO_2

		Phenazoprifine			Propranolol			Methimazole		
T	EOS	Mixing rules			Mixing rules			Mixing rules		
308		vdW1	vdW2	AD	vdW1	vdW2	AD	vdW1	vdW2	AD
	PR	23.62	14.46	13.06	29.82	9.78	9.30	12.06	5.17	7.58
	Pazuki	5.11	2.50	6.44	10.56	6.37	18.71	8.75	7.56	7.55
	MPR	19.29	14.21	12.81	30.16	9.22	8.87	11.81	5.11	8.43
318	PR	34.24	20.34	20.13	29.81	18.26	18.06	23.45	5.49	4.62
	Pazuki	4.78	2.02	4.78	5.29	4.72	11.93	18.20	4.89	7.35
	MPR	25.59	18.43	18.32	30.19	17.89	17.79	23.33	5.40	5.08
328	PR	42.11	20.72	20.69	33.80	27.72	26.63	22.14	8.82	8.81
	Pazuki	9.80	3.41	3.34	19.99	10.48	10.78	23.88	7.93	7.86
	MPR	29.10	20.88	23.10	34.63	26.94	26.75	22.12	8.88	8.86
338	PR	35.38	23.08	22.71	30.20	29.51	29.02	25.10	25.10	25.09
	Pazuki	7.86	3.39	3.39	10.10	5.63	5.61	22.59	22.58	22.19
	MPR	29.28	22.71	22.40	30.72	29.00	28.95	25.08	25.08	25.08
348	PR	34.18	17.78	17.41	51.61	31.41	30.69	50.49	39.18	39.06
	Pazuki	12.38	7.40	7.36	16.35	14.08	13.75	43.65	34.42	34.41
	MPR	38.82	18.46	18.45	44.32	32.24	28.28	50.48	38.82	38.64

In addition, the overall error calculated by the vdW2 mixing rule with the three EOSs is better than the other two mixing rules. The EOS-vdW1 set has one interaction parameter (k_{ij}) and EOS-vdW2 and EOS-AD sets have two interaction parameters (l_{ij} for vdW2 and l_{ij} and λ_{ij} for AD). The parameter l_{ij} is used for parameter b and λ_{ij} for parameter a. The l_{ij} parameter for parameter b is more efficient than the λ_{ij} for parameter a. The AARD% error for the three EOSs and the three mixing rules that have been used for methimazol + CO_2 systems is greater than the propranolol + CO_2 and phenazopyridine + CO_2 systems.

Properties of pure compounds are important in precisely predicting drug substance solubility in SC-CO_2. For example, the molar volume of a solid drug is very important in solubility precision. v_s data for propranolol and phenazopyridine have been calculated by using density experimental data; nevertheless, the value of v_s for methimazole has been calculated as an adjustable parameter of regression of solubility results due to lack of density. The higher error of the binary system of methimazole-CO_2 can be attributed to this matter.

In general, the calculated values of AARD% in the mentioned study increase for all systems as the temperature increases. This difference can be due to the lack of compatibility of experimental solubility data at high temperatures.

3.2.6 Peng-Robinson-Stryjeck-Vera (PRSV)

In this section, solubility of salicylic acid in CO_2 + 1-propanol has been studied using the PR EOS by Stryjeck and Vera (PRSV EOS) [17]. The main form of PRSV EOS is similar to PR EOS and only difference between them is due to the $\alpha(T_r)$ parameter of PRSV EOS. The $\alpha(T_r)$ parameter of PRSV EOS is written as [37]

$$\alpha(T_r) = \left[1 + \kappa\left(1 - \sqrt{T_r}\right)\right]^2 \tag{56}$$

$$\kappa = \kappa_0 + \kappa_1\left(1 + T_r^{0.5}\right)(0.7 - T_r) \tag{57}$$

$$\kappa_0 = 0.378893 + 1.4897153\omega - 0.17131848\omega^2 + 0.0196554\omega^3 \tag{58}$$

In this equation, κ_1 is a constant obtained for each pure compound using saturation-pressure fitting. The vdW2 mixing rule is used to examine the phase behavior of drug-CO_2 system. For calculation fugacity coefficient of solid soluble compound, equation related to fugacity coefficient from the PR EOS is used. However, for parameter a, the PRSV EOS will be used. Since EOSs are used to investigate the phase behavior of gases, the PRSV is not applicable to the investigation of phase behavior of solid phase. Therefore, a modified equation is used in order to assess solid phase behavior [20]:

$$\ln\varphi^s = \ln\varphi^L + \frac{\Delta h_{tp}}{R}\left(\frac{1}{T_{tp}} - \frac{1}{T}\right) + \frac{v_{tp}}{RT}\left(P - P_{tp}\right) + \frac{\Delta c_p}{R}\left(\frac{T_{tp}}{T} - 1\right) - \frac{\Delta c_p}{R}\ln\frac{T_{tp}}{T} \tag{59}$$

By omitting the last two terms of the equation, it will be simplified as follows:

$$\ln\varphi^s = \ln\varphi^L + \frac{\Delta h_{tp}}{R}\left(\frac{1}{T_{tp}} - \frac{1}{T}\right) + \frac{v_{tp}}{RT}\left(P - P_{tp}\right) \tag{60}$$

The investigated system is in the form of solid (salicylic acid)-vapor-liquid (SVL) equilibrium. The equilibrium equations (SLV) for the aforementioned system are written as

$$\frac{\hat{\varphi}_i^L}{\hat{\varphi}_i^V}x_i = y_i,\quad i = 1,2,3\ (VLE),\quad \frac{\varphi_3^S}{\hat{\varphi}_3^L} = x_3\ (SLE) \tag{61}$$

The equilibrium experimental data for the salicylic acid + CO_2 + 1-propanol system was correlated with the proposed model applying PRSV EOS. The interaction parameters of vdW2 mixing rule can be obtained from the optimization of experimental data for each temperature. The obtained AARD% is equal to 0.391.

3.2.7 PR EOS for solvent/CO_2/drug ternary system

Solubility of acetaminophen drug substance in ethanol-CO_2 system has been modeled by PR EOS [38]. The PR EOS along with vdW1 mixing rule is applied estimate fugacity coefficient. Solid solubility in a solvent mixture can be calculated from the following equation [38]:

$$x_s = \frac{f_s(P^{sub})}{\hat{\varphi}_s P}\exp\left[\int_{P^{sub}}^{P}\frac{v_s^s}{RT}dP\right] = \frac{K(T)}{\hat{\varphi}_s P}\exp\left[\int_{P^{sub}}^{P}\frac{v_s^s}{RT}dP\right] \tag{62}$$

For a binary system, fugacity of pure solid is considered the solid sublimation pressure ($f_s(P^{sub})$). The $K(T)$ value can be calculated from solubility of drug in a pure solvent at $P = 1$ atm, as well as the fugacity coefficient of solid at sublimation pressure. The proposed model has three interaction parameters. The parameter required for CO_2-solvent system (k_{12}) can be either obtained from the literature or from binary experimental data. The parameter for CO_2-solute system (k_{13}) is often obtained or considered as fitting parameter. The parameter for solvent-solute system (k_{23}) can be found. In this work, k_{23} is assumed zero, and the unpressurized solution is assumed in solvent as reference state. The effect of pressure on the pure solvent solubility is obtained from the following equation:

$$\frac{dx_s}{dP} = \left(v_s^{sol} - \overline{V}_s\right)\frac{x_s}{RT} \tag{63}$$

In which $\overline{V}_s$ is solid partial molar volume. For this system, the interaction parameter is available for solvent-CO_2 (k_{12}). Therefore, the interaction parameter is obtained for the solute-CO_2 system (k_{23}) by minimizing the AARD function.

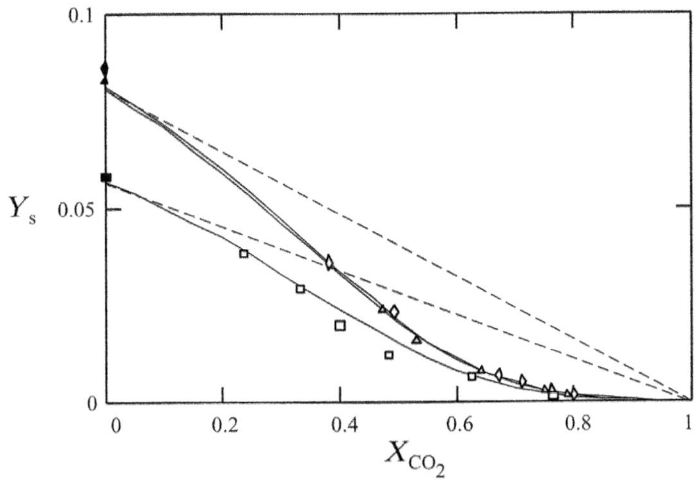

Fig. 7 Solubility of acetaminophen in ethanol and CO_2 at various temperatures, 295 K (□), 315 K (Δ), and 317 K (◊) at a pressure of 7, 10, and 15 MPa [38].

The value of $K(T)$ parameter is obtained from Y_s^o data. The a_s parameter (PR EOS for solid) and k_{13} can be obtained from fitting experimental data. Experimental data and the theoretical results for acetaminophen in ethanol and CO_2 at three temperatures of 295, 315, and 317 K are shown in Fig. 7.

If concentrations of CO_2 are low, the trend of solubility curve is similar to ideal state. This phenomenon shows that the affinity of solvent is not effect on solids and system is at dilute state. When more antisolvent was added, solubility becomes less than the ideal dilution line since antisolvent decreases the dissolution ability of the solvent. The results show that the acetaminophen solubility in ethanol is reduced with. Results also showed that the pressure is less effective than the temperature and CO_2 concentration in the solution. When compositions are near the critical point, the pressure effect becomes clearer due to a reduction of the solid volume in the solution. For the studied system, the results of the predictions of EOS indicate that the solubility of acetaminophen in ethanol decreases under pressure. This is due to the large size of partial molar volume of ethanol in comparison to the solid molar volume.

3.3 Solubility prediction of drug component using noncubic EOS

As stated in the previous section, cubic EOSs have low accuracy to estimate drug solubility components in SC-CO_2. Cubic EOSs require thermodynamic information of drug components that usually cannot be found in the literatures. Therefore, it is necessary to calculate these properties from the thermodynamic equations and then the solubility of the drug component using EOSs. This will reduce the prediction accuracy of the EOSs. In this section, we will review the behavior of the noncubic EOSs.

3.3.1 Leonhard-Kraska (EOS)

In the following section, the solubility of salicylic acid, RS-ibuprofen, S-naproxen, and phytosterol in SC-CO_2, which was investigated in the previous section applying PR as a cubic EOS, will be investigated using the Leonhard-Kraska (LK EOS) [39]. The common cubic EOSs have been developed based on the van der Waals equation, which did not have the ability to estimate the thermodynamic properties in the near-critical region. This lack of ability is due to the fact that these types of EOSs give analytical results, while in the near-critical region liquids have nonanalytic behavior due to the low difference in density fluctuation. Therefore, cubic EOSs are not suitable candidates for predicting thermodynamics properties in the near-critical regions as well as supercritical region. One of the EOSs that have a very high accuracy to predict the behavior of liquids in near-critical point is the noncubic equation proposed by Leonhard and Kraska (LK), which is written as below:

$$P = P_{ref}(a, b) + P_{pert}(a, b, \sigma), \quad \sigma = \sigma(P, T) \tag{64}$$

The first part of the equation is the reference section that shows the effects of repulsion and the second part of the equation is the perturbation section obtained from the following equation [40]:

$$P_{ref} = \frac{RT}{v}\left(1 + \frac{4\zeta - 2\zeta^2}{(1-\zeta)^3}\right) - \frac{8RbT^*}{v^2}\psi(T) \tag{65}$$

$$\zeta = \frac{b}{v} = b\rho \tag{66}$$

$$\psi(T) = 1 + f_T^* \frac{T_c - T}{T_c} \tag{67}$$

$$f_T^* = \begin{cases} f_T & \text{for} \quad T \leq T_c \\ 0.25 f_T & \text{for} \quad T > T_c \end{cases} \tag{68}$$

In which b and T^* are the parameters of the equation. f_T is a temperature-dependent parameter and ζ is packing fraction. The perturbation term is obtained from the following equation:

$$P_{pert} = RT \sum_{i=2}^{n_{max}} \left((i+2)(i-1)b^{i-1} \sum_{k=1}^{\text{int}[i/2]} \left(\sigma^{2k}\rho^{i-2k} C_{ik}\right) \right) - 8RbT^*\psi(T)\sigma^2 \tag{69}$$

$$C_{i,k} = \frac{i!}{(i-2k)!k!2^k} \tag{70}$$

In low densities $n_{max}=20$ and in high densities $n_{max}=40$. σ is also calculated from the following equation:

$$\sigma=\sigma_c\sigma_\zeta(\zeta)\sigma(T) \tag{71}$$

$$\sigma_\zeta(\zeta)=\zeta\left(\frac{1}{\zeta_c}-\frac{B_\zeta}{\zeta_c^2}y+\sum_{i=2}^{n_{coeff}}C_i y^i\right) \tag{72}$$

$$y=\frac{\zeta-\zeta_c}{B_\zeta} \tag{73}$$

$$\sigma_T=\exp\left[-\left(\frac{\Delta T}{B_T}\right)^2\right] \tag{74}$$

The C_i coefficients for $i=2$ to 11 have been reported in the literature [40]. Therefore, the noncubic EOS has six adjustable parameters including T^*, b, σ_c, B_ζ, σ_T, B_T.

The noncubic LK EOS has been used for solubility modeling of salicylic acid, RS-ibuprofen, S-naproxen, and phytosterol in SC-CO_2 along with the PR EOS. The critical data point for pure drug is not required because parameters of LK EOS were fitted. The results of the solubility correlation obtained by LK EOS and PR EOS are presented in Table 16.

The AARD% results reported in Table 16 emphasized that the LK EOS has a better accuracy than PR-EOS. Especially for the phytosterol drug, the LK EOS equation has a better accuracy, while the PR-EOS seems to be weak.

3.3.2 *SAFT of variable range EOS*

Cubic EOS is not precise enough to predict drug solubility in SC-CO_2 due to the simplicity as well as nonconsideration of intermolecular interactions such as hydrogen bonding effects between components in the system. The SAFT equations are a preferable alternative for cubic EOSs. Since these EOSs consider various intermolecular effects, such as hard sphere, chain effects, dispersion, and hydrogen bonding effects, they have a good accuracy for complex mixture containing association fluids [41].

Table 16 Comparison between the AARD% values for PR and LK EOSs [39]

Drug	PR-EOS AARD (%)	LK-EOS AARD (%)
Salicylic acid	10–44	5
RS-Ibuprofen	16–33	9
S-Naproxen	14–33	10
Phytosterol	37–77	7

SAFT of variable range (SAFT-VR) which is coupled with Yukawa potential function is a suitable alternative model among the SAFT EOSs. Accordingly, a predictive model among SAFT EOSs is proposed to estimate pharmaceutical compounds solubility in SC-CO_2 [42]. Based on the theory of the SAFT-VR EOS, the residual Helmholtz energy consists of three parts mono, chain, and assoc, as following equation [42]:

$$\frac{A^{res.}}{NkT} = \frac{A^{mono.}}{NkT} + \frac{A^{chain.}}{NkT} + \frac{A^{assoc.}}{NkT} \tag{75}$$

In which $A^{mono.}$ is the energy of monomer, $A^{chain.}$ is the energy of chain formation, and $A^{assoc.}$ is the association energy between the associating sites.

The thermodynamic functions that are needed to calculate the phase equilibria and the solubility such as chemical potential and fugacity of species i in a mixture are as follows:

$$\mu_i = \left(\frac{\partial nA}{\partial n_i}\right)_{T,V,n_j} \tag{76}$$

$$\ln \hat{\varphi}_i = \frac{\mu_i - \mu_i^{i.d}}{RT} - \ln Z_m \tag{77}$$

where in Eq. (77), Z_m is compressibility factor, which is computed from the following equation:

$$Z_m = 1 + \rho\left(\frac{\partial A^{ref}}{\partial \rho}\right)_{T,n} \tag{78}$$

Similar to cubic EOSs in which mixing rules are used for multicomponent systems, the SAFT EOSs also use a mixing rule for the CO_2-drug system. The Lorentz-Bertholt combining rule has been used in this work.

$$\sigma_{ij} = \frac{\sigma_{ii} + \sigma_{jj}}{2} \tag{79}$$

$$\varepsilon_{ij} = \left(\varepsilon_{ii}\varepsilon_{jj}\right)^{1/2} \tag{80}$$

$$\lambda_{ij} = \frac{\lambda_{ii}\sigma_{ii} + \lambda_{jj}\sigma_{jj}}{\sigma_{ii} + \sigma_{jj}} \tag{81}$$

As CO_2 is a nonassociating compound, the energy and volume association is not required for combining rule. In addition, since the experimental data of the drug for sublimation pressure are not available, this property should be estimated. As it was previously mentioned, there are two methods to calculate sublimation pressure. The sublimation pressure is obtained using the group contribution and correlation methods. The group contribution method is a predictive model based on the drug functional group, while pressure is considered as an adjustable parameter in correlation method. Also in the correlation method, the sublimation pressure method is considered as a function of temperature and its constants are tuned by solubility data. Here, the SAFT-VR EOS is coupled with both methods. The following equation is considered in the group contribution method based on the Clausius-Clapeyron equation:

$$\frac{d\ln P^{sub}}{dT} = \frac{\Delta h_V}{\Delta Z R T^2} \tag{82}$$

In which Δh_V is the heat of vaporization and ΔZ is the difference of compressibility factors of solid and vapor phases. In this method, $\Delta h_V / \Delta Z$ was considered to be independent of temperature and the value of Δh_V is calculated from the following relation:

$$\Delta h_V = \Delta h_{Vb}\left(\frac{1 - T/T_c}{1 - T_b/T_c}\right)^m \tag{83}$$

With the approximation of $T_c \approx \frac{3}{2} T_b$, the above equation can be written as

$$\Delta h_V = \Delta h_{vb}\left(3 - 2T_{\rho b}\right)^m, \quad T_{\rho b} = \frac{T}{T_b} \tag{84}$$

As a result, the sublimation pressure equation is written as follows:

$$\ln P^{sub} = \frac{\Delta h_{vb}}{\Delta Z_b R T_b}\left[1 - \frac{\left(3 - 2T_{\rho b}\right)^m}{T_{\rho b}} - 2m\left(3 - 2T_{\rho b}\right)^{m-1} \ln T_{\rho b}\right] \tag{85}$$

In this method, $\Delta Z_b = 0.47$ and Δh_{vb} is heat of vaporization at the normal boiling point, which is calculated by the following equation:

$$\frac{\Delta h_{vb}}{T_b} = K_F(8.75 + R\ln T_b) \tag{86}$$

In the correlation method, the sublimation pressure is considered as the following in terms of temperature:

$$\ln P^{sub} = a - \frac{b}{T} \tag{87}$$

For any drug component, a and b are constants obtained from the correlation of drug solubility data in SC-CO_2. For calculation solubility, the SAFT-VR EOS contains parameters such as λ, m, ε/k, σ which should be specified for the presented components in the system. In the beginning, the parameters of SAFT-VR EOS for CO_2 have to be calculated. These parameters were obtained from optimization of pressure-volume-temperature (PVT) data for pure CO_2. Two methods are used to predict drugs solubility in SC-CO_2. First, authors used group contribution method.

Table 17 The values of AARD% for correlation in two different methods [42]

Drug	*T* (K)	*P* (bar)	First approach	Second approach
Ketoprofen	312.5	100–220	11.53	11.68
Methimazole	308	122–355	0.58	0.46
Nabumetone	308	100–220	1.81	2.27
Nimesulide	313	130–220	4.45	4.44
phenazopyridine	308	122–355	3.83	5.22
Phenylbutazone	308	100–220	4.57	4.24
Piroxicam	312	100–220	11.18	11.09
Propranolol	308	122–334	7.38	6.85
Salicylic acid	313	96–250	15.28	14.33
Salicylamide	308	100–220	0.76	0.97
Overall AAD%			6.13	6.15

The main purpose of the research is to evaluate the predictive capability of SAFT-VR EOS. Accordingly, to calculate the SAFT-VR EOS parameters for drug components, only the first set of available solubility data has been used for correlation and the rest of the data has been used for validating the predicted solubility data using the proposed EOS. Therefore, the proposed model is called a semicorrelative-semipredictive model. In the second method, sublimation pressure data are considered as correlation. A comparison between the results of correlation in two methods is reported in Table 17.

By comparing the results reported in Table 17, we can state that the SAFT-VR EOS along with the two proposed methods have the same accuracy for correlating drug components solubility.

Later ability of prediction of this equation will be assessed using the SAFT-VR parameters. The AARD% error values for predicting pharmaceuticals solubility in SC-CO_2 are represented in Table 18. It should be noted that at this point the correction of the additional adjustable parameters has not been used.

Table 18 The AARD% values for prediction in two different methods [42]

Drug	*T* (K)	*P* (bar)	First approach	Second approach
Ketoprofen	331.5	115–220	55.46	64.02
Methimazole	318–328	122–355	12.61	12.89
Nalometone	318–328	100–220	26.78	36.05
Nimesulide	333	130–220	37.82	41.36
phenazopyridine	318–328	122–355	26.33	58.53
Phenylbutazone	318–328	100–220	32.08	31.21
Piroxicam	331	130–220	42.11	44.95
Propranolol	318–328	122–355	35.27	24.44
Salicylic acid	328	101–203	42.82	36.49
Salicylamide	318–328	100–220	35.95	37.91
Overall AAD%			34.72	38.78

As the values of overall AARD% indicate, the SAFT-VR EOS with the first method (first approach) has better accuracy. However, the overall error of the second method is also acceptable.

3.3.3 Perturebed-chain polar statistical associating fluid theory

A version of SAFT EOS called perturebed-chain polar statistical associating fluid theory (PCP-SAFT) has been used to model ketotifen fumarate (KTF) solubility in SC-CO_2 [43]. Despite of nonpolar nature of CO_2, it has a quadrupole momentum. Therefore, quadrupole-quadrupole (QQ) intermolecular effects are essential for the thermodynamic modeling of mixtures containing CO_2. Due to consideration of the intermolecular effects of (QQ), the SAFT EOS is called PCP-SAFT. Similar to the other versions of SAFT EOS, the dimensionless residual Helmhotz energy can be calculated as

$$\frac{A^{res.}}{NkT}=\frac{A^{hc.}}{NkT}+\frac{A^{disp.}}{NkT}+\frac{A^{QQ.}}{NkT} \tag{88}$$

In the above equation, "*hc*" represents the hard chain as reference term, "*disp*" represents the term related to dispersion and "QQ" represents the intermolecular effects of quadrupole-quadrupole. The PCP-SAFT EOS has four parameters as m,σ, ε and Q, for which thermodynamic calculations are required to be specified. The following combining rules are considered for multicomponent system:

$$\sigma_{ij}=\frac{1}{2}\left(\sigma_i+\sigma_j\right) \tag{89}$$

$$\varepsilon_{ij}=\sqrt{\varepsilon_{ii}\varepsilon_{jj}}\left(1-k_{ij}\right) \tag{90}$$

In which $k_{ij}$is interaction parameter. In addition, the following combining rules were used for chain length:

$$m_{ij}=\left(m_i m_j\right)^{\frac{1}{2}} \tag{91}$$

$$m_{ijk}=\left(m_i m_j m_k\right)^{\frac{1}{3}} \tag{92}$$

In order to model KTF solubility, parameters of the PCP-SAFT EOS should be specified for CO_2. In the current study, the parameters of the EOS for CO_2 are obtained from the literature, and the model parameters for the drug are calculated by nonlinear regression between the experimental solubility data and the results obtained from proposed model. However, it is worth noting that solubility modeling by PCP-SAFT is calculated considering $k_{12}=0$ and $k_{12}\neq 0$. The values of AARD% for the PCP-SAFT EOS are compared for $k_{12}=0$ and $k_{12}\neq 0$, in Table 19.

The values of AARD% errors indicate that $k_{12}\neq 0$ is a suitable assumption which reduces the amount of error by 43.5%. The experimental solubility data and the results of PCP-SAFT EOS in various temperatures were compared in Fig. 8.

Table 19 The comparison between the results of AARD% of PCP-SAFT EOS for two different states of $k_{12}=0$ and $k_{12}\neq 0$ [43]

k_{12}	AARD%
0	64.5
$-0.8247+\frac{255.8845}{T}$	20.7

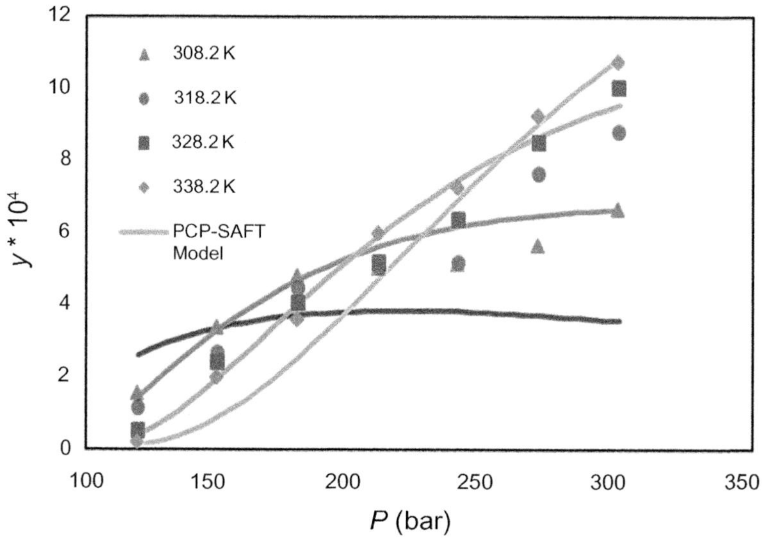

Fig. 8 Comparison between experimental and correlated solubility of KTF obtained from PCP-SAFT EOS [43].

As Fig. 8 shows, the error decreases with increasing temperature, and at higher temperatures there is good correlation between experimental data and PCP-SAFT EOS. Experimental results indicate that with increasing pressure (temperature is tuned constant), higher levels drug solubility were obtained. The PCP-SAFT EOS for $T=308.2$ K and $P=219$ bar predicted the maximum solubility. Based on the results, it can be inferred that PCP-SAFT EOS results can meet experimental data at low temperatures, but the accuracy of the results is low.

3.3.4 Association-SRK EOS with quadruple effect (qCPA) EOS

Association-SRK EOS with quadruple effect (quadruple-Cubic-Plus-Association [qCPA]-EOS) as a noncubic EOS as well as PCP-SAFT EOS has been applied for estimating loratadine solubility as an antihistamine drug in SC-CO_2 [9].

Since van der Waals cubic EOSs do not consider all the intermolecular effects, they do not have enough precision to predict the phase behavior of compounds such as CO_2, which have quadrupole-quadrupole (QQ) intermolecular effects. In the current research, The SRK cubic EOS is corrected by including the QQ intermolecular effects. Accordingly, the reduced residual Helmholtz energy has been modified as [9]

$$\frac{A^{res.}}{NkT}=\frac{A^{SRK.}}{NkT}+\frac{A^{Assoc.}}{NkT}+\frac{A^{QQ.}}{NkT} \quad (93)$$

The parameters of those two methods were not available for this drug. Therefore, the parameters of the two models were obtained applying the experimental data of sublimation pressure and optimization of a suitable objective functions. The objective function can be written as below:

$$AARD\%=\frac{100}{N}\sum_{i=1}^{N}\left|\frac{P_i^{sub}-P_i^{sub,cal}}{P_i^{sub}}\right| \quad (94)$$

Both EOSs that have been used have four parameters for the drug that were obtained from the above objective function. Also, binary interaction parameter k_{ij} ($i=CO_2$, $j=loratadine$) is obtained for these two EOSs by fitting solubility data. The results for two states of $k_{12}=0$ and $k_{12}\neq 0$ for solubility of loratadine are represented in Table 20.

According to the results reported in Table 20, the two noncubic EOSs are not able to correlation loratadine solubility. The AARD% values for both EOSs are higher than 100%. By considering the binary interaction parameter (k_{12}) as a function of temperature, the AARD% values for both EOSs became almost equal. The results of experimental solubility and calculated results obtained from both EOSs considering $k_{12}\neq 0$ are shown in Fig. 9.

According to the figure, the solubility correlation results for the two EOSs are not the same. The mean values of AARD% for qCPA and PCP-SAFT EOSs are 15.17 and 15.02 at various temperatures. The lowest value of AARD% was seen at 308.15 K. Moreover, by increasing temperature at any point, qCPA EOS underestimates the drug solubility in SC-CO_2. While PCP-SAFT EOS overestimates the solubility. Authors can be concluded that ability of two EOSs is similar.

Table 20 The AARD% values for qCPA and PCP-SAFT with and without interaction parameter [9]

qCPA		PCP-SAFT	
k_{12}	AARD%	k_{12}	AARD%
0	147.2	0	168.6
$-0.2788+\frac{100.493}{T}$	15.18	$-0.2977+\frac{102.575}{T}$	15.03

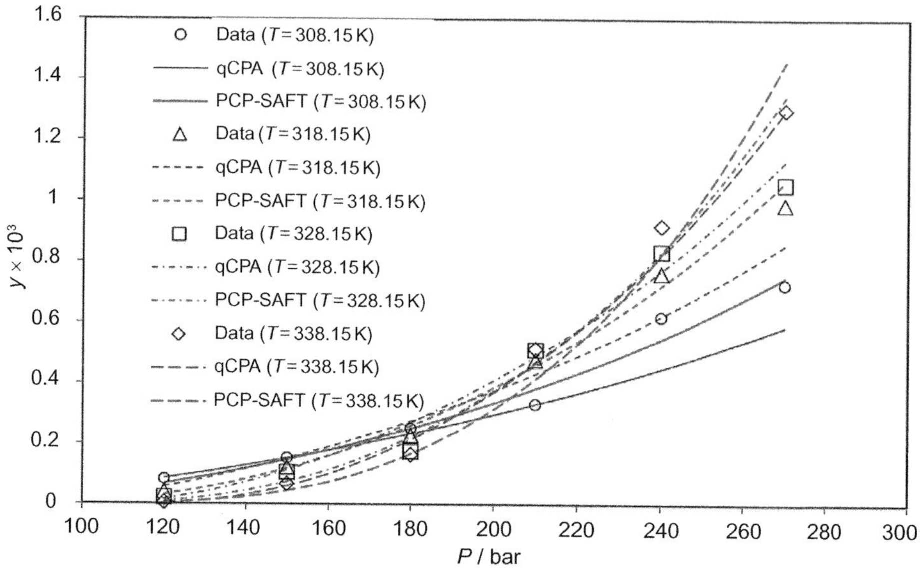

Fig. 9 Comparison between experimental and correlated solubility for loratadine/SC-CO_2 solution using qCPA and PCP-SAFT EOSs applying interaction parameter [9].

3.4 Solubility parameter-based models

There is a basic principle in chemistry which says "like dissolves like." It means that compounds with similar structures dissolve in each other. Solubility parameter is a thermodynamic parameter that expresses this principle quantitatively. This parameter can help us to find a suitable solvent in various industries including the pharmaceutical industry. The solubility parameter was first used by Hildebrand and Scott. The Hilderbrand solubility parameter (HSP) is defined in the following equation [44]:

$$\delta = \left(\frac{\Delta u^{vap}}{v}\right)^{1/2} \tag{95}$$

In this equation, Δu^{vap} is the cohesive energy density and v is the molar volume of pure solvent. The solubility parameter can be used to calculate the solubility of a component in the desired solvent. The parameter $\Delta = \delta_1 - \delta_2$ is considered as a criterion for the solubility of the components (1) and (2) in each other. However, higher values of Δ means that the solubility of these two components seem to be unfavorable. The solubility of these two components decrease by increasing the value of Δ. On the other hand, if Δ becomes smaller, the two component can dissolve better in each other. Since solubility parameters can be calculated for different components, models that are based on the solubility parameters are predictive models.

3.4.1 Solution model theory

A new method has been proposed for solid component solubility especially drug components in SC-CO_2 using activity coefficient [45]. This model is suggested based on the solution model. Accordingly, the SCF is assumed to be an expanded liquid. The component (1) represents a SCF and component (2) represents a solid solute. At equilibrium state, the following equation is available [45]:

$$y_2 = \frac{f_2^s}{\gamma_2 f_2^l} \tag{96}$$

In which, γ_2 is activity coefficient of component (2). Due to the fact that the value of component (2) in the solution is very low, activity coefficient at infinite dilution (γ_2^∞) can be used instead of (γ_2). Therefore, the above equation is obtained as follows:

$$y_2 = \frac{f_2^s}{\gamma_2^\infty f_2^l} \tag{97}$$

In this study, along with the Flory-Huggins equation, the modified regular solution model is being used for determination of γ_2^∞.

$$\ln \gamma_2^\infty = \left(\frac{v_2}{RT}\right)(\delta_1 - \delta_2)^2 + 1 - \left(\frac{v_2}{v_1}\right) + \ln\left(\frac{v_2}{v_1}\right) \tag{98}$$

In this equation, δ is the solubility parameter and v is the molar volume. Using Eqs. (97) and (98) with assumption $C_P^S = C_P^{SCF}$, drug substance solubility in SCF is obtained from the following equation:

$$\ln y_2 = \left(\frac{\Delta h_2^f}{R}\right)\left(\frac{1}{T_{2,m}} - \frac{1}{T}\right) - \left(\frac{v_2}{RT}\right)(\delta_1 - \delta_2)^2 - 1 + \left(\frac{v_2}{v_1}\right) - \ln\left(\frac{v_2}{v_1}\right) \tag{99}$$

The value of δ_1 is calculated from PR EOS. δ_2 is obtained using molar volume of drug and the internal energy change of vaporization ΔU_2 that calculated by group contribution method. In this equation, v_2 is considered as an adjustable parameter, which is calculated by minimization of the AARD% as an objective function:

The sublimation pressure of the drug component is estimated by fusion experimental data with the experimental data for heat of vaporization, which is calculated using the PR EOS. A simple linear relationship between v_2 and ρ_1 (density of CO_2) is given as follows:

$$\ln v_2 = \alpha \ln \rho_1 + \beta \tag{100}$$

In this equation α and β are two temperature-independent parameters which are obtained for each drug substance. α and β parameters for a pharmaceutical component

Table 21 Correlations results of the proposed model for solubility of drug components in SC-CO_2 using solution model method [45]

Drug	Temperature range (K)	Pressure range (MPa)	N_P	AARD%[a]	AARD%[b]
Nifedipine	333.15–373.15	12.6–29.6	29	15.6	29.69
Ketorprofen	312.50–331.50	10.0–22.0	10	20.00	26.67
Nimesulide	312.50–331.50	13.0–22.0	8	21.05	21.64
Piroxicam	312.50–331.50	10.0–22.0	9	8.50	15.86
Average				16.28	23.46

[a]Two-parameter method.
[b]One parameter method.
NP: The number of experimental data points.

indicate that β has a constant value for all components and is considered as -13.26. Therefore, based on the drug solubility data, the only parameter which is going to be fitted for each drug-CO_2 system is α. The results of AARD% for the solubility of nifedipine, ketorprofen, nimesulide, and piroxicam in SC-CO_2 have been reported for both the one-parameter and the two-parameter methods in Table 21.

The average AARD% values for the two-parameter and one-parameter methods are 16.28% and 23.46%, respectively. These results indicate that the accuracy of the two-parameter method is more accurate than one-parameter method. However, one-parameter method error values are also acceptable. Furthermore, α is obtained as a linear relationship based on Δh_2^{vap} by the following equation:

$$\alpha = m + n\Delta h_2^{vap} \tag{101}$$

where $m = -0.9517$ and $n = -2.1729 \times 10^{-3}$. The average of AARD% value is 48.84 for predicting the solubility of the four specified drugs in SC-CO_2 using the solution model by the generalized correlation method. By comparing this error with the two previous models, it can be concluded that the error of two-parameter and one-parameter models is better than the generalized correlation.

3.5 Mathematical models

In the three previous sections, the solubility of pharmaceutical components was modeled by cubic and noncubic EOSs, as well as thermodynamic models based on the solubility parameter. In this section, there are other models that are used to model the solubility of drug components. The models used in this section are association-based model, conductor-like screening model activity coefficient (COSMO-SAC), neural network, which will be discussed below.

3.5.1 Association model for drug-CO_2 system

A simple and modified association model has been proposed, which is used for solubility of drug compounds including steroids, antibiotics, and antiinflammatory in SC-CO_2 [46]. The following associate reaction is considered for molecules A and B:

$$A + kB \rightarrow AB_k \tag{102}$$

The equilibrium constant of this reaction is written in terms of fugacity:

$$K_f = \frac{\left(\hat{f}_{AB_k}/f^*_{AB_k}\right)_{SCP}}{\left(\hat{f}_A/f^*_A\right)_S \left(\hat{f}_B/f^*_B\right)^k_{SCP}} \tag{103}$$

In which SCP, S and f^* represent the supercritical phase, solute phase, and fugacity, respectively. The fugacity of each component is calculated from the following relationships:

$$\hat{f}_i = y_i \hat{\varphi}_i P \tag{104}$$

$$f^*_i = \varphi^*_i P^* \tag{105}$$

where $i = A, B, AB_k$. In this model, it is assumed that the liquid-phase compounds are not soluble in solid phase, and the solid is pure. Suppose that the standard state of solute i is considered as pure solute at the desired temperature and pressure. Thus,

$$\hat{f}^*_i = f_i \ , \quad i = solute \tag{106}$$

For which the fugacity of the pure component A is calculated from the following equation [46]:

$$f_A = P^{sub}_A \exp\left[\frac{v_A}{RT}\left(P - P^{sub}_A\right)\right] \tag{107}$$

By replacing Eq. (104) in Eq. (103) the following equation can be obtained as

$$K_f = \frac{\left(y\hat{\varphi}_{AB_k} P / \varphi^*_{AB_k} P^*\right)}{\left(P^{sub}_A \exp\left[\frac{v_A}{RT}\left(P - P^{sub}_A\right)\right] / \varphi^*_A P^*\right)\left(y_B \hat{\varphi}_B P / \varphi^*_B P^*\right)^k} \tag{108}$$

The equilibrium constant is obtained from the following relation:

$$\ln(K_f) = \frac{\Delta h_s}{RT} + q_s \tag{109}$$

$$\frac{v_A P}{RT} = \frac{Z v_A \rho}{M} \tag{110}$$

The sublimation pressure can be considered as the following relation:

$$\ln\left(P_A^{sub}\right) = A_1 - \frac{B_1}{T} \tag{111}$$

where A_1 and B_1 are temperature independent and $\frac{\nu_A P_A^{sub}}{RT}$ is very small (10^{-9}). As the solubility of the drug components including steroids, antibiotics, and statin in SC-CO_2 is very low, $y_B = 1$, the equation is simplified as follows:

$$y = \left(\frac{P}{P^*}\right)^{(k-1)} \exp\left(\frac{a}{T} + b\rho + c\right) \tag{112}$$

where:

$$a = \left(\frac{\Delta h_s}{R}\right) - B_1 \tag{113}$$

$$b = \frac{Z\nu_A}{M} \tag{114}$$

$$c = \ln\left(\left(\varphi_A^*\right)\left(\frac{\hat{\varphi}_{AB_k}}{\varphi_{AB_k}^*}\right) / \left(\frac{\hat{\varphi}_B}{\varphi_B^*}\right)^k\right) - \ln P^* + q_s + A_1 \tag{115}$$

The solubility of drug components including steroids, antibiotics, and statin in SC-CO_2 has been calculated using the association model. The parameters of this model are computed by minimizing the AARD% objective function. The AARD% results for these drugs are represented in Table 22.

According to the results obtained, the model successfully correlates the drug component data in SC-CO_2 with an error about 7%. Therefore, it can be inferred that proposed model predicts solubility of various drugs with small error.

3.5.2 Association model for drug-cosolvent-CO_2 system

An association model has been developed for correlating pharmaceutical component solubility in SC-CO_2 with cosolvent [47]. In accordance with what has been stated for drug-SCF system in the previous section, A is associated with SCF (B) and cosolvent (E). As a result, the following reaction represents this system:

$$A + kB + \gamma E \rightarrow AE_\gamma B_k \tag{116}$$

The equilibrium constant based on the fugacity is written as follows:

$$K_f = \frac{\left(\hat{f}_{AE_\gamma B_k} / f_{AE_\gamma B_k}^*\right)_{SCF}}{\left(\hat{f}_A / f_A^*\right)_S \left(\hat{f}_B / f_B^*\right)_{SCF}^k \left(\hat{f}_E / f_E^*\right)_{SCF}^\gamma} \tag{117}$$

Table 22 Physical properties and AARD% values of the association model for drug components

No.	Drug	*T* (K)	*P* (bar)	N_P	AARD%
1	Anastrazole	308–348	120–355	45	6.9
2	Antipyrine	308–328	100–220	21	3.9
3	Artemisinin	310–338	100–270	36	6.3
4	Aspirin	308–328	120–250	24	5.0
5	Atorvastatin	308–348	120–355	45	6.4
6	Benzocaine	308–318	80–250	20	9.9
7	Climbazole	313–333	100–400	24	5.6
8	Codeine	308–348	120–360	45	11.9
9	Diazepam	308–348	120–360	45	9.2
10	Flurbiprofen	303–323	80–250	27	6.5
11	Fluvastatin	308–348	120–355	45	11.7
12	Ibuprofen	308–318	80–220	29	10.2
13	Ketoprofen	313–332	90–250	10	7.0
14	Lamotrigine	318–348	120–355	36	3.8
15	Lovastatin	308–348	120–355	45	4.8
16	Methimazole	308–348	120–355	40	9.4
17	Nabumetone	308–328	100–220	21	7.7
18	Naproxen	313–333	80–200	18	4.9
19	Nimesulide	313–333	120–220	8	11.0
20	Penicillin V	313–335	80–300	24	11.2
21	Paraxetin	308–328	80–220	24	4.7
22	Phenacetin	308-328	90–190	16	5.6
23	Phenazopyridine	308–348	120–360	45	7.7
24	Piroxicam	313–331	100–220	9	5.4
25	Procaine	308–318	80–250	18	3.6
26	Rosuvastatin	308–348	120–355	45	5.0
27	Salicylamide	308–328	100–220	21	6.1
28	Salicylic acid	313–328	90–250	11	4.4
29	Sulfamethazine	313–333	130–480	20	5.4
Average					6.9

The fugacity of species *i* is calculated as below:

$$\hat{f}_i = \gamma_i \hat{\varphi}_i P, \quad f_i^* = \varphi_i^* P^* \tag{118}$$

Since fugacity of solid can be calculated from Henry's law:

$$\hat{f}_{AE_\gamma B_k} = \gamma_{AE_\gamma B_k} H_{AE_\gamma B_k} \tag{119}$$

In this research, it is assumed that

$$\gamma_{AE_\gamma B_k} = \gamma_2, \quad H_{AE_\gamma B_k} = H_2^{eff} \tag{120}$$

Considering the relation between fugacity of pure solid with temperature and pressure [Eq. (107)], the equilibrium constant is obtained from the following equation:

$$K_f = \frac{\left(y_2 H_2^{eff} / \varphi^*_{AE_\gamma B_k} P^*\right)}{\left(P_A^{sub} \exp\left(\frac{v_A\left(P - P_A^{sub}\right)}{RT}\right) / \varphi^*_A P^*\right) \left(\frac{y_B \hat{\varphi}_B P}{\varphi^*_B P^*}\right)^k \left(\frac{y_E \hat{\varphi}_E P}{f_E^*}\right)^\gamma} \tag{121}$$

The Henry constant for cosolvent-SCF-solid system is related to density and cosolvent concentration as [47]

$$\ln H_2^{eff} = \ln f_s + \frac{(D + E'\rho + F y_E)}{RT} \tag{122}$$

In this equation, f_s is the fugacity of the mixed solvent and y_E is the cosolvent mole fraction. The equilibrium constant is similar to Eq. (109). By replacing $\frac{v_A P}{RT}$ with $\frac{Z v_A \rho}{M}$, the above equation can be written as

$$\begin{aligned} \ln y_2 = & k \ln(y_B) + (k + \gamma - 1) \ln\left(\frac{P}{P^*}\right) + \gamma \ln(y_E) - \ln\left(\frac{\hat{\varphi}_s}{\varphi^*_{AE_\gamma B_k}}\right) + \ln\left(\frac{P_A^{sub}}{P^*}\right) + \frac{Z v_A \rho}{M} \\ & - \frac{v_A P_A^{sub}}{RT} + k \ln\left(\frac{\hat{\varphi}_B}{\varphi^*_B}\right) + \gamma \ln\left(\frac{\hat{\varphi}_E}{\varphi^*_E}\right) - \ln\left(\varphi^*_A\right) + \frac{\Delta h_s}{RT} + q_s + \frac{(D + E'\rho + F y_B)}{RT} \end{aligned} \tag{123}$$

According to the Clausius-Clapeyron equation for sublimation pressure and the fact that $y_B \approx 1.0$ and also by omitting $\left(\frac{v_A P_A^{sub}}{RT}\right)$, we have

$$\ln y_2 = (k + \gamma - 1) \ln\left(\frac{P}{P^*}\right) + \gamma \ln(y_E) + \frac{(A_1 + A_2\rho + A_3 y_E)}{T} + A_4\rho + A_5 \tag{124}$$

In which

$$\begin{aligned} & A_1 = ((\Delta h_s - D)/R) - B, \; A_2 = E'/R, \; A_3 = F/R, \; A_4 = Z v_A / M \\ & A_5 = q_s + A - \ln\left(\left(\frac{\varphi^*_A \hat{\varphi}_s}{\varphi^*_{AE_\gamma B_k}}\right)\left(\frac{\hat{\varphi}_B}{\varphi^*_B}\right)^{-k}\left(\frac{\hat{\varphi}_E}{\varphi^*_E}\right)^{-\gamma}\right) \end{aligned} \tag{125}$$

Relation (124) is used to calculate the solubility of pharmaceutical components based on the solvent composition, temperature, pressure, and supercritical-phase density. k and γ are the constants of the equation. Also, A_1 to A_5 are constants of the equation.

The density of CO_2 required for the proposed model is obtained from the Span and Wagner EOS [48]. The adjustable parameters of the modified solution model are obtained from optimization of AARD%. The results for the ternary system of cosolvent-CO_2-drug have been reported in Table 23.

Table 23 The AARD% values for the association model in correlating solubility of drug components in cosolvent-CO_2 systems [47]

System	T/K	P/MPa	Cosolvent mole fraction	AARD%
Aspirin-acetone-SC-CO_2	318–328	10–20	0.03	5.53
O-Hydroxybenzoic acid-acetone-SC-CO_2	318–328	8.6–20.1	0.035	2.78
O-Hydroxybenzoic acid-methanol- SC-CO_2	318–328	9.1–20.1	0.035	2.29
Naproxan-acetone-SC-CO_2	318–333	8.96–17.3	0.0175–0.0525	10.41
Naproxan-ethyl acetone-SC-CO_2	333	8.96–17.93	0.0175–0.0525	12.19
Average				6.64

As the results show, the association model can correlate solubility of drug in cosolvent-CO_2 system. Although, the accuracy of model decreases as the mole fraction of cosolvent increases.

3.5.3 PR-COSMO-SAC model for drug solubility in SC-CO_2 system

One of the weaknesses of common cubic EOSs such as PR or SRK to estimate drug compounds solubility is to estimate the parameters of these equations for drug components. These parameters are the function of the critical properties and the acentric factor, which are not available experimentally for drugs and usually are obtained from equations such as group contribution method. The PR-COSMO-SAC EOS has been applied for this propose [49]. Parameters of PR EOS were obtained from results of the first principle of solvation. The solubility of solid drug is calculated from isofugacity equation between solid and fluid phases:

$$f_i^S(T,P) = \hat{f}_i^F(T,P,x) \tag{126}$$

However, it is worth mentioned that the solid phase is pure. By solving above equation, x will be calculated. Solid fugacity is calculated by

$$\frac{f_i^S(T,P)}{f_i^L(T,P)} = \exp\left[\frac{\left(v_i^S - v_i^L(T)\right)\left(P - P_i^t\right)}{RT} + \frac{\Delta h_i^t}{RT_i^t}\left(1 - \frac{T_i^t}{T}\right)\right] \tag{127}$$

Since the pressure is high, $P - P_i^t \cong P$, solid volume is assumed to be equal to parameter b of the EOS. T_i^t and Δh_i^t have been obtained from the literature and $v_i^L(T)$ and f_i^L were calculated from EOS. $a(T,x)$ is calculated using the solvation free energy:

$$a(T,x) = \frac{b(x)}{C_{PR}} \Delta g^{*chg}(T,x) \tag{128}$$

In which $C_{PR} = -0.623$. Furthermore, $b(x)$ is obtained from the following equation:

$$b(x) = \sum_i x_i b_i \tag{129}$$

In which b_i is the molar cavity. The charging free energy can be obtained as

$$\Delta g^{*chg}(T, x) = \sum_i x_i \Delta g_{i/s}^{*chg}(T, x) \tag{130}$$

The following equation is applied to calculate Δg:

$$\Delta g_{i/s}^{*chg}(T, X) = \underbrace{\Delta g_i^{*is}(T)}_{\text{ideal solvation}} + \underbrace{\Delta g_i^{*cc}(T)}_{\text{charge averaging correlation}} + \underbrace{\Delta g_i^{*res}(T)}_{\text{restoring}} + \underbrace{\Delta g_i^{*dsp}(T)}_{\text{dispersion}} \tag{131}$$

The different terms of Eq. (131) are obtained using COSMO solvations calculations [49]. The required parameters for calculating the various contributions of Eq. (131) have been reported in the literature [49]. By calculating PR EOS parameter for drug + SCF mixture, fugacity of species i in the mixture is obtained by

$$\ln\frac{\hat{f}_i(T, P, x)}{x_i P} = \frac{\Delta g_{i/s}^{*chg}}{RT} - \ln\left(1 - \frac{bP}{ZRT}\right) + \frac{b_i P}{ZRT - bP} - \ln Z \tag{132}$$

The fugacity of the solid component is calculated from Eq. (132). The algorithm needed for calculate drug components solubility in SC-CO_2 is reported in Fig. 10.

The proposed model has a parameter (w) for drug components, which modifies the calculations for ring-containing compounds. The mentioned parameter for PR-COSMO-SAC EOS is equal to 1 and its optimum value is 2.14. In this case, the EOS is called PR-COSMO-SAC^{opt}. The optimal value of this parameter is obtained by fitting. A list of predicted solubility results of 15 different drug components in SC-CO_2 has been reported in Table 24.

Average AARD% values of 392 data points of solubility in SC-CO_2 for PR-COSMO-SAC and PR-COSMO-SAC^{opt} EOSs are 1.18 and 0.55, respectively. As the results show, the PR-COSMO-SAC^{opt} EOS with an optimized w-parameter has a better accuracy to predict the SC-CO_2 solubility than the PR-COSMO-SAC EOS. The results of ALD-x indicate that error of PR-COSMO-SAC EOS is twice as much as PR-COSMO-SAC^{opt}.

3.5.4 PR-COSMO model for drug-cosolvent-CO_2 system

Further, in accordance with the previous work, the solubility of drug components in SC-CO_2 along with an organic cosolvent has been predicted according to the PR-COSMO-SAC EOS [50]. Solubility of five drug components including

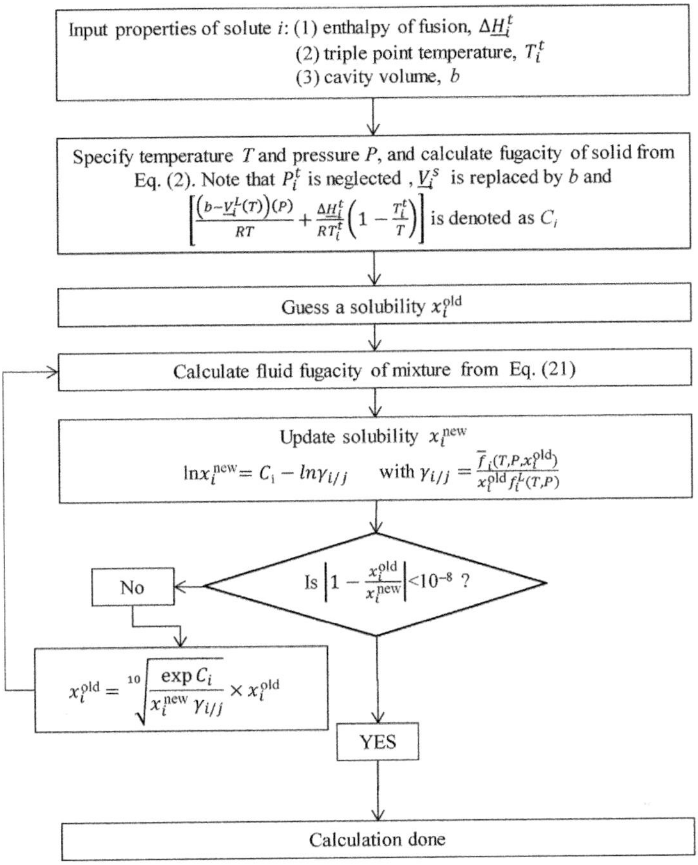

Fig. 10 Algorithm of drug solubility calculation in SC-CO_2 [49].

theophylline, naproxen, aspirin, paracetamol, and hydroquinone in SC-CO_2 along with organic cosolvents such as acetone, ethyl acetate, methanol, ethanol, 1-propanol, and 2-propanol has been predicted using this EOS. The experimental solubility of naproxen with acetone and aspirin with ethanol in SC-CO_2 which was obtained from PR-COSMO-SAC and PR-COSMO-SAC^{opt} EOSs has been compared in Fig. 11.

Fig. 11 indicates that solubility increases sharply at low pressures with increasing pressure and the high-pressure range, it increases slowly. It can also be said that the results of the two EOSs are different than from experimental results, however, the results of PR-COSMO-SAC^{opt} EOS have a better accuracy than the PR-COSMO-SAC EOS. Drug solubility in SC-CO_2 with cosolvents is important for selecting a cosolvent.

Naproxen solubility behavior in SC-CO_2 with cosolvents was studied at various conditions and shown in Fig. 12. The experimental solubility sequence is 1-propanol > 2-propanol > ethanol > methanol > acetone > ethyl acetone, however, the predicted sequence of PR-COSMO-SAC EOS is ethyl acetone > 1-propanol > 2-propanol > acetone > ethanol > methanol.

Table 24 Comparison of the results for prediction of drug solubility for 15 drug substances from PR+COSMOSAC EOS [49]

Drug	N_P	ALD-x	
		PR-COSMO-SAC	PR-COSMO-SAC^{opt}
Aspirin	24	1.34	0.45
Ketoprofen	25	1.95	0.43
Naproxen	58	1.11	0.67
Salicylic acid	11	1.22	0.87
Carbamazepine	39	0.59	0.39
Theophylline	23	0.86	0.43
Diazepam	45	1.62	0.57
Nifedipine	29	0.61	0.53
Piroxicam	9	0.92	0.29
Codiene	45	0.81	0.51
Flurbiprofen	27	1.57	0.90
Phenacetin	16	0.80	0.50
Progesterone	11	1.96	0.48
Sulfamerazine	18	0.63	0.21
Tebuconazole	12	1.72	1.05
Average	392	1.18	0.55

$ALD-x=\frac{1}{M}\sum_{k=1}^{M}\sum_{i=1}^{N}\frac{\log\left|x^{scf}_{exp,i}/x^{scf}_{cal,i}\right|}{N_k}$

M: The total binary systems.

N_k: The number of data points for each binary systems k.

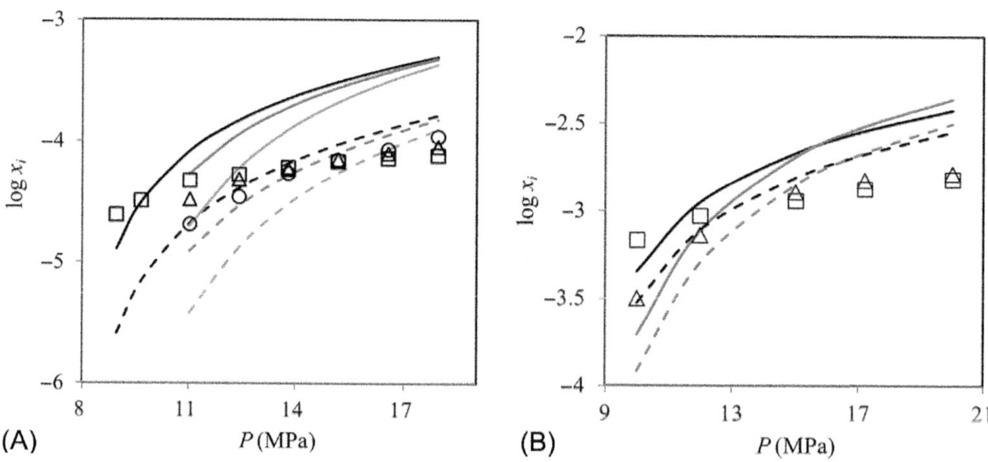

(A) (B)

Fig. 11 Comparison between experimental and predicted solubility of drugs in SC-CO_2 using PR-COSMO-SAC (—) and PR-COSMO-SAC^{opt} (- - -) EOSs. (A) Naproxen and acetone. (B) Aspirin and ethanol [50].

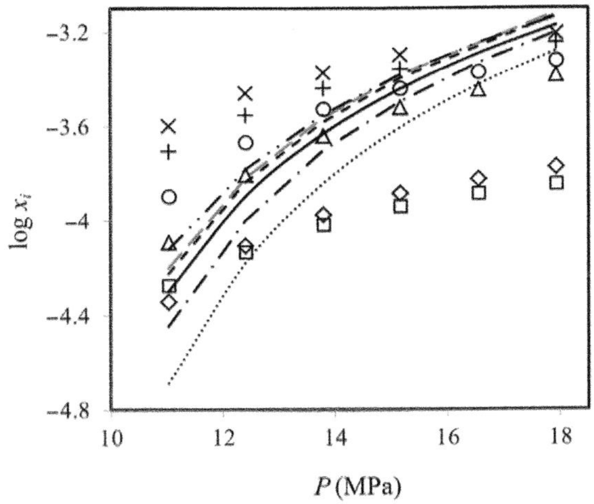

Fig. 12 Experimental solubility of naproxen in SC-CO_2 with six different cosolvents at 333.1 K, PR-COSMO-SAC EOS (—), acetone (◊), methanol (Δ), ethanol (○), 1-propanol (×), 2-propanol (+), and ethyl acetate (□) [50].

Although the solubility prediction of PR + COSMOSAC is not entirely correct, but naproxen solubility in the SC-CO_2 + alcohols (1-propanol > 2-propanol > ethanol > methanol) systems also in the SC-CO_2 + C3-solvents systems (1-propanol > 2-propanol > acetone) is correlated using real results, which shows the importance of predicted results. In other words, an EOS such as PR-COSMO-SAC, by taking the temperature, pressure, and mole fraction, can be used to choose the best solvent for increasing solubility of drug in SC-CO_2.

3.5.5 Activity coefficient model based on the COSMO method

As indicated previously, one of the methods for predicting drug component solubility is to use activity coefficient of drug at infinite dilution state. For this purpose, the drug activity coefficient has been calculated using the COSMO-SAC model [51]. In this model, a molecule can be divided into surface charge segments. Segment probability can be obtained from the following equation [51]:

$$P(\sigma_m) = \frac{\sum_i^N x_i A_i P_i(\sigma_m)}{\sum_i^N x_i A_i} \quad (133)$$

$$P_i(\sigma_m) = \frac{A_i(\sigma_m)}{A_i} \quad (134)$$

In this equation, A_i represents the surface area of molecule i. Activity coefficient of segment value can be obtained from the following equation [51]:

$$\ln\Gamma^{SAC}(\sigma_m) = -\ln\left\{\sum_{n=1}^{k} P(\sigma_n)\Gamma^{SAC}(\sigma_n)\exp\left[-\frac{\Delta w(\sigma_m\sigma_n)}{kT}\right]\right\} \tag{135}$$

Δw is energy required for transfer for segment pair (m, n) from neutral pair where it can be calculated as

$$\Delta W(\sigma_m, \sigma_n) = \frac{0.64 \times 0.3 \times a_{eff}^{3/2}}{2\varepsilon_{\circ}}(\sigma_m + \sigma_n) + c_{nb}\max[0, \sigma_{ac} - \sigma_{hb}]\min[0, \sigma_{do} + \sigma_{hb}] \tag{136}$$

c_{nb} is constant and σ_{hb} represents the cut-off value for the hydrogen bonding interactions. In addition, $\sigma_{do} = \min(\sigma_m, \sigma_n)$ and $\sigma_{ac} = \max(\sigma_m, \sigma_n)$. Drug activity coefficient (2) in SCF at infinite dilution state has been calculated using the following relation:

$$\ln\gamma_2^{\infty} = \ln\gamma_2^{c,\infty} + \frac{A_2}{\sigma_{eff}}\sum_{\sigma_m}^{N} P_2(\sigma_m)\left[\ln\Gamma^{\infty,SAC}(\sigma_m) - \ln\Gamma^{o}(\sigma_m)\right] \tag{137}$$

In which the infinite dilute condition and pure solute is replaced by superscripts ∞ and o. $\ln\gamma_2^c$ represents the difference in molecular size and shape. Since COSMO model is not able to predict supercritical and gas-phase thermodynamic properties, CASMO-Vac model which is based on the vacancy assumption is used. In COSMO-SAC EOS, possibility of finding segment pair is obtained from the following equation:

$$P(\sigma_m, \sigma_n) = \frac{\sum_{k}^{N} n_k}{n_{vac} + \sum_{k}^{N} n_k} \times \frac{2n_{mn}}{\sum_{k}^{N} n_k} = (1-\alpha)\frac{2n_{mn}}{\sum_{k}^{N} n_k} \tag{138}$$

α can be defined as

$$\alpha = \frac{n_{vac}}{n_{vac} + n_{solv}} \tag{139}$$

Given that $n_{solv} = \frac{A_{solv}}{a_{eff}}$ and $n_{vac} = \frac{A_{vac}}{a_{eff}}$ are the solvent molecule surface area and vacancy, respectively. Parameter α is calculated as

$$\alpha = \frac{n_{vac}}{n_{vac} + n_{solv}} \tag{140}$$

In which

$$A_{vac} = \pi d_1^2 - A_{solv} \tag{141}$$

In this equation, d_1 is molecular diameter. The A_{solv} value is obtained from the COSMO calculations. The segment activity coefficient in solution that contains vacancy has been calculated using the following equation [51]:

$$\ln \Gamma^{vac}(\sigma_m) = - \ln\left\{ \sum_{n=1}^{k} P(\sigma)(1-\alpha)\Gamma^{vac}(\sigma_n)\exp\left[-\frac{\Delta W(\sigma_m, \sigma_n)}{kT}\right]\right\} \tag{142}$$

In which:

$$\Delta W(\sigma_m, \sigma_n) = E(\sigma_m, \sigma_n) - E(0,0) \tag{143}$$

$E(0,0)$ is the energy of the system consists of neutral segments. The drug activity coefficient at infinite dilution state is calculated from the following equation:

$$\ln \gamma_2^{\infty} = \ln \gamma_2^{C\infty} + \frac{A_2}{\sigma_{eff}} \sum_{\sigma_m}^{k} P_2(\sigma_m)[\ln \Gamma^{vac,\infty}(\sigma_m) - \ln \Gamma^{\circ}(\sigma_m)] \tag{144}$$

The combinatorial part of activity coefficient sentence is calculated from as below:

$$\ln \gamma_i^c = \ln \frac{\varphi_i}{x_i} + \frac{z}{2} q_i \ln \frac{\theta_i}{\varphi_i} + l_i - \frac{\varphi_i}{x_i} \sum_j x_j l_j \tag{145}$$

$$\theta_i = \frac{x_i q_i}{\sum_j^N x_j q_j}, \quad \phi_i = \frac{x_i r_i}{\sum_j^N x_j r_j}, \quad l_i = \frac{z}{2}[(r_i - q_i) - (r_i - 1)] \tag{146}$$

The results of solubility prediction for drug component in SC-CO_2 using the COSMO-Vac model are reported in Table 25.

The results reported in Table 25 pointed that the activity coefficient model based on the COSMO method as a predictive model can calculate the drugs solubility in SC-CO_2 successfully.

3.5.6 ANN system

The solubility of five drug components including phenazopyridine, propranolol, methyl salycilate, benzocain, and aspirin in SC-CO_2 has been modeled using an ANN system [52]. The three-layer feed-forward neural network (TFFNN) is applied to model with sigmoid (hyperbolic) transfer function. The schematic structure of this network has been shown in Fig. 13.

Table 25 The results of solubility prediction of drug substances using the COSMO-vac [51]

Drug	N_p	*T* (K)	$\Delta \log(y)$
Aspirin	24	308.15–328.15	0.36
Carbamazepine	31	308–338	0.61
Codeine	27	308–328	0.83
Diazepam	45	308–348	0.86
Flurbiprofen	27	303–323	0.57
Ketoprofen	15	313, 328	0.46
Naproxen	39	308–348	0.66
	18	313.1–333.1	0.66
Nifedipine	8	333.15	0.79
Piroxicam	10	312.5, 333.15	0.92
Salicylic acid	11	313, 328	0.34
Theophylline	23	313–353	0.37
$\Delta \log(y) = \lvert \log y_2^{\exp} - \log y_2^{calc} \rvert$		$\Delta \log(y)^{ave}$	0.61

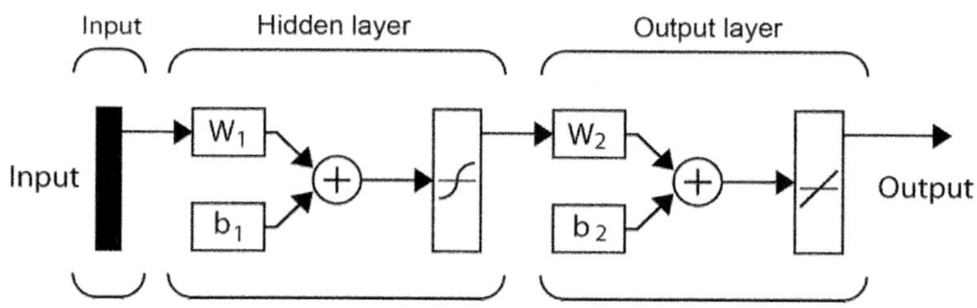

Fig. 13 The structure of a TFFNN artificial neural network [52].

Accordingly, the neural network includes input and output variables that are related to the following relationship:

$$output(i) = \omega_2 Tanh[\omega_1 input(i) + b_1] + b_2 \tag{147}$$

In this equation, ω_i is the weighting factor and b_i is bias. Neural network design requires a set of experimental data which will be used in three steps. The training stage where the network parameters (ω_i, b_i) are computed, the validation step, and the test step which will be used to verify the validity of the proposed network consistency. The input variables for this network are T_c, P_c, ω, v_s, T, and P. Therefore, the number of input variables is six. The output variable is solids solubility in SC-CO_2.

However, output variable is considered as the logarithm drug solubility in SC-CO_2. In the next step, experimental data are divided into three parts. In all, 90% of the data are considered for the training process, 5% of the data for validation, and 5% for the test.

The Levenberg-Marquardt optimization method has been used to optimize the neural network. The AARD% error value for solubility of the intended drug in SC-CO_2 is

Table 26 AARD% values for the ANN system for correlation of solubility of pharmaceutics [52]

Drug	*T* (K)	*P* (MPa)	N_p	AARD%
Phenazopyridine	308–348	12.2–35.5	45	8
Propranolol	308–348	12.2–35.5	83	9.8
Methyl salycilate	343.15–423.15	9.0–31.0	44	13.56
Benzocain	308–348	8.4–35.5	52	33.16
Aspirin	308.15–328.15	12.0–25.0	24	9.15
Average				14.73

reported in Table 26. The average AARD% values indicate that the ANN system has a good accuracy to estimate drug components solubility in SC-CO_2.

3.5.7 Molecular dynamics simulation

The solubility of drug component including acetaminophen and ibuprofen has been modeled using molecular dynamics simulation in terms of intermolecular forces [53]. Solubility of a drug component in SC-CO_2 is calculated from the following equation [53]:

$$y_2 = \frac{Z P_2^{sub} \exp\left[\beta v^s \left(P - P^{sub}\right)\right]}{P \exp\left(\Delta g_2^{solv}\right)} \tag{148}$$

Solvation free energy based on the thermodynamic interaction is obtained from the following equation [54]:

$$\Delta g_2^{solv} = \int_0^1 << \frac{\partial H(\lambda)}{\partial \lambda} >> d\lambda \tag{149}$$

where H is Hamiltonian and λ is the coupling parameter. Solvation free energy can be computed during NVT Ensemble (NVT) runs. In this research, Chemistry at Harvard Macromolecular Mechanics (CHARMM) force field was used for calculations:

$$U_{ij} = \sum k_b (r - r_\circ)^2 + k_\theta (\theta - \theta_\circ)^2 + k_x [1 + \cos(nx - \delta)] + k_\varphi (\varphi - \varphi_\circ)^2 + 4\varepsilon_{ij} \left[\left(\frac{\sigma_{ij}}{r_{ij}}\right)^{12} - \left(\frac{\sigma_{ij}}{r_{ij}}\right)^{6} \right] + \frac{q_i q_j}{r_{ij}} \tag{150}$$

The parameters of Eq. (150) are defined in Ref. [53].

A comparison between the experimental results of solvation free energies of acetaminophen and ibuprofen in SC-CO_2 and the results of stimulation in force field including Zhang and Duan [54], Harris and Yung (EPM2) model [55], and Potoff (TraPPE) model [56] is represented in Table 27.

Table 27 Comparison between experimental and simulated results of solvation free energies of acetaminophen and ibuprofen [53]

	Solvation free energy (Kcal/mol)			
Density (Kmol/m^3)	TraPPE	EPM2	Zhang	Exp
Acetaminophen ($T = 313$ K)				
14.30	−9.06	−8.68	−8.30	−8.84
16.61	−9.60	−9.16	−8.83	−9.21
18.50	−9.94	−9.47	−9.13	−9.44
19.98	−10.04	−9.58	−9.26	−9.57
Ibuprofen ($T = 308$ K)				
16.24	−9.26	−9.10	−8.80	−9.24
17.46	−9.46	−9.30	−9.04	−9.48
18.81	−9.60	−9.36	−9.10	−9.62
20.03	−9.43	−9.25	−9.01	−9.60

Table 28 Solubility of acetaminophen and ibuprofen in SC-CO_2 obtained from various simulation force fields [53]

	Mole fraction × 10^6			
Density (Kmol/m^3)	TraPPE	EPM2	Zhang	Exp
Acetaminophen ($T = 313$ K)				
14.30	0.53	0.24	0.16	0.37
16.61	1.24	0.61	0.36	0.65
18.50	2.44	1.14	0.66	1.08
19.98	3.80	1.82	1.08	1.78
Ibuprofen ($T = 308$ K)				
16.24	1.40	1.08	0.66	1.35
17.46	2.07	1.59	1.04	2.13
18.81	3.15	2.13	1.39	3.23
20.03	3.33	2.48	1.08	4.41

According to the reported results in Table 27, different force forces predict different outcomes for solvation free energy. The results of experimental acetaminophen and ibuprofen solubility with the simulated results by force fields, TraPPE, EPM2, and Zhang have been reported in Table 28.

As the reported results in Table 28 indicate the solubility of drug components depend on the model. As can be inferred from Table 28, EPM2 model has a very good accuracy for predicting the solubility of acetaminophen and TraPPE model has a very good precision for predicting ibuprofen solubility in SC-CO_2.

3.6 Empirical correlations

One of the methods for estimation of solubility of solid drug components is empirical correlations. In most of correlations, it is assumed that logarithm of drug component solubility is related to logarithm of solvent density at a given temperature and pressure. The common correlations that used for prediction of drug component in SC-CO_2 are summarized in Table 29.

As can be seen from Table 29, in Chrastil [57], Adachi and Lu (Chrastil-A-L) [58], Sung and Shim (Chrastil-S-S) [59], Sparks et al. [60], Kumar and Johnston (K-J) [61], del Valle and Aguilera (dV-A) [62], Garlapati and Madras (G-M) [63], and Bian et al. [64], drug solubility in SC-CO_2 (y_2, S) is dependent to two variables including temperature (T) and supercritical-phase density (ρ) while in Bartle, Gordillo et al., Mendez-Santiago and Teja (M-S-T), Jouyban et al., Ch and Madras (C-M), Hozhabar et al., Keshmiri et al., and Khansary et al. drug solubility in SC-CO_2 (y_2) is dependent to three variables, temperature (T), supercritical-phase density (ρ), and pressure (P). It should be noted that

Table 29 Emperical relations for predicting the drug solubility in SC-CO_2

Model	Correlation	Ref
Chrastil	$\ln S = a_0 + a_1 \ln\rho + \frac{a_2}{T}$	[57]
Modified Chrastil (Adachi and Lu), (Chrastil-A-L)	$\ln S = (a_0 + a_1\rho + a_2\rho)\ln\rho + \frac{a_3}{T} + a_4$	[58]
Modified Chrastil (Sung and Shim), (Chrastil-S-S)	$\ln y_2 = a_0 + \frac{a_1}{T} + \left(a_2 + \frac{a_3}{T}\right)\ln\rho$	[59]
de Valle and Aguilera (dV-A)	$\ln S = a_0 + a_1 \ln\rho + \frac{a_2}{T} + \frac{a_3}{T^2}$	[62]
Bartle	$\ln\left(\frac{y_2 P}{P_{ref}}\right) = a_0 + a_1\left(\rho - \rho_{ref}\right) + \frac{a_2}{T}$	[97]
Gordillo et al.	$\ln S = a_0 + a_1 P + a_2 P^2 + a_3 PT + a_4 T + a_5 T^2$	[98, 99]
Kumar and Johnston (K-J)	$\ln y_2 = a_0 + \frac{a_1}{T} + a_2\rho$	[61]
Mendez-Santiago and Teja (M-S-T)	$T\ln(y_2 P) = a_0 + a_1\rho + a_2 T$	[100]
Jouyban et al.	$\ln y_2 = a_0 + a_1 P + a_2 P^2 + a_3 PT + \frac{a_4 T}{P} + a_5 \ln\rho$	[101]
Sparks et al.	$\ln y_2 = a_0 + \frac{a_1}{T} + \frac{a_2}{T^2} + (a_3 + a_4\rho)\ln\rho$ $\ln y_2 = a_0 + \frac{a_1}{T} + \frac{a_2}{T^2} + (a_3 + a_4\rho + a_5\rho^2)\ln\rho$	[60]
Garlapati and Madras (G-M)	$\ln y_2 = a_0 + \frac{a_1}{T} + a_2 \ln\rho + a_3\rho \ln\rho + a_4 \ln(\rho T)$	[63]
Ch and Madras (CM)	$\ln y_2 = (a_0 - 1)\ln\left(\frac{P}{P_{ref}}\right) + a_1 + a_2\rho + \frac{a_3}{T}$	[46]
Hozhabr et al.	$\ln y_2 = a_0 + \frac{a_1}{T} + \frac{a_2\rho}{T} - a_3 \ln P$	[102]
Keshmiri et al.	$\ln y_2 = a_0 + \frac{a_1}{T} + a_2 P^2 + \left(a_3 + \frac{a_4}{T}\right)\ln\rho$	[90]
Khansary et al.	$\ln y_2 = \frac{a_0}{T} + a_1 P + \frac{a_2 P^2}{T} + (a_3 + a_4 P)\ln\rho$	[103]
Bian et al.	$\ln y_2 = a_0 + \frac{a_1}{T} + \frac{a_2\rho}{T} + (a_3 + a_4\rho)\ln\rho$	[64]
Wubbolts et al.	$y_s^{eq} = y_s^{\circ}(1 - x_{CO_2})^{A + Bx_{CO2}} + y_{s_{CO2}} x_{CO_2}$	[38]
Yu et al.	$y_2 = a_0 + a_1 P + a_2 P^2 + a_3 PT(1 - y_2) + a_4 T + a_5 T^2$	[104]
Jafari Nejad et al.	$\ln y_2 = a_0 + a_1 P^2 + a_2 T^2 + a_3 \ln\rho$	[105]

Wubbolts et al. model is applied for drug solubility in supercritical CO_2-cosolvent system. In Bartle model, $P_{ref*} = 0.1$ MPa and $\rho_{ref} = 700$ kg/m^3.

For some of these models, coefficients have physical meanings. In Chrastil model, a_0 is a function of solute molecular weights and SCF, a_1 is association number (average number of SCF molecules in the solvated complex), and $a_2 = \Delta h = \Delta h_{vap} + \Delta h_{solv}$ is the sum of the heat vaporization and the heat of solvation of the solute. In Bartle model, $a_2 = -\Delta h_{vap}/R$.

In Wubbolts et al. model, solubility is dependent to CO_2 mole fraction as an antisolvent and an organic solvent. In the following, application of the empirical correlations for prediction of drug solubility in SC-CO_2 is studied and investigated.

3.6.1 Comparative study 1

The ability of empirical correlations in order to prediction solubility of 19 drug components has been compared [65]. Characteristics of drugs and variation range for temperature and pressure are reported in Table 30. Also, values of AARD% for empirical correlations are compared in Table 31. The lowest value of AARD% for each drug is specified in Table 31 with a star.

With comparison the values of average AARD% for 19 drug components it can be said that Yu model has high error for solubility prediction of drug components in SC-CO_2.

Table 30 Characteristics of the studied drugs and variation range of *T* and *P* [65]

No.	Drug	N_p	*P* (bar)	*T* (K)
1	Astaxanthin	26	100–400	313–333
2	Cyproterone acetate	40	120–350	303–348
3	Exemestane	45	120–350	303–348
4	Fluvastatin	45	120–350	303–348
5	Letrozole	45	120–350	303–348
6	Lovastatin	45	120–350	303–348
7	Medroxyprogesterone acetate	40	120–350	303–348
8	Methimazole	40	120–350	303–348
9	Naproxen	18	90–195	313–333
10	Nimesulide	8	130–220	312–329
11	Nitrendipine	42	100–300	333–373
12	Paclitaxel	12	140–345	312–329
13	Penicillin G	18	100–350	313–333
14	Penicillin V	24	80–280	314–334
15	Phenazopyridine	45	120–350	308–348
16	Propanolol	45	120–350	308–348
17	Retionol	20	200–350	313–353
18	Simvastatin	45	120–350	308–348
19	Taxol	12	205–475	308–318

Table 31 The AARD% values for 19 drug components obtained from various empirical relations [65]

No.	Chrastil	VA	AL	Sparks	MST	KJ	Bartle	Yu	Gordillo
1	52.98	55.67	41.33	43.07	52.72	49.56	56.29	33.06	25.45*
2	24.74	24.73	15.57	16.21	12.91	19.77	23.40	22.92	5.52*
3	30.58	18.60	36.43	14.88*	32.68	30.22	32.06	98.55	35.99
4	14.73	14.59	9.31*	10.64	14.30	11.81	11.42	56.85	20.47
5	38.47	38.99	28.79	29.37	20.49	21.98	17.38*	54.27	27.11
6	5.90	5.78	4.38	4.27	5.37	9.80	4.19*	8.55	12.53
7	18.91	17.10	11.12	10.57	16.23	14.63	17.24	15.70	6.90*
8	12.69	12.34	10.43	10.95	10.99	9.56*	12.43	15.29	11.90
9	9.77	9.74	8.93	8.58	10.41	11.36	11.80	6.27*	15.63
10	14.08	14.06	6.35	5.63	12.59	8.59	14.58	4.72*	14.51
11	15.35	14.13	14.01	12.58*	14.91	18.51	15.11	21.53	25.20
12	30.46	29.37	18.97	18.90	31.23	26.60	31.92	4.66*	7.04
13	25.43	26.00	21.17	20.47	25.22	24.36	25.90	23.43	13.39*
14	18.17	18.13	5.99*	6.24	16.59	11.49	16.17	6.68	16.10
15	15.46	15.71	21.40	20.92	9.21	8.14*	10.94	17.97	13.09
16	22.43	18.14	19.68	18.17	21.69	21.08	20.75	71.02	12.99*
17	7.96	7.97	6.66	5.32	8.30	7.18	8.95	6.73	3.90*
18	13.06	12.74	14.80	8.93*	10.95	13.68	9.15	95.20	52.79
19	4.34	3.58	4.36	3.13	3.53	3.81	4.71	3.12	1.50*
Average	19.76	18.78	15.77	14.14	17.91	16.95	18.12	29.76	16.96

However, Gordillo model is the modified version of Yu model. The average AARD% for Gordillo model is 19.96%. It is shown that the performance of Yu model is improved. The order of AARD% for the existing models from the most error is as below:

$$\text{Yu} > \text{Chrastil} > \text{VA} > \text{Bartle} > \text{MST} > \text{Gordillo} > \text{KJ} > \text{AL} > \text{Sparks}$$

The results show that both Gordillo and KJ models have the same error. Gordillo model has six adjustable parameters while KJ model has three adjustable parameters. This indicates that KJ model with fewer parameters has good accuracy for prediction drug solubility. In addition, Sparks model that has good performance has six adjustable parameters. Generally, it should be mention that the accuracy of model increases with increasing the number of adjustable parameters. Also, with comparison between Bartle, KJ, Chrastil, and MST three-parameters models, it is found that KJ model has better performance than either three models.

3.6.2 Comparative study 2

Drug compound solubility in SC-CO_2 is correlated by using correlation models [64]. The models used for predicting solubility of drugs can be classified into three classes as:

1. $\ln y_2 = m \ln \rho_{SCF} + n, \quad T = \text{constant}.$
2. $\ln y_2 = m(\rho_{SCF}/T) + n, \quad T = \text{constant}.$

Table 32 The characteristics of drug components and temperature and pressure ranges as well as density of SC-CO_2 [64]

No.	Drug	N_P	*P* (MPa)	*T* (K)	ρ_1 (kg/m^3)
1	Finasteride	45	12.2–35.5	308–348	327–955
2	Fenofibrate	21	10.01–22.02	308.2–328.2	498.3–881
3	Gemfibrozil	21	10.01–22.02	308.2–328.2	498.3–881
4	Levonorgestrel	18	10–18	313–323	384.3–819.5
5	Ketoprofen	28	16–40	308.15–338.15	592.3–972.3
6	Piroxicam	28	16–40	308.15–338.15	592.4–972.3
7	Cephalexin	27	16–40	308.15–338.15	681.1–972.3
8	Ergosterol	15	12–24	318.15–333.15	442–849.4
9	Levonorgestrel	36	12.2–35.5	308–339	396–955
10	Niflumic acid	24	12–36	313–333	436.3–939.8
11	Clofenamic acid	24	12–36	313–333	436.3–939.8
12	Tolfenamic acid	24	12–36	313–333	436.3–939.8

3. $\ln y_2 = m(\rho_{SCF} \ln \rho_{SCF}) + n, \quad T = \text{constant}.$

The characteristics of drugs as well as range of temperature, pressure, and CO_2 density are represented in Table 32 and the values of AARD% of 12 drug components for 15 correlations are represented in Table 33.

The lowest value of AARD% for each drug is specified in Table 33 with star. Comparing the AARD% values shows that Bian correlation with five parameters has very low errors in correlating drug solubility in SC-CO_2. The five-parameter correlation proposed by Keshmiri has good performance. Three parameters correlations such as MST and Chrastil have the highest errors. Generally, it should be emphasized that the accuracy of model increases by increasing the number of adjustable parameters of model. Of course, this is not always true. For instance, Sparks model has six parameters but its accuracy is worse than Bian correlation as a five-parameter correlation.

4. Conclusion

In this chapter, the applications of SC-CO_2 as a green solvent in pharmaceutical industry-related processes as well as the experimental methods for measurement of drug solubility in SC-CO_2 are investigated. In addition, mathematical models for solubility correlation of drugs in SC-CO_2 have been studied. The mathematical models can be classified to EOSs, solubility parameter-based model, mathematical models such as ANN and COSMO models, and empirical correlations. Most of the EOSs are two-parameter cubic EOSs coupled with mixing rules. The disadvantage of these EOSs is that the parameters of cubic EOSs are dependent to critical properties and acentric factor of species which are not experimentally obtained for drug components. In order to conquer this problem,

Table 33 The AARD% values for 12 drug components from 15 empirical correlations [64]

No.	1	2	3	4	5	6	7	8	9	10	11	12	13	14	15
1	24.37	16.43	24.37	18.85	16.42	14.83	15.28	18.85	20.85	21.02	25.58	20.39	19	21.23	13.14*
2	25.69	21.64	19.83	18.82	17.08	21.25	19.20	18.84	29.14	29.26	32.05	22.91	9.74*	30.21	10.38
3	31.00	27.15	20.82	20.31	18.77	23.57	21.81	20.48	34.01	34.00	38.52	28.41	11.02*	35.83	12.90
4	7.88	6.31	7.89	6.69	6.23	6.64	6.15*	6.69	6.75	6.80	7.92	6.95	7.44	6.93	6.28
5	5.67	5.39	5.71	5.49	5.51	8.63	13.12	5.49	5.15	7.99	5.53	5.37	5.24	5.12*	5.28
6	17.81	16.03	16.92	14.11	14.29	15.33	15.17	14.04	16.50	16.36	18.03	17.28	14.42	14.52	13.61*
7	8.99	8.64	8.99	8.96	8.67	11.96	12.32	8.96	8.12	10.11	11.59	8.97	7.41	10.23	7.35*
8	7.18	6.93	6.82	6.01	6.12	5.55	3.48	6.01	7.33	7.98	8.09	7.02	3.81	5.05	2.44*
9	15.00	9.64	15.03	10.00	10.02	12.72	11.02	9.99	9.60	11.26	16.44	9.62	9.51*	10.17	9.65
10	3.33	3.00	3.33	3.32	3.00	12.81	5.59	3.32	3.30	7.12	3.60	3.34	2.96*	3.35	3.16
11	5.37	5.00	5.36	5.38	5.02	16.45	6.76	5.38	5.42	8.68	4.82*	5.30	4.99	5.52	5.05
12	5.40	4.50	5.34	4.71	4.43	11.23	7.91	4.71	4.46	6.12	6.54	4.99	4.32	4.77	3.74*
Average	13.14	10.84	11.70	10.22	9.63	13.41	11.48	10.23	12.55	13.89	14.89	11.71	8.32	12.74	7.47

1: Chrastil, 2: AL, 3: dV, 4: Sparks1, 5: Sparks2, 6: Gordillo, 7: Jouyban, 8: GM, 9: CM, 10: KJ, 11: MST, 12: Hozhabr, 13: Keshmiri, 14: Khansary, 15: Bian.

T_c,P_c as well as ω have been calculated by using group contribution method. The results showed that the accuracy of EOSs increased as the number of parameters of EOS and binary interaction parameters of mixing rule increased. Also, SAFT EOS has been used for predicting the solubility of drugs in SC-CO_2. In this chapter, SAFT-VR and PCP-SAFT versions of SAFT EOS as well as dipole-dipole-association-SRK (qCPA) as non-cubic EOSs have been applied. Also, authors concluded that SAFT-VR EOS has good accuracy as a correlative model. PCP-SAFT and qCPA EOSs have low accuracy where the accuracy of models increases with consideration of binary interaction parameters. In addition, the solubility parameter-based model (solution model) along with Flory-Huggins activity coefficient model has been used for prediction of drug component solubility. The results of the solution model show that this model has good error with less parameters. The charging component of the solvation free energy so-called COSMO model has been studied. The COSMO model is based on the quantum chemistry. The results indicate that the COSMO model as a predictive model has good accuracy in predicting the solubility of drugs in SC-CO_2. Also, ANN system has been used for correlation and prediction of drug solubility. Although ANN system has less error among other models, this model needs a lot of experimental data and there is not a simple and suitable relation between solubility of drugs in SC-CO_2 and input variables. Finally, the molecular dynamic simulation has been applied. It should be pointed out that the molecular dynamic simulation is a predictive model. Of course, this method is needed high-speed computers and it is not a suitable alternative for traditional thermodynamics models. With comparison between the results of the studied model, it can be said that the cubic EOSs and empirical correlations as well as simple models have been widely used for estimation solubility of drugs in SC-CO_2.

References

[1] K. Moribe, Y. Tozuka, K. Yamamoto, Supercritical carbon dioxide processing of active pharmaceutical ingredients for polymorphic control and for complex formation, Adv. Drug Deliv. Rev. 60 (2008) 328–338.
[2] T. Yasuji, H. Takeuchi, Y. Kawashima, Particle design of poorly water-soluble drug substances using supercritical fluid technologies, Adv. Drug Deliv. Rev. 60 (2008) 388–398.
[3] L.S. Daintree, A. Kordikowski, P. York, Separation processes for organic molecules using SCF technologies, Adv. Drug Deliv. Rev. 60 (2008) 351–372.
[4] M. Mukhopadhyay, Natural Extracts Using Supercritical Carbon Dioxide, CRC Press, 2000.
[5] A. Mortazavi, N. Hadidi, M. Saffari, A. Gaffari, A review on characteristics and manufacturing methods of powders: particle engineering and design of novel generation of pharma excipients, published by Andishe Asr, Tehran, 2008.
[6] H. Sim Yeoh, G. Hean Chong, N. Mohd Azahan, R. Abdul Rahman, T.S. Yaw Choong, Solubility measurement method and mathematical modeling in supercritical fluids, Eng. J. 17 (2013) 67–78.
[7] S. Bristow, B.Y. Shekunov, P. York, Solubility analysis of drug compounds in supercritical carbon dioxide using static and dynamic extraction systems, Ind. Eng. Chem. Res. 40 (2001) 1732–1739.
[8] A.Z. Hezave, A. Mowla, F. Esmaeilzadeh, Cetirizine solubility in supercritical CO_2 at different pressures and temperatures, J. Supercrit. Fluids 58 (2011) 198–203.

[9] G. Sodeifian, F. Razmimanesh, S.A. Sajadian, H. Soltani Panah, Solubility measurement of an antihistamine drug (Loratadine) in supercritical carbon dioxide: assessment of qCPA and PCP-SAFT equations of state, Fluid Phase Equilib. 472 (2018) 147–159.

[10] H. Asiabi, Y. Yamini, F. Latifeh, A. Vatanara, Solubilities of four macrolide antibiotics in supercritical carbon dioxide and their correlations using semi-empirical models, J. Supercrit. Fluids 104 (2015) 62–69.

[11] S.A. Shojaee, A.Z. Hezave, S. Aftab, M. Lashkarbolooki, F. Esmaeilzadeh, Solubility of gabapentin in supercritical carbon dioxide, J. Supercrit. Fluids 78 (2013) 1–6.

[12] C. Tang, Y.-X. Guan, S.-J. Yao, Z.-Q. Zhu, Solubility of dexamethasone in supercritical carbon dioxide with and without a cosolvent, J. Chem. Eng. Data 59 (2014) 3359–3364.

[13] Z. Huang, W.D. Lu, S. Kawi, Y.C. Chiew, Solubility of aspirin in supercritical carbon dioxide with and without acetone, J. Chem. Eng. Data 49 (2004) 1323–1327.

[14] S.S.T. Ting, D.L. Tomasko, N.R. Foster, S.J. Macnaughton, Solubility of naproxen in supercritical carbon dioxide with and without cosolvents, Ind. Eng. Chem. Res. 32 (1993) 1471–1481.

[15] I.A. Cuadra, A. Cabañas, J.A.R. Cheda, F.J. Martínez-Casado, C. Pando, Pharmaceutical co-crystals of the anti-inflammatory drug diflunisal and nicotinamide obtained using supercritical CO_2 as an antisolvent, J. CO_2 Util. 13 (2016) 29–37.

[16] I. Kikic, P. Alessi, F. Eva, M. Moneghini, B. Perissutti, Supercritical antisolvent precipitation of atenolol: the influence of the organic solvent and of the processing approach, J. Supercrit. Fluids 38 (2006) 434–441.

[17] A. Shariati, C.J. Peters, Measurements and modeling of the phase behavior of ternary systems of interest for the GAS process: I. The system carbon dioxide + 1-propanol + salicylic acid, J. Supercrit. Fluids 23 (2002) 195–208.

[18] E. Reverchon, I. De Marco, Supercritical antisolvent micronization of Cefonicid: thermodynamic interpretation of results, J. Supercrit. Fluids 31 (2004) 207–215.

[19] R. Hartono, G.A. Mansoori, A. Suwono, Prediction of solubility of biomolecules in supercritical solvents, Chem. Eng. Sci. 56 (2001) 6949–6958.

[20] J.M. Prausnitz, R.N. Lichtenthaler, E.G. de Azevedo, Molecular Thermodynamics of Fluid-Phase Equilibria, Pearson Education, 1998.

[21] P. Coimbra, C.M.M. Duarte, H.C. de Sousa, Cubic equation-of-state correlation of the solubility of some anti-inflammatory drugs in supercritical carbon dioxide, Fluid Phase Equilib. 239 (2006) 188–199.

[22] R.F. Fedors, A method for estimating both the solubility parameters and molar volumes of liquids, Polym. Eng. Sci. 14 (1974) 147–154.

[23] B.E. Poling, J.M. Prausnitz, J.P. O'Connell, The Properties of Gases and Liquids, McGraw-Hill, 2001.

[24] A. Zeinolabedini Hezave, M.H. Khademi, F. Esmaeilzadeh, Measurement and modeling of mefenamic acid solubility in supercritical carbon dioxide, Fluid Phase Equilib. 313 (2012) 140–147.

[25] F. Esmaeilzadeh, M. Roshanfekr, A new cubic equation of state for reservoir fluids, Fluid Phase Equilib. 239 (2006) 83–90.

[26] A.Z. Hezave, S. Shahnazar, H. Rajaei, M. Lashkarbolooki, F. Esmaeilzadeh, Solubility of spironolactone in supercritical carbon dioxide: experimental and modeling approaches, Fluid Phase Equilib. 355 (2013) 130–134.

[27] C. Garlapati, G. Madras, Temperature independent mixing rules to correlate the solubilities of antibiotics and anti-inflammatory drugs in $SCCO_2$, Thermochim. Acta 496 (2009) 54–58.

[28] T.Y. Kwak, G.A. Mansoori, Van der waals mixing rules for cubic equations of state. Applications for supercritical fluid extraction modelling, Chem. Eng. Sci. 41 (1986) 1303–1309.

[29] G. Sodeifian, S.A. Sajadian, F. Razmimanesh, Solubility of an antiarrhythmic drug (amiodarone hydrochloride) in supercritical carbon dioxide: Experimental and modeling, Fluid Phase Equilib. 450 (2017) 149–159.

[30] G.R. Pazuki, S. Mansouri, F. Feyzi, Modified cubic equation of state for prediction VLE phase behavior of fluids: Pure and mixture, Sep. Sci. Technol. 42 (2007) 1883–1899.

[31] J. Marrero, R. Gani, Group-contribution based estimation of pure component properties, Fluid Phase Equilib. 183-184 (2001) 183–208.

[32] M.R. Housaindokht, M.R. Bozorgmehr, Calculation of solubility of methimazole, phenazopyridine and propranolol in supercritical carbon dioxide, J. Supercrit. Fluids 43 (2008) 390–397.
[33] A. Dashtizadeh, G.R. Pazuki, V. Taghikhani, C. Ghotbi, A new two-parameter cubic equation of state for predicting phase behavior of pure compounds and mixtures, Fluid Phase Equilib. 242 (2006) 19–28.
[34] K.A.M. Gasem, W. Gao, Z. Pan, R.L. Robinson, A modified temperature dependence for the Peng–Robinson equation of state, Fluid Phase Equilib. 181 (2001) 113–125.
[35] Y. Adachi, H. Sugie, A new mixing rule—modified conventional mixing rule, Fluid Phase Equilib. 28 (1986) 103–118.
[36] E. Neau, S. Garnier, L. Avaullée, A consistent estimation of sublimation pressures using a cubic equation of state and fusion properties, Fluid Phase Equilib. 164 (1999) 173–186.
[37] R. Stryjek, J.H. Vera, PRSV: an improved Peng-Robinson equation of state for pure compounds and mixtures, Can. J. Chem. Eng. 64 (1986) 323–333.
[38] F.E. Wubbolts, O.S.L. Bruinsma, G.M. van Rosmalen, Measurement and modelling of the solubility of solids in mixtures of common solvents and compressed gases, J. Supercrit. Fluids 32 (2004) 79–87.
[39] M. Türk, M. Crone, T. Kraska, A comparison between models based on equations of state and density-based models for describing the solubility of solutes in CO2, J. Supercrit. Fluids 55 (2010) 462–471.
[40] K. Leonhard, T. Kraska, An equation of state describing the critical region: extension to high pressure, J. Supercrit. Fluids 16 (1999) 1–10.
[41] G.M. Kontogeorgis, G.K. Folas, Thermodynamic Models for Industrial Applications: From Classical and Advanced Mixing Rules to Association Theories, Wiley, 2010.
[42] M. Hosseini Anvari, G. Pazuki, A study on the predictive capability of the SAFT-VR equation of state for solubility of solids in supercritical CO_2, J. Supercrit. Fluids 90 (2014) 73–83.
[43] G. Sodeifian, N. Saadati Ardestani, S.A. Sajadian, H.S. Panah, Measurement, correlation and thermodynamic modeling of the solubility of Ketotifen fumarate (KTF) in supercritical carbon dioxide: Evaluation of PCP-SAFT equation of state, Fluid Phase Equilib. 458 (2018) 102–114.
[44] C.M. Hansen, Hansen Solubility Parameters: A User's Handbook, second ed., CRC Press, 2007.
[45] J.-S. Cheng, M. Tang, Y.-P. Chen, Correlation of solid solubility for biological compounds in supercritical carbon dioxide: comparative study using solution model and other approaches, Fluid Phase Equilib. 194–197 (2002) 483–491.
[46] R. Ch, G. Madras, An association model for the solubilities of pharmaceuticals in supercritical carbon dioxide, Thermochim. Acta 507–508 (2010) 99–105.
[47] S.N. Reddy, G. Madras, A new semi-empirical model for correlating the solubilities of solids in supercritical carbon dioxide with cosolvents, Fluid Phase Equilib. 310 (2011) 207–212.
[48] R. Span, W. Wagner, A new equation of state for carbon dioxide covering the fluid region from the triple-point temperature to 1100 K at pressures up to 800 MPa, J. Phys. Chem. Ref. Data 25 (1996) 1509–1596.
[49] L.-H. Wang, S.-T. Lin, A predictive method for the solubility of drug in supercritical carbon dioxide, J. Supercrit. Fluids 85 (2014) 81–88.
[50] Y.-H. Ting, C.-M. Hsieh, Prediction of solid solute solubility in supercritical carbon dioxide with organic cosolvents from the PR + COSMOSAC equation of state, Fluid Phase Equilib. 431 (2017) 48–57.
[51] Y. Shimoyama, Y. Iwai, Development of activity coefficient model based on COSMO method for prediction of solubilities of solid solutes in supercritical carbon dioxide, J. Supercrit. Fluids 50 (2009) 210–217.
[52] F. Gharagheizi, A. Eslamimanesh, A.H. Mohammadi, D. Richon, Artificial neural network modeling of solubilities of 21 commonly used industrial solid compounds in supercritical carbon dioxide, Ind. Eng. Chem. Res. 50 (2011) 221–226.
[53] J. Noroozi, C. Ghotbi, J.J. Sardroodi, J. Karimi-Sabet, M.A. Robert, Solvation free energy and solubility of acetaminophen and ibuprofen in supercritical carbon dioxide: Impact of the solvent model, J. Supercrit. Fluids 109 (2016) 166–176.
[54] T. Merker, J. Vrabec, H. Hasse, Comment on "An optimized potential for carbon dioxide" [J. Chem. Phys. 122, 214507 (2005)], J. Chem. Phys. 129 (2008) 087101.
[55] J.G. Harris, K.H. Yung, Carbon dioxide's liquid-vapor coexistence curve and critical properties as predicted by a simple molecular model, J. Phys. Chem. 99 (1995) 12021–12024.

[56] J.J. Potoff, J.I. Siepmann, Vapor–liquid equilibria of mixtures containing alkanes, carbon dioxide, and nitrogen, AICHE J. 47 (2001) 1676–1682.
[57] J. Chrastil, Solubility of solids and liquids in supercritical gases, J. Phys. Chem. 86 (1982) 3016–3021.
[58] Y. Adachi, B.C.Y. Lu, Supercritical fluid extraction with carbon dioxide and ethylene, Fluid Phase Equilib. 14 (1983) 147–156.
[59] H.-D. Sung, J.-J. Shim, Solubility of C. I. Disperse red 60 and C. I. Disperse blue 60 in supercritical carbon dioxide, J. Chem. Eng. Data 44 (1999) 985–989.
[60] D.L. Sparks, R. Hernandez, L.A. Estévez, Evaluation of density-based models for the solubility of solids in supercritical carbon dioxide and formulation of a new model, Chem. Eng. Sci. 63 (2008) 4292–4301.
[61] S.K. Kumar, K.P. Johnston, Modelling the solubility of solids in supercritical fluids with density as the independent variable, J. Supercrit. Fluids 1 (1988) 15–22.
[62] J.M. Del Valle, J.M. Aguilera, An improved equation for predicting the solubility of vegetable oils in supercritical carbon dioxide, Ind. Eng. Chem. Res. 27 (1988) 1551–1553.
[63] C. Garlapati, G. Madras, New empirical expressions to correlate solubilities of solids in supercritical carbon dioxide, Thermochim. Acta 500 (2010) 123–127.
[64] X.-Q. Bian, Q. Zhang, Z.-M. Du, J. Chen, J.-N. Jaubert, A five-parameter empirical model for correlating the solubility of solid compounds in supercritical carbon dioxide, Fluid Phase Equilib. 411 (2016) 74–80.
[65] A. Tabernero, E.M.M. del Valle, M.Á. Galán, A comparison between semiempirical equations to predict the solubility of pharmaceutical compounds in supercritical carbon dioxide, J. Supercrit. Fluids 52 (2010) 161–174.
[66] C.-C. Tsai, H.-m. Lin, M.-J. Lee, Solubility of niflumic acid and celecoxib in supercritical carbon dioxide, J. Supercrit. Fluids 95 (2014) 17–23.
[67] M. Asghari-Khiavi, Y. Yamini, Solubility of the drugs bisacodyl, methimazole, methylparaben, and iodoquinol in supercritical carbon dioxide, J. Chem. Eng. Data 48 (2003) 61–65.
[68] S.J. Macnaughton, I. Kikic, N.R. Foster, P. Alessi, A. Cortesi, I. Colombo, Solubility of anti-inflammatory drugs in supercritical carbon dioxide, J. Chem. Eng. Data 41 (1996) 1083–1086.
[69] M.A. Sabegh, H. Rajaei, F. Esmaeilzadeh, M. Lashkarbolooki, Solubility of ketoprofen in supercritical carbon dioxide, J. Supercrit. Fluids 72 (2012) 191–197.
[70] M. Ashraf-Khorassani, M.T. Combs, L.T. Taylor, F.K. Schweighardt, P.S. Mathias, Solubility study of sulfamethazine and sulfadimethoxine in supercritical carbon dioxide, fluoroform, and subcritical freon 134A, J. Chem. Eng. Data 42 (1997) 636–640.
[71] L. Manna, M. Banchero, Solubility of tolbutamide and chlorpropamide in supercritical carbon dioxide, J. Chem. Eng. Data 63 (2018) 1745–1751.
[72] R. Heryanto, E.C. Abdullah, M. Hasan, Solubility of isoniazid in supercritical carbon dioxide, J. Chem. Eng. Data 55 (2010) 2306–2309.
[73] M. Türk, T. Kraska, Experimental and theoretical investigation of the phase behavior of naproxen in supercritical CO_2, J. Chem. Eng. Data 54 (2009) 1592–1597.
[74] D. Suleiman, L.A. Estévez, J.C. Pulido, J.E. García, C. Mojica, Solubility of anti-inflammatory, anti-cancer, and anti-HIV drugs in supercritical carbon dioxide, J. Chem. Eng. Data 50 (2005) 1234–1241.
[75] R.D. Weinstein, K.R. Muske, J. Moriarty, E.K. Schmidt, The solubility of benzocaine, lidocaine, and procaine in lquid and supercritical carbon dioxide, J. Chem. Eng. Data 49 (2004) 547–552.
[76] Y. Yamini, J. Hassan, S. Haghgo, Solubilities of some nitrogen-containing drugs in supercritical carbon dioxide, J. Chem. Eng. Data 46 (2001) 451–455.
[77] O. Guney, A. Akgerman, Solubilities of 5-fluorouracil and β-estradiol in supercritical carbon dioxide, J. Chem. Eng. Data 45 (2000) 1049–1052.
[78] J.W. Hampson, R.J. Maxwell, S. Li, R.J. Shadwell, Solubility of three veterinary sulfonamides in supercritical carbon dioxide by a recirculating equilibrium method, J. Chem. Eng. Data 44 (1999) 1222–1225.
[79] M. Ko, V. Shah, P.R. Bienkowski, H.D. Cochran, Solubility of the antibiotic penicillin V in supercritical CO_2, J. Supercrit. Fluids 4 (1991) 32–39.
[80] C.-S. Su, Y.-P. Chen, Measurement and correlation for the solid solubility of non-steroidal anti-inflammatory drugs (NSAIDs) in supercritical carbon dioxide, J. Supercrit. Fluids 43 (2008) 438–446.

[81] H.C. de Sousa, M.S. Costa, P. Coimbra, A.A. Matias, C.M.M. Duarte, Experimental determination and correlation of meloxicam sodium salt solubility in supercritical carbon dioxide, J. Supercrit. Fluids 63 (2012) 40–45.
[82] M. Khamda, M.H. Hosseini, M. Rezaee, Measurement and correlation solubility of cefixime trihydrate and oxymetholone in supercritical carbon dioxide (CO_2), J. Supercrit. Fluids 73 (2013) 130–137.
[83] A. Zeinolabedini Hezave, H. Rajaei, M. Lashkarbolooki, F. Esmaeilzadeh, Analyzing the solubility of fluoxetine hydrochloride in supercritical carbon dioxide, J. Supercrit. Fluids 73 (2013) 57–62.
[84] H. Rajaei, A.Z. Hezave, M. Lashkarbolooki, F. Esmaeilzadeh, Representing experimental solubility of phenylephrine hydrochloride in supercritical carbon dioxide and modeling solute solubility using semi-empirical correlations, J. Supercrit. Fluids 75 (2013) 181–186.
[85] M. Lashkarbolooki, A.Z. Hezave, Y. Rahnama, R. Ozlati, H. Rajaei, F. Esmaeilzadeh, Solubility of cyproheptadine in supercritical carbon dioxide; experimental and modeling approaches, J. Supercrit. Fluids 84 (2013) 13–19.
[86] M. Lashkarbolooki, A.Z. Hezave, Y. Rahnama, H. Rajaei, F. Esmaeilzadeh, Solubility of chlorpheniramine maleate in supercritical carbon dioxide, J. Supercrit. Fluids 84 (2013) 29–35.
[87] C.-A. Lee, M. Tang, S.-L. Ho, Y.-P. Chen, Solubilities of chlormezanone, metaxalone and methocarbamol in supercritical carbon dioxide, J. Supercrit. Fluids 85 (2014) 11–16.
[88] A.B.S. Rosa, S. Marceneiro, M.E.M. Braga, A.M.A. Dias, H.C. de Sousa, Solubility of all-trans retinoic acid in supercritical carbon dioxide, J. Supercrit. Fluids 98 (2015) 70–78.
[89] N. Lamba, R.C. Narayan, J. Modak, G. Madras, Solubilities of 10-undecenoic acid and geraniol in supercritical carbon dioxide, J. Supercrit. Fluids 107 (2016) 384–391.
[90] K. Keshmiri, A. Vatanara, Y. Yamini, Development and evaluation of a new semi-empirical model for correlation of drug solubility in supercritical CO_2, Fluid Phase Equilib. 363 (2014) 18–26.
[91] X.-Y. Gong, X.-J. Cao, Measurement and correlation of solubility of artemisinin in supercritical carbon dioxide, Fluid Phase Equilib. 284 (2009) 26–30.
[92] S. Marceneiro, P. Coimbra, M.E.M. Braga, A.M.A. Dias, H.C. de Sousa, Measurement and correlation of the solubility of juglone in supercritical carbon dioxide, Fluid Phase Equilib. 311 (1-8) (2011).
[93] R. Chim, S. Marceneiro, M.E.M. Braga, A.M.A. Dias, H.C. de Sousa, Solubility of norfloxacin and ofloxacin in supercritical carbon dioxide, Fluid Phase Equilib. 331 (2012) 6–11.
[94] J.-l. Li, J.-s. Jin, Z.-t. Zhang, Y.-b. Wang, Measurement and correlation of solubility of benzamide in supercritical carbon dioxide with and without cosolvent, Fluid Phase Equilib. 307 (2011) 11–15.
[95] M. Sauceau, J.J. Letourneau, B. Freiss, D. Richon, J. Fages, Solubility of eflucimibe in supercritical carbon dioxide with or without a co-solvent, J. Supercrit. Fluids 31 (2004) 133–140.
[96] A.R.C. Duarte, P. Coimbra, H.C. de Sousa, C.M.M. Duarte, Solubility of flurbiprofen in supercritical carbon dioxide, J. Chem. Eng. Data 49 (2004) 449–452.
[97] K.D. Bartle, A.A. Clifford, S.A. Jafar, G.F. Shilstone, Solubilities of solids and liquids of low volatility in supercritical carbon dioxide, J. Phys. Chem. Ref. Data 20 (1991) 713–756.
[98] M.D. Gordillo, M.A. Blanco, A. Molero, E. Martinez de la Ossa, Solubility of the antibiotic penicillin G in supercritical carbon dioxide, J. Supercrit. Fluids 15 (1999) 183–190.
[99] M.D. Gordillo, C. Pereyra, E.J. Martı´nez de la Ossa, Measurement and correlation of solubility of Disperse Blue 14 in supercritical carbon dioxide, J. Supercrit. Fluids 27 (2003) 31–37.
[100] J. Méndez-Santiago, A.S. Teja, The solubility of solids in supercritical fluids, Fluid Phase Equilib. 158–160 (1999) 501–510.
[101] A. Jouyban, H.-K. Chan, N.R. Foster, Mathematical representation of solute solubility in supercritical carbon dioxide using empirical expressions, J. Supercrit. Fluids 24 (2002) 19–35.
[102] S.B. Hozhabr, S.H. Mazloumi, J. Sargolzaei, Correlation of solute solubility in supercritical carbon dioxide using a new empirical equation, Chem. Eng. Res. Des. 92 (2014) 2734–2739.
[103] M. Asgarpour Khansary, F. Amiri, A. Hosseini, A. Hallaji Sani, H. Shahbeig, Representing solute solubility in supercritical carbon dioxide: A novel empirical model, Chem. Eng. Res. Des. 93 (2015) 355–365.
[104] Z.-R. Yu, B. Singh, S.S.H. Rizvi, J.A. Zollweg, Solubilities of fatty acids, fatty acid esters, triglycerides, and fats and oils in supercritical carbon dioxide, J. Supercrit. Fluids 7 (1994) 51–59.
[105] S. Jafari Nejad, H. Abolghasemi, M.A. Moosavian, M.G. Maragheh, Prediction of solute solubility in supercritical carbon dioxide: A novel semi-empirical model, Chem. Eng. Res. Des. 88 (2010) 893–898.

CHAPTER 11

Decaffeination using supercritical carbon dioxide

Giovani L. Zabot
Laboratory of Agroindustrial Processes Engineering (LAPE), Federal University of Santa Maria (UFSM), Cachoeira do Sul, Brazil

Contents

1. Introduction

Caffeine (alkaloid) is one of the most well-known psychoactive compounds worldwide. Since the last five decades, the per capita consumption of caffeine in the world has increased by approximately 100% as a consequence of the increase in consumption of teas, energy drinks, and coffee. Based on this fact, several scientific studies have been performed to evaluate the physiological actions of caffeine. Up to now, it has not been possible to reach a consensus on its positive and negative effects. The controversial information eventually led to an increase in consumption of decaffeinated products, generally by medical advice.

The decaffeination is an alternative to provide products without caffeine or with a reduced content of caffeine. Some methods use organic solvents, such as alcohol, dichloromethane, chloroform, ethyl acetate, or acetone. Other methods use water and supercritical CO_2. The CO_2 under supercritical conditions is preferable because it can

Green Sustainable Process for Chemical and Environmental Engineering and Science
https://doi.org/10.1016/B978-0-12-817388-6.00011-8

retain most of the compounds that give taste and flavor. Decaffeination by supercritical CO_2 is a relatively recent technology that is expanding in several countries.

Based on this context, this chapter has the objective to present and discuss the main characteristics of decaffeination by supercritical CO_2. Even though after performing the extraction processes two fractions are obtained, the emphasis is put on decaffeinated products (coffee beans, tea leaves, cocoa beans, etc.) instead of the extract. The constructive discussion is focused on the solid material without or with a reduced content of caffeine. The aim is to overview the contribution of the supercritical technology using a green and renewable solvent (CO_2) toward making the realistic high-pressure decaffeination process widespread and even more feasible.

2. Carbon dioxide as a green supercritical fluid

A supercritical fluid has simultaneously properties of gas and liquid, especially a high solubility and a low viscosity, respectively. Pure substances submitted to pressure and temperature higher than the critical conditions (critical temperature and critical pressure—critical point) are called supercritical fluids. In the phase diagram, the visible differences between liquid and gas phases disappear in this region and the fluid interchanges between liquid-like/gas-like characteristics. This phenomenon of interchanging is more pronounced in the region near to the critical point because there is a high solvation power, where the physicochemical characteristics are highly sensitive to changes in the parameters of pressure and temperature.

CO_2 (critical pressure of approximately 7.4 MPa and critical temperature of approximately 31°C) is a fluid used to fulfill supercritical fluid extraction (SFE) because it permits the operation at moderate temperature. Overall, supercritical CO_2 is commonly applied for extracting nonpolar and moderate polar phytochemicals. The density of CO_2 can be altered by modifying the system conditions (especially pressure or temperature), especially in the region near to the critical point. Therefore, the selective extraction of different phytochemicals can be reached. CO_2 has other advantages, as the large availability in the environment, the nontoxic properties, and the easy separation from the products (liquid extracts and solid coproducts) by decompression at room pressure (approximately 0.1 MPa). This last one advantage can make the extracted material a solvent-free product to be applied in several industrial fields, as in the biochemical, pharmaceutical, chemical, materials, cosmetic, and food-related fields.

The solubility of a diversified quantity of phytochemicals in supercritical CO_2 or other supercritical fluids can change significantly because it is closely related to solvent density. Consequently, the solubility of different bioactive compounds depends on pressure and temperature. Typically, the solubility of CO_2 increases near the critical point. Thereafter, it can decrease if the system is compressed, leading to dominant repulsive CO_2-solute-cosolvent interactions. Otherwise, increased solubility can occur when

Table 1 Solubility data of caffeine in supercritical CO_2 with or without a cosolvent

Pressure (MPa)	Temperature (°C)	Cosolvent	Solubility (mmol/kmol solvent)[a]	Ref.
15	40	–	61.0	[31]
15	60	–	106.3	[31]
25	40	–	64.8	[31]
25	60	–	125.3	[31]
22	40	14% methanol (molar basis)	360.0	[32]
10	40	–	140.0	[33]
20	40	–	220.0	[33]
28	40	–	360.0	[33]
15	40	Water (103.7 mol/m^3)	3150.0	[34]
35	50	–	87.0	[35]
35	60	–	66.6	[35]
30	40	Water (saturated)	7100.0	[36]
24	70	5% Ethanol (molar basis)	1600.0	[37]
18	40	5% Ethanol (molar basis)	1200.0	[37]
15	50	5% Isopropanol (molar basis)	250.0	[37]

[a]The solvent means the supercritical CO_2 with the addition of modifier (if applicable).

attractive CO_2-solute-cosolvent interactions take place, which is a feature of small chain and highly volatile molecules. It is desirable to fulfill the extractions as far as possible from the equilibrium solubility. The best choice is to use an excess of CO_2 to have a nonequilibrium condition. However, there is a limit value of excess of solvent imposed by the CO_2 characteristics and operational procedures (capacities of pump, pipes, valves, and heat exchangers, among others) that prevent the use of a larger excess of CO_2. In general, Table 1 presents a brief compilation of solubility data of caffeine in CO_2 found in some solid matrices.

3. Why extracting caffeine?

Caffeine ($C_8H_{10}N_4O_2$, molar mass of 194.1906 g/mol, also known as 1H-Purine-2,6-dione, 3,7-dihydro-1,3,7-trimethyl- or 1,3,7-trimethyl-2,6-dioxopurine) is a methylxanthine alkaloid found in several parts of plants, like beans, nuts, seeds, and leaves. It is a stimulant and the form I of caffeine is metastable. This form is only stable at temperatures ranging from 155°C to 237°C. The form II is stable at room conditions, as 20–25°C. One well-known source of caffeine is the coffee bean and the well-known products containing caffeine are drinks as coffee, tea, energy drinks, and chocolate, among others.

The extraction of caffeine from the plants aims at presenting a decaffeinated product, as decaffeinated coffee or decaffeinated tea.

Producing and processing coffee is one of the industrial areas most economically relevant around the world. Brazil, Colombia, Indonesia, and Mexico are the main producers of coffee, which can account more than 3 million tons per year (almost 60% of the worldwide production). The current annual worldwide production is larger than 5.5 million tons. The ingestion of caffeine favors oxidation of fat and speeds up metabolism. Furthermore, caffeine can block adenosine receptors on the central nervous system. Otherwise, caffeine intake can cause some adverse effects, such as palpitations, gastrointestinal disturbances, anxiety and malaise, mood changes, insomnia and sleep problems, and increased blood pressure. When the consumption of beverages containing caffeine is relatively high, some consumers change the comfort of drinking the beverages to the discomfort. For some people, especially who have health problems, the pleasure of drinking caffeinated products is not outweighed by the caffeine-fueled negatives.

Therefore, extraction of caffeine is a goal of companies that sell a decaffeinated product ready for consumption. For example, coffee needs to have more than 97% of the caffeine extracted for being called decaffeinated. Such kind of coffee is another product available to consumers who want to taste the aroma of coffee without experiencing the stimulant effects of caffeine. The decaffeination process is used to remove harshness and irritation compounds. Unfortunately, some processes leave a bland coffee beverage with low flavor. Coffee has up to 1500 different chemical compounds, including alkaloids, phenolics, and terpenes, among others. Therefore, the extraction of only caffeine is a hard task because approximately 2% (mass basis) of this compound should be removed without altering significantly the original flavor and characteristics of the product. Furthermore, caffeine extracted from coffee beans, tea leaves, cocoa beans, and cola nuts can be used in the production of cosmetics (for the treatment of localized excess fat), soft drinks, cola-type drinks, energy drinks, and psychoactive drugs (e.g., for treating neuralgia and headache), among other products.

The water solubility of caffeine is approximately 16 mg/mL (approximately 1.48 mol/kmol water) [1]. Regarding supercritical CO_2 (with or without some modifiers), the solubility of caffeine in this solvent is presented in Table 1. The levels of caffeine in some common products are presented in Table 2. It is estimated that approximately 10% of the current consumption of coffee-based beverages (e.g., coffee, energy drinks, and chocolate drinks) in the United States and Europe are decaffeinated beverages.

4. Decaffeination by supercritical technology

Decaffeination is the extraction of caffeine from yerba mate, coffee, guarana, tea, cocoa, and other plants by different methods. Commonly, the decaffeination can be fulfilled by water immersion, water-ethyl acetate immersion, methylene chloride (or dichloromethane)

Table 2 Average levels of caffeine in coffee drinks, tea, chocolate, and sodas

Soft drink	Caffeine (mg/100 mL)
Express coffee	250–330
Decaffeinated coffee	1–5
Soluble coffee	30–120
Instantaneous tea	25–50
Chocolate	2–20
Sodas	2–20

extraction, and supercritical CO_2 extraction. In the water immersion, the caffeine is dissolved in water and can be removed by activated carbon or other adsorbents as agricultural solid coproducts (e.g., rice husks). In the water-ethyl acetate immersion, the caffeine is extracted from the seeds, beans, nuts, or leaves by ethyl acetate, which the solvent flows through the solid matrix embedded in water. In the methylene chloride extraction, the caffeine is removed from the solid biomass by this solvent using mild temperatures (generally near to 40°C). In such case, the moisture of the solid matrix should be high. Until the mid-1970s, the dichloromethane was extensively used for extraction of caffeine. Although the residual amount of dichloromethane in the solid biomass is not too high (below the limit defined by the Food and Drug Administration), recent concerns about the problems to humans have risen due to its high toxic effects.

Supercritical carbon dioxide extraction (SFE-CO_2) uses CO_2 above its critical conditions of pressure and temperature to solubilize the caffeine. The SFE-CO_2 process has some advantages whether compared to the other methods because uses a nontoxic solvent at mild conditions (especially temperature). The consumption of energy is generally lower than those associated with conventional processes and the CO_2 can be easily separated from the products by a simple decompression. Indeed, the main process of removing caffeine and maintaining most of the other flavor substances from coffee, tea, cocoa, and yerba mate, among others, is CO_2. This solvent practically does not attack the carbohydrates (e.g., sugars and starch) and proteins. This fact is important because several compounds that give aroma and flavor are converted from proteins after roasting. Since 1970, the worldwide use of SFE-CO_2 is being expanded. Supercritical technology is considered green and sustainable, especially for decaffeination. Therefore, this section is intended to provide the main characteristics of processing plants containing caffeine with supercritical CO_2.

4.1 Batch process

CO_2 has some positive characteristics for the extraction of caffeine. Besides the advantages previously cited, CO_2 has a good dissolving ability and a good mass transfer performance (rapid penetration ability). SFE-CO_2 of caffeine is generally more efficient because lower aromatic substances can be lost during the decaffeination process. One operational mode of extracting caffeine is by a batch process, which the biomass (e.g., coffee, tea, yerba mate, cocoa, etc.) and the CO_2 are mixed in a fixed bed during a static time under a defined values of pressure and temperature. After the static extraction time, which generally occurs in a few hours, the bed is depressurized, the extract is collected, and the solid matrix is recovered.

Typically, large-scale extraction comprises vessels of volume higher than 20 m^3. According to Lack and Seidlitz [2], decaffeination of coffee beans by SFE-CO_2 under a batch and nonisobaric process presents different cost estimates, as € 1.10/kg feed for a capacity of 3500 tons/year and € 0.75/kg feed for a capacity of 7000 tons/year. In fact, the decaffeination of coffee by SFE-CO_2 under a batch and isobaric process is commonly more economically feasible. For the capacities of 3500 tons/year and 7000 tons/year, the costs estimates are € 0.85/kg feed and € 0.55/kg feed, respectively.

The batch process is commonly performed in a single-stage mode since the solids (beans, leaves, etc.) are difficult to process in pressurized vessels by a continuous mode due to their particulate forms [3]. Furthermore, there is no harmful residue in the decaffeinated material (solid material) because the CO_2 is a solvent that can be completely and easily removed without a further treatment. A schematic flowchart of decaffeination in a batch process is shown in Fig. 1.

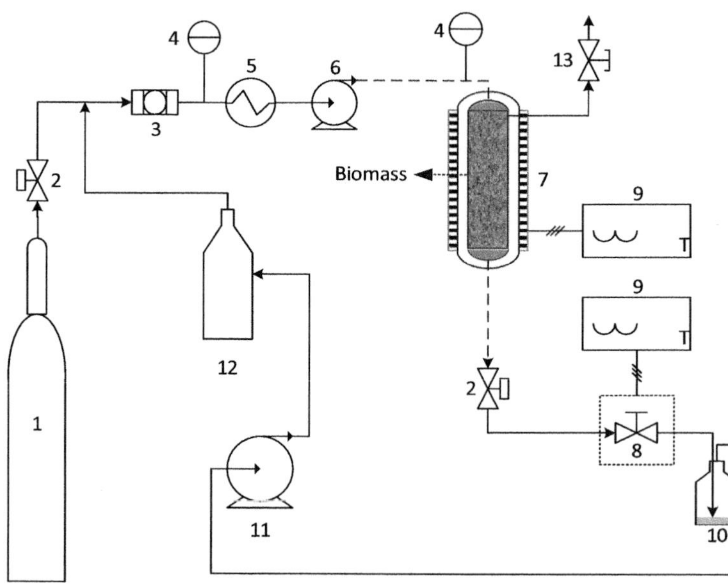

1	CO_2 reservoir
2	Blocking valve
3	CO_2 filter
4	Pressure gauge
5	Cooling exchanger
6	High-pressure pump
7	Extraction vessel
8	Micrometering valve
9	Temperature controller
10	Extract separator
11	Compressor
12	Buffer tank
13	Safety valve

Fig. 1 Typical flowchart of decaffeination by SFE-CO_2 in a batch process; *dashed lines* indicate a noncontinuous flow.

4.2 Semicontinuous and continuous processes

Other modes of extracting caffeine are performed by semicontinuous (also known as semi-batch) and continuous processes. In the semicontinuous process, the solid biomass is fed into a fixed bed and the CO_2 flows continuously to the bed under the defined values of temperature and pressure. Caffeine is recovered overtime at the extraction vessel outlet (Fig. 2). In the continuous process, the solid biomass and the CO_2 are mixed and pressurized continuously by a pump to the bed under the defined conditions of temperature and pressure. At the vessel outlet, caffeine and a decaffeinated solid material are separated (Fig. 3).

A recent study dealing with SFE-CO_2 under a semicontinuous mode for extraction of caffeine was performed with black tea [4]. At the beginning of the process, 500 g black tea with particle size ranging from 600 to 780 μm and 2.16% (mass basis) caffeine content was mixed with 300 mL water and loaded into the extractor (10 L). After evaluating different values of temperature and pressure, the mass extraction yield ranged from 16.2% to 100%. Commonly, in the extraction processes the biomass is saturated with water to increase the solubility of caffeine. It is important to mention that the critical overview and discussion of the effect of process parameters on the decaffeination is provided into Section 4.3.

In most of the cases, the semicontinuous mode is applied. This mode allows using pumps that pressurize fluids (pure substances or mixtures) without particles, like CO_2. As presented in Fig. 2, the steps consist of cooling CO_2 to change its phase from gas to liquid using a cooler, pressurizing CO_2 to the target pressure using a high-pressure

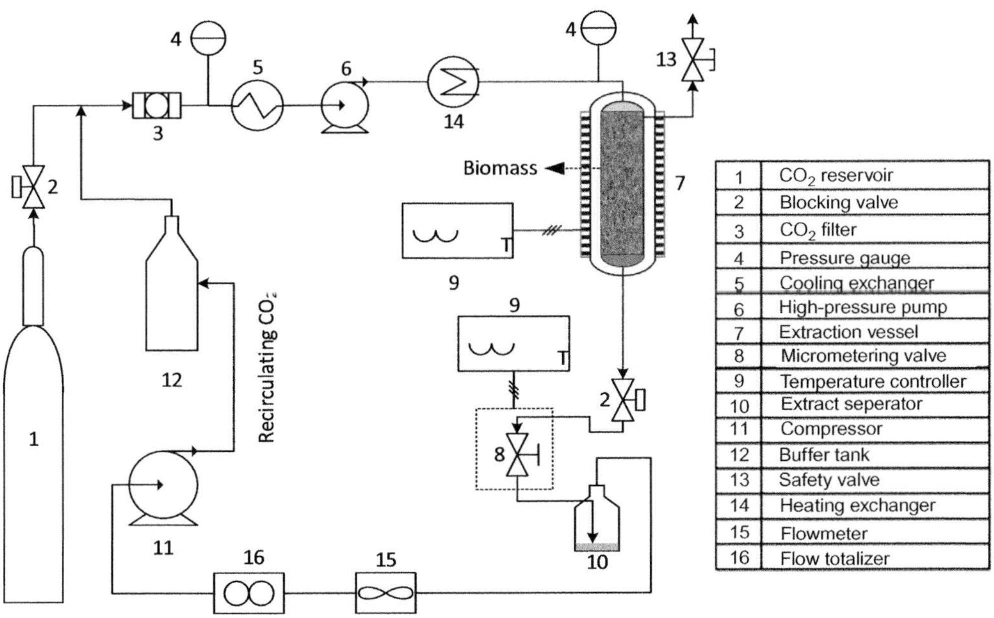

Fig. 2 Typical flowchart of decaffeination by SFE-CO_2 in a semicontinuous process.

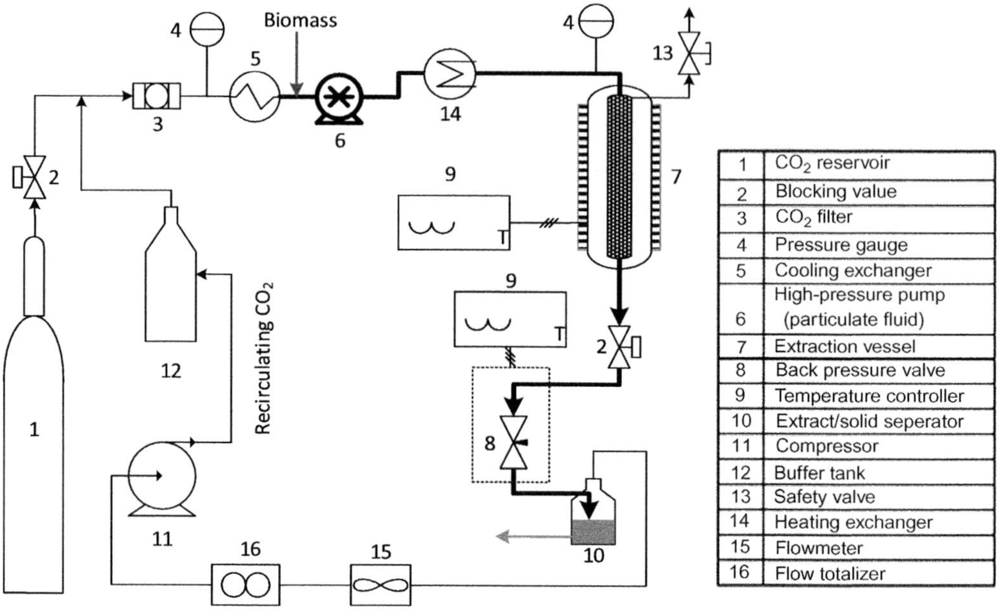

Fig. 3 Typical flowchart of decaffeination by SFE-CO_2 in a continuous process.

pump, heating CO_2 to the desired temperature using a jacketed heater, pumping CO_2 into the vessels containing the fixed bed (biomass mixed with water), and collecting the extracted caffeine at the exit of the vessels (in the separators) because CO_2 flowing through the fixed-bed solubilizes caffeine. Thereafter, the decaffeinated biomass is unloaded from the vessel for drying. The moisturizing step in the beginning of the process is important for decaffeination regardless of the solvent. The main characteristics of the systems operating in semicontinuous mode include a short CO_2 residence time, a long solid residence time, a possibility of allowing different sizes of particles (in the ranges of micrometers until a few centimeters), an adjustable solvent to feed mass ratio, and a medium process control.

A supercritical laboratory scale system from Thar Technologies Inc. (Pittsburgh, United States) having a vessel of 1 L was used in experimental assays for decaffeination of orthodox Kangra tea (particles ranging from 7.1 to 9.1 mm) in a semicontinuous mode. SFE-CO_2 with cosolvent (water) was performed at different temperatures (35–85°C) and pressures (9–17 MPa) [5]. The semicontinuous mode was performed with a fixed bed (900 g of orthodox Kangra black tea) and a continuous CO_2 flow rate (60 g/min) for 90 min. The first separator was maintained at a pressure of 8–9 MPa and a temperature range from −10°C to −9°C. The second separator was maintained at a pressure ranging from 2 to 3 MPa and a temperature ranging from −5°C to 0°C. At 90°C and 17 MPa (in the extraction vessel), approximately 48% (mass basis) was removed, reducing the caffeine content from 1.71% to 0.89%.

The continuous mode needs pumps that pressurize particulate fluids, that is, a mixture of solvent and particles of biomass. As presented in Fig. 3, the steps consist of cooling CO_2 to change its phase from gas to liquid using a cooler, mixing CO_2 with milled biomass, pressurizing the mixture of CO_2 and solid biomass to the target pressure using a high-pressure pump, heating the mixture to the desired temperature using a jacketed heater, pumping continuously the mixture into the vessels for having the extraction, and collecting the bulk extract mixed with biomass at the outlet of the vessels (in the separators). The main characteristics of the systems operating in continuous mode include a short CO_2 residence time, a short solid residence time, the necessity of using really small particles (generally micrometric particles), a high solvent to feed mass ratio, and a high process control.

Unfortunately, decaffeination in continuous mode is uncommon because some challenges should be overcome. The main challenge refers to pumping the mixture of moistened solids and CO_2. There are a limited number of high-pressure pumps that can pressurize particulate fluids. The particles can block the piping and cause a high-pressure drop. Therefore, it is preferable to perform the procedures in a semicontinuous mode with various extraction vessels disposed of in parallel, which can represent a pseudo-continuous mode. This strategy can be based on the methodological approaches presented by Moraes et al. [6], where a successful pseudo-continuous operating SFE equipment was validated.

4.3 Process parameters

There are some process parameters that influence the solubility and yields of extracted substances in processes operating with supercritical fluids. For decaffeination process, the design of extraction vessels in supercritical processes comprises the knowledge of phase equilibrium among the CO_2, the caffeine to be extracted, and the insoluble solids (in some cases, soluble solutes can be included herein). The main process parameters are temperature, pressure, cosolvent type and concentration, extraction time, and solvent to feed mass ratio.

4.3.1 Temperature

The temperature applied in the decaffeination using supercritical CO_2 commonly ranges from 35°C to 70°C. In some cases, this parameter can reach higher values, such as 120°C, because its increase can favor mass transfer. Overall, thermal labile compounds can be affected and degraded at mild to high temperatures while hydrolysis of compounds and other reactions can take place [7].

The temperature is one of the process parameters that influence the solvent density and determines the solvating power of CO_2. In the decaffeination of black tea using a pilot-scale equipment with supercritical CO_2 as solvent, three levels of temperature (55°C, 62.5°C, and 70°C) were evaluated in the process [4]. According to the authors, the caffeine extraction yield was higher when the temperature increased at constant pressure of 25 MPa. Besides, no significant effect of this parameter was observed at constant pressure of 50 MPa. In fact, typically, there is an interaction between some process parameters, especially temperature, pressure, and cosolvent. In this referenced decaffeination of black tea, the temperature-modifier interaction presented a different impact on decaffeination yield than other interactions did. As the concentration of modifier (cosolvent) increased, the yield increased when 62.5°C was used if compared to 55°C. However, the yield decreased when 70°C was used if compared to 62.5°C.

Tea (black, green, etc.) is one of the most common beverages consumed in the world. SFE-CO_2 process for caffeine removal from black tea leaves was reported [8]. In such study, three temperatures (40°C, 60°C, and 80°C) were evaluated. After a statistical analysis, the parameter temperature (principal effects) had no statistical significant effect ($P > .05$) on caffeine recovery. Consequently, considering energy savings and technical responses, lower values of temperature could be selected to extract caffeine. However, still regarding the decaffeination of black tea leaves, a synergistic effect between temperature and other parameters (pressure, ethanol concentration, and modifier flow rate) was seen. After applying the Response Surface Methodology to define the decaffeination conditions, the average mass decaffeination efficiency was 99.8% and the loss of phenolic compounds was only 3.3% when the following conditions were used: 53°C, 30 MPa, 0.7 mL/min of modifier (ethanol as cosolvent) and 87% ethanol concentration (volumetric basis). According to Bahar et al. [8], it is of remarkable importance for the decaffeination from black tea (it could be extended to other caffeinated raw materials) to remove caffeine while maintaining the maximum quantity of phenolics (and all aromatic compounds in general). Therefore, the caffeine removal and the amount of phenolic compounds remaining in black tea are important responses that indicate the yield of the process from tea by supercritical technology using CO_2.

Although the decaffeination is generally applied to coffee or tea, some other caffeine-rich natural matrices can be decaffeinated by SFE-CO_2. One example of another matrix is guarana, which is a Brazilian native plant containing up to 6% (mass basis) of caffeine [9]. Guarana seeds and shells are often used in soft drinks (named in Brazil as "guarana") and as an ingredient of some medicinal products. Based on this context, aiming at obtaining decaffeinated guarana seeds, a study was performed where the SFE-CO_2 process was applied. Experimental data indicated the reduction of the caffeine content from guarana seeds with supercritical CO_2 saturated with water [10]. In that study, two levels of temperature were evaluated: 40°C and 70°C. Approximately 98% (mass basis) of caffeine was removed from seeds at 70°C for 4 h, indicating that SFE-CO_2 is a successful method for the decaffeination of such seeds. As reported by the authors [10], the decaffeination at lower levels of pressure or temperature needed extra time and higher amounts of CO_2 to obtain the same relative yield than that obtained at 70°C, for instance.

From the foregoing outcomes, we infer that higher temperatures are recommendable to remove caffeine from caffeinated samples. However, an integrated evaluation should be performed, which should take into account the contents of phenolics and other aromatic compounds. There is a trend of solubilizing more phenolic acids at higher temperatures, especially when a larger content of a polar cosolvent (as modifier) is used. In some cases, the reduction of volatile compounds is more evidenced if more caffeine is extracted. Therefore, it is preferable to perform the processes with temperatures not higher than 70°C.

4.3.2 Pressure

The pressure applied for the decaffeination using supercritical CO_2 commonly ranges from 10 to 35 MPa. The increase in pressure raises the solvation power of CO_2 as a consequence of the increase in density. In some cases, pressures larger than 50 MPa are applied, including in systems that a modifier is used. Some experimental data of decaffeination of different caffeinated plants is presented in Table 3, where different process conditions have been used and different findings have been obtained.

The decaffeination of coffee and tea is the most common practice worldwide regarding the decaffeination process. The semicontinuous mode is the main process applied, where the CO_2 circulates through the bed and caffeine is extracted from the leaves or beans. One alternative in the separation system is using activated carbon, which can retain caffeine as an adsorbent. In fact, the extracted caffeine can be separated from the supercritical CO_2 by a pressure reduction to the room conditions (approximately 0.1 MPa), consequently decreasing its solubility in CO_2, which becomes a gas in lower pressures.

Table 3 Experimental data for decaffeination using supercritical CO_2

Biomass	*T* (°C)	*P* (MPa)	Time (h)	*S/F* (–)	Cosolvent	Yield (wt.%)	Ref.
Green tea	50	40	5	28.1	Water	59.8	[38]
Green tea	50	30	5	28.1	Water	43.0	[38]
Green tea	40	40	5	28.1	Water	54.0	[38]
Guarana seeds	70	40	4	273.6	Water	98.0	[10]
Black tea	62.5	37.5	5	1008	Ethanol	100	[4]
Black tea	62.5	37.5	5	1512	Ethanol	100	[4]
Black tea	55	37.5	3	604.8	Ethanol	95.4	[4]
Black tea	62.5	37.5	1	201.6	Ethanol	56.5	[4]
Coffee beans	50	13.8	3	–	Water	80.0	[12]
Cocoa beans	70	40	1	100	Water	66.4	[13]
Yerba mate	60	17	–	20.0	NA	33.9	[39]
Green tea	70	20	1	51.0	Ethanol	82.4	[27]
Black tea	85	9	1.5	6.0	Water	40.4	[5]

NA, not applicable; *P*, pressure; *S/F*, solvent to feeds mass ratio (g solvent/g biomass); *T*, temperature.

A study proposed a decaffeination of coffee beans using saturated CO_2 [11]. After maintaining the CO_2 in contact with the beans at relatively high temperatures (up to 100°C) and high pressures (up to 30 MPa), there is a fast pressure drop. In the sequence, the beans are washed for up to 2 h. Afterward, the CO_2-saturated green coffee extract containing caffeine is decaffeinated by SFE-CO_2. Franca [11] also reports that liquid CO_2 submitted at lower pressures (generally <7 MPa) can be used in decaffeination processes. Green coffee beans (moisture ranging from 45% to 55%) are mixed with water-saturated CO_2 at such pressures (e.g., 6.5 MPa) in order to retain most of the flavors during the process. One drawback of this processing line is the extraction time, which could be longer than 4 days.

Some former studies that demonstrated the possibility of decaffeination have been developed at the end of the last century. In one of these studies, the influence of pressure on the decaffeination of coffee beans was demonstrated by Peker et al. [12]. The authors have tested four pressures: 10.3, 13.8, 16.5, and 19.3 MPa. According to the findings, there is an increase of the extraction rate with increasing the pressure. As seen, at the highest pressures and temperatures (main process parameters), the data depended on changes in operational conditions. Consequently, it means that larger variations of extraction rates were observed in the pressures of 16.5 and 19.3 MPa. Furthermore, based on results obtained at large-scale commercial facilities, the flavor compounds released during roasting are not recovered in the decaffeination process, which is a positive and promising outcome.

More recently, the influence of pressure was also evaluated in the SFE-CO_2 of caffeine from black tea leaves [4]. Caffeine recovery yield was high with high pressure. According to the Response Surface Methodology, the interactions between the parameters pressure/extraction time and pressure/flow rate presented similarities on the extraction yield of caffeine. The yield of caffeine demonstrated a slight increase as pressure increased, achieving 97.8% of decaffeination. Still considering black tea leaves, pressures between 15 and 45 MPa have been tested for the decaffeination of such raw material [8]. Caffeine extraction was high when the pressure increased up to approximately 30 MPa. Thereafter, the pressure presented a negative influence on extraction yield.

Not only coffee and teas are decaffeinated by SFE-CO_2, but also cocoa beans [13]. In such case, the authors evaluated the influence of 20 and 40 MPa on the extraction of caffeine and theobromine. The initial amount of caffeine in cocoa beans was 0.26 wt.%. The supercritical CO_2 was also compared with supercritical ethane. In all decaffeination conditions, CO_2 provided higher yields than supercritical ethane (C_2H_6) did, which justifies its preference in most of the processes. When the levels of pressure were compared at 70°C, SFE-CO_2 at 40 MPa yielded 66.4 wt.% of decaffeination while SFE-CO_2 at 20 MPa yielded 37.4 wt.%.

Based on this context, we infer that pressures near to 20–40 MPa are recommendable to remove caffeine from caffeinated samples. In addition, saturating the caffeinated samples with water is important if large extraction rates are aimed at obtaining by supercritical technology. In such conditions, extraction rates increase as a consequence of the high partition coefficient for caffeine between CO_2 and water.

4.3.3 Time

The extraction time is one of the parameters that ultimately determine the capacity the SFE-CO_2 plant to process different quantities of biomass, including the caffeinated ones. The extraction time depends on the raw material properties (e.g., physiology) and the CO_2 mass flow rate (Table 3). Overall, a decaffeination process can follow some steps during the extraction time. The steps are comprised of an initial constant extraction rate period, an intermediary falling extraction rate period that diffusion and convection of CO_2 predominate, and a final diffusion-controlled rate period. These steps are also presented in the extraction of other bioactive components, as terpenoids and tocotrienols [14, 15]. Overall, the major modifications in the caffeinated samples occur at the beginning of the extractions. Park et al. [16] reported that the extraction of caffeine was larger at the beginning of SFE-CO_2 process. More than 75.7, 93.9, and 97.8 wt.% of caffeine were recovered after 40, 80, and 120 min, respectively. Consequently, more time and CO_2 were needed to extract the remained solutes as a consequence of the internal mass transfer resistance.

The decaffeination of black tea was also reported [4]. The increase of time and the solvent mass flow rate presented a linear effect on the decaffeination yield, while higher values of extraction time and modifier (cosolvent) concentration presented a logarithmic influence on the decaffeination yield. This behavior indicates an interaction between the extraction time and these other parameters (e.g., modifier concentration and mass flow rate).

Obviously, the decaffeination rates increase with the increase of extraction time. However, longer times are not feasible because they indicate a high CO_2 consumption and a high cost of manufacturing (COM) [17]. In the decaffeination of coffee and teas, it is common to perform the process during 60–180 min. Notoriously, there are extraction times longer than 600 min. In fact, the extraction time also depends on the operating mode, which the batch process is generally more prolonged. Furthermore, optimization of temperature, pressure, flow rate or solvent to feed mass ratio, and cosolvent concentrations in an integrated way shortens the extraction time and the efforts spent in the SFE runs.

4.3.4 Solvent to feed mass ratio

Solvent to feed mass ratio is a process parameter straightly associated with the extraction time. This parameter is used instead of CO_2 flow rate because it is a dimensionless parameter. Even though the CO_2 flow rate depends on the size of the equipment (laboratory, pilot, or industrial scales), the solvent to feed mass ratio does not depend on the size and can be used for scale up purposes.

Generally, the increase of the solvent to feed mass ratio increases the mass ratio of solute (e.g., caffeine) in the supercritical phase at the vessel exit up to a maximum value and decreases afterward. This behavior is associated to the solubility of caffeine in the supercritical CO_2. The solubility increases at the beginning of extraction time as a consequence of the initial constant extraction rate period and decreases afterward as a consequence of the intermediary falling rate period and the final diffusion-controlled period. The thermodynamic equilibrium is not a goal in the SFE-CO_2 because the solvent saturation would never be reached under any practical industrial application. Indeed, the technological and economic viability analyses would define the most appropriate value of solvent to feed mass ratio to be applied in the decaffeination of raw materials. Typically, this ratio is close to the end of the constant extraction rate period. In some cases, this ratio can be near the end of falling rate period as well. However, if more than 99% of decaffeination is needed, this ratio could be higher, which the extraction can be in the diffusion-controlled period.

It is important to remember that caffeine is linked to phenols. Therefore, removing the total content of caffeine can lead to reduced levels of phenols poor taste (low-quality products). Generally, decaffeinated products are not completely caffeine-free. Such products are called decaffeinated products, even they can contain traces or low concentrations of caffeine (generally $<3\%$ from the initial fresh mass content).

When processing cocoa beans [13], approximately 60 wt.% of decaffeination was reached when using 100 kg CO_2/kg cocoa beans. When using 700 kg CO_2/kg cocoa beans, the decaffeination increased only to 66.4 wt.%. The findings presented in that work highlight the previous inference that it is preferable to operate an SFE-CO_2 process near to the end of the constant extraction rate or falling rate periods instead of the end of the diffusion-controlled period.

When processing black tea [4], 97.2 wt.% of decaffeination was reached when using approximately 600 kg CO_2/kg leaves. When using approximately 1000 kg CO_2/kg leaves, the decaffeination was 98.1 wt.%. We can see that the ratios are relatively higher. The CO_2 is recycled and can be used again in the extractions. However, too high ratios can cause some operational troubles or can demand high-pressure pumps with extremely high capacities of flow rate. Indeed, these limitations can impact high investment and operational costs. For example, if an extraction vessel is filled with 500 kg of black tea leaves and the extraction time is 180 min, the CO_2 flow rate should be approximately 1667 kg/min to satisfy the ratio of 600 kg CO_2/kg leaves. Therefore, for a practical application, all process parameters (temperature, pressure, cosolvent, etc.) should be systematically optimized to avoid using too high solvent to feed mass ratios.

4.3.5 Cosolvent type and concentration

Regarding cosolvent type and concentration, a cosolvent is used together with CO_2 to improve the solubility of compounds as caffeine (a polar compound). Supercritical CO_2 is properly used to recover nonpolar substances. Cosolvents, also known as modifiers, can increase the solubility of target compounds in the mixture. The modifiers used in SFE-CO_2 process have specific properties that are indicated for specific substances (e.g., caffeine) or a class of substances (e.g., alkaloids). In a sustainable concept and in order to maintain SFE-CO_2 as a clean technology, the cosolvent type (or cosolvents) should be carefully chosen. In the decaffeination, the water and ethanol are recommended. If possible, methanol should be avoided.

In the decaffeination using supercritical CO_2, water is the main modifier used in the extractions. In fact, water is not necessarily pumped together with CO_2. In most of the cases, the solid biomass is soaked with water before being loaded in the extraction vessels to have a moistened biomass. This procedure was performed by Joshi et al. [5], where 900 g Kangra orthodox tea was soaked with 100 mL water and loaded in a vessel of 1 L. The objective of using water as cosolvent stood for improving the solubility of caffeine, according to the behavior presented in Table 1. Since water increases the solubility of caffeine, its single use (without CO_2) is not recommendable because it removes soluble solids (aromatic compounds) and secondary metabolites, which is a disadvantage.

Water as cosolvent was also used in the supercritical CO_2 decaffeination of unroasted coffee beans [18]. Water was used to moisten unroasted beans to a moisture content of 50% and extracted by SFE-CO_2 in semicontinuous mode at 30 MPa and 150°C until reaching a final caffeine level of 0.16 wt.%. Another work was performed to evaluate the extraction conditions of caffeine from black tea while maintaining the maximum content of phenolics [8]. In such study, a cosolvent mixture of ethanol-water was added to supercritical CO_2 at different ratios applying a high-pressure pump and all assays were performed for 1 h. The mixture ethanol-water (cosolvent) was evaluated in the following mass proportions (%): 75:25, 87.5:12.5, and 100:0. The following cosolvent volumetric concentrations (based on CO_2) were used: 0.25%, 0.375%, and 0.5%. The authors [8] reported that the extraction with a mixture of 87.5% ethanol and 12.5% water (87.5:12.5) at a cosolvent concentration of 0.25% removed almost all caffeine from black tea. The caffeine was reduced from 1283 mg/100 g leaves to <1 mg/100 g leaves. Therefore, the ethanol (cosolvent) concentration had significant influences on both caffeine and phenolics (P value $< .01$). In fact, the recovery yield of caffeine was low when the ethanol concentration was high. Otherwise, reducing the ethanol content in ethanol-water mixture had also decreased the content of phenolics in the leaves. This behavior is attributed to water, which is a better cosolvent for caffeine, but also removes other phenolic substances.

From the foregoing facts, the decaffeination using supercritical CO_2 is preferable to be performed with a certain content of modifiers (cosolvents). Even a low concentration (e.g., 0.25%) can be very suitable because the polarity of the mixture is changed to favor caffeine solubilization. The effects of cosolvents (types and concentration) are attributed to a specific chemical (e.g., acid-base interactions) or physical interactions (e.g., dipole-dipole interactions) between the modifier and target solute, and also to a likely solvent-modifier interaction, which can influence the solvent-solute interactions. Overall, such effects can accelerate the extractions and can make the processes more feasible.

4.4 Economic approach

The costs related to a processing plant operated by SFE-CO_2 should be estimated before the economic viability of the supercritical process can be assessed [7]. The variables that compose the COM are fixed costs, direct costs (related to manufacturing), and general expenses. The main fixed costs include the costs of purchasing the equipment, acquiring the accessories, and building the factory. Depreciation, local taxes, and insurance also compose fixed costs. The main direct manufacturing costs include the costs of raw materials (biomass, solvent, cosolvent, etc.), waste treatment, operational labor, and utilities (electricity, water for washing, cleaning material, etc.). In the same trend, the main general expenses refer to security, yard improvement, research and development, distribution, and administration costs, among others. All these costs are extended to decaffeination factories.

The people who are consuming totally decaffeinated or partially decaffeinated products generally are looking for quality and healthy products. Therefore, they do not hesitate in acquiring a more expensive product. Decaffeinated products are a few more expensive because there are further processing steps to reach the required characteristics. The decaffeination processes, especially the SFE-CO_2, needs energy and robust equipment to reduce the concentration of caffeine. Even though the CO_2 can be recycled, a percent fraction (generally 2%–5%) is lost during the recirculation, which is a considerable fraction when larger scales are considered in the evaluation. In addition, the increase in the solvent to feed mass ratio increases the mass flow rates. Therefore, as aforementioned, the capacities of high-pressure pumps should be increased. Clearly, the increase in capacities leads to an increase in the total production costs.

Besides being technologically feasible, a supercritical CO_2 process should evidence economic viability. Regarding the subject overviewed in this chapter, some optimization studies were disclosed in terms of technological responses (e.g., decaffeination yield). However, to the far of our knowledge, we could not find a study in the main database (e.g., Scopus) that evaluated systematically the production costs of decaffeinated products. Overall, based on the experience, we can infer that decaffeination of several solid matrices (tea leaves, coffee beans, guarana seeds, cocoa beans, yerba mate leaves, etc.) by SFE-CO_2 can be economically feasible because the investment cost is not the main cost that influences the total COM. A decaffeination plant using supercritical CO_2 generally is commercially feasible at larger capacities, such as 5000 tons/year. Of course, supercritical equipment is more expensive than conventional equipment. However, the contribution of this cost rarely exceeds 30% of the total cost because the equipment is built to last for more than 10 years without needing significant maintenance. In fact, the main cost refers to raw materials, such as biomass and solvents. Consequently, the main alternatives are to reduce the purchasing costs of raw materials and to avoid losses of solvents during the processes.

4.5 Life-cycle assessment approach

SFE-CO_2 is considered as eco-friendly technology. Even having this feature, it is necessary to evaluate the environmental emissions of SFE-CO_2 during the decaffeination. One form to assess the environmental impact of a process in a quantitative description is using life-cycle assessment (LCA) analysis [19].

The LCA approach has been applied to the decaffeination of coffee with a 60/40 Arabica/Robusta mass blend [19]. The beans were moistened and the decaffeination process consisted of SFE-CO_2 at 90°C and 25 MPa. During the process, the caffeine + CO_2 mixture was accumulated in a holding tank until completing the procedure. Thereafter, the vessel started a downtime. A 97% of decaffeination rate was obtained for 11.5 h of extraction. Regarding the LCA approach, its goal was to evaluate the environmental impacts of decaffeination of coffee using CO_2. After an in-depth description and assessment of impacts, the main real factors that contribute to the environmental impact are agricultural stages (use of fertilizers), transportation, and caffeine extraction steps.

When emphasizing the decaffeination step, an improved and suggested solution to this impact refers to replacing a portion of the electricity by sustainable photovoltaic panels. Considering a solution for these three impacts (agricultural stages—reduce fertilizers; transportation—use renewable fuels; caffeine extraction steps—use alternative sources of energy), a reduction of the environmental impact can be 17.6% in terms of human health, 16.1% in terms of resource availability, and 10.3% in terms of ecosystem diversity.

In another work [20], an LCA analysis was also performed for the decaffeination of coffee beans by supercritical technology. Data were obtained from an industrial company, which uses supercritical CO_2 to recover caffeine from coffee beans at 90°C and 25 MPa for 5 h with energy, CO_2, water, and fuel supply. Transportation is one of the main factors that impact negatively the environment due to the high consumption of fuels. Regarding the decaffeination process, its impact on the nonrenewable energy consumption (e.g., electricity from thermoelectric plants) is related to the higher consumption of energy to heat and pressurize the systems. Even though, SFE-CO_2 is still very positive if compared to other methods because it does not have a big drawback (Fig. 4).

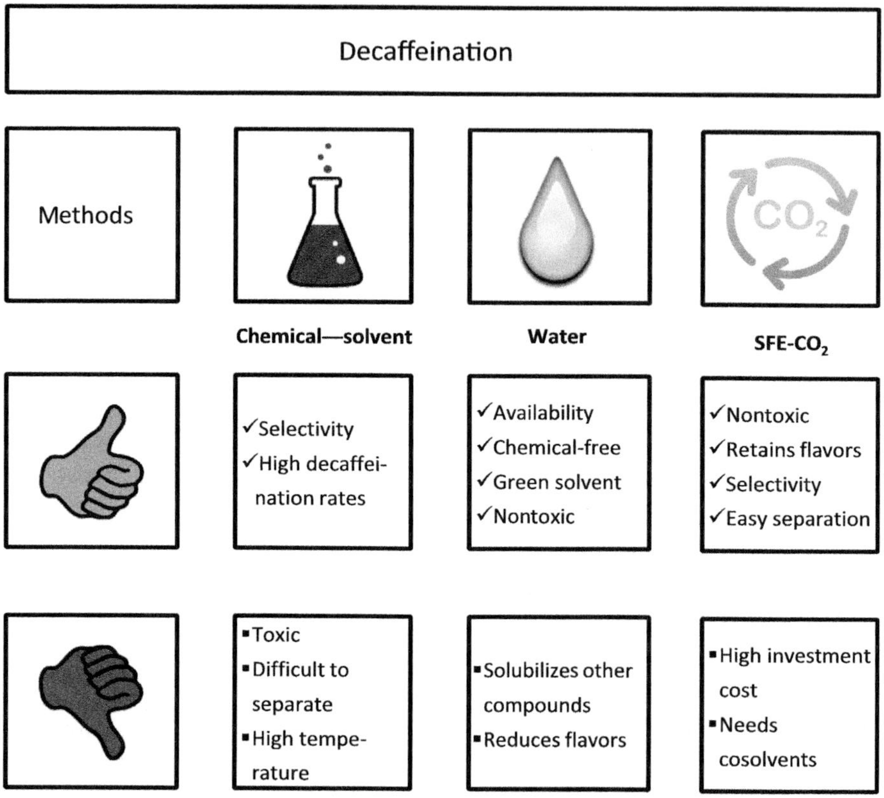

Fig. 4 Simply versed descriptions of the main decaffeination methods, including pros and cons of each method.

5. Decaffeination of coffee

Coffee is a beverage consumed worldwide because of its target bitter taste. The effect of coffee on human physiology is different from person to person (age, gender, etc.) and also on the quantity and quality of coffee consumed. Coffee beans have different contents of caffeine depending on the species. For example, arabica coffee contains approximately 1.2 wt.% while canephora (robusta) coffee and the hybrid arabusta coffee contain approximately 2.2 wt.% and 1.7 wt.%, respectively. Overall, agricultural factors demonstrated to have little effect on caffeine content in coffee beans [21].

Decaffeinated coffee is a product which caffeine has been extracted from the beans, corresponding to roasted and ground coffee. The decaffeination process is usually fulfilled on beans prior to roasting. For avoiding extraction of aromatic compounds, the extraction process is commonly fulfilled with green beans. For example, roasted beans have up to 2% (mass basis) of caffeine. The decaffeination is carried out to remove approximately 97% of the caffeine to meet the requirements of a decaffeinated product. Therefore, for instance, a cup of 150 mL of decaffeinated coffee has approximately 5 mg or less of caffeine. Otherwise, a cup of 150 mL of regular caffeinated coffee generally contains from 60 to 180 mg of caffeine [11].

For most of people, a fatal dose of caffeine is approximately 10 g or approximately 50–100 cups of regular coffee (depending on volume and strength). This excessive consumption can cause excitement, restlessness, insomnia, and mild delirium, demonstrating a hard influence on the respiratory and central nervous systems. Most of people can suffer from sleepiness (most common), nervousness, heart stimulation, and other effects after drinking larger quantities of caffeine [21]. Therefore, since genetically decaffeinated coffee beans are not commercially available (i.e., ongoing scientific studies), the decaffeination of coffee is a positive alternative to provide a decaffeinated product for coffee addicted people.

Decaffeination of coffee by SFE-CO_2 is the most promising process. The removal of caffeine can be done in batch, semicontinuous (or semibatch), or continuous modes (Figs. 1–3). Typically, the semicontinuous mode is preferable. Previously to the extractions, the coffee beans are generally moistened (20%–50%). Cosolvents as ethanol, water or a mixture of them can be used as polarity modifiers.

Decaffeination of coffee beans is a subject of several patents since the end of the last century [22–24]. The SFE-CO_2 was patented for the first time by Zosel in the mid-1970 for removing the caffeine from green coffee beans [25]. For instance, an invention reports a method for the decaffeination of green coffee beans without affecting in an adverse manner the acidity or the taste and flavor of coffee beverages [26]. According to the inventors, prior to or simultaneously with decaffeination process, green coffee beans can be contacted with an acidic fluid formed by an organic acid (preferably an aqueous citric acid solution) that is not soluble in CO_2. Considering that citric acid can be dissociated in aqueous solution and that it can be inexpensively obtained in food grades for decaffeination purposes, such acid is favorite to the decaffeination. Furthermore, according to the invention, the process conditions of SFE-CO_2 are extraction at 90°C and 28 MPa for 8 h and separation using 30–40°C and 7–8 MPa.

When using water or an aqueous solution, coffee beans can swell, the pores usually open, and the caffeine can diffuse out of the beans. Consequently, CO_2 at supercritical condition flowing in the bed solubilizes caffeine. After the extraction time, the beans can be unloaded from the vessel for drying and for further use. The key factor of operation is to optimize the extraction conditions, especially regarding the parameters discussed in Section 4. Therefore, the operators and assistants should have the know-how for controlling the high-pressure facilities of decaffeination. Also, an excellent technological infrastructure for maintaining the plant is needed. Although the supercritical process has a large complexity than the processes fulfilled with some liquid solvents, the clean and positive physicochemical characteristics of supercritical CO_2 expand its acceptance by consumers. As a consequence of the high selectivity, it is asserted that decaffeinated products obtained by SFE-CO_2 have a better physicochemical and sensorial quality.

Overall, the main advantages of decaffeination of coffee by SFE-CO_2 are informed in the sequence. Firstly, there is no harmful residue in the decaffeinated coffee because CO_2 becomes a gas at room conditions (e.g., 25°C and 0.1 MPa) and naturally separates from the solids after the expansion. There are no problems if some traces of CO_2 remain in the products because it is nontoxic and safe at low concentrations. Secondly, decaffeinated coffee using supercritical CO_2 has superior quality than the products obtained from other decaffeination methods with appearance and flavor similar to that of caffeinated coffee. Thirdly, the extraction of other compounds (phenols, amino acids, etc.) is quite low, resulting in a greater product.

6. Decaffeination of tea

Recent innovations in food-related areas have improved the use of novel technologies for satisfying the increased consumers' requirements. Low-fat products, low-sugar products, low-sodium additives, and more fibrous ingredients, among others, are examples of healthy food products. In this trend, decaffeinated teas are also included as differentiated products. Tea is a beverage obtained from the shoots and leaves of *Camellia sinensis* plants. Decaffeinated tea products have a high commercial market as a consequence their interesting healthy benefits.

Consequently, to obtain a decaffeinated tea, the leaves should undergo an extraction procedure. As in coffee decaffeination, SFE-CO_2 is preferable because it yields decaffeinated teas with the retained content of taste and flavors. The supercritical CO_2 can penetrate into the swollen pores of leaves, thus solubilizing caffeine. It is not recommendable to use a high concentration of water or other polar cosolvents because some phenolic compounds can be solubilized in the mixture, which is an undesired effect.

Regarding some reports in the scientific literature, a black tea with particle size ranging from 600 to 780 μm was processed in a pilot-scale supercritical plant with two vessels, each one having 10 L [4]. Extraction time, pressure, solvent to feed mass ratio and modifier quantity (as process parameters) had a high impact on tea decaffeination yield. From an initial content of 2.16 wt.% caffeine, a 100% decaffeination was reached at 37.5 MPa, 62.5°C, 300 min, 2.5% (molar basis) of ethanol as molar concentration and solvent to feed mass ratio of approximately 1512 kg/kg tea leaves (Table 3). Considering that supercritical CO_2 was utilized as a solvent, there was no need to remove CO_2 from decaffeinated tea because it separates naturally from the extract at low pressures. The authors also mention that it was important to attain maximum decaffeination at a relatively low temperature (62.5°C) in a short time (300 min) in terms of applicability of the results at industrial scale.

Black tea has also been studied for obtaining a decaffeinated product at a pilot-scale supercritical plant with two extractors of 5 L [5]. However, in that study, the maximum rate of decaffeination was only 40.4 wt.%. The findings of low decaffeination yield are attributed to the lower extraction time (1.5 h) and solvent to feed mass ratio (6.0 kg/kg tea leaves) if compared to the findings reported by Ilgaz et al. [4]. The temperature was higher (85°C), which could have influenced the composition of aldehydes, ketones, alcohols, esters, and other compounds. For example, in the decaffeination by SFE-CO_2 at 85°C and 17 MPa, the concentration of α-pinene reduced to 0.10 wt.% if compared to the fresh black tea (0.33 wt.%) [5].

Green tea was decaffeinated by SFE-CO_2 in a laboratory system containing a 100 mL extraction vessel [27]. The highest decaffeination yield was 82.4 wt.% at 70°C and 20 MPa. Ethanol (3%, mass basis) was used as a cosolvent. When either the pressure or temperature was maintained constant, the decaffeination was high when the quantity of modifier was high. In fact, among the parameters evaluated, the modifier concentration had the highest influence on the decaffeination efficiency. However, for a processing line at industrial scale, further processes such as recovering chlorophylls and using them in the reconstitution of chlorophyll contents in decaffeinated green tea can be needed to improve its quality.

7. Decaffeination using carbon dioxide at industrial scale

In 1903, Ludwig Roselius proposed the first commercially successful decaffeination process. In such invention, coffee beans were initially steamed with a brine solution and the caffeine was extracted with benzene. Nowadays, benzene cannot be used as a solvent in decaffeination process because it is toxic. Decaffeination using CO_2 has been proposed later by a scientist of the Max Plank Institute (Kurt Zosel) in mid-1970 [25]. Currently, the SFE-CO_2 at industrial scale is performed in high-volumes reactors, generally higher than 1000 L. The process is done at large scale because the demand for decaffeinated products is high and there are several tons of raw materials to be processed per day around the world. Overall, this process is primarily used to decaffeinate large amount of commercial-grade and less-exotic coffee found in grocery stores. It is important to highlight that supercritical processes for decaffeination of coffee and tea were one of the first commercialized applications of supercritical technology at the large industrial scale.

Some companies around the world are using supercritical CO_2 for decaffeination of caffeinated biomass. One example is the Brazilian Canadian Coffee Co. Ltd. (note that we do not have the intention of marketing any company) [28]. This company produces decaffeinated coffee by SFE-CO_2. Although the processing conditions and the total capacities are not disclosed, the process incorporates only water (to moisturize the coffee beans and thus making the caffeine accessible for removal) and CO_2 to remove the caffeine and to leave the flavor. Once the coffee is 99.9% caffeine free, it is dried and ready for roasting.

Another company is Queen Bean Coffee Company [29]. Colombian CO_2 decaf, an "Especiale" bean, is shade-grown in Colombia's Bucaramanga region. The "Especiale" beans are naturally 99.5 wt.% decaffeinated by a patented process using supercritical CO_2 that permits the beans to retain higher amount of taste and aroma than most decaffeination processes. According to the company, these beans can be softer and more delicate than other decaffeinated coffees.

Regarding decaffeinated tea, the Hälssen and Lyon Company [30] offers different decaffeination processes, including SFE-CO_2. The company has many years (since 1935) of experience with the decaffeination of tea. The decaffeinated teas by SFE-CO_2 are produced more recently, which the process is suitable for all leaf grades and teas, including organic teas. Once the products have been decaffeinated, they are dried back to the original 5%–12% water content.

Although manufacturing processes may slightly differ from one factory to another, the main steps aforementioned and presented in Figs. 1–3 (especially the semicontinuous mode) are followed. Overall, the caffeine-rich CO_2 stream that exits the extraction vessels is either directed through a bed of activated charcoal (solid adsorption) or through a water bath tower (gas absorption) to absorb the caffeine. At industrial scale, the CO_2 should be recycled back to the vessel. Indeed, decaffeination by SFE-CO_2 is capital-cost intensive, but it usually provides higher yields. Interestingly, the extracted caffeine is further processed and sold to food-related industries.

8. Future outlooks

In the whole text, we have discussed the characteristics of SFE-CO_2 in order to obtain decaffeinated products able to be consumed after the preparation of energy drinks, chocolate drinks and teas, among others. Supercritical CO_2 is a naturally occurring nontoxic solvent, which justifies its preference in decaffeination processes. Indeed, any decaffeination process involves some challenges, especially related to avoiding the loss of aroma and flavor since coffee has more than 1500 compounds. Therefore, as future outlooks in this field, it is expected to have more decaffeinated products (coffee, tea, chocolate, soda, etc.) with and increased similarity to the caffeinated ones in terms of taste and flavor. As people are becoming more used to quality coffee and tea, this is forcing coffee-making and tea-making companies to find ways of enhancing the flavor even in decaffeinated products.

Regarding the technological perspectives, it is expected to have a complete decaffeination using lower solvent to feed mass ratios. Consequently, it can result in smaller extraction times and lower CO_2 flow rates, which can reduce the operating costs. Also, it is expected to maintain most of the aromatic and bioactive compounds, as phenols, ketones, acids (e.g., chlorogenic acid), and aldehydes, among others. Although there are several studies on the chemical composition of whole coffee and teas, little is still known about the composition of decaffeinated coffee and teas. Therefore, another outlook refers to the development of systematic studies to disclose the complete composition of decaffeinated products.

Acknowledgments

The author thanks the Coordination for the Improvement of Higher Education Personnel (CAPES), National Council of Technological and Scientific Development (CNPq), and the Research Support Foundation of the State of Rio Grande do Sul (FAPERGS; 16/2551-0000522-2; 17/2551-0000 893-6) for the financial support. G.L. Zabot thank CNPq for the productivity grant (304882/2018-6).

References

[1] A. Beri, P.K. Banipal, T.S. Banipal, Effect of potassium chloride on the solvation behavior of caffeine, theophylline and theobromine: Volumetric, viscometric, calorimetry and spectroscopic approach, Food Chem. 266 (2018) 110–118.

[2] E. Lack, H. Seidlitz, Economics of high pressure processes, in: A. Bertucco, G. Vetter (Eds.), High Pressure Process Technology, Elsevier, Amsterdan, 2001.

[3] G. Brunner, Supercritical fluids: technology and application to food processing, J. Food Eng. 67 (2005) 21–33.

[4] S. Ilgaz, I.G. Sat, A. Polat, Effects of processing parameters on the caffeine extraction yield during decaffeination of black tea using pilot-scale supercritical carbon dioxide extraction technique, J. Food Sci. Technol. 55 (2018) 1407–1415.

[5] R. Joshi, G.D.K. Babu, A. Gulati, Effect of decaffeination conditions on quality parameters of Kangra orthodox black tea, Food Res. Int. 53 (2013) 693–703.

[6] M.N. Moraes, G.L. Zabot, M.A.A. Meireles, Extraction of tocotrienols from annatto seeds by a pseudo continuously operated SFE process integrated with low-pressure solvent extraction for bixin production, J. Supercrit. Fluids 96 (2015) 262–271.

[7] G.L. Zabot, M.N. Moraes, M.A.A. Meireles, Supercritical fluid extraction of bioactive compounds from botanic matrices: experimental data, process parameters and economic evaluation, Recent Pat. Eng. 6 (2012) 182–206.

[8] B. Bahar, E. Pelvan, I. Hasbay, C. Alasalvar, Decaffeinated black tea: process optimization and phenolic profiles, J. Supercrit. Fluids 82 (2013) 116–121.

[9] F. Belliardo, A. Martelli, M.G. Valle, HPLC determination of caffeine and theophylline in *Paullinia cupana* Kunth (guarana) and Cola spp. samples, Z. Lebensm. Unters. Forsch. 180 (1985) 398–401.

[10] M.D.A. Saldaña, C. Zetzl, R.S. Mohamed, G. Brunner, Decaffeination of guarana seeds in a microextraction column using water-saturated CO_2, J. Supercrit. Fluids 22 (2002) 119–127.

[11] A.S. Franca, Coffee: Decaffeination, in: B. Caballero, P. Finglas, F. Toldra (Eds.), Encyclopedia of Food and Health, Academic Press, Cambridge, 2016.

[12] H. Peker, M.P. Srinivasan, J.M. Smith, B.J. McCoy, Caffeine extraction rates from supercritical carbon coffee beans with dioxide, AICHE J. 38 (1992) 761–770.

[13] R.S. Mohamed, M.D.A. Saldaña, P. Mazzafera, C. Zetzl, G. Brunner, Extraction of caffeine, theobromine, and cocoa butter from brazilian cocoa beans using supercritical CO_2 and ethane, Ind. Eng. Chem. Res. 41 (2002) 6751–6758.

[14] G.L. Zabot, M.N. Moraes, M.A.A. Meireles, Process integration for producing tocotrienols-rich oil and bixin-rich extract from annatto seeds: a techno-economic approach, Food Bioprod. Process. 109 (2018) 122–138.

[15] P.I.N. Carvalho, J.F. Osorio-Tobón, G.L. Zabot, M.A.A. Meireles, Spatial and temporal temperature distributions in fixed beds undergoing supercritical fluid extraction, Innovative Food Sci. Emerg. Technol. 47 (2018) 504–516.

[16] H.S. Park, H.-K. Choi, S.J. Lee, K.W. Park, S.-G. Choi, K.H. Kim, Effect of mass transfer on the removal of caffeine from green tea by supercritical carbon dioxide, J. Supercrit. Fluids 42 (2007) 205–211.

[17] J.C.F. Johner, T. Hatami, G.L. Zabot, M.A.A. Meireles, Kinetic behavior and economic evaluation of supercritical fluid extraction of oil from pequi (*Caryocar brasiliense*) for various grinding times and solvent flow rates, J. Supercrit. Fluids 140 (2018) 188–195.

[18] Y. Chen, P.H. Brown, K. Hu, R.M. Black, R.L. Prior, B. Ou, Y.F. Chu, Supercritical CO_2 decaffeination of unroasted coffee beans produces melanoidins with distinct NF-κB inhibitory activity, J. Food Sci. 76 (2011) 182–186.
[19] I.D. Marco, S. Riemma, R. Iannone, Life cycle assessment of supercritical CO_2 extraction of caffeine from coffee beans, J. Supercrit. Fluids 133 (2018) 393–400.
[20] I.D. Marco, S. Riemma, R. Iannone, Supercritical carbon dioxide decaffeination process: a life cycle assessment study, Chem. Eng. Trans. 57 (2017) 1699–1704.
[21] K. Ramalakshmi, B. Raghavan, Caffeine in coffee: its removal. why and how? Crit. Rev. Food Sci. Nutr. 39 (1999) 441–456.
[22] S.N. Katz, J.E. Spence, M.J. O'Brien, R.H. Skiff, G.J. Vogel, R. Prasad, Decaffeination of coffee, 1991. Patent number EP0 424 579 A1.
[23] M. Gehring, O. Vitzthum, H. Wienges, Decaffeination of roast coffee, 1984. Patent number DE 33 03 679 A1.
[24] S.N. Katz, Decaffeination Process, 1984. Patent number US 4 472 442.
[25] K. Zosel, Process for recovering caffeine, 1974. Patent number US 3806619.
[26] P.T. Kazlas, R.D. Novak, R.J. Robey, Supercritical carbon dioxide decaffeination of acidified coffee, 1994. Patent number US5288511A.
[27] H.S. Park, N.G. Im, K.H. Kim, Extraction behaviors of caffeine and chlorophylls in supercritical decaffeination of green tea leaves, Food Sci. Technol. 45 (2012) 73–78.
[28] Braziliancoffeeco, Decaffeination Process, Available from: http://braziliancoffeeco.com, 2018.
[29] Queenbean, Natural Decaffeination: Swiss Water and CO_2 Processes, Available from: https://www.thequeenbean.com, 2018.
[30] Haelssen-lyon, Decaffeinated tea, Available from: https://www.haelssen-lyon.com, 2018.
[31] P.V. Gadkari, M. Balaraman, Mass transfer and kinetic modelling of supercritical CO_2 extraction of fresh tea leaves (*Camellia sinensis* L.), Braz. J. Chem. Eng. 34 (2017) 799–810.
[32] V.A. Golubev, M.Y. Nikiforov, G.A. Alper, Solubility of caffeine in the supercritical CO_2-methanol binary solvent, Russ. J. Phys. Chem. 9 (2014) 1054–1058.
[33] B. Li, W. Guo, W. Song, E.D. Ramsey, Interfacing supercritical fluid solubility apparatus with supercritical fluid chromatography operated with and without on-line post-column derivatization: determining the solubility of caffeine and monensin in supercritical carbon dioxide, J. Supercrit. Fluids 115 (2016) 17–25.
[34] Y. Iwai, H. Nagano, G.S. Lee, M. Uno, Y. Arai, Measurement of entrainer effects of water and ethanol on solubility of caffeine in supercritical carbon dioxide by FT-IR spectroscopy, J. Supercrit. Fluids 38 (2006) 312–318.
[35] P.V. Gadkari, M. Balaraman, Solubility of caffeine from green tea in supercritical CO_2: a theoretical and empirical approach, J. Food Sci. Technol. 52 (2015) 8004–8013.
[36] H. Lentz, M. Gehrig, J. Schulmeyer, Dynamic solubility measurements of caffeine in carbon dioxide and in carbon dioxide saturated with water, Phys. B+C 139-140 (1986) 70–72.
[37] U. Kopcak, R.S. Mohamed, Caffeine solubility in supercritical carbon dioxide/co-solvent mixtures, J. Supercrit. Fluids 34 (2005) 209–214.
[38] W.-J. Kim, J.-D. Kim, J. Kim, S.-G. Oh, Y.-W. Lee, Selective caffeine removal from green tea using supercritical carbon dioxide extraction, J. Food Eng. 89 (2008) 303–309.
[39] E. Cassel, R.M.F. Vargas, G.W. Brun, D.E. Almeida, L. Cogoi, G. Ferraro, R. Filip, Supercritical fluid extraction of alkaloids from *Ilex paraguariensis* St. Hil, J. Food Eng. 100 (2010) 656–661.

CHAPTER 12

Supercritical fluids for the extraction of oleoresins and plant phenolics

Wan-Jun Lee, Norhidayah Suleiman, Noor Hadzuin Nik Hadzir, Gun-Hean Chong
Faculty of Food Science and Technology, Universiti Putra Malaysia, Serdang, Selangor, Malaysia

Contents

List of abbreviations

Ace acetone
ACN acetonitrile
AcOH acetic acid
ar-TUR aromatic-turmerone
CA carnosic acid
CL carnosol
CO_2 carbon dioxide
CUR curcumin
CU Curcuminoids
DAD diode array detection
EoS equation of state
ETAC ethyl acetate

Green Sustainable Process for Chemical and Environmental Engineering and Science
https://doi.org/10.1016/B978-0-12-817388-6.00012-X

EtOH	ethanol
FA	formic acid
FR	carbon dioxide flow rate
GC	gas chromatography
H	hydrogen
HHP	high hydrostatic pressure
H_2O	water
HPLC	high-performance liquid chromatography
IPA	isopropyl alcohol
MAE	microwave-assisted extraction
MeOH	methanol
MS	mass spectrometer
NIST	National Institute of Standards and Technology
NR	not reported
PET	petroleum ether
PLE	pressurized liquid extraction
RSLDE	rapid solid-liquid dynamic extraction
SAF	supercritical antisolvent fractionation
SAFT-EoS	statistical associating fluid theory equation of state
SCF	supercritical fluid
$scCO_2$	supercritical carbon dioxide
SFE	supercritical fluid extraction
T	temperature
t	time
TBME	*t*-butyl methyl ether
TPC	total phenolic content;
UAE	ultrasound-assisted extraction
UMB	umbelliferon
UP	ultrasonic power

1. Introduction

The world is moving toward the use of materials extracted from natural sources in order to replace the synthetically produced ingredients such as bioactive components, coloring agent, flavoring agent, etc. Plants are the richest known source for the extraction of these valuable components and the extract is often in the form of oleoresins containing targeted groups of phenolic compounds. There are several reported methods used for the extraction of oleoresins and plant phenolics. Among these, supercritical fluid extraction (SFE) is one of the most studied techniques. One of the major motivations behind the application of supercritical fluid (SCF) in oleoresin and plant phenolic extraction is due to the implementation of the green chemistry principles at the beginning of the 1990s. The 12 principles of the green chemistry serve as the guidelines to maximize efficiency and minimize the negative impacts of chemistry research and industry on human health and environment and these requirements can be achieved by employing supercritical fluid (SCF) extraction technologies. Although there are other types of SCFs, the supercritical carbon dioxide ($scCO_2$) is usually preferred in oleoresin and plant phenolic extractions due to the low critical temperature and pressure (31.1°C, 7.39 MPa) and high purity grade at relatively low cost. Also, $scCO_2$ is nontoxic, noncorrosive, nonflammable, and listed as a substance that is generally recognized as safe by the food and drug administration. Hence, the use of $scCO_2$ is in fact, environmental-friendly and safe for human consumption complying with the green chemistry principles. Besides, SFE is a very useful technique for extraction since the extraction conditions including temperature, pressure, carbon dioxide (CO_2) flow rate, time, and the addition of cosolvent can be modified in order to allow the variation in density and viscosity, hence, changing the solvent power of CO_2 to achieve a selective extraction process. The short extraction time, high selectivity of the $scCO_2$, and easy removal of the gas at the end of the extraction process allow for the high recovery of oleoresins and plant phenolics, which contributed to the high efficacy of the SFE.

2. Oleoresins and plant phenolics

2.1 Types of oleoresins

Being one of the richest sources of bioactive components, oleoresins find a wide application in the food and pharmaceutical industries. Oleoresins consists of two fractions, (1) volatile oil and (2) nonvolatile components (resins). The resins are made up of waxes, lipophilic components, color, etc. The oleoresins can be extracted from different parts of the plant, such as the bulb, bark, fruits, seeds, roots, and rhizomes. Some examples of oleoresins extracted from the different plant parts and their characteristics are listed in Table 1.

Oleoresins can be a replacement for ground spices or herbs for the following reasons:

- oleoresins exhibit similar aroma and flavor properties as spices and herbs;
- the concentrated form allows for ease of handling, transportation, and storage;
- spice oleoresins are available throughout the year compared to some spices which are seasonal;
- longer shelf life (minimal oxidative degradation, minimal loss of flavor, and free of microorganism) due to the concentrated form and they are virtually moisture-free;
- highly coveted as flavoring agents in the food processing industry as they are more economical regarding dosage and are able to replace large quantities of traditional raw spice;
- better quality consistency with a standardized taste and aroma that can be tailored as per the requirement of the product.

Oleoresins have been extracted from fruits, spices, herbs, and some from the agricultural and food industries by-products (agri-food-by-product). Extractions from the agri-food-by-products are being encouraged due to the availability of the material in large quantities since there is no competition with the food industries. Besides, the production of wastes such as peels, seeds, leaves, stems, roots, and bark can be minimized. Moreover,

Table 1 Various oleoresins extracted from different parts of plants and their characteristics

Parts	Oleoresins	Botanical name	Characteristics
Aril	Mace	*Myristica fragrans*	Reddish yellow, free flowing liquid with aroma of mace and a warm taste
Bark	Cinnamon	*Cinnamomum verum*	Dark yellow to dark brown liquid with flavor of cinnamon
Bud	Clove	*Syzygium aromaticum*	Yellow to brown viscous liquid with pungent taste
Bulb	Garlic	*Allium sativum*	Brown viscous liquid with pungent and spicy aroma and taste
	Onion	*Allium cepa*	Dark brown viscous liquid with a characteristic onion odor and a slight sweet taste
Flower	Marigold	Various species	Light to dark colored viscous liquid
	Pyrethrum	*Chrysanthemum cinerariifolium*	Dark yellowish-brown viscous liquid
	Saffron	*Crocus sativus*	Orange red liquid with pungent taste
Fruit	Amla	*Phyllanthus emblica*	Light greenish-brown and free-flowing liquid with a refreshing aroma
	Black pepper	*Piper nigrum*	Green viscous liquid with characteristic aroma of pepper and pungent taste
	Capsicum	*Capsicum annuum*	Dark red liquid with strong spicy taste
	Tamarind	*Tamarindus indica*	Dark brown viscous liquid with acidic-sweet flavor
	Tomato	*Solanum lycopersicum*	Dark red viscous liquid
Leaf	Basil	*Ocimum basilicum*	Yellowish green liquid with basil-minty odor
	Curry tree	*Murraya koenigii*	Greenish brown liquid, sweet and spicy with bitter tone characteristic of curry leaf
	Oregano	*Origanum vulgare*	Greenish-brown liquid with a floral odor with a slightly bitter taste
	Parsley	*Petroselinum crispum*	Dark brownish greenish liquid with a fresh green garden herb characteristic
	Rosemary	*Rosemarinus officianalis*	Light green liquid with an aroma characteristic of rosemary
	Sage	*Salvia officinalis*	Dark greenish brown viscous liquid with herb characteristic
	Thyme	*Thymus vulgaris*	Dark greenish brown viscous liquid with a herbaceous aroma, slight bitterness and a faint clovish aftertaste
	Tarragon	*Artemisia dracunculus*	Dark green viscous liquid

Table 1 Various oleoresins extracted from different parts of plants and their characteristics—cont'd

Parts	Oleoresins	Botanical name	Characteristics
Rhizome	Ginger	*Zingiber officinale*	Dark brown viscous liquid, pungent taste with characteristic aroma of ginger
	Turmeric	*Curcuma longa*	Yellowish orange viscous liquid with aroma characteristic of turmeric
Pods	Vanilla	*Vanilla planifolia*	Dark brown viscous liquid
Seed	Ajowan	*Trachyspermum ammi*	Pale green liquid with a sharp taste
	Anise	*Pimpinella anisum*	Dark brown liquid with flavor of anise
	Celery	*Apium graveolens*	Greenish yellow liquid with a characteristic aroma of celery with a bitter taste
	Cardamom	*Elettaria cardamomum*	*Dark brown liquid with a sweet-spicy, warming fragrance*
	Coriander	*Coriandrum sativum*	Dark to golden-brown liquid, sweet, aromatic flavors, with musty and citrus-like notes
	Caraway	*Carum carvi*	Greenish-brown viscous liquid with a fresh, herbaceous, carvone odor
	Cumin	*Cuminum cyminum*	Brown liquid with the aroma characteristic of cumin
	Fennel	*Foeniculum vulgare*	Dark brown liquid has the characteristic odor and flavor of sweet fennel, spicy sweet scent
	Fenugreek	*Trigonella foenum-graecum*	Dark brown viscous liquid, bitter, pungent
	Mustard	*Brassica nigra*	Pale yellow liquid, Pungent odor and is lachrymatory
	Nutmeg	*Myristica fragrans*	Yellowish red to light brown viscous liquid with pungent smell

the by-products contain health beneficial bioactive compounds such as carotenoids, tocochromanols, and phytosterols that could be extracted and utilized. The fruit processing sectors often generate great quantities of seeds which are then discarded as wastes. These fruit seeds were reported to contain a high concentration of bioactive compounds [1, 2] and some studies showed the content of bioactive compounds in the seeds was higher compared to that of fruit pulp and skin. For example, the total phenolic content of the grape seed was higher (2178.8 mg/g gallic acid equivalent) compared to the skin (374.6 mg/g) and pulp (23.8 mg/g) and higher antioxidant capacity values of 281.3 μM TEAC/g for seed followed by 12.8 μM TEAC/g for skin and 2.4 μM TEAC/g for pulp [3]. Chandra and Ramalingam [4] reported the high levels of lycopene and phenolic contents in the skin comparing to that of pulp in different tomato cultivars. Therefore, the fruit seeds and skin have a good potential to be a source for the extraction of bioactive compounds.

2.2 Conventional extraction methods for oleoresins

Recovery of oleoresins from their plant sources can be carried out by maceration with organic solvents, Soxhlet, and solvent extraction processes. These conventional extraction methods are two-step processes, which entail the use of organic solvents such as ethyl acetate, alcohols, acetone, and hexane to extract the oleoresin, followed by a solvent removal step. Reviews on the different types of solvent and solvent extraction conditions have been published by Muhammad and Dewettinck [5] and Melgar-Lalanne et al. [6] for cinnamon and capsicum oleoresins. Other recent examples include the Soxhlet extraction of basil leaves oleoresin with ethyl acetate [7] and solvent extraction of oleoresin extraction from *Aloe vera* flowers with acetone, ethyl acetate, hexane, and petroleum ether [8]. Although the aforementioned methods are commonly used, several drawbacks should be taken into consideration such as the long extraction time and more importantly, solvent residues in the final product. Accumulative consumption on the product and the exposure to the solvent during the preparation could still pose risks to the environment and human health. Constant contact with the organic solvents upturns the risk for serious medical, neurological, and neuropsychological impairments [9]. The allowable limit of different solvent residues in oleoresins that is safe for human consumption is shown in Table 2. No complete removal can be guaranteed and hence, a trace amount of chemicals will be left in the food. Chronic health effects after prolonged consumption might be inevitable. Instead of working toward more effective separation process, the use of toxic chemicals must be avoided.

2.3 Classification of plant phenolics

Phenolic compounds are secondary metabolites produced in the shikimic acid of plants and pentose phosphate through phenylpropanoid metabolization [10]. Approximately 8000 types of phenolic compounds have been identified in various vegetables, fruits, and cereals [11]. The molecular structures of phenolics range from simple molecules

Table 2 Tolerance level of various solvents in oleoresin that is safe for human consumption

Solvent	Tolerance level (ppm)
Acetone	30
Ethylene dichloride	30
Isopropyl alcohol	50
Methyl alcohol	50
Methylene chloride	30
Hexane	25
Trichloroethylene	30

Adapted from Food and Drug Administration, CFR-Code of Federation Regulations Title 21, volume 3, Subchapter B-Food for human consumption, Part 173, secondary direct food additives permitted in food for human consumption, Subpart C, Solvents, lubricants, release agents and related substances, revised as of April 1, 2018.

to highly polymerized compounds, consisting of benzene rings with one or more hydroxyl substituents [10]. The three major classes of plant phenolics are the phenolic acids, flavonoids, and nonflavonoid polyphenols (Table 3). As shown in Table 3, the plant phenolics are classified according to their basic carbon skeleton and plant sources where the phenolics are found. All these plant phenolics have been reported to be prominent antioxidants and showed different biological properties such as antiinflammatory, antimicrobial, anticancer, and some hepaprotective properties [12–15]. Some phenolics are capable of preventing and controlling aging-associated diseases including Alzheimer

Table 3 Categories of plant phenolics, the basic carbon skeleton, examples, and plant sources

Major category	Carbon skeleton	Examples	Plant sources
Phenolic acids			
Hydroxybenzoic acids	C_6-C_1	• *p*-hydroxybenzoic acid • Protocatechuic acid • Vanillic acid • Veratric acid • Gallic acid • Gentisic acid • Syringic acid • 5-sulphosalicylic acid	Onions, black radish, red fruits, wheat, sorghum, rye, oat, barley, potato
Hydrocinnamic acids	C_6-C_3	• Coumaric acid • *p*-coumaric acid • Caffeic acid • Ferulic acid • Sinapic acid	Oat, wheat, sorghum, rye, barley, tomato, sweet potato, potato, bell pepper, mushroom, lettuce, corn, cauliflower, carrot, cabbage, broccoli, beet
Flavanoids			
Flavones	C_6-C_3-C_6	• Tricin • Baicalin • Luteolin • Apigenin • Nobiletin • Tengeretin • Sinensetin • Isosinensitin • Polymethoxylated flavones	Sweet orange, grapefruit, lemon, celery, parsley, blue passion flower, garlic

Continued

Table 3 Categories of plant phenolics, the basic carbon skeleton, examples, and plant sources—cont'd

Major category	Carbon skeleton	Examples	Plant sources
Isoflavones	C_6-C_3-C_6	• Genistin • Daidzin • Glycitin • Pretensein • Puerarin • Biochanin A	Soybeans, alfalfa sprouts, red clover, pueraria
Flavonols	C_6-C_3-C_6	• Myricetin • Fisetin • Kaempferol • Isorhamnetin • Quercetin	Olive, onion, garlic, apple, pear, grapes, grapefruit, berries, tea, red wine, broccoli, radish, galangal root, tomato, spinach, lettuce, cauliflower
Monomeric flavanols	C_6-C_3-C_6	• Catechin, • Epicatechin • Epicatechin gallate • Epigallocatechin • Epigallocatechin gallate	Black tea, green tea cocoa, red wines, apple, grape, pear, peach
Polymeric flavanols-proanthocyanidins	$(C_6$-C_3-$C_6)_n$	• Polyepicathechin • Polyepigallocathechins • Polyepiafzelechin	Wine, tea, cocoa, legume seeds, cereal grains, fruits
Anthocyanins	C_6-C_3-C_6	• Cyanidin • Delphinidin • Peonidin • Pelargonidin • Petunidin • Malvidin	Blueberry, blackcurrant, cranberry, red wine, strawberry, sweet orange, grapes, cherry, peach, plum, peony
Flavanones	C_6-C_3-C_6	• Naringenin • Hesperetin • Silibinin • Eriodictyol	Grapefruit, orange peel, lemons, milk thistle
Flavanonol	C_6-C_3-C_6	• Taxifolin	Citrus fruit, onion, grape

Table 3 Categories of plant phenolics, the basic carbon skeleton, examples, and plant sources—cont'd

Major category	Carbon skeleton	Examples	Plant sources
Nonflavanoid polyphenols			
	C_6-C_7-C_6	• Curcumin	Turmeric
	C_6-C_1	• Ellagic acid	Pomegranate, black raspberry, raspberry, strawberry, cranberry, pecans, walnut, black raspberry seed
	C_6-C_{10}	• Gingerol	Ginger
	C_6-C_6-C_6	• Rosmarinic acid	Rosemary
Stilbenes	C_6-C_2-C_6	• Resveratrol • Viniferins	Wines, grapes
Lignans	$(C_6$-$C_3)_2$	• Sesamin • Sesamolin • Pinerisol	Barley, oat, wheat, sesame seed, flax seed

and coronary heart diseases [16]. Hence, the extraction of phenolic compounds from plants is a great effort to the search for alternative, cheap sources of natural antioxidants as well as to the development of functional food products.

2.4 Identification and quantification of plant phenolics

Extraction is a very important step for the isolation and identification of phenolic compounds [17] and it is vital to develop selective extraction techniques to get an efficient extraction process with high extraction yield. Various techniques including Soxhlet, reflux, cold pressing, and maceration by organic solvents have been used by researchers to extract phenolic compounds from plant sources [18]. Table 4 lists some examples of plant phenolics extracted from different plant sources using conventional methods.

Identification and quantification of phenolic compounds have been successfully carried out via (a) spectrophotometric and (b) chromatographic methods. Spectrophotometric methods provide a simple and cost-effective way to quantify phenolic compounds. One of the established methods for determination of total phenolic content is the Folin-Ciocalteu assay [31, 32] which uses gallic acid as the standard. A colorimetric assay based on aluminum complex formation has been employed to evaluate total flavonoid content without the interference of other phenolic compounds [33, 34].

Table 4 Plant phenolics extracted from various plant sources using conventional methods

Plant	Phenolic compound	Extraction method	Solvent	Experimental parameters			References
				T (°C)	*t* (Min)	UP (W)	
Citrus pomade	• *p*-Coumaric acid • Hesperidin • Caffeic acid • Chlorogenic acid	• UAE	H_2O	RT–50	10–60	150, 200	[19]
Indian herbs (Unani system of medicine)	• Total phenolic content	• UAE • MAE • Soxhlet • Maceration	EtOH	60 60 NR 70	60 20 300–600 1440	NR 150 – –	[20]
Pomegranate peel	• Ellagic acid • Gallic acid • Punicalagin • Punicalin	• UAE	EtOH	20–80	10–60	NR	[21]
Mandarin peel	• Hesperidin	• UAE	Ace	30–50	20–40	30–60	[22]
Red sorghum bran	• Procyanidin • Quercetin-diglucoside • Taxifolin • Apigenin • Epicatechin • Apigenin-glucoside • Rutin • Naringenin	• UAE • solvent extraction	EtOH	30–70 –	5–40 21	100–300 –	[23]
Soybean seed	• Isoflavones	• aqueous extraction	H_2O	25–40	6–1440	–	[24]
Green tea	• Epigallocatechin gallate • Epicatechin gallate • Epigallocatechin • Epicatechin	• UAE • aqueous extraction	H_2O EtOH	66–94 20–70 72–88	18–102 10–110 5–55	2000 2000	[25]

Red grape skin	• Malvidin-3-*O*-glucoside • Quercetin • Rutin • Catechin • Epicatechin	• UAE • MAE • solvent extraction	EtOH, H_2O	28 NR 30	1–9 1–30 1–30	1000 1000 –	[26]
Rosemary	Total phenolic content	• UAE	EtOH	50	15–180	NR	[27]
Lapins cherry	• Cyanidin-3-rutinoside • Peonidin-3-rutinoside • Cyanidin-3-glucoside • Pelargonidin-3-rutinoside	• Solvent extraction	Acidified MeOH, EtOH	4–70	30–1440	–	[28]
							[29]
Leek	• Total phenolic content	• Solvent extraction	MeOH	22	30–90	–	
Rice	• Total phenolic content	• UAE	EtOH, MeOH, IPA	37	60	240	[30]

TBME: *t*-butyl methyl ether; PET: petroleum ether; ETAC: ethyl acetate; Ace: acetone; H_2O: water; EtOH: ethanol; MeOH: methanol; IPA: isopropanol; UAE: ultrasound-assisted extraction; MAE: microwave-assisted extraction; *T*: temperature; *t*: time; UP: ultrasonic power.

Quantification of total proanthocyanidin can be performed via acid-butanol, vanillin, and 4-dimethylaminocinnamaldehyde assay [35–37]. Total flavanols can be determined using *p*-dimethylaminocinnamaldehyde assay [38]. Chromatographic methods used for the detection, identification, and quantification of individual phenolic compounds include high performance liquid chromatography (HPLC) and gas chromatography (GC). The primary concern in the application of GC technique is the low volatility of the most phenolic compounds. Thus, phenolic compounds need to be converted to volatile derivatives prior to chromatography via methylation, conversion into trimethylsilyl derivatives, or derivatization with *N*-(tert-butyldimethylsilyl)-*N*-methyltrifluoroacetamide [39]. HPLC is the preferred method for plant phenolics analyses due to its precision, versatility, and low-cost. Phenolic compounds have been commonly detected using reversed-phase column with diode array detection (DAD) and/or mass spectrometry (MS) as shown in Table 5.

Table 5 Identification and quantification of various plant phenolic using high-performance liquid chromatography and *ultrahigh-performance liquid chromatography systems

Sample	Examples of detected phenolic compounds	Chromatography system	References
Thymus algeriensis Boiss. and *Reut.* *Ephedra alata Decne.*	• Rosmarinic acid • Salvianolic acid K, lithospermic acid • Kaempferol-*O*-glucuronide • Apigenin-6,8-*C*-dihexoside • Apigenin-7-*O*-glucuronide • Myricetin-C-hexoside • Hydroxypuerarin • 5,5′-dihydroxi-methoxy-isoflavone-*O*-glucoside	Column: Spherisorb S3 ODS-2 C18 (150 × 4.6 mm, 3 μm) Mobile phase: H_2O, 0.1% FA/CAN Detector: ESI MSn DAD	[40]
Jabuticaba fruits	• Derivatives of quercetin • Derivatives of myricetin derivatives of methyl ellagic acid • Derivatives of ellagic acid	Column: Zorbax Eclipse XDB-C18 (150 x 2.1 mm, 3.5 μm) Mobile phase: ACN/H_2O/FA: 3/88.5/8.5 v/v/v ACN/ H_2O/FA: 50/41.5/8.5/v/ vvMeOH/H_2O/ FA:90/1.5/8.5 v/v/v Detector: DAD-ESI/MS^n	[41]

Table 5 Identification and quantification of various plant phenolic using high-performance liquid chromatography and *ultrahigh-performance liquid chromatography systems—cont'd

Sample	Examples of detected phenolic compounds	Chromatography system	References
**Phoenix dactylifera L.* *Palm date fruits*	• Caffeoylglucoside-formic acid • Caffeic acid-formic acid • Salicylic acid • rutin • luteolin • Caffeoylshikimic acid • *p*-coumaric acid • Apigenin pentosyl hexoside • Ferulic acid-o-hexoside • Luteolin rhamnosyl dihexoside • Quercetin-7-glucoside • phloridzin • *O*-p-coumaroylshikimic acid • Caffeic acid-*O*-(sinapoyl-O-hexoside) • Isorhamentin hexoside • Caffeoylshikimic acid	Column: Hypersil Gold C18 (100 × 2.1 mm, 1.9 μm) Mobile phase: ACN/(H_2O, 0.1% FA) Detector: DAD-ESI/MS^n	[42]
Physalis alkekengi	• 3-Caffeoylquinic acid • 5-caffeoylquinic acid • Protocatechuic acid, *p*-coumaric acid • Caffeic acid, suberic acid • 1,3-dicaffeoylquinic acid • 4,5-dicaffeoylquinic acid • Vitexin, luteolin, apigenin, chrysoeriol • Quercetin *O*-rhamnoside • Quercetin 3-*O*-glucoside • Luteolin 7-*O*-glucoside • Apigenin *O*-glucoside • Diosmetin O-glucoside • Luteolin *O*-glucoside	Column: Zorbax SB C18 (150 × 4.6 mm, 5 μm) Mobile phase: ACN/(H_2O, 0.1% FA) Detector: DAD-ESI-QTOF-MS	[43]
**Thymus capitatus (L.) Hoff. et Link.*	• Rosmarinic acid • Salvianolic acid A • Salvianolic acid E • Hesperidin • Eriodictyol • Naringenin • Apigenin-*C*-di-hexoside • Gallocatechin • Taxifolin	Column: Hypersil Gold C18 (100 × 2.1 mm, 1.9 μm) Mobile phase: ACN, 0.1% FA Detector: DAD-ESI/MS^n	[44]

AcOH: acetic acid; ACN: acetonitrile; FA: formic acid; MeOH: methanol; H_2O: water; DAD: diode array detector.

3. Supercritical extraction of oleoresins

SFE has emerged as a promising green technique to substitute conventional methods to extract oleoresins from plant sources. There are several aspects that need to be considered for optimal extractions as discussed in the following subsections.

3.1 Solubility

Solubility is defined as the weight fraction of a substrate in the SCF, which is in equilibrium with the bulk substrate. The solubility of a substrate can be determined using the dynamic or static method. It is vital to study the solubility of oleoresins in $scCO_2$ as it serves to evaluate the compatibility of the SCF in the extraction process and for a proper design of the extraction process. Besides, the solubility data is important for optimizing the SFE, allowing for the selection of optimum conditions such as the operating pressures, temperatures, and CO_2 flow rates [45, 46]. Fernández-Ronco et al. [47] compared the solubility of paprika oleoresin and capsicum oleoresin in $scCO_2$ as a preliminary step to determine the suitable type of oleoresin for the subsequent SFE process. It was found that capsicum oleoresin had a higher solubility in $scCO_2$ (7.53 kg/m^3 CO_2) compared to the solubility of paprika oleoresin (3.68 kg/m^3 CO_2) and therefore, the capsicum oleoresin was selected for the following analyses. The solubility of vanilla oleoresin and selectivity study on vanillin were conducted at 14 and 18 MPa, respectively at temperatures of 37.35–75.15°C and the results were correlated with a modified version of del Valle-Aguilera model [48]. At highest operating pressures and temperatures, the vanilla oleoresin had the highest solubility in $scCO_2$; nevertheless, the selectivity of vanillin from the extracted oleoresin decreased at high temperature. Hence, the solubility studies are important for the selection of operating parameters to enable the optimal extraction efficacy depending on the desired results (high oleoresin extraction yield vs high phytonutrient concentration).

3.2 Sample pretreatment

SFE is a mass transfer operation in which convection in the supercritical solvent phase is generally the main transport mechanism. The SFE of oleoresins from plant materials usually involves certain pretreatment on the samples such as reducing the particle size and moisture content of the samples and frequently, a combination of both these treatments are needed for optimal extraction yield. For SFE of oleoresins, the samples are normally subjected to the drying steps followed by particle size reduction prior to the extraction process.

3.2.1 Particle size

The particle size of samples is known to be a prominent factor affecting the SFE as it influences the extraction kinetics. When the particle size of a sample is reduced, the contact surface area between the sample and the scCO_2 is increased, hence, increasing the mass transfer rate. This phenomenon can be explained on the basis of convection and diffusion processes that occur during the extraction. In the convection mass transfer theory, small particles tend to possess higher contact surface to bed volume, whereby the fluid-phase convection is enhanced [49]. According to Fick's laws of diffusion, materials having smaller particle diameters have smaller diffusive paths to solvent and extract [50] and thus, the decrease in both internal and external mass transfer resistances, with an associated increase in extraction rate. Furthermore, during the particle size reduction process such as grinding, the cell walls of the plant materials which acted as mass transfer barriers are destroyed [51] and this process enables higher permeability for the scCO_2 and subsequently higher extraction yield. The SFE of grounded tomato peels (0.3mm) recorded higher oleoresin extraction yield of 4.86% compared to ungrounded tomato peels (1.0 mm) with extraction yield of 3.97% [52]. The SFE yields of red pepper (*Capsicum frutescens*) oleoresins from samples with three different particle sizes of 0.43, 0.93, 1.42mm were investigated [49] and as anticipated, a sample with the smallest particle size of 0.43mm had the highest oleoresin yield. However, it should be noted that if the particle size is reduced too much, this will cause an increase in the packing density in the extraction vessel causing the channeling effect. Ultimately, the mass transfer rate of scCO_2 will be affected, resulting in a lower extraction yield. The scCO_2 flow rate could be inconsistent when the samples particle size is too small and hence affect the optimization of SFE. For instance, Said et al. [53] reported the optimum size for SFE of oleoresin from ginger rhizomes as 0.25mm, further reduction in the particle size decreased the ginger oleoresin extraction yield. The sample particle size for SFE was normally less than 1.0mm and some examples are listed as follows:

- tomato peel: <0.1mm and tomato seed:<1.0mm [54]
- tomato-by-products: 0.20–0.3mm [55, 56]
- seeds from pomegranate, tomato, and grape: <1.0mm [57]
- flesh from watermelon and gac fruit: < 0.5mm [58]
- citrus fruits peel: <0.8mm [59]
- marigold flowers: 0.175–1.3mm [60]
- black pepper: <0.4–1.0mm [61, 62]
- capsicum: 0.51–0.88mm [51]

3.2.2 Moisture content

When the moisture ratio of the solute is changed, the rate of mass transfer and the efficacy of the SFE are affected, either positively, or negatively. The moisture content in the material acts as a barrier that hinders the diffusion of scCO_2 into the plant matrix and lowers the extraction yield. However, in certain cases, the moisture does not act as barricade but as cosolvent that enhanced the extraction yield, especially for polar compounds [63]. Hence, the functionality of moisture content in affecting the efficiency of the SFE process is dependent on the targeted outcomes.

In the case of SFE of oleoresin, it is suggested that samples with lower moisture content would yield oleoresins with high fat content and high fat-soluble bioactive components; whereas oleoresins with low fat, low resin content, and high polar phenolic content are expected when the moisture content of the sample is high. It is crucial to dehydrate the samples, especially fruit samples which are very high in moisture content in order to have an efficient SFE of oleoresins with high lipophilic compounds. The mass transfer of solute in the solid phase could be affected differently by varying the moisture content of the samples. Vasapollo et al. [64] compared the extraction yield of lycopene from sun-dried tomato (containing 50–60% moisture content) and vacuum-dried tomato (containing 6% moisture content). It was found that the lycopene extraction yield increased from approximately 15% in sun-dried tomato compared to approximately 35% for the vacuum-dried samples. The reported moisture contents of samples for optimal SFE extraction have been 3%–18% as evident from Table 6.

Since the drying process is energy intensive, different extraction dynamics are produced by various drying methods. The particle structure (size and porosity) varies with different drying methods, hence resulting in different oleoresin extraction yields. For instance, the oleoresin yields obtained from freeze-dried and oven-dried capsicums were

Table 6 The final moisture content in various samples acquired by different drying techniques for supercritical fluid extraction of oleoresins

Sample	Drying method	Final moisture content (%)	References
(a) Gac fruit aril **(b)** Watermelon flesh **(c)** Tomato flesh and seed	Freeze-dried	**(a)** 4.0 **(b)** 3.5 **(c)** 3.1	[58]
Tomato by-products	Sun-dried	5.0	[52]
Marigold flower	Air-dried	11.8	[60]
Ginger	Oven-dried	12	[53]
Capsicum peppers	**(a)** Freeze-dried **(b)** Oven-dried	**(a)** 6.18 **(b)** 2.05	[51]

12.1% and 13.6%, respectively [51] due to the differences in the final moisture content (Table 6) and particle diameter (freeze-dried: 0.88 mm; oven-dried: 0.51 mm) resulted from these two different drying techniques.

3.2.3 *Enzyme treatment*

Enzymes are known as bio-catalysts, most of which are proteins and some are ribozymes. There are six classes of enzymes, namely, hydrolase, isomerase, lyase, oxidoreductase, synthetases, and transferase, with each class providing different functionalities. In SFE, enzymes could be used to break down the cellular structures of the plant material that potentially hinder the extraction process. The loss of cell integrity allows for better solvent penetration and hence, higher extraction yield. Enzyme α-amylase (source: *Bacillus licheniformis*) has been used to hydrolyse the black pepper coating, which is made up of 30% starch before performing the SFE at 30 MPa, 60°C. The results showed that the extraction yield of black pepper oleoresin increased from 15% to 88% after the enzymatic treatment using 0.4–4.0 mg of α-amylase. Lenucci et al. [65] explored the use of different plant cell-wall glycosidases in enhancing the yield of tomato oleoresin from the SFE process. The tomato matrix treated with different enzymes such as Cellulast, Novozyme, and Viscozyme, showed higher extracted oleoresins compared to the untreated sample at SFE conditions of 50 MPa, and 86°C.

3.2.4 *Cosolvent*

Cosolvent is added to a primary solvent at a relatively smaller amount to increase the solubility of the component of lower solubility. In the SFE process, the $scCO_2$ is the primary solvent and in certain experiments, an organic solvent is added as cosolvent with the purpose of enhancing the extraction efficiency. There are two main factors to be considered when using cosolvent to aid in the SFE; (1) the type of organic solvent and (2) the amount to be added into the extraction process.

(1) Type of organic solvent:

Organic solvents have different properties and polarities. Two of the most widely used cosolvents for SFE of oleoresins are methanol (MeOH) and ethanol (EtOH) which aid in improving the polarity of the SFE system. The hydroxyl groups in both MeOH and EtOH are able to form hydrogen bonds with the polar molecules present in the samples, hence, increase the extraction yields of polar compounds and oleoresin. Other than the polarity, the interaction between the cosolvent and the solute is another factor affecting the ability of the cosolvent to aid in the SFE process [61].

(2) Amount of cosolvent:
When cosolvent is added, the density and the solvation power of the $scCO_2$ increase, allowing for higher extraction yield. It is anticipated that when a higher amount of co-solvent is used, the SFE extraction yield will increase correspondingly.

EtOH as cosolvent has been used in different amounts for the SFE of oleoresins from cyanobacteria powder (*Arthrospira platensis*) [66]. Under identical SFE conditions, the oleoresin yield was higher when extracted with 53% of EtOH compared to when extraction was done with 26% of cosolvent. Nagavekar and Singhal [61] have examined the effect of different cosolvents such as MeOH, EtOH, and acetone (Ace), added in different amounts (10%, 20%, 30%, v/w) on the black pepper oleoresin yield. As anticipated, when the volume of added cosolvent was high (30%), the oleoresin extraction yield was maximum for all the three types of cosolvents. From the results, the extraction yield of black pepper oleoresin was the highest for MeOH (108.51 mg/g), followed by EtOH (106.65 mg/g) and finally Ace (86.26 mg/g) as affected by the properties of each type of organic solvent. Even though a higher percentage of cosolvent contributes to higher extraction yield, it should be taken in consideration that only a minimal amount should be used for SFE. This is to comply with the green chemistry principle which is also one of the main purposes of SFE as an alternative method that does not require an organic solvent.

3.3 Operating parameters

The SFE process is governed by four important parameters that will largely affect the extraction efficiencies and the extraction yield; (1) pressure; (2) temperature; (3) extraction time; and (4) carbon dioxide flow rate.

3.3.1 Pressure and temperature

In SFE of oleoresins, the process pressures normally ranged between 10 and 35 MPa, and in certain cases, the higher pressures of 40–55 MPa have been used. The extraction pressure has a positive relation with the extraction yield. It is anticipated that when the SFE is conducted at high-pressure conditions, the extraction yield will be high. The density as well as the solvation power of $scCO_2$ increased at high operating pressure (Fig. 1), which enables higher solubility and hence, higher extraction yields.

Nonetheless, it is possible for the extraction yield to decrease if the operating pressure is too high. The interaction and the mass transfer between the samples and the $scCO_2$ are affected as the result of the solid matrix becoming too packed and causes chaneling issues at very high operating pressures. The solubility of the oleoresin is then affected due to the

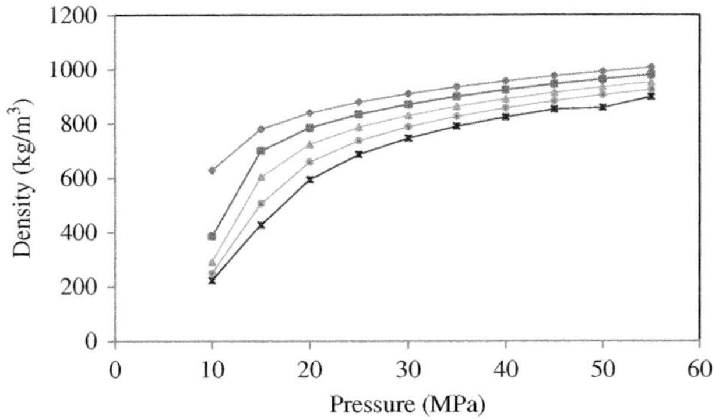

Fig. 1 The change in the $scCO_2$ at different pressures and temperatures of 40°C, 50°C, 60°C, 70°C, 80°C. *The densities of carbon dioxide were adapted from "Thermophysical Properties of Fluid System," NIST ChemistryWebbook, NIST Standard Reference Database Number 69.*

internal and external mass transfer resistances and hence $scCO_2$ does not penetrate easily into the solid matrix.

The temperature commonly used for oleoresin SFE has been in the range of 32–80°C. Low extraction temperatures are desirable in order to reduce the possibility of destroying thermolabile components that are present in the samples. From Fig. 2, it is evident that at constant pressure, when the temperature increases, the $scCO_2$ density decreases. It has been observed that with the reduction in $scCO_2$ solvation power, the oleoresin extraction yield lowered at high temperatures.

Nevertheless, the effect of temperature is complex due to the occurrence of crossover effect and this effect varies for different samples. The crossover occurs due to the competing effect between the $scCO_2$ density and solute vapor pressure. Generally, the extraction yield decreased with increasing temperature due to the effect of reducing $scCO_2$ density. At a certain point, the crossover effect took over and causes the extraction yield to increase with temperature, although the $scCO_2$ density is lower. The solute vapor pressure increases at high temperature because of the kinetic energy of the molecules increased, enabling more molecules to transition into the vapor state and solubilized in the $scCO_2$, hence higher extraction yield. The occurrence of crossover is different for each type of sample; the solubility of vanilla oleoresin in $scCO_2$ increased when the temperature increased from 55 to 75°C at 18 MPa [48]. However, at 12 MPa, the horsetail (*Equisetum giganteum* L.) oleoresin yield decreased with increasing temperature from 29 to 49°C, but increased with temperature at pressure 15–25 MPa [67]. At temperature of 26–40°C, crossover occurred at 14–16 MPa for marigold oleoresin [68] and similarly, another research group reported the crossover for marigold oleoresin

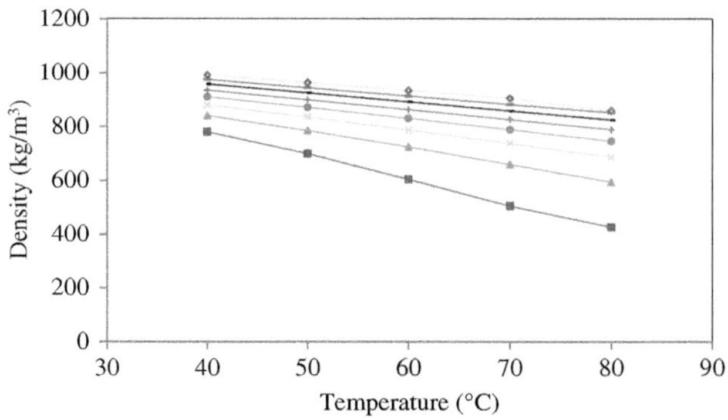

Fig. 2 The change in the density of $scCO_2$ when at a various temperature at fixed operating pressure (the lowest line represents 15–50 MPa). *The densities of carbon dioxide were adapted from "Thermophysical Properties of Fluid System," NIST ChemistryWebbook, NIST Standard Reference Database Number 69.*

occurred at 15 MPa [69]. The crossover effect was found to be below 15 MPa for Brazilian herb (*Polygala cyparissias*) oleoresin [70]. Although the crossover effect is solute dependent, it can be concluded that the crossover region for oleoresin is generally below 20 MPa, as crossover pressures for substrates are commonly in the range of 15–20 MPa [71].

3.3.2 Time

The extraction curve for the effect of extraction time consists of three stages (Fig. 3). The 1st stage: convective mass transfer occurs between the sample surface and $scCO_2$ and the oleoresin yield increased rapidly as expressed by a linear increasing plot. The 2nd stage: falling rate period, whereby the extractions yield increased slowly (rate started to decrease), indicating a stronger hindering of the mass transfer of the more deeply embedded oleoresin within particles [72]. The 3rd stage: completion of the extraction process, whereby the rate of extraction remained constant as indicated by the plateau state. In most of the oleoresin extractions, a preliminary static extraction step (approximately 10–20 min) is carried out to increase the $scCO_2$ contact time and higher penetrate into the plant matrix following dynamic extraction. For instance, Vallecilla-Yepez and Ciftci [54] and Urbonaviciene and Viskelis [55] subjected the tomato samples to 20 and 10 min static extractions, respectively, prior to the dynamic extraction step for tomato oleoresin extraction.

Time is one of the main factors for the evaluation of extraction efficiency and often, the extraction time is reliant on: (1) the cell structure and physicochemical properties of the plant materials, the extraction vessel's dimension and packing density and the $scCO_2$-

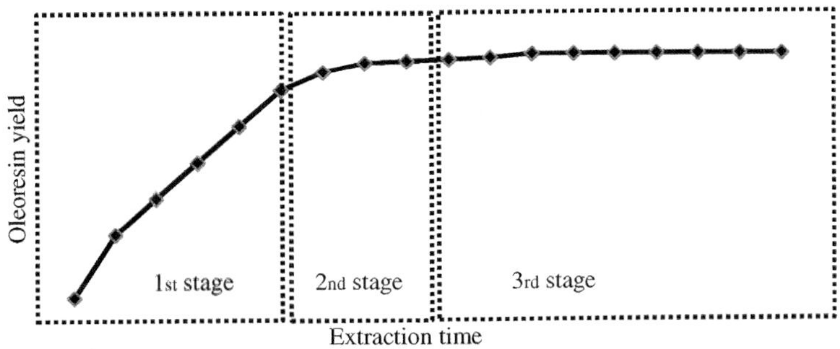

Fig. 3 Extraction curve represented by three different stages obtained by plotting yield against extraction time.

solute interaction; (2) process parameters such as the pressure, temperature, and CO_2 flow rate. Incomplete extraction process could arise when the extraction time is too short, and if the extraction time is too long over the required period, it could be time-consuming and requires high production cost.

One of the factors that influence the extraction time is the particle size of the substrate. When the particle size of the material is small, the intra-particle diffusion resistance is smaller and hence high mass transfer rate which reduces the required extraction time. On the contrary, a substrate with a large particle size will need a longer extraction time, which may consequently influence the extraction efficiency and yield. It is evident that the oleoresin extraction rate and yield of grounded tomato peel were higher compared to the ungrounded samples at similar SFE conditions of 40 MPa, 50°C, and CO_2 flow rate of 4 g/min [52]. Under high-pressure and or high-temperature conditions, the extraction rate is faster due to the higher mass transfer rate and hence less time is required to achieve the equilibrium state. At high CO_2 flow rate, the extraction time is reduced. Urbonaviciene and Viskelis [55] examined the effects of extraction pressure and temperature on time needed to achieve the completion of the tomato oleoresin extraction. It was observed that most of the oleoresins have been extracted during the first 120 min of the extraction time at a higher pressure of 55 MPa. At 37.5 MPa, the plateau stage was achieved after 160 min of extraction time. As regards the effect of temperature, the completion of the extraction was achieved faster at a higher temperature of 80°C compared to that of 60°C followed by 40°C.

3.3.3 Carbon dioxide flow rate

Optimal CO_2 flow rate is crucial to obtain an optimized oleoresin SFE process. The extraction rate is controlled by the scCO_2 flow rate; a higher flow rate results in a higher extraction rate with shorter extraction time. The increase of velocity causes a retraction of the mass boundary layer that surrounds the particles. As a consequence, the external mass transfer resistance is reduced and the extraction rate increases. Following the same trend, the convective mass transfer coefficient also increased with solvent flow rate. Therefore, the constant extraction rate period decreased [49]. However, the flow rate should not be too high as it might cause the sample to be flushed out which results in a higher yield than it should be. The used flow rate should allow for sufficient contact time to saturate the CO_2 with the solute, achieving thermodynamic equilibrium in the bulk of solid and fluid phase.

3.4 Supercritical fluid extraction of oleoresin from selected plant samples

SFE has been employed to extract oleoresins from different plant samples, ranging from spices, herbs, vegetables, fruits, and many more. The following section covers a few selected types of oleoresins extracted with SFE and the comparison with conventional extraction technologies.

3.4.1 Tomato oleoresin

Tomato oleoresin consists of bioactive components such as lycopene, ascorbic acid, vitamin E, carotenoids, and flavonoids. With the intense red color, tomato oleoresin finds its application in the food industry as a substitute for tomato fruit, coloring and flavoring agent. The capability of using tomato oleoresin in enhancing the oxidative stability of edible oils such as refined olive oil and sunflower oils in replacing synthetic preservative has been reported [73].

The SFE of tomato oleoresin has been carried out at pressures of 20-55 MPa, temperatures of 40–80°C, at various CO_2 flow rate and extraction times for obtaining oleoresin yield of 5%–33% (Table 7). The oleoresin yield varied depending on the different parts of tomato that were used for extraction, i.e., flesh, seed, peel, pulp, and the extraction conditions. Besides, the extraction yield was also influenced by sample pretreatment and the presence of co-matrix or cosolvent. Lenucci et al. [65] reported the effect of enzymatic pretreatment and addition of co-matrix on the extraction yield of oleoresin and lycopene from tomato purée. Tomato samples were digested with different combinations of food-grade enzymes, namely, Celluclast, Novozyme 188, Viscozyme L, and Flavourzyme 500 L before extraction. Although the substrate load was increased after the digestion, the SFE was affected due to the channeling effect caused by the smaller particle size and denser microstructure. Hence, hazelnut seed as co-matrix was added to the tomato samples to enhance the diffusion of $scCO_2$ into the tomato matrix besides of co-extracting lipids.

Table 7 Examples of supercritical fluid and conventional extractions of tomato oleoresin from various tomato samples

Tomato	SFE conditions				Co-matrix/ enzyme treatment	Oleoresin yield (%)	Lycopene (%) from oleoresin	References
	P (MPa)	*T* (°C)	FR	*t* (min)				
Peel	30–50	50–80	3–6 g/min	105	–	4.58–6.31	32–60 (% from peel)	[52]
Seed	35	60	4 mL/min	120	–	7.8	-	[57]
Pulp	45	80	600 kg/h	120	–	NR	10.9	[74]
	40–45	60–70	18–20 kg/h	420	Hazelnut	–	0.12–0.58	[75]
	33.5–45	45–70	8–20 kg/h	480	Grape seed	–	0.37–2.85	[76]
Purée	50	86	4 mL/min	270	Celluclast, novozyme, viscozyme, hazelnut	5.3–12.4	18.24	[65]
	45	65–70	18–20 kg/h	180	Hazelnut	4.9–33	0.81–2.98	[77]
Peel +seed	30,50	40,80	1 L/min	240	–	24.6	0.02–0.26	[54]
	40	80	4 g/min	120	–	5.79	0.4	[56]
Peel, pulp, seed	20,37.5, 55	40,60,80	NR	120–240	–	10.8–25.61	2.4–8.5	[55]

P: pressure; *T*: temperature; FR: carbon dioxide flow rate; *t*: extraction time; NR: not reported

The use of co-matrix in enhancing the SFE of tomato oleoresin was also reported, with the aid from hazelnut [75, 77] and grape seed [76]. However, under the same SFE conditions (50 MPa, 86°C), it was found that the oleoresin extraction yield for the untreated sample was similar to that of enzymatically treated tomato matrix (purée + hazelnut seed) but the total extracted lycopene was approximately three times higher from the enzyme-treated sample. Vallecilla-Yepez and Ciftci [54] compared the oleoresin extraction yields from tomato peel and seed by SFE (50 MPa, 40°C and 80°C, at a $scCO_2$ flow rate of 1 L/min for 240 min) and solvent extraction (hexane, 360 min). Using SFE, the extracted tomato oleoresin was 24.6% with lycopene of 0.26% as compared to solvent extraction whereby lower oleoresin of 19.3% and 0.24% of lycopene were extracted. Other extraction techniques including maceration using hexane for overnight reported oleoresin yield of 2.66% [73] and sonication with acetone reported oleoresin yield of 1.43% and lycopene yield of 0.065% [78]. Thus, the higher yield and shorter extraction time suggested SFE to be a better extraction technique for tomato oleoresin.

3.4.2 Flower oleoresin

3.4.2.1 Marigold oleoresin

Marigold is a plant of the daisy family that is used as nutritional and medicinal supplements. Marigold flowers have been reported for cancer and cardiovascular diseases prevention and protection against oxidant-induced cell damage properties [79, 80]. The marigold species can be classified into:

Genus	Species
Calendula	*Calendula officinalis*
Tagetes	*Tagetes patula*
	Tagetes erecta
	Tagetes lucida
	Tagetes tenuifolia
Arctotheca	*Arctotheca calendula*
Baileya	*Baileya multiradiata*
Caltha	*Caltha palustris*
Glebionis	*Glebionis segetum*
Tithonia	*Tithonia diversifolia*

Due to the presence of lutein in the marigold flowers, the extracted oleoresin is yellowish-orange in color and is widely used as a natural coloring and flavoring agent in the food industry. The SFE of marigold flowers was conducted under the operating pressure of 12–40 MPa, 20–70°C, at a various CO_2 flow rate and reported oleoresin yield of 2.1%–7.4%. López-Padilla et al. [60] extracted oleoresin from *Calendula officinalis* at pressures of 14, 24, 34 MPa, 40°C and at a CO_2 flow rates of 15, 30, and 45 g/min CO_2 for 4.5 h. The extraction was carried out using different extractors. The small, medium, and large scale extraction yields of oleoresins were 4.56%–6.28%, 3.08%–4.15%, and 4.77%–7.38%, respectively. Oleoresin extraction from *Calendula officinalis* determined at 12–20 MPa, 20–40°C, at a CO_2 flow rate of 0.79–1.67 g/min for 6 h [69], was approximately 2%. The SFE data obtained were then used by Mezzomo et al. [81] to estimate the operation cost of using SFE to extract marigold oleoresin by using the software Tech analysis, considering the specific costs to be equal to the market value of the oils. Under the conditions studied, it was reported that the use of SFE was not competitive in comparison to the commercial extraction method. The low SFE yield was reported by Danielski et al. [68] whereby the oleoresin yield from cold maceration with hexane was 6.94%, comparing to SFE ***(P=12–20 MPa; T=20–40°C; CO_2 flow rate=0.79–1.67 g/min) with yield of 2.08%–3.54%. Extraction of marigold oleoresin from the *Tagetes erecta* species also showed the lower yield of 5.89% (P=20–40 MPa; T=30–70°C; CO_2 flow rate=250 g/min) compared to Soxhlet extraction oleoresin yield of 6.1% with hexane [82].

3.4.2.2 Pyrethrum oleoresin

Pyrethrum, commercially known as chrysanthemum (*Chrysanthemum cinerariifolium*), contains pyrethrins which have been used as food compatible insecticides. The SFE of pyrethrum oleoresin was conducted by Kiriamiti et al. [83], investigating the effect of pressure (8.10 MPa) and temperature (29.40°C) at a fixed CO_2 flow rate of 0.3 kg/h. (Gallo et al. [84] compared the extraction yield of pyrethrum oleoresin and pyrethrins using three extraction processes such as the traditional maceration with organic solvents, SFE, and rapid solid-liquid dynamic extraction (RSLDE). The SFE was conducted at 40 MPa, 40°C, the CO_2 flow rate of 1.155×10^{-4} kg/s for an extraction time of 210 min. The SFE yield was reported to be 4.2% in comparison to maceration (hexane: 3.03%, ethanol: 5.77%), and RSLDE (hexane: 3.26%, ethanol: 6.31%). The RSLDE method showed advantages over the SFE with higher yield and shorter extraction time of 120 min compared to SFE which was carried out for 210 min.

3.4.2.3 Chamomile oleoresin

Chamomile is one of the medicinal plants used in the pharmaceutical industry because of its antiinflammatory, antiseptic and antispasmodic properties, besides the use in reducing stress, anxiety, and insomnia [85, 86]. The extracts are also used in the cosmetology industry to impart flavor to personal care products. Some examples of the SFE and conventional extraction methods available in the literature on the two different species of chamomile, namely *Matricaria chamomilla* and *Chamomilla recutita* [L.] Rauschert, are listed in Table 8.

Table 8 Examples of SFE and other extraction techniques used to obtain chamomile oleoresin

Chamomile	Part	SFE conditions: *P* (MPa)	*T* (°C)	FR Kg/h	*t* (min)	Oleoresin yield (%)	Bioactive component (mg/100 g)	Conventional method	Oleoresinyield (%)	Bioactive component (mg/100 g)	References
Matricaria chamomilla	A	30	40	2	90	1.57	UMB:ND H: 13.08	Soxhlet extraction: Hexane, 8 h	4.6	UMB:0.50 H: 103.9	[87]
								Maceration: Aqueous EtOH, 50%, 5 d	20.85	UMB: 5.59 H: 47.45	
								Hydrodistillation: 4 h	0.41	UMB: ND H: ND	
	B	30	40	2	90	3.64	UMB:0.33 H: 37.05	Soxhlet extraction: Hexane,8 h	4.98	UMB:0.00 H: 20.22	
								Maceration: Aqueous EtOH, 50%, 5 d	22.30	UMB:4.78 H: 45.54	
								Hydrodistillation: 4 h	0.62	UMB:ND H:ND	
	C	25	40	0.13–0.17	–	3.81	M: 0.28	Steam distillation: 4 h	0.6	M: 0.0014 C: 0.60	[88]
								Maceration: EtOH, 48 h	8.9	M: 0.0101 C: ND	
								Soxhlet extraction: EtOH, 6 h	10	M: 0.0108 C: 0.01	
Chamomilla recutita [L.] Rauschert	–	10–38	60,70	8	180	2.29–16.3	Apigenin-7-glucoside: 0.0466 – 0.432	–	–	–	[89]
	–	10–20	30,40	0.017	600	4.33	–	–	–	–	[90]
	–	24	40	0.124	300	2.25 – 3.09	–	–	–	–	[91]

T: pressure; *T*: temperature; FR: carbon dioxide flow rate; *t*: time; ND: not detected; EtOH: ethanol; (A) unprocessed flower; (B) processed flower (C) flower head; UMB: umbelliferon;

3.4.3 Turmeric oleoresin

Turmeric (*Curcuma longa* L.) is a type of spice rhizome widely used in cooking and added in various food preparations and condiments as flavoring and coloring material. Turmeric has also found its applications in the pharmaceuticals industry and as traditional medicine because of its role as antiinflammatory, antimicrobial, antiviral, and is able to treat digestive tract disorders and skin disorders [92]. Turmeric oleoresin is known to be rich in curcuminoids which are a family of active compounds consisting of curcumin, demethoxycurcumin, and bisdemethoxycurcumin [93]. The studies performed on SFE of turmeric oleoresin and the integration of SFE with other technologies such as ultrasound and pressurized liquid extraction (PLE) are presented in Table 9. Chhouk et al. [99] integrated the SFE with ultrasound technology (ultrasound-assisted SFE), reporting the ability of ultrasound (45 kHz and 600 W) in improving the SFE of turmeric oleoresin by generating the cavitation and breaking the cell wall of the turmeric samples. Investigation on the effect of the integration of SFE and pressurized liquid extraction (PLE) was conducted by Osorio-Tobón et al. [100] to enhance the extraction yield of curcuminoids. The turmeric samples were first subjected to SFE (10–35 MPa; 40–80°C) and instantaneously after the SFE process, without removing the turmeric samples from the extractor, the PLE was performed to extract the curcuminoids using ethanol. The extraction yield of the phenolic compounds was improved by this SFE—PLE process and has also been confirmed by Martinez-Correa et al. [98]. The manufacturing cost of turmeric oleoresin and its bioactive compound ar-turmerone obtained using SFE was reported to be economically feasible by Carvalho et al. [92] and Priyanka and Khanam [94] have reported the feasibility of using SCF to extract turmeric at industrial scale as a profitable method.

3.4.4 Rosemary oleoresin

Rosemary oleoresin extracted from the rosemary herbs (*Rosmarinus officinalis*) has been used as a culinary spice and also as natural preservatives due to its high antioxidant and antimicrobial properties. The bioactive compounds present in rosemary that are responsible for the antioxidative effects are mainly the rosmarinic acid, carnosic acid, and carnosol. Studies on using rosemary oleoresins as natural preservatives in different food items have been reported. For instance, application of rosemary oleoresin in enhancing the stability of fresh pork sausages [101], pork batter formulation [102], sunflower oil [103], and ground pork [104] are some examples to cite. The different SFE processes and conventional extraction methods used for rosemary oleoresin extractions are listed in Table 10. One of the interesting SFE technologies used for rosemary extraction was the supercritical antisolvent fractionation (SAF) by A. P. Sánchez-Camargo et al. [111]. Optimization of the single-step SAF of rosemary oleoresin that was previously obtained after pressurized liquid extraction process with ethanol and water as extracting solvent was conducted to produce rosemary oleoresin rich in both carnosic acid and

Table 9 Extraction of turmeric oleoresins with supercritical fluid, various conventional methods and integration of SFE with other technologies.

SFE conditions										
P (MPa)	*T* (°C)	FR	*t* (min)	Co-solvent	Oleoresin yield (%)	Bioactive component (%)	Conventional method	Oleoresin yield (%)	Bioactive component (%)	References
20–40	40–60	5–15 g/min	260	EtOH	2–5.3	–	Soxhlet extraction: Hexane, 24 h	5.954	–	[94]
26	40	5 L/min	150	–	6.98	α-β-ar-TUR: 67.7	–	–	–	[95]
30	60,70	5.2, 20.8 g/min	150	–	6.47	ar-TUR: 40.5 mg/g	–	–	–	[93]
20,30	30	4.2×10^{-5} kg/s	360	EtOH	5.8	CU: 0.0005–0.0006	Hydrodistillation: 3.5 h	2.1–3.1	CU:~0.0002	[96]
				IPA	5.7–8.0	CU:0.0007–0.0041	Low pressure solvent extraction: IPA/ EtOH,6 h, 30°C	7.5–13	CU:0.34–3.1	
				EtOH/ IPA	5.9–7.8	CU: 0.0083–0.0152	Soxhlet: EtOH, 2.5 h	21–31	CU:0.38–1.4	
–	–	–	–	–	–	–	Soxhlet extraction: EtOH/ IPA, 2.5 h	9.4–21	CU:3.7–8.43	
25, 30	40,44	11–14 g/min	85–107	–	4.5–6.51	CU:~3.75	–	–	–	[97]
25,30	40,44	6–9 g/min	95	EtOH	13–22	CU:~6.5	–	–	–	
10–35	40–60	8.6 g/min	–	–	0.4–6.4	ar-TUR:0.07–1.14	–	–	–	[92]
40	60	0.04 g/s	360	–	5.8	CUR:52.3 mg/g	Solvent extraction : EtOH,42 h, 25°C	9.5	CUR:354 mg/g	[98]
							Water extraction: 10 min, 60°C	18.5	CUR:2.2 mg/g	

Two steps extraction, SFE followed by EtOH extraction (42 h, 25°C)										
					6.9	CUR:565 mg/g				
Two steps extraction, SFE followed by water extraction (10 min, 60°C)										
					16.8	CUR:3.7 mg/g				
Ultrasound assisted SFE (ultrasound power of 45 kHz and 600 W)										
15–25	40–60	2–4 mL/min	30–120	EtOH	7.17	CUR:1.69	Soxhlet extraction EtOH, 6 h	~11.03	CUR:1.62	[99]
–	–	–	–	–	–	–	Soxhlet extraction hexane, 6 h	~4.96	CUR:~0.07	
Integrated SFE and pressurized liquid extraction										
10–35	40–60 60–80	8.6 g/min	–	–	6–15	CU: 2.2–4.4	Soxhlet extraction: EtOH 6 h, 80°C	12	CU: 4.2	[100]
							Low-pressuresolvent extraction: EtOH, 3 h, 40°C	12	CU: 3.7	

SFE: supercritical fluid extraction; *P*: pressure; *T*: temperature; FR: carbon dioxide flow rate; *t*: time; CUR: curcumin; CU: Curcuminoids; ar-TUR: aromatic-turmerone; EtOH: ethanol; IPA: isopropyl alcohol.

Table 10 Extraction of rosemary oleoresins with supercritical fluid, various conventional methods and integration of SFE with other technologies

SFE conditions										
P (MPa)	T (°C)	FR	t (min)	Co-solvent	Oleoresin yield (%)	Bioactive component (mg/g)	Conventional method	Oleoresin yield (%)	Bioactive component (mg/g)	Ref.
30,40	50	0.5 L/min	–	–	1.9–2.4	TPC: 36.8–45.2	Maceration: EtOH (4 h)	30	TPC:123.9	[105]
30,40	50	0.5 L/min	–	EtOH	7.6	TPC:102.0	Maceration: Hexane (4 h)	8.8	TPC:52.9	
10,35	40,60	25 g/min	60	EtOH	0.41–24.01	–	–	–	–	[106]
20,30	40	60 g/min	300	–	3.14–4.45	CA: 52.88–94.84 CL:6.81 β-carotene: 0.64–0.92	Pressurized liquid extraction: Hexane, 10 min, 100°C, 150°C	9.87–15.63	CA: 104.20–161.49 CL: 9.62–13.64 β-carotene: 0.17–0.23	[107]
30	40	60 g/min	360	–	1.62	CA: 180 CL: 16	–	–	–	[108]
15	40	60 g/min	180	EtOH	7.26	CA: 256 CL: 38	–	–	–	
Single-step SFE										
15	40	60 g/min	300	EtOH	6.74	–	PLE: $EtOH/H_2O$ 20 min,10 MPa, 50°C	38.46	–	[109]

Two-step sequential SFE										
Step 1										
30	40	60 g/min	60	–	4.68	–	–	–	–	
Step 2										
15	40	60 g/min	120	EtOH	–		–	–		
Supercritical antisolvent fractionation (SAF)										
10 15	40 40	10 L/min 60 g/min	60–180 180	EtOH EtOH	5.74–20.65 7.26–13.44	– CA: 14.18–25.66	– –	– –	– –	[110]
Fractionation										
Step 1										
15–30	40	60 g/min	60–360	–	1.42–4.93	2–30.69	–	–	–	
Step 2										
15	40	60 g/min	120–180	EtOH						

SFE: supercritical fluid extraction; *P*: pressure; *T*: temperature; FR: carbon dioxide flow rate; *t*: time; TPC: total phenolic content; CA: carnosic acid; CL: carnosol.

carnosol. This SAF process was achieved by the continuous contact between $scCO_2$ and the polar liquid mixture dissolved in EtOH and H_2O in a pressurized separation chamber. The polar fraction that was insoluble in $scCO_2$ was precipitated whereas the nonpolar fraction (soluble in $scCO_2$) was recovered by downstream pressure reduction [112].

4. Supercritical extraction of plant phenolics

The SFE has been introduced as an alternative method for the extraction of plant phenolics. By using SFE, extracts containing a high concentration of plant phenolics could be obtained. For instance, the higher recovery of selected secondary metabolites of 25.5%–84.9% from *Syzygium campanulatum* Korth using SFE with cosolvent, as compared to the conventional solvent extraction (0.92%–66%) has been reported [113]. Besides, the high selectivity of $scCO_2$ is desired for the extraction of targeted plant phenolics from plant matrices [114]. The low operating temperature of the SFE technique is another added parameter for the extraction of thermolabile phenolic compounds. The issue on the presence of solvent residue in the final product can also be avoided since SCF involves the use of very less amount of organic solvents in the form of cosolvents for plant phenolic extraction. Table 11 shows some representative applications of SFE for the extraction of phenolic compounds from different plants at different operating pressures and temperatures.

4.1 Operational conditions of supercritical fluid extraction of phenolic compounds

In the SFE of plant phenolics, several variables are investigated and evaluated before the extraction process for understanding the effects of all variables on the extraction yield and selectivity of each targeted compound and plant matrices [127, 128]. The phase equilibrium and kinetics of the extraction process are largely affected by the operational conditions [129]. Operational conditions such as the pressure and temperature are among the crucial factors for the optimal extraction of plant phenolics. By altering the extraction pressures and temperatures, the phase equilibrium and extraction kinetics are affected due to the change in solvent density and solvating power of $scCO_2$ [130]. Sanjaya et al. [122] reported the SFE of several polyphenol and flavonoid compounds including gallic acid, catechin, ferulic acid, caffeic acid, p-coumaric acid, quercetin, luteolin, and kaempferol from ant-nest-tuber at pressures ranged between 9–22.5 MPa and temperature of 40–70°C. Maran et al. [119] reported the extraction of phenolic compounds including total phenolics, flavonoids and tannins from tea leaves using $scCO_2$ at pressure ranging from 10 to 20 MPa, temperature of 40–60°C, and ethanol as a cosolvent at a flow rate of 1–3 g/min. The pressure and cosolvent flow rate were found to significantly affect the responses with the optimum condition (18.8 MPa, 50°C, and 2.94 g/min) for the recovery of maximum total phenolic compound [119]. Cássia et al. [131] investigated

Table 11 Representative applications involving the use of SFE for phenolic compounds published

Matrix	*T* (°C)/*P* (MPa)	*t* (min)	Compound of interest	References
Soybean residues	35,40/40	–	• Total phenolics • Flavonoids	[33]
Olive leaves	35/15	30–40	• Oleuropein	[115]
Euterpre edulis Mart.	60/20	46	• Phenolic • Anthocyanins	[116]
	50/22		• Anthocyanins	[117]
Moringa oleifera	50/15	180	• Total phenolics • Total flavonoids	[118]
Camellia sinensis L.	50/18.8	60	• Total phenolics • Flavonoids • Tannin	[119]
Clinacanthus nutans lindau	60/35	120	• Phenols • Flavonoids	[120]
Capsium frutescens L.	40/ 15	–	• Capsaicinoids	[117]
Vaccinium myrtillus L.	40/20	–	• Phenolic compounds • Anthocyanins	[121]
Myrmecodia pendans	70/22.5	420	• Flavonoids • Polyphenol	[122]
Eruca sativa	75/30	60	• Glucosinolates	[123]
Sasa palmata	95/20	180	• DL-alanine • Gluconic acid • Phosphoric acid • β-amyrene • A 87-amyrin acetate Friedelin	[124]
Rosmarinus officinalis	40/40	300	• Rosmarinic acid • Carnosic acid	[125]
Horchata by-products	40/30,40	120	• Isohydroxymatairesinol	[126]

P: pressure; *T*: temperature; *t*: time.

the phenolic compounds and anthocyanin contents of acai berry oil using scCO_2 extraction. The total phenolic compounds obtained from lyophilized acai berry pulp at 60°C, 19 MPa were comparable to that found in pulp sample and the best conditions for obtaining anthocyanins compound were 50°C and, 15 MPa [131]. Rodríguez-Pérez et al. [118] reported under the optimum condition of 15 MPa and 50°C, the recovery of different fractions with varied compositions of bioactive compounds from *Moringa oleifera* leaves using scCO_2. Table 11 shows some representative applications of SFE for the extraction of phenolic compounds from different plants at different operating pressures and temperatures.

4.1.1 Pressure

The pressure used for the extraction of plant phenolic for different plant materials varied greatly in the range of 5–60.2 MPa as reviewed by Roberto et al. [132] and Tyśkiewicz et al. [133]. An increase of pressure results in an increase of the density as well as the solvating power of $scCO_2$ = and thus improving the interactions between the SCF and solid matrices allowed higher extraction yield [134]. When the operating pressure increased from 30 to 40 MPa, the extraction yield of phenol from palm kernel shell oil was also increased [135]. A similar observation was reported by Bimakr et al. [136] whereby the extraction yield of flavonoids from spearmint (*Mentha spicata* L.) was increased when the pressure increased from 10 to 20 MPa. However, when the pressure was raised beyond 20 MPa, an unforeseen reduction in the flavonoids yield was observed. The extraction yield of phenolic compounds from propolis increased at pressures of 15–25 MPa but decreased at 25 MPa [137], suggesting the selectivity of $scCO_2$ for phenolic compounds was high at low pressures. At higher pressure, more recovery of volatile fractions and less recovery for nonvolatile fractions were obtained [138]. Therefore, in certain cases such as antioxidant extraction, the selectivity was reduced with an increase of pressure because of the co-extraction of compounds that reduce the purity of the compound, color alteration, and prooxidant action to SCF extracts [139]. Therefore, it is important to optimize the extraction pressure since the relationship of extraction yield does not always increase linearly with pressure.

4.1.2 Temperature

Another important variable in SFE is temperature, where the selection of operating range is dependent on the $scCO_2$ solvent power, thermal stability of solutes, the vapor pressure of the solute, and the properties of the matrix. The effect of temperature on the extraction yield of plant phenolics is also dependent on the change in the $scCO_2$ solvent density. The density of $scCO_2$ decreased at high temperatures and hence, lower extraction yield is anticipated. Also, the effect of temperature on the solute solubility is different at pressures in the critical range. At the critical pressures, the fluid density greatly varies with a small change of temperature. However, beyond the crossover pressure region, high extraction temperatures result in high extraction yield as the vapor pressure, diffusion, and mass transfer rate of the phenolic compounds to be extracted are increased [140]. Hence, the increase of extraction temperature will lead to an increase in the extraction yield, unless too high a temperature is employed, thermolabile components might be degraded. The temperature needs to be set in the vicinity of the critical point, but, as low as possible to prevent degradation of targeted compounds [141]. Hence, a moderate temperature in the ranges of 40–60°C is the frequent temperature used to extract antioxidants from the plant, particularly, phenolic acids [142], flavonoids and terpenoid [143–145], carotenoids [146], and tocopherols [147, 148].

4.1.3 Other factors

The CO_2 flow rate, the particle size of the materials, and the extraction time are important parameters that should be considered for optimal extraction yield.CO_2 flow rate plays an important role in the extraction process and is controlled by an external mass transfer resistance or by phase equilibrium. Meanwhile, the particle size significantly influences the extraction efficiency as the process is controlled by an internal mass transfer resistances [149]. For instance, smaller mean particle size increases the surface of the plant material that is in contact with the solvent, hence, reduces the length of diffusion of the solvent and the mass transfer rate is enhanced [150]. In the extraction of phenolic compounds from *Origanum vulgare* leaves, highest phenolic contents were recovered from the sample with the smallest particle size [151]. The common particle diameter ranged between 0.25 and 2.0 mm for optimal extraction. The particle size of plant materials should not be too small to avoid channeling inside the extraction bed [152]. Extraction time is another parameter that contributes to the optimum extraction process. However, it is related to the CO_2 flow rate and particle size of the sample. For example, a smaller particle size enhances the recovery of phenolic compounds at a shorter extraction time [153].

4.2 Supercritical fluid extraction with cosolvent for phenolics compound

Even though $scCO_2$ is a promising alternative for a sustainable green extraction, it has a limitation in the extraction of polar substrates. Generally, plant phenolics are polar in nature, hence, not completely soluble in the nonpolar $scCO_2$ [154]. This causes the difficulty to extract the covalently or hydrophobically entrapped phenolics (bound forms) [154]. To overcome this limitation either high extraction pressure is employed or a polar modifier/cosolvent is added to improve the solubility of polar organic compounds or their interaction with the solutes [155–157]. Technically, the cosolvents should have a higher polarity than CO_2, thus, expanding the range of compounds attainable [158]. In this regard, the cosolvent increases the affinity [159] and solubility of polar compounds, hence, resulting in a greater yield [160, 161]. The addition of cosolvent is able to enhance the extraction of polar substances due to: (1) change in polarity, density, and viscosity of the extraction fluid; (2) miscibility of the modifier and solvent and the solute solubility; (3) the interaction between $scCO_2$ and the matrix; (4) disruption of the bonding between solutes and the solid matrix. It is important to note that the changes of solubility that are affected by the addition of cosolvent are still depended on the pressure and temperature employed in the extraction process.

The cosolvent used in the extraction process can be a pure solvent or a mixture of solvents. In the food and pharmaceutical applications, ethanol (polarity = 5.2) and water (polarity = 9.0) have been widely used cosolvents, particularly, to improve the extraction efficiency of phenolic compounds [133]. Both ethanol and water are considered as green solvents [162] and can be easily removed from the final product. The cosolvent can be either mixed with CO_2 before being pumped to the extractor or mixed with the raw material in SFE. The mixing of the cosolvent into the CO_2 stream is known as a sequential, gradual, or continuous cosolvent addition. This type of operation has been used for the extraction of phenolic compounds from yeast biomass, herbs, ginger, leaves, rye bran, carrots, and mushroom.

Water has been commonly chosen as cosolvent to use in several prominent industries such as $scCO_2$ extraction processes of nicotine, caffeine, and vanillin [163]. Solana et al. [123] have reported the effect of different cosolvents for the $scCO_2$ extraction of phenolic compounds from the *Eruca sativa* leaves extract. The results showed that water (8%, w/w) was the most efficient cosolvent for the extraction of phenolic compounds at 30 MPa and 75°C. The extraction yield was enhanced by the addition of water as cosolvent due to the ability of water to swell the solid sample which increased the solute diffusivity. It happens due to reduction in interactions between the solute and the matrix as a result of adsorption of water onto the polar sites and the interactions of functional groups of the oxygenated compounds with water [163]. However, the increased polarity of CO_2 would be disadvantageous for extracting nonpolar components at high pressure because of swelling limitations due to compression [129]. Besides that, some limitations might arise, including: (i) the formation of ice blockages during expansion; (ii) reduced solubility and extractability of ionizable compounds; and (iii) reduced the shelf life of the product [163].

Ethanol also has been widely used as a cosolvent in the extraction of phenolic compounds because of its ability to form hydrogen bonding and dipole-dipole interactions with phenols. Thus, ethanol is compatible as a cosolvent with CO_2 to extract the phenolic compounds [159]. Castro-Vargas et al. [164] found ethanol at concentration level of 10% (w/w) as the best cosolvent for the extraction of phenolic fraction from guava seeds at 30 MPa and 50°C using $scCO_2$. Santos et al. [165] also reported that ethanol (20%) as a cosolvent for the extraction of phenolic compounds from *Eucalyptus globulus* provided the best extraction yield, high total phenolic content, and antioxidant activity.

The effectiveness of the SFE using the combination of water and ethanol mixtures as cosolvent in obtaining high phenolic extraction yield has been reported [166–168]. Paes et al. [121] evaluated the phenolic compounds and anthocyanins from the blueberry residues using $scCO_2$ at the operating conditions of 15–25 MPa and 40°C with a cosolvents of ethanol, water, and acidified water. They suggested a moderate pressure of 20 MPa with a combination of ethanol and water as cosolvents for satisfactory results for the extraction of phenolic compounds from blueberry residues.

4.3 Benefits and limitations of SFE in phenolic compounds extraction

Lang and Wai [169] had discussed the advantages, as summarized below, of using SCF as a tool for the extraction of the phenolic compound.

(1) The unique properties of the SCF such as high mass diffusivity, low viscosity, and surface tension favored the effective penetration of SCF into the sample matrices for enhancing the mass transfer rate [170]. As a result, the extraction time reduced compared to conventional methods. However, the nonpolar characteristic of the $scCO_2$ was not favorable for the extraction of the polar phenolic compounds [126]. The study conducted by Castro-Vargas et al. [164] found out that the phenolic fraction was lower using SFE when compared to the Soxhlet extraction although the total extraction yield was higher with SFE.

(2) The naturally occurring separation of the CO_2 solvent and the extract due to the high volatility of CO_2 eliminated the need for further separation process which means complete extraction through the repeated reflux of SCF to the sample. The fluid changed to gas and separated from the extracts with the pressure drop.

(3) The selectivity of SCF was higher than liquid solvent [150].

(4) The dissolving power of the SCF depending on its density can be varied continuously by manipulating the temperature and pressure of the system, particularly, in the vicinity of the critical point [171, 172]. It is important to know that a small isothermal raise in pressure near the critical point resulted in significant increase in the fluid density [172], which suggest that the changing of the conditions may be necessary for some applications to ensure the phenolic compounds completely dissolve in the SCF.

(5) SFE was suitable for thermolabile compound since it operated at a moderate temperature. For instance, $scCO_2$ has a critical temperature of 31.1 °C.

(6) The minimum consumption of organic solvent in SFE has contributed to saving the environment. Even though the modifiers are needed in the extraction of phenolic compounds, however, it still considers low consumption of organic solvent compared to the conventional extraction. Ethanol as a modifier in SFE was successfully used to recovered polyphenol from grape [173] and antioxidants from lees of rice wine [174, 175].

(7) SFE can be modified by integrating with other devices such as online coupling of SFE with chromatographic for highly volatile compounds [176].

(8) The SCF can be recycled and thus the waste generation is reduced [177].

(9) The specific doses from a few milligram samples in the laboratory to tons of sample in industries can be applied by arranging the SFE scale [178].

(10) The optimization extraction process can be done by manipulating the experimental conditions of the SFE process [119].

4.4 Perspective and future direction for SFE of phenolic compounds (assisted supercritical fluids technology)

SFE, particularly $scCO_2$ is emerging as a green technique to extract and isolate the high-valuable compounds and phytochemicals from natural matrices. The superiority of this technique to recover relatively high purity of compounds has made it special for the application in functional food, pharmaceutical, nutraceutical, and cosmeceutical industries [179]. Combining supercritical fluids with other processes especially for the extraction and isolation of high-valuable compounds is a further area of interest that is expected to gain momentum and be the focus of wider research projects and collaborations. Herein, directions of trends that can be employed to overcome the challenges and gaps in the SFC process, particularly for phenolic compounds extraction are discussed.

As per discussion, plant phenolics being polar have low solubility in the nonpolar $scCO_2$. This limitation can be overcome with the addition of a small volume of cosolvent such as ethanol to improve the polarity and enhance the extractability. However, it is hardly achievable to recover the covalent and hydrophobic entrapped phenolics (bound forms) [154]. The extractability of plant phenolic compounds can be enhanced with the integration of SFE with other technologies.

4.4.1 Enzyme-assisted supercritical fluid extraction

Enzyme-assisted SFE can be one of the techniques to enhance the extraction efficiency of the phenolic compound. Previously, the role of the enzyme in assisting the extraction of phenolic compounds using the solvent extraction and cold pressing has been investigated. The findings showed that the use of enzymes had improved the extraction yield of polysaccharides [180], edible and nonedible oils [181], protein [182], and phenolics [183]. This is because an enzyme can hydrolyze the cellulosic composite structure of plant cell wall, hence, enhance the recovery of both bound and free phenolics [154]. In the enzyme-assisted SFE of plant phenolics, the sample is pretreated with the enzyme before going through the $scCO_2$ extraction. Extraction of antioxidant phenolics using enzyme-assisted SFE from pomegranate peel doubled the recovery of crude extracts with an increased level of phenolic compounds [154].

4.4.2 Ultrasound-assisted supercritical fluid extraction

Ultrasound-assisted supercritical fluid extraction (US-SFE) integrates both the ultrasound and SFE technologies. The principle of ultrasound-assisted extraction (UAE) is the sonochemical phenomenon associated with acoustic cavitation and the formation of microbubbles when a large pressure is applied to a liquid [184]. These formed bubbles collapsed after growing up to an unstable size with the release of intense local energy due to important chemical and mechanical effects [185]. UAE has emerged as a promising technique for extraction with various benefits. The main advantage of the ultrasound application is that it allows the extraction of both polar and nonpolar compounds [186].

4.4.3 High hydrostatic pressure supercritical fluid extraction

The combination of high hydrostatic pressure (HHP) and SFE (HPP-SFE) is a promising technology to enhance the extraction efficiency of the process. The HHP is a nonthermal process, mainly to extend the shelf life of products through maintaining the quality and organoleptic characteristics of foods [187]. It has been previously used in the extraction of bioactive compounds from several fruits and products since it can increase the cell permeability and enables the diffusion of metabolites [188]. Therefore, the use of HHP as a pretreatment process may improve the efficiency of the extraction process. Torres-Ossandón et al. [189] reported the extraction of bioactive compounds with high antioxidant capacity (phenolics and β-carotene) from Cape gooseberry using HHP-SFE. The observed results showed that the high pressure in the SFE promoted active ingredient release in plant matrices, hence, increased the antioxidant activity. In addition, the recovery of sulfur bioactive compounds from Allium species using the HHP-SFE technology produced a superior quality compared to thermal extraction methods [190].

5. Conclusion

The extraction of oleoresins and plant phenolics using supercritical fluid extraction (SFE) technologies has been discussed. In order to optimize the extraction efficiency, there are several important parameters that need to be considered: (1) the operating parameters including the pressure, temperature, carbon dioxide flow rate, and extraction time and (2) the state of raw material, such as the particle size and moisture content. Besides, enzymatic pretreatment of the sample and the addition of cosolvent in the SFE process could also enhance the efficacy of the SFE for the oleoresins and plant phenolics. With the introduction of SFE integrated technologies, namely enzyme-assisted SFE, ultrasound assisted SFE, and high hydrostatic pressure SFE, the extraction yield of plant phenolics has increased.

References

[1] A.R.N. Raihana, J.M.N. Marikkar, I. Amin, M. Shuhaimi, A review on food values of selected tropical fruits seeds, Int. J. Food Prop. 18 (2015) 2380–2392, https://doi.org/10.1080/10942912.2014.980946.
[2] Y.-Y. Soong, P.J. Barlow, Antioxidant activity and phenolic content of selected fruit seeds, Food Chem. 88 (2004) 411–417, https://doi.org/10.1016/J.FOODCHEM.2004.02.003.
[3] E. Pastrana-Bonilla, C.C. Akoh, S. Sellappan, G. Krewer, Phenolic content and antioxidant capacity of muscadine grapes, J. Agric. Food Chem. 51 (2003) 5497–5503, https://doi.org/10.1021/jf030113c.

[4] H.M. Chandra, S. Ramalingam, Antioxidant potentials of skin, pulp, and seed fractions of commercially important tomato cultivars. Food Sci. Biotechnol. 20 (2011) 15–21, https://doi.org/10.1007/s10068-011-0003-z.
[5] D.R.A. Muhammad, K. Dewettinck, Cinnamon and its derivatives as potential ingredient in functional food—A review. Int. J. Food Prop. 20 (2017) 2237–2263, https://doi.org/10.1080/10942912.2017.1369102.
[6] G. Melgar-Lalanne, A.J. Hernández-Álvarez, M. Jiménez-Fernández, E. Azuara, Oleoresins from Capsicum spp.: extraction methods and bioactivity. Food Bioprocess. Technol. 10 (2017) 51–76, https://doi.org/10.1007/s11947-016-1793-z.
[7] R. Tambun, R.R.H. Purba, H.K. Ginting, in: Extraction of basil leaves (ocimum canum) oleoresin with ethyl acetate solvent by using soxhletation method, IOP Conf. Ser. Mater. Sci. Eng, 2017. https://doi.org/10.4324/9781315519135.
[8] J. López-Cervantes, D.I. Sánchez-Machado, P. Cruz-Flores, M.F. Mariscal-Domínguez, G. Servín de la Mora-López, O.N. Campas-Baypoli, Antioxidant capacity, proximate composition, and lipid constituents of Aloe vera flowers. J. Appl. Res. Med. Aromat. Plants 10 (2018) 93–98, https://doi.org/10.1016/j.jarmap.2018.02.004.
[9] J. da Silva, DNA damage induced by occupational and environmental exposure to miscellaneous chemicals, Mutat. Res. 770 (2016) 170–182.
[10] D. Lin, M. Xiao, J. Zhao, Z. Li, B. Xing, X. Li, M. Kong, L. Li, Q. Zhang, Y. Liu, H. Chen, W. Qin, H. Wu, S. Chen, An overview of plant phenolic compounds and their importance in human nutrition and management of type 2 diabetes. Molecules 21 (2016), https://doi.org/10.3390/molecules21101374.
[11] A. Altemimi, N. Lakhssassi, A. Baharlouei, D. Watson, D. Lightfoot, Phytochemicals: extraction, isolation, and identification of bioactive compounds from plant extracts. Plants 6 (2017) 42, https://doi.org/10.5912/jcb722.
[12] L. Abenavoli, A.A. Izzo, N. Milić, C. Cicala, A. Santini, R. Capasso, Milk thistle (Silybum marianum): a concise overview on its chemistry, pharmacological, and nutraceutical uses in liver diseases. Phyther. Res. 32 (2018) 2202–2213, https://doi.org/10.1002/ptr.6171.
[13] Y. Bi, M. Min, W. Shen, Y. Liu, Genistein induced anticancer effects on pancreatic cancer cell lines involves mitochondrial apoptosis, G0/G1cell cycle arrest and regulation of STAT3 signalling pathway. Phytomedicine 39 (2018) 10–16, https://doi.org/10.1016/J.PHYMED.2017.12.001.
[14] S. Chen, Z. Zeng, N. Hu, B. Bai, H. Wang, Y. Suo, Simultaneous optimization of the ultrasound-assisted extraction for phenolic compounds content and antioxidant activity of Lycium ruthenicum Murr. fruit using response surface methodology. Food Chem. 242 (2018) 1–8, https://doi.org/10.1016/j.foodchem.2017.08.105.
[15] H.-J. Jang, H.-J. Lee, D.-K. Yoon, D.-S. Ji, J.-H. Kim, C.-H. Lee, Antioxidant and antimicrobial activities of fresh garlic and aged garlic by-products extracted with different solvents. Food Sci. Biotechnol. 27 (2018) 219–225, https://doi.org/10.1007/s10068-017-0246-4.
[16] D.Y. Choi, Y.J. Lee, J.T. Hong, H.J. Lee, Antioxidant properties of natural polyphenols and their therapeutic potentials for Alzheimer's disease. Brain Res. Bull. 87 (2012) 144–153, https://doi.org/10.1016/j.brainresbull.2011.11.014.
[17] T.T. Garmus, L.C. Paviani, C.L. Queiroga, P.M. Magalhães, F.A. Cabral, Extraction of phenolic compounds from pitanga (Eugenia uniflora L.) leaves by sequential extraction in fixed bed extractor using supercritical CO2, ethanol and water as solvents. J. Supercrit. Fluids 86 (2014) 4–14, https://doi.org/10.1016/j.supflu.2013.11.014.
[18] C. Da Porto, E. Porretto, D. Decorti, Comparison of ultrasound-assisted extraction with conventional extraction methods of oil and polyphenols from grape (Vitis vinifera L.) seeds. Ultrason. Sonochem. 20 (2013) 1076–1080, https://doi.org/10.1016/j.ultsonch.2012.12.002.
[19] K. Papoutsis, P. Pristijono, J.B. Golding, C.E. Stathopoulos, M.C. Bowyer, C.J. Scarlett, Q. V. Vuong, Screening the effect of four ultrasound-assisted extraction parameters on hesperidin and phenolic acid content of aqueous citrus pomace extracts. Food Biosci. 21 (2018) 20–26, https://doi.org/10.1016/J.FBIO.2017.11.001.
[20] S.H. Nile, A.S. Nile, Y.-S. Keum, Total phenolics, antioxidant, antitumor, and enzyme inhibitory activity of Indian medicinal and aromatic plants extracted with different extraction methods. 3 Biotech. 7 (2017) 76, https://doi.org/10.1007/s13205-017-0706-9.

[21] J. Živković, K. Šavikin, T. Janković, N. Ćujić, N. Menković, Optimization of ultrasound-assisted extraction of polyphenolic compounds from pomegranate peel using response surface methodology. Sep. Purif. Technol. 194 (2018) 40–47, https://doi.org/10.1016/J.SEPPUR.2017.11.032.
[22] S. Nipornram, W. Tochampa, P. Rattanatraiwong, R. Singanusong, Optimization of low power ultrasound-assisted extraction of phenolic compounds from mandarin (Citrus reticulata Blanco cv. Sainampueng) peel. Food Chem. 241 (2018) 338–345, https://doi.org/10.1016/J.FOODCHEM.2017.08.114.
[23] X. Luo, J. Cui, H. Zhang, Y. Duan, D. Zhang, M. Cai, G. Chen, Ultrasound assisted extraction of polyphenolic compounds from red sorghum (Sorghum bicolor L.) bran and their biological activities and polyphenolic compositions. Ind. Crops Prod. 112 (2018) 296–304, https://doi.org/10.1016/J.INDCROP.2017.12.019.
[24] A. Lante, G. Barion, S. Zannoni, M.R. Pastore, F. Tinello, C. Dal Cortivo, T. Vamerali, G. Mosca, An ecofriendly procedure to extract isoflavones from soybean seeds. J. Clean. Prod. 170 (2018) 1102–1110, https://doi.org/10.1016/J.JCLEPRO.2017.09.218.
[25] S. Saklar Ayyildiz, B. Karadeniz, N. Sagcan, B. Bahar, A.A. Us, C. Alasalvar, Optimizing the extraction parameters of epigallocatechin gallate using conventional hot water and ultrasound assisted methods from green tea. Food Bioprod. Process. 111 (2018) 37–44, https://doi.org/10.1016/J.FBP.2018.06.003.
[26] T.W. Caldas, K.E.L. Mazza, A.S.C. Teles, G.N. Mattos, A.I.S. Brígida, C.A. Conte-Junior, R.G. Borguini, R.L.O. Godoy, L.M.C. Cabral, R.V. Tonon, Phenolic compounds recovery from grape skin using conventional and non-conventional extraction methods. Ind. Crop. Prod. 111 (2018) 86–91, https://doi.org/10.1016/J.INDCROP.2017.10.012.
[27] Y.-Z. Wang, S.-G. Fu, S.-Y. Wang, D.-J. Yang, Y.-H.S. Wu, Y.-C. Chen, Effects of a natural antioxidant, polyphenol-rich rosemary (Rosmarinus officinalis L.) extract, on lipid stability of plant-derived omega-3 fatty-acid rich oil. LWT 89 (2018) 210–216, https://doi.org/10.1016/J.LWT.2017.10.055.
[28] M.L. Blackhall, R. Berry, N.W. Davies, J.T. Walls, Optimized extraction of anthocyanins from Reid Fruits' Prunus avium 'Lapins' cherries. Food Chem. 256 (2018) 280–285, https://doi.org/10.1016/J.FOODCHEM.2018.02.137.
[29] I. Strati, G. Kostomitsopoulos, F. Lytras, P. Zoumpoulakis, C. Proestos, V. Sinanoglou, I.F. Strati, G. Kostomitsopoulos, F. Lytras, P. Zoumpoulakis, C. Proestos, V.J. Sinanoglou, Optimization of polyphenol extraction from allium ampeloprasum var. porrum through response surface methodology. Foods 7 (2018) 162, https://doi.org/10.3390/foods7100162.
[30] M. Zubail, F. Anwar, S.A. Shahid, Effect of extraction solvents on phenolics and antioxidant activity of selected varieties of Pakistani rice (Oryza sativa), Int. J. Agric. Biol. 14 (2012) 935–940.
[31] J. Ito, K. Hara, T. Someya, T. Myoda, Y. Sagane, T. Watanabe, R.G.S. Wijesekara, K. Toeda, S. Nojima, Data on the inhibitory effect of traditional plants from Sri Lanka against tyrosinase and collagenase. Data Brief 20 (2018) 573–576, https://doi.org/10.1016/j.dib.2018.08.143.
[32] Q.U. Le, H.L. Lay, M.C. Wu, Antioxidant activities and HepG2 cells growth inhibitory capacity of whole plant ethanol extracts (Eclipta alba Hassk and Mesona procumbens Hemsl). J. Food Biochem. 42 (2017) e12454. https://doi.org/10.1111/jfbc.12454.
[33] M.V. Alvarez, S. Cabred, C.L. Ramirez, M.A. Fanovich, Valorization of an agroindustrial soybean residue by supercritical fluid extraction of phytochemical compounds. J. Supercrit. Fluids 143 (2019) 90–96, https://doi.org/10.1016/j.supflu.2018.07.012.
[34] F.S. Al Amri, M.A. Hossain, Comparison of total phenols, flavonoids and antioxidant potential of local and imported ripe bananas, Egypt. J. Basic Appl. Sci. (2018), https://doi.org/10.1016/J.EJBAS.2018.09.002.
[35] P.J. Correia, T. Saavedra, F. Gama, M. da Graça Miguel, A. de Varennes, M. Pestana, Biologically active compounds available in Ceratonia siliqua L. grown in contrasting soils under Mediterranean climate. Sci. Hortic. (Amsterdam) 235 (2018) 228–234, https://doi.org/10.1016/J.SCIENTA.2018.03.010.
[36] M. Sintara, L. Li, D.G. Cunningham, R.L. Prior, X. Wu, T. Chang, Single-laboratory validation for determination of total soluble proanthocyanidins in cranberry using 4-dimethylaminocinnamaldehyde. J. AOAC Int. 101 (2018) 805–809, https://doi.org/10.5740/jaoacint.17-0288.

[37] R. Tegelberg, V. Virjamo, R. Julkunen-Tiitto, Dry-air drying at room temperature - a practical pretreatment method of tree leaves for quantitative analyses of phenolics?. Phytochem. Anal. 29 (2018) 493–499, https://doi.org/10.1002/pca.2755.
[38] D. Šamec, I. Pavlović, I. Radojčić Redovniković, B. Salopek-Sondi, Comparative analysis of phytochemicals and activity of endogenous enzymes associated with their stability, bioavailability and food quality in five Brassicaceae sprouts. Food Chem. 269 (2018) 96–102, https://doi.org/10.1016/J.FOODCHEM.2018.06.133.
[39] C.D. Stalikas, Extraction, separation, and detection methods for phenolic acids and flavonoids. J. Sep. Sci. 30 (2007) 3268–3295, https://doi.org/10.1002/jssc.200700261.
[40] B.E.C. Ziani, S.A. Heleno, K. Bachari, M.I. Dias, M.J. Alves, L. Barros, I.C.F.R. Ferreira, Phenolic compounds characterization by LC-DAD- ESI/MSn and bioactive properties of Thymus algeriensis Boiss. & Reut. and Ephedra alata Decne. Food Res. Int. (2018), https://doi.org/10.1016/j.foodres.2018.08.041.
[41] N. de A. Neves, P.C. Stringheta, S. Gómez-Alonso, I. Hermosín-Gutiérrez, Flavonols and ellagic acid derivatives in peels of different species of jabuticaba (Plinia spp.) identified by HPLC-DAD-ESI/MSn. Food Chem. 252 (2018) 61–71, https://doi.org/10.1016/j.foodchem.2018.01.078.
[42] S. Hachani, C. Hamia, S. Boukhalkhal, A.M.S. Silva, A. Djeridane, M. Yousfi, Morphological, physico-chemical characteristics and effects of extraction solvents on UHPLC-DAD-ESI-MSn profiling of phenolic contents and antioxidant activities of five date cultivars (Phoenix dactylifera L.) growing in Algeria. NFS J. 13 (2018) 10–22, https://doi.org/10.1016/J.NFS.2018.10.001.
[43] Q. Zhang, X.F. Hu, M. man Xin, H.B. Liu, L.J. Sun, S.L. Morris-Natschke, Y. Chen, K.H. Lee, Antidiabetic potential of the ethyl acetate extract of Physalis alkekengi and chemical constituents identified by HPLC-ESI-QTOF-MS. J. Ethnopharmacol. 225 (2018) 202–210, https://doi.org/10.1016/j.jep.2018.07.007.
[44] R. Jaouadi, S.M. Cardoso, A.M.S. Silva, I.B.H. Yahia, M. Boussaid, Y. Zaouali, Variation of phenolic constituents of Tunisian Thymus capitatus (L.) Hoff. et Link populations. Biochem. Syst. Ecol. 77 (2018) 10–15, https://doi.org/10.1016/j.bse.2017.12.009.
[45] M.H. Zuknik, N.A. Nik Norulaini, W.S.W.N. Dalila, N.R. Ali, A.K.M. Omar, Solubility of virgin coconut oil in supercritical carbon dioxide, J. Food Eng. 168 (2016) 240–244.
[46] S. Zhao, D. Zhang, An experimental investigation into the solubility of Moringa oleifera oil in supercritical carbon dioxide, J. Food Eng. 138 (2014) 1–10.
[47] M.P. Fernández-Ronco, C. Ortega-Noblejas, I. Gracia, A. De Lucas, M.T. García, J. F. Rodríguez, Supercritical fluid fractionation of liquid oleoresin capsicum: Statistical analysis and solubility parameters. J. Supercrit. Fluids 54 (2010) 22–29, https://doi.org/10.1016/j.supflu.2010.03.011.
[48] G. Romero De La Vega, M.A. Salgado Cervantes, M.A. Garcia Alvarado, A. Romero-Martínez, P.E. Hegel, Fractionation of vanilla oleoresin by supercritical CO_2 technology. J. Supercrit. Fluids 108 (2016) 79–88, https://doi.org/10.1016/j.supflu.2015.10.022.
[49] L.P.S. Silva, J. Martínez, Mathematical modeling of mass transfer in supercritical fluid extraction of oleoresin from red pepper. J. Food Eng. 133 (2014) 30–39, https://doi.org/10.1016/j.jfoodeng.2014.02.013.
[50] R.B. Bird, Transport phenomena. Appl. Mech. Rev. 55 (2002) R1, https://doi.org/10.1115/1.1424298.
[51] A.C. De Aguiar, L.P. Sales, J.P. Coutinho, G.F. Barbero, H.T. Godoy, J. Martínez, Supercritical carbon dioxide extraction of Capsicum peppers: global yield and capsaicinoid content. J. Supercrit. Fluids 81 (2013) 210–216, https://doi.org/10.1016/j.supflu.2013.05.008.
[52] M. Kehili, M. Kammlott, S. Choura, A. Zammel, C. Zetzl, I. Smirnova, N. Allouche, S. Sayadi, Supercritical CO_2 extraction and antioxidant activity of lycopene and β-carotene-enriched oleoresin from tomato (Lycopersicum esculentum L.) peels by-product of a Tunisian industry. Food Bioprod. Process. 102 (2017) 340–349, https://doi.org/10.1016/j.fbp.2017.02.002.
[53] P.P. Said, O.P. Arya, R.C. Pradhan, R.S. Singh, B.N. Rai, Separation of oleoresin from ginger rhizome powder using green processing technologies. J. Food Process Eng. 38 (2015) 107–114, https://doi.org/10.1111/jfpe.12127.
[54] L. Vallecilla-Yepez, O.N. Ciftci, Increasing cis-lycopene content of the oleoresin from tomato processing byproducts using supercritical carbon dioxide. LWT 95 (2018) 354–360, https://doi.org/10.1016/j.lwt.2018.04.065.

[55] D. Urbonaviciene, P. Viskelis, The cis-lycopene isomers composition in supercritical CO2 extracted tomato by-products. LWT Food Sci. Technol. 85 (2017) 517–523, https://doi.org/10.1016/j.lwt.2017.03.034.

[56] M. Kehili, L.M. Schmidt, W. Reynolds, A. Zammel, C. Zetzl, I. Smirnova, N. Allouche, S. Sayadi, Biorefinery cascade processing for creating added value on tomato industrial by-products from Tunisia. Biotechnol. Biofuels 9 (2016) 1–12, https://doi.org/10.1186/s13068-016-0676-x.

[57] M. Durante, A. Montefusco, P.P. Marrese, M. Soccio, D. Pastore, G. Piro, G. Mita, M.S. Lenucci, Seeds of pomegranate, tomato and grapes: an underestimated source of natural bioactive molecules and antioxidants from agri-food by-products. J. Food Compos. Anal. 63 (2017) 65–72, https://doi.org/10.1016/j.jfca.2017.07.026.

[58] A. Bruno, M. Durante, P.P. Marrese, D. Migoni, M.N. Laus, E. Pace, D. Pastore, G. Mita, G. Piro, M.S. Lenucci, Shades of red: Comparative study on supercritical CO2 extraction of lycopene-rich oleoresins from gac, tomato and watermelon fruits and effect of the α-cyclodextrin clathrated extracts on cultured lung adenocarcinoma cells' viability. J. Food Compos. Anal. 65 (2018) 23–32, https://doi.org/10.1016/j.jfca.2017.08.007.

[59] M.H. Chen, T.C. Huang, Volatile and nonvolatile constituents and antioxidant capacity of oleoresins in three Taiwan citrus varieties as determined by supercritical fluid extraction. Molecules 21 (2016) 1735, https://doi.org/10.3390/molecules21121735.

[60] A. López-Padilla, A. Ruiz-Rodriguez, G. Reglero, T. Fornari, Supercritical carbon dioxide extraction of *Calendula officinalis*: kinetic modeling and scaling up study, J. Supercrit. Fluids 130 (2017) 292–300.

[61] N. Nagavekar, R.S. Singhal, Enhanced extraction of oleoresin from Piper nigrum by supercritical carbon dioxide using ethanol as a co-solvent and its bioactivity profile. J. Food Process Eng. 41 (2017) 1–12, https://doi.org/10.1111/jfpe.12670.

[62] L. Trung, H. Chi, M. City, T.P. Ward, H. Chi, M. City, Optimization of supercritical CO_2 extraction of oleoresin from black pepper (Piper nigrum L.) and antioxidant capacity of the oleoresin, Int. Food Res. J. 21 (2014) 1489–1493.

[63] A. Mouahid, C. Crampon, S.A.A. Toudji, E. Badens, Effects of high water content and drying pre-treatment on supercritical CO2extraction from Dunaliella salina microalgae: experiments and modelling. J. Supercrit. Fluids 116 (2016) 271–280, https://doi.org/10.1016/j.supflu.2016.06.007.

[64] G. Vasapollo, L. Longo, L. Rescio, L. Ciurlia, Innovative supercritical CO_2 extraction of lycopene from tomato in the presence of vegetable oil as co-solvent, J. Supercrit. Fluids 29 (2004) 87–96.

[65] M.S. Lenucci, M. De Caroli, P.P. Marrese, A. Iurlaro, L. Rescio, V. Böhm, G. Dalessandro, G. Piro, Enzyme-aided extraction of lycopene from high-pigment tomato cultivars by supercritical carbon dioxide. Food Chem. 170 (2015) 193–202, https://doi.org/10.1016/j.foodchem.2014.08.081.

[66] D.A. Esquivel-Hernández, V.H. López, J. Rodríguez-Rodríguez, G.S. Alemán-Nava, S.P. Cuéllar-Bermúdez, M. Rostro-Alanis, R. Parra-Saldívar, Supercritical carbon dioxide and microwave-assisted extraction of functional lipophilic compounds from Arthrospira platensis. Int. J. Mol. Sci. 17 (2016) https://doi.org/10.3390/ijms17050658.

[67] E.M.Z. Michielin, L.F.V. Bresciani, L. Danielski, R.A. Yunes, S.R.S. Ferreira, Composition profile of horsetail (Equisetum giganteum L.) oleoresin: Comparing SFE and organic solvents extraction. J. Supercrit. Fluids 33 (2005) 131–138, https://doi.org/10.1016/j.supflu.2004.07.004.

[68] L. Danielski, L.M.A.S. Campos, L.F.V. Bresciani, H. Hense, R.A. Yunes, S.R.S. Ferreira, Marigold (Calendula officinalis L.) oleoresin: solubility in scCO2 and composition profile, Chem. Eng. Process. Process Intensif. 46 (2007) 99–106.http://www.sciencedirect.com/science/article/pii/S0255270106001218. Accessed 8 March 2015.

[69] L.M.A.S. Campos, E.M.Z. Michielin, L. Danielski, S.R.S. Ferreira, Experimental data and modeling the supercritical fluid extraction of marigold (Calendula officinalis) oleoresin. J. Supercrit. Fluids 34 (2005) 163–170, https://doi.org/10.1016/j.supflu.2004.11.010.

[70] T. de S. Weinhold, L.F.V. Bresciani, C.W. Tridapalli, R.A. Yunes, H. Hense, S.R.S. Ferreira, Polygala cyparissias oleoresin: Comparing CO2 and classical organic solvent extractions. Chem. Eng. Process. Process Intensif. 47 (2008) 109–117, https://doi.org/10.1016/j.cep.2007.08.007.

[71] J.M. Del Valle, C. Mena, M. Budinich, Extraction of garlic with supercritical CO2 and conventional organic solvents. Braz. J. Chem. Eng. 25 (2008) 535–542, https://doi.org/10.1590/S0104-66322008000300011.

[72] L. Ciurlia, M. Bleve, L. Rescio, Supercritical carbon dioxide co-extraction of tomatoes (Lycopersicum esculentum L.) and hazelnuts (Corylus avellana L.): A new procedure in obtaining a source of natural lycopene. J. Supercrit. Fluids 49 (2009) 338–344, https://doi.org/10.1016/j.supflu.2009.03.003.

[73] M. Kehili, S. Choura, A. Zammel, N. Allouche, S. Sayadi, Oxidative stability of refined olive and sunflower oils supplemented with lycopene-rich oleoresin from tomato peels industrial by-product, during accelerated shelf-life storage. Food Chem. 246 (2018) 295–304, https://doi.org/10.1016/j.foodchem.2017.11.034.

[74] M. Honda, T. Higashiura, T. Fukaya, Safety assessment of a natural tomato oleoresin containing high amounts of *Z*-isomers of lycopene prepared with supercritical carbon dioxide. J. Sci. Food Agric. 97 (2017) 1027–1033, https://doi.org/10.1002/jsfa.7830.

[75] C. Longo, L. Leo, A. Leone, Carotenoids, fatty acid composition and heat stability of supercritical carbon dioxide-extracted-oleoresins. Int. J. Mol. Sci. 13 (2012) 4233–4254, https://doi.org/10.3390/ijms13044233.

[76] A. Leone, R. Zefferino, C. Longo, L. Leo, G. Zacheo, Supercritical CO2-extracted tomato oleoresins enhance gap junction intercellular communications and recover from mercury chloride inhibition in keratinocytes. J. Agric. Food Chem. 58 (2010) 4769–4778, https://doi.org/10.1021/jf1001765.

[77] M.S. Lenucci, A. Caccioppola, M. Durante, L. Serrone, R. Leonardo, G. Piro, G. Dalessandro, Optimisation of biological and physical parameters for lycopene supercritical CO2 extraction from ordinary and high-pigment tomato cultivars. J. Sci. Food Agric. 90 (2010) 1709–1718, https://doi.org/10.1002/jsfa.4006.

[78] J. López-Cervantes, D.I. Sánchez-Machado, K.P. Valenzuela-Sánchez, J.A. Núñez-Gastélum, A.A. Escárcega-Galaz, R. Rodríguez-Ramírez, Effect of solvents and methods of stirring in extraction of lycopene, oleoresin and fatty acids from over-ripe tomato. Int. J. Food Sci. Nutr. 65 (2014) 187–193, https://doi.org/10.3109/09637486.2013.839630.

[79] J.M. Fernández-Sevilla, F.G. Acién Fernández, E. Molina Grima, Biotechnological production of lutein and its applications. Appl. Microbiol. Biotechnol. 86 (2010) 27–40, https://doi.org/10.1007/s00253-009-2420-y.

[80] J. Vechpanich, A. Shotipruk, Recovery of free lutein from tagetes erecta: determination of suitable saponification and crystallization conditions. Sep. Sci. Technol. 46 (2011) 265–271, https://doi.org/10.1080/01496395.2010.506904.

[81] N. Mezzomo, J. Martínez, S.R.S. Ferreira, Economical viability of SFE from peach almond, spearmint and marigold. J. Food Eng. 103 (2011) 473–479, https://doi.org/10.1016/j.jfoodeng.2010.10.032.

[82] Y. Gong, S. Plander, H. Xu, B. Simandi, Y. Gao, Supercritical CO 2 extraction of oleoresin from marigold (Tagetes erecta L.) flowers and determination of its antioxidant components with online HPLC-ABTS·+ assay. J. Sci. Food Agric. 91 (2011) 2875–2881, https://doi.org/10.1002/jsfa.4537.

[83] H.K. Kiriamiti, S. Camy, C. Gourdon, J.S. Condoret, Pyrethrin exraction from pyrethrum flowers using carbon dioxide. J. Supercrit. Fluids 26 (2003) 193–200, https://doi.org/10.1016/S0896-8446(02)00165-1.

[84] M. Gallo, A. Formato, D. Ianniello, A. Andolfi, E. Conte, M. Ciaravolo, V. Varchetta, D. Naviglio, Supercritical fluid extraction of pyrethrins from pyrethrum flowers (Chrysanthemum cinerariifolium) compared to traditional maceration and cyclic pressurization extraction. J. Supercrit. Fluids 119 (2017) 104–112, https://doi.org/10.1016/j.supflu.2016.09.012.

[85] J. Sarris, A. Panossian, I. Schweitzer, C. Stough, A. Scholey, Herbal medicine for depression, anxiety and insomnia: a review of psychopharmacology and clinical evidence. Eur. Neuropsychopharmacol. 21 (2011) 841–860, https://doi.org/10.1016/j.euroneuro.2011.04.002.

[86] J.D. Amsterdam, Y. Li, I. Soeller, K. Rockwell, J.J. Mao, J. Shults, A randomized, double-blind, placebo-controlled trial of oral matricaria recutita (chamomile) extract therapy for generalized anxiety disorder. J. Clin. Psychopharmacol. 29 (2009) 378–382, https://doi.org/10.1097/JCP.0b013e3181ac935c.

[87] M. Molnar, N. Mendešević, D. Šubarić, I. Banjari, S. Jokić, Comparison of various techniques for the extraction of umbelliferone and herniarin in Matricaria chamomilla processing fractions. Chem. Cent. J. 11 (2017) 1–8, https://doi.org/10.1186/s13065-017-0308-y.

[88] P. Kotnik, M. Škerget, Ž. Knez, Supercritical fluid extraction of chamomile flower heads: Comparison with conventional extraction, kinetics and scale-up, J. Supercrit. Fluids 43 (2007) 192–198.
[89] C.S. Kaiser, H. Römpp, P.C. Schmidt, Supercritical carbon dioxide extraction of chamomile flowers: Extraction efficiency, stability, and in-line inclusion of chamomile-carbon dioxide extract in β-cyclodextrin. Phytochem. Anal. 15 (2004) 249–256, https://doi.org/10.1002/pca.775.
[90] N.P. Povh, M.O.M. Marques, M.A.A. Meireles, Supercritical CO2extraction of essential oil and oleoresin from chamomile (Chamomilla recutita [L.] Rauschert). J. Supercrit. Fluids 21 (2001) 245–256, https://doi.org/10.1016/S0896-8446(01)00096-1.
[91] Z.P. Zekovic, Chamomile ligulate flowers in supercritical CO2-extraction. J. Essent. Oil Res. 12 (2000) 85–93, https://doi.org/10.1080/10412905.2000.9712049.
[92] P.I.N. Carvalho, J.F. Osorio-Tobón, M.A. Rostagno, A.J. Petenate, M.A.A. Meireles, Techno-economic evaluation of the extraction of turmeric (Curcuma longa L.) oil and ar-turmerone using supercritical carbon dioxide. J. Supercrit. Fluids 105 (2015) 44–54, https://doi.org/10.1016/j.supflu.2015.03.020.
[93] L.H. Chang, T.T. Jong, H.S. Huang, Y.F. Nien, C.M.J. Chang, Supercritical carbon dioxide extraction of turmeric oil from Curcuma longa Linn and purification of turmerones. Sep. Purif. Technol. 47 (2006) 119–125, https://doi.org/10.1016/j.seppur.2005.06.018.
[94] S. Priyanka, Khanam, Influence of operating parameters on supercritical fluid extraction of essential oil from turmeric root. J. Clean. Prod. 188 (2018) 816–824, https://doi.org/10.1016/j.jclepro.2018.04.052.
[95] L. Kao, C.R. Chen, C.M.J. Chang, Supercritical CO2extraction of turmerones from turmeric and high-pressure phase equilibrium of CO2+ turmerones. J. Supercrit. Fluids 43 (2007) 276–282, https://doi.org/10.1016/j.supflu.2007.06.007.
[96] M.E.M. Braga, P.F. Leal, J.E. Carvalho, M.A.A. Meireles, Comparison of yield, composition, and antioxidant activity of turmeric (Curcuma longa L.) extracts obtained using various techniques. J. Agric. Food Chem. 51 (2003) 6604–6611, https://doi.org/10.1021/jf0345550.
[97] A.L. Chassagnez-me, Supercritical CO2 Extraction of Curcumins and Essential Oil from the Rhizomes of Turmeric (Curcuma longa L.), Ind. Eng. Chem. Res. (2000) 4729–4733.
[98] H.A. Martinez-Correa, J.T. Paula, A.C.A.V. Kayano, C.L. Queiroga, P.M. Magalhães, F.T. M. Costa, F.A. Cabral, Composition and antimalarial activity of extracts of Curcuma longa L. obtained by a combination of extraction processes using supercritical CO2, ethanol and water as solvents. J. Supercrit. Fluids 119 (2017) 122–129, https://doi.org/10.1016/j.supflu.2016.08.017.
[99] K. Chhouk, W. Wahyudiono, H. Kanda, M. Goto, Comparison of conventional and ultrasound assisted supercritical carbon dioxide extraction of curcumin from turmeric (Curcuma longa L.). Eng. J. 21 (2017) 53–65, https://doi.org/10.4186/ej.2017.21.5.53.
[100] J.F. Osorio-Tobón, P.I.N. Carvalho, M.A. Rostagno, A.J. Petenate, M.A.A. Meireles, Extraction of curcuminoids from deflavored turmeric (Curcuma longa L.) using pressurized liquids: process integration and economic evaluation. J. Supercrit. Fluids 95 (2014) 167–174, https://doi.org/10.1016/j.supflu.2014.08.012.
[101] D. Georgantelis, I. Ambrosiadis, P. Katikou, G. Blekas, S.A. Georgakis, Effect of rosemary extract, chitosan and α-tocopherol on microbiological parameters and lipid oxidation of fresh pork sausages stored at 4°C. Meat Sci. 76 (2007) 172–181, https://doi.org/10.1016/J.MEATSCI.2006.10.026.
[102] E. Hernández-Hernández, E. Ponce-Alquicira, M.E. Jaramillo-Flores, I. Guerrero Legarreta, Antioxidant effect rosemary (Rosmarinus officinalis L.) and oregano (Origanum vulgare L.) extracts on TBARS and colour of model raw pork batters. Meat Sci. 81 (2009) 410–417, https://doi.org/10.1016/J.MEATSCI.2008.09.004.
[103] S. Urbančič, M.H. Kolar, D. Dimitrijević, L. Demšar, R. Vidrih, Stabilisation of sunflower oil and reduction of acrylamide formation of potato with rosemary extract during deep-fat frying. LWT Food Sci. Technol. 57 (2014) 671–678, https://doi.org/10.1016/J.LWT.2013.11.002.
[104] S. Huang, B. Liu, D. Ge, J. Dai, Effect of combined treatment with supercritical CO2and rosemary on microbiological and physicochemical properties of ground pork stored at 4°C. Meat Sci. 125 (2017) 114–120, https://doi.org/10.1016/j.meatsci.2016.11.022.
[105] I. Vieitez, L. Maceiras, I. Jachmanián, S. Alborés, Antioxidant and antibacterial activity of different extracts from herbs obtained by maceration or supercritical technology. J. Supercrit. Fluids 133 (2018) 58–64, https://doi.org/10.1016/j.supflu.2017.09.025.

[106] J.S. García-Pérez, S.P. Cuéllar-Bermúdez, A. Arévalo-Gallegos, C. Salinas-Salazar, J. Rodríguez-Rodríguez, R. de la Cruz-Quiroz, H.M.N. Iqbal, R. Parra-Saldívar, Influence of supercritical CO_2 extraction on fatty acids profile, volatile compounds and bioactivities from Rosmarinus officinalis. Waste Biomass Valorization (2018), https://doi.org/10.1007/s12649-018-0408-5.
[107] E. Vázquez, M.R. García-risco, L. Jaime, G. Reglero, T. Fornari, The journal of supercritical fluids simultaneous extraction of rosemary and spinach leaves and its effect on the antioxidant activity of products. J. Supercrit. Fluids 82 (2013) 138–145, https://doi.org/10.1016/j.supflu.2013.07.004.
[108] E. Arranz, L. Jaime, M.R. García-Risco, T. Fornari, G. Reglero, S. Santoyo, Anti-inflammatory activity of rosemary extracts obtained by supercritical carbon dioxide enriched in carnosic acid and carnosol. Int. J. Food Sci. Technol. 50 (2015) 674–681, https://doi.org/10.1111/ijfs.12656.
[109] A.D.P. Sánchez-Camargo, V. García-Cañas, M. Herrero, A. Cifuentes, E. Ibáñez, Comparative study of green sub- and supercritical processes to obtain carnosic acid and carnosol-enriched rosemary extracts with in vitro anti-proliferative activity on colon cancer cells. Int. J. Mol. Sci. 17 (2016) 1–18, https://doi.org/10.3390/ijms17122046.
[110] G. Vicente, S. Molina, M. González-vallinas, M.R. García-risco, T. Fornari, G. Reglero, A. Ramírez, D. Molina, The journal of supercritical fluids supercritical rosemary extracts, their antioxidant activity and effect on hepatic tumor progression. J. Supercrit. Fluids 79 (2013) 101–108, https://doi.org/10.1016/j.supflu.2012.07.006.
[111] A.P. Sánchez-Camargo, J.A. Mendiola, A. Valdés, M. Castro-Puyana, V. García-Cañas, A. Cifuentes, M. Herrero, E. Ibáñez, Supercritical antisolvent fractionation of rosemary extracts obtained by pressurized liquid extraction to enhance their antiproliferative activity. J. Supercrit. Fluids 107 (2016) 581–589, https://doi.org/10.1016/j.supflu.2015.07.019.
[112] O.J. Catchpole, J.B. Grey, K.A. Mitchell, J.S. Lan, Supercritical antisolvent fractionation of propolis tincture. J. Supercrit. Fluids 29 (2004) 97–106, https://doi.org/10.1016/S0896-8446(03)00033-0.
[113] A.H. Memon, M.S.R. Hamil, M. Laghari, F. Rithwan, S. Zhari, M.A.A. Saeed, Z. Ismail, A.M.S. A. Majid, A comparative study of conventional and supercritical fluid extraction methods for the recovery of secondary metabolites from Syzygium campanulatum Korth. J. Zhejiang Univ. Sci. B 17 (2016) 683–982, https://doi.org/10.1631/jzus.B1600019.
[114] I.J. Seabra, M.E.M. Braga, M.T.P. Batista, H.C. de Sousa, Fractioned high pressure extraction of anthocyanins from elderberry (Sambucus nigra L.) pomace. Food Bioprocess. Technol. 3 (2010) 674–683, https://doi.org/10.1007/s11947-008-0134-2.
[115] L. Baldino, G. Della Porta, L.S. Osseo, E. Reverchon, R. Adami, Concentrated oleuropein powder from olive leaves using alcoholic extraction and supercritical CO2assisted extraction. J. Supercrit. Fluids 133 (2018) 65–69, https://doi.org/10.1016/j.supflu.2017.09.026.
[116] M. del P. Garcia-Mendoza, F.A. Espinosa-Pardo, A.M. Baseggio, G.F. Barbero, M.R. Maróstica Jr., M.A. Rostagno, J. Martínez, Extraction of phenolic compounds and anthocyanins from juçara (Euterpe edulis Mart.) residues using pressurized liquids and supercritical fluids. J. Supercrit. Fluids 119 (2017) 9–16, https://doi.org/10.1016/j.supflu.2016.08.014.
[117] C. De Cássia Rodrigues Batista, M.S. De Oliveira, M.E. Araújo, A.M.C. Rodrigues, J.R.S. Botelho, A.P.D.S.S. Filho, N.T. Machado, R.N. Carvalho, Supercritical CO2extraction of açaí (Euterpe oleracea) berry oil: Global yield, fatty acids, allelopathic activities, and determination of phenolic and anthocyanins total compounds in the residual pulp. J. Supercrit. Fluids 107 (2015) 364–369, https://doi.org/10.1016/j.supflu.2015.10.006.
[118] C. Rodriguez-Perez, J. Mendiola, R. Quirantes-Piné, E. Ibáñez, A. Segura-Carretero, Green downstream processing using supercritical carbon dioxide, CO2-expanded ethanol and pressurized hot water extractions for recovering bioactive compounds from Moringa oleifera leaves. J. Supercrit. Fluids 116 (2016) 90–100, https://doi.org/10.1016/j.supflu.2016.05.009.
[119] J.P. Maran, S. Manikandan, B. Priya, P. Gurumoorthi, Box-Behnken design based multi-response analysis and optimization of supercritical carbon dioxide extraction of bioactive flavonoid compounds from tea (Camellia sinensis L.) leaves. J. Food Sci. Technol. 52 (2015) 92–104, https://doi.org/10.1007/s13197-013-0985-z.
[120] A.N. Mustapa, Á. Martin, R.B. Mato, M.J. Cocero, Extraction of phytocompounds from the medicinal plant Clinacanthus nutans Lindau by microwave-assisted extraction and supercritical carbon dioxide extraction. Ind. Crop. Prod. 74 (2015) 83–94, https://doi.org/10.1016/j.indcrop.2015.04.035.

[121] J. Paes, R. Dotta, G.F. Barbero, J. Martínez, Extraction of phenolic compounds and anthocyanins from blueberry (Vaccinium myrtillus L.) residues using supercritical CO2and pressurized liquids. J. Supercrit. Fluids 95 (2014) 8–16, https://doi.org/10.1016/j.supflu.2014.07.025.

[122] R.E. Sanjaya, Y.Y. Tedjo, A. Kurniawan, Y.H. Ju, A. Ayucitra, S. Ismadji, Investigation on supercritical CO_2 extraction of phenolic-phytochemicals from an epiphytic plant tuber (Myrmecodia pendans). J. CO_2 Util. 6 (2014) 26–33, https://doi.org/10.1016/j.jcou.2014.03.001.

[123] M. Solana, I. Boschiero, S. Dall'Acqua, A. Bertucco, Extraction of bioactive enriched fractions from Eruca sativa leaves by supercritical CO2technology using different co-solvents. J. Supercrit. Fluids 94 (2014) 245–251, https://doi.org/10.1016/j.supflu.2014.08.022.

[124] Z.D. Zulkafli, H. Wang, F. Miyashita, N. Utsumi, K. Tamura, Cosolvent-modified supercritical carbon dioxide extraction of phenolic compounds from bamboo leaves (Sasa palmata). J. Supercrit. Fluids 94 (2014) 123–129, https://doi.org/10.1016/j.supflu.2014.07.008.

[125] M. Herrero, M. Plaza, A. Cifuentes, E. Ibáñez, Green processes for the extraction of bioactives from Rosemary: Chemical and functional characterization via ultra-performance liquid chromatography-tandem mass spectrometry and in-vitro assays. J. Chromatogr. A 1217 (2010) 2512–2520, https://doi.org/10.1016/j.chroma.2009.11.032.

[126] P. Garcia-Salas, A. Morales-Soto, A. Segura-Carretero, A. Fernández-Gutiérrez, Phenolic-compound-extraction systems for fruit and vegetable samples. Molecules 15 (2010) 8813–8826, https://doi.org/10.3390/molecules15128813.

[127] M. Herrero, L. Jaime, P.J. Martín-Álvarez, A. Cifuentes, E. Ibáñez, Optimization of the extraction of antioxidants from Dunaliella salina microalga by pressurized liquids. J. Agric. Food Chem. 54 (2006) 5597–5603, https://doi.org/10.1021/jf060546q.

[128] B. Díaz-Reinoso, A. Moure, H. Domínguez, J.C. Parajó, Supercritical CO2 extraction and purification of compounds with antioxidant activity. J. Agric. Food Chem. 54 (2006) 2441–2469, https://doi.org/10.1021/jf052858j.

[129] J.C. Parajo, H. Dominguez, A. Moure, B. Diaz-Reinoso, Obtaining antioxidants by supercritical fluid extraction, in: Extr. Bioact. Compd. Food Prod., CRC Press Taylor & Francis Group, Boca Raton, 2009, , pp. 288–305.

[130] K.Y. Khaw, M.O. Parat, P.N. Shaw, J.R. Falconer, Solvent supercritical fluid technologies to extract bioactive compounds from natural sources: a review. Molecules 22 (2017), https://doi.org/10.3390/molecules22071186.

[131] C. De Cássia, R. Batista, M. Santana, D. Oliveira, M. Emmi, A.M.C. Rodrigues, J. Rafael, S. Botelho, A. Pedro, S. Filho, N.T. Machado, R.N. Carvalho, Supercritical CO2 extraction of açaí (Euterpe oleracea) berry oil: Global yield, fatty acids, allelopathic activities, and determination of phenolic and anthocyanins total compounds in the residual pulp. J. Supercrit. Fluids 107 (2016) 364–369, https://doi.org/10.1016/j.supflu.2015.10.006.

[132] M. Roberto, M. Junior, A.V. Leite, N. Romanelli, V. Dragano, Supercritical fluid extraction and stabilization of phenolic compounds from natural sources—review (supercritical extraction and stabilization of phenolic compounds). Open Chem. Eng. J. 4 (2010) 51–60, https://doi.org/10.2174/1874123101004010051.

[133] K. Tyśkiewicz, M. Konkol, E. Rój, The application of supercritical fluid extraction in phenolic compounds isolation from natural plant materials. Molecules 23 (2018) 2625, https://doi.org/10.3390/molecules23102625.

[134] Z. Wimmer, M. Zarevúcka, A review on the effects of supercritical carbon dioxide on enzyme activity. Int. J. Mol. Sci. 11 (2010) 233–253, https://doi.org/10.3390/ijms11010233.

[135] Y.H. Chan, S. Yusup, A.T. Quitain, Y.H. Chai, Y. Uemura, S.K. Loh, Extraction of palm kernel shell derived pyrolysis oil by supercritical carbon dioxide: evaluation and modeling of phenol solubility. Biomass Bioenergy 116 (2018) 106–112, https://doi.org/10.1016/j.biombioe.2018.06.009.

[136] M. Bimakr, R.A. Rahman, A. Ganjloo, F.S. Taip, L.M. Salleh, M.Z.I. Sarker, Optimization of supercritical carbon dioxide extraction of bioactive flavonoid compounds from spearmint (Mentha spicata L.) leaves by using response surface methodology. Food Bioprocess. Technol. 5 (2012) 912–920, https://doi.org/10.1007/s11947-010-0504-4.

[137] L.C. Paviani, C. Dariva, M.C. Marcucci, F.A. Cabral, Supercritical carbon dioxide selectivity to fractionate phenolic compounds from the dry ethanolic extract of propolis. J. Food Process Eng. 33 (2010) 15–27, https://doi.org/10.1111/j.1745-4530.2008.00256.x.

[138] P.B. Gomes, V.G. Mata, A.E. Rodrigues, Production of rose geranium oil using supercritical fluid extraction. J. Supercrit. Fluids 41 (2007) 50–60, https://doi.org/10.1016/j.supflu.2006.08.018.
[139] A.F. Gouveia, C. Duarte, M.L. Beirão Da Costa, M.G. Bernardo-Gil, M. Moldão-Martins, Oxidative stability of olive oil flavoured by Capsicum frutescens supercritical fluid extracts. Eur. J. Lipid Sci. Technol. 108 (2006) 421–428, https://doi.org/10.1002/ejlt.200500273.
[140] B.E. Richter, B.A. Jones, J.L. Ezzell, N.L. Porter, N. Avdalovic, C. Pohl, Accelerated solvent extraction: a technique for sample preparation. Anal. Chem. 68 (1996) 1033–1039, https://doi.org/10.1021/ac9508199.
[141] A. Mokrani, K. Madani, Effect of solvent, time and temperature on the extraction of phenolic compounds and antioxidant capacity of peach (Prunus persica L.) fruit, Elsevier B.V. Sep. Purif. Technol. (2016), https://doi.org/10.1016/j.seppur.2016.01.043.
[142] R.L. Smith Jr., R.M. Malaluan, W.B. Setianto, H. Inomata, K. Arai, Separation of cashew (Anacardium occidentale L.) nut shell liquid with supercritical carbon dioxide. Bioresour. Technol. 88 (2003) 1–7, https://doi.org/10.1016/S0960-8524(02)00271-7.
[143] E. Vági, E. Rapavi, M. Hadolin, K. Vásárhelyiné Perédi, A. Balázs, A. Blázovics, B. Simándi, Phenolic and Triterpenoid Antioxidants from Origanum majorana L. Herb and Extracts Obtained with Different Solvents. J. Agric. Food Chem. 53 (2005) 17–21, https://doi.org/10.1021/jf048777p.
[144] R.N. Carvalho, L.S. Moura, P.T.V. Rosa, M.A.A. Meireles, Supercritical fluid extraction from rosemary (Rosmarinus officinalis): kinetic data, extract's global yield, composition, and antioxidant activity. J. Supercrit. Fluids 35 (2005) 197–204, https://doi.org/10.1016/j.supflu.2005.01.009.
[145] C. Yang, Y.-R. Xu, W.-X. Yao, Extraction of pharmaceutical components from Ginkgo biloba leaves using supercritical carbon dioxide. J. Agric. Food Chem. 50 (2002) 846–849, https://doi.org/10.1021/jf010945f.
[146] A.P.R.F. Canela, P.T.V. Rosa, M.O.M. Marques, M.A.A. Meireles, Supercritical Fluid Extraction of Fatty Acids and Carotenoids from the Microalgae Spirulina maxima, Ind. Eng. Chem. Res. 41 (2002) 3012–3018. doi: 10.1021/ie010469i.
[147] E. Ibáñez, J. Palacios, F.J. Señoráns, G. Santa-María, J. Tabera, G. Reglero, Isolation and separation of tocopherols from olive by-products with supercritical fluids. J. Am. Oil Chem. Soc. 77 (2000) 187–190, https://doi.org/10.1007/s11746-000-0030-8.
[148] Y. Ge, Y. Ni, H. Yan, Y. Chen, T. Cai, Optimization of the supercritical fluid extraction of natural vitamin E from wheat germ using response surface methodology. J. Food Sci. 67 (2002) 239–243, https://doi.org/10.1111/j.1365-2621.2002.tb11391.x.
[149] H.S. Park, H.K. Choi, S.J. Lee, K.W. Park, S.G. Choi, K.H. Kim, Effect of mass transfer on the removal of caffeine from green tea by supercritical carbon dioxide. J. Supercrit. Fluids 42 (2007) 205–211, https://doi.org/10.1016/j.supflu.2007.03.002.
[150] A. Oreopoulou, D. Tsimogiannis, V. Oreopoulou, Extraction of Polyphenols From Aromatic and Medicinal Plants: An Overview of the Methods and the Effect of Extraction Parameters. second ed., Elsevier Inc., 2019. https://doi.org/10.1016/B978-0-12-813768-0.00025-6
[151] M. Majeed, A.I. Hussain, S.A.S. Chatha, M.K.K. Khosa, G.M. Kamal, M.A. Kamal, X. Zhang, M. Liu, Optimization protocol for the extraction of antioxidant components from Origanum vulgare leaves using response surface methodology. Saudi J. Biol. Sci. 23 (2016) 389–396, https://doi.org/10.1016/j.sjbs.2015.04.010.
[152] E. Reverchon, I. De Marco, Supercritical fluid extraction and fractionation of natural matter. J. Supercrit. Fluids 38 (2006) 146–166, https://doi.org/10.1016/j.supflu.2006.03.020.
[153] G.D.A.R. Oliveira, A.E. De Oliveira, E.C. Da Conceição, M.I.G. Leles, Multiresponse optimization of an extraction procedure of carnosol and rosmarinic and carnosic acids from rosemary. Food Chem. 211 (2016) 465–473, https://doi.org/10.1016/j.foodchem.2016.05.042.
[154] M. Mushtaq, B. Sultana, F. Anwar, A. Adnan, S.S.H. Rizvi, Enzyme-assisted supercritical fluid extraction of phenolic antioxidants from pomegranate peel. J. Supercrit. Fluids 104 (2015) 122–131, https://doi.org/10.1016/j.supflu.2015.05.020.
[155] S. Soltani, S.H. Mazloumi, A new empirical model to correlate solute solubility in supercritical carbon dioxide in presence of co-solvent. Chem. Eng. Res. Des. 125 (2017) 79–87, https://doi.org/10.1016/j.cherd.2017.07.006.

[156] A. Queiroz, J. Cajaiba, A sustainable process for (-)-α-bisabolol extraction from Eremanthus erythropappus using supercritical CO2and ethanol as co-solvent. J. Supercrit. Fluids 110 (2016) 39–46, https://doi.org/10.1016/j.supflu.2015.12.011.

[157] D. Trabelsi, A. Aydi, A.W. Zibetti, G. Della Porta, M. Scognamiglio, V. Cricchio, E. Langa, M. Abderrabba, A.M. Mainar, Supercritical extraction from Citrus aurantium amara peels using CO2with ethanol as co-solvent. J. Supercrit. Fluids 117 (2016) 33–39, https://doi.org/10.1016/j.supflu.2016.07.003.

[158] M. Herrero, A. del P. Sánchez-Camargo, Plants, seaweeds, microalgae and food by-products as natural sources of functional ingredients obtained using pressurized liquid extraction and supercritical fluid extraction. TrAC Trends Anal. Chem. 71 (2015) 26–38, https://doi.org/10.1016/j.trac.2015.01.018.

[159] R. Murga, R. Ruiz, S. Beltrán, J.L. Cabezas, Extraction of natural complex phenols and tannins from grape seeds by using supercritical mixtures of carbon dioxide and alcohol. J. Agric. Food Chem. 48 (2000) 3408–3412, https://doi.org/10.1021/jf9912506.

[160] H.S. Park, H.J. Lee, M.H. Shin, K.W. Lee, H. Lee, Y.S. Kim, K.O. Kim, K.H. Kim, Effects of cosolvents on the decaffeination of green tea by supercritical carbon dioxide. Food Chem. 105 (2007) 1011–1017, https://doi.org/10.1016/j.foodchem.2007.04.064.

[161] A.S. Zarena, N.M. Sachindra, K. Udaya Sankar, Optimisation of ethanol modified supercritical carbon dioxide on the extract yield and antioxidant activity from *Garcinia mangostana* L, Food Chem. 130 (2012) 203–208.https://doi.org/10.1016/j.foodchem.2011.07.007.

[162] Y.M. Monroy, R.A.F. Rodrigues, A. Sartoratto, F.A. Cabral, Influence of ethanol, water, and their mixtures as co-solvents of the supercritical carbon dioxide in the extraction of phenolics from purple corn cob (Zea mays L.). J. Supercrit. Fluids 118 (2016) 11–18, https://doi.org/10.1016/j.supflu.2016.07.019.

[163] G. Leeke, F. Gaspar, R. Santos, Influence of water on the extraction of essential oils from a model herb using supercritical carbon dioxide. Ind. Eng. Chem. Res. 41 (2002) 2033–2039, https://doi.org/10.1021/ie010845z.

[164] H.I. Castro-Vargas, L.I. Rodríguez-Varela, S.R.S. Ferreira, F. Parada-Alfonso, Extraction of phenolic fraction from guava seeds (Psidium guajava L.) using supercritical carbon dioxide and co-solvents. J. Supercrit. Fluids 51 (2010) 319–324, https://doi.org/10.1016/j.supflu.2009.10.012.

[165] S.A.O. Santos, J.J. Villaverde, C.M. Silva, C.P. Neto, A.J.D. Silvestre, Supercritical fluid extraction of phenolic compounds from Eucalyptus globulus Labill bark. J. Supercrit. Fluids 71 (2012) 71–79, https://doi.org/10.1016/j.supflu.2012.07.004.

[166] C. Da Porto, D. Decorti, A. Natolino, Water and ethanol as co-solvent in supercritical fluid extraction of proanthocyanidins from grape marc: a comparison and a proposal. J. Supercrit. Fluids 87 (2014) 1–8, https://doi.org/10.1016/j.supflu.2013.12.019.

[167] M. Solana, I. Boschiero, S. Dall'Acqua, A. Bertucco, A comparison between supercritical fluid and pressurized liquid extraction methods for obtaining phenolic compounds from Asparagus officinalis L. J. Supercrit. Fluids 100 (2015) 201–208, https://doi.org/10.1016/j.supflu.2015.02.014.

[168] J.L. Pasquel Reátegui, A.P.D.F. Machado, G.F. Barbero, C.A. Rezende, J. Martínez, Extraction of antioxidant compounds from blackberry (Rubus sp.) bagasse using supercritical CO2assisted by ultrasound. J. Supercrit. Fluids 94 (2014) 223–233, https://doi.org/10.1016/j.supflu.2014.07.019.

[169] Q. Lang, C.M. Wai, Supercritical fluid extraction in herbal and natural product studies—a practical review, Talanta 53 (2001) 771–782.

[170] R.P.F.F. da Silva, T.A.P. Rocha-Santos, A.C. Duarte, Supercritical fluid extraction of bioactive compounds. TrAC Trends Anal. Chem. 76 (2016) 40–51, https://doi.org/10.1016/j.trac.2015.11.013.

[171] K.K. Liong, P.A. Wells, N.R. Foster, Diffusion coefficients of long-chain esters in supercritical carbon dioxide. Ind. Eng. Chem. Res. 30 (1991) 1329–1335, https://doi.org/10.1021/ie00054a039.

[172] M. Sihvonen, E. Järvenpää, V. Hietaniemi, R. Huopalahti, Advances in supercritical carbon dioxide technologies.pdf. Trends Food Sci. Technol. 10 (1999) 217–222, https://doi.org/10.1016/S0924-2244(99)00049-7.

[173] A.M. Farías-Campomanes, M.A. Rostagno, M.A.A. Meireles, Production of polyphenol extracts from grape bagasse using supercritical fluids: yield, extract composition and economic evaluation. J. Supercrit. Fluids 77 (2013) 70–78, https://doi.org/10.1016/j.supflu.2013.02.006.

[174] J.J. Wu, J.C. Lin, C.H. Wang, T.T. Jong, H.L. Yang, S.L. Hsu, C.m.J. Chang, Extraction of antioxidative compounds from wine lees using supercritical fluids and associated anti-tyrosinase activity. J. Supercrit. Fluids 50 (2009) 33–41, https://doi.org/10.1016/j.supflu.2009.04.010.

[175] M.M.R. de Melo, A.J.D. Silvestre, C.M. Silva, Supercritical fluid extraction of vegetable matrices: Applications, trends and future perspectives of a convincing green technology. J. Supercrit. Fluids 92 (2014) 115–176, https://doi.org/10.1016/j.supflu.2014.04.007.
[176] A. del P. Sánchez-Camargo, F. Parada-Alfonso, E. Ibáñez, A. Cifuentes, On-line coupling of supercritical fluid extraction and chromatographic techniques. J. Sep. Sci. 40 (2017) 213–227, https://doi.org/10.1002/jssc.201601040.
[177] A.M. Farías-Campomanes, M.A. Rostagno, J.J. Coaquira-Quispe, M.A.A. Meireles, Supercritical fluid extraction of polyphenols from lees: overall extraction curve, kinetic data and composition of the extracts. Bioresour. Bioprocess. 2 (2015), https://doi.org/10.1186/s40643-015-0073-5.
[178] J. Azmir, I.S.M. Zaidul, M.M. Rahman, K.M. Sharif, A. Mohamed, F. Sahena, M.H.A. Jahurul, K. Ghafoor, N.A.N. Norulaini, A.K.M. Omar, Techniques for extraction of bioactive compounds from plant materials : A review. J. Food Eng. 117 (2013) 426–436, https://doi.org/10.1016/j.jfoodeng.2013.01.014.
[179] M. Herrero, A. Cifuentes, E. Iban, Sub- and supercritical fluid extraction of functional ingredients from different natural sources: Plants , food-by-products, algae and microalgae A review. Food Chemistry 98 (2006) 136–148, https://doi.org/10.1016/j.foodchem.2005.05.058.
[180] X. Yin, Q. You, Z. Jiang, Optimization of enzyme assisted extraction of polysaccharides from Tricholoma matsutake by response surface methodology. Carbohydr. Polym. 86 (2011) 1358–1364, https://doi.org/10.1016/j.carbpol.2011.06.053.
[181] S. Shah, A. Sharma, M.N. Gupta, Extraction of oil from Jatropha curcas L. seed kernels by enzyme assisted three phase partitioning. Ind. Crops Prod. 20 (2004) 275–279, https://doi.org/10.1016/j.indcrop.2003.10.010.
[182] Y.W. Sari, M.E. Bruins, J.P.M. Sanders, Enzyme assisted protein extraction from rapeseed, soybean, and microalgae meals. Ind. Crop. Prod. 43 (2013) 78–83, https://doi.org/10.1016/j.indcrop.2012.07.014.
[183] M. Mushtaq, B. Sultana, H.N. Bhatti, M. Asgher, Optimization of enzyme-assisted revalorization of sweet lime (citrus limetta risso) peel into phenolic antioxidants. BioResources 9 (2014) 6153–6165, https://doi.org/10.15376/biores.9.4.6153-6165.
[184] J.L. Capelo-Martinez, Ultrasound In Chemistry, Wiley-VCH, Weinheim, 2009.
[185] T.J. Mason, Sonochemistry, Oxford University Press, Oxford, New York, 1999.
[186] M. Koubaa, H. Mhemdi, J. Fages, Recovery of valuable components and inactivating microorganisms in the agro-food industry with ultrasound-assisted supercritical fluid technology. J. Supercrit. Fluids 134 (2018) 71–79, https://doi.org/10.1016/j.supflu.2017.12.012.
[187] N. Kaushik, B.P. Kaur, P.S. Rao, H.N. Mishra, Effect of high pressure processing on color, biochemical and microbiological characteristics of mango pulp (Mangifera indica cv. Amrapali). Innov. Food Sci. Emerg. Technol. 22 (2014) 40–50, https://doi.org/10.1016/j.ifset.2013.12.011.
[188] G. Arroyo, ENGINEERING / PROCESSING MS 5707 High Hydrostatic Pressure Effects, J. Food Sci. 63 (1998) 1–4.
[189] M.J. Torres-Ossandón, A. Vega-Gálvez, J. López, K. Stucken, J. Romero, K. Di Scala, Effects of high hydrostatic pressure processing and supercritical fluid extraction on bioactive compounds and antioxidant capacity of Cape gooseberry pulp (Physalis peruviana L.). J. Supercrit. Fluids 138 (2018) 215–220, https://doi.org/10.1016/J.SUPFLU.2018.05.005.
[190] M.M. Poojary, P. Putnik, D. Bursać Kovačević, F.J. Barba, J.M. Lorenzo, D.A. Dias, A. Shpigelman, Stability and extraction of bioactive sulfur compounds from Allium genus processed by traditional and innovative technologies. J. Food Compos. Anal. 61 (2017) 28–39, https://doi.org/10.1016/j.jfca.2017.04.007.

CHAPTER 13

Applications of supercritical carbon dioxide in textile industry

Andrew N. Amenaghawon[a], Chinedu L. Anyalewechi[a], Heri Septya Kusuma[b], Mahfud Mahfud[b]
[a]Department of Chemical Engineering, Faculty of Engineering, University of Benin, Benin City, Nigeria
[b]Department of Chemical Engineering, Faculty of Industrial Technology, Institut Teknologi Sepuluh Nopember, Surabaya, Indonesia

Contents

Green Sustainable Process for Chemical and Environmental Engineering and Science
https://doi.org/10.1016/B978-0-12-817388-6.00013-1

1. Introduction

Large volumes of water-soluble dyes (synthetic) are typically used up by the textile industries during conventional dyeing operations and hence huge volumes of effluents are generated. The increasing difficulty in waste water treatment, environmental concerns, and the high cost of treating these effluents have facilitated the application of carbon dioxide as supercritical fluid to the textile industry for dyeing operations with several benefits [1].

The use of supercritical carbon dioxide ($scCO_2$) in the textile industry as a substitute to water-based dyeing processes has continued to be a subject of significant interest among researchers in recent years due to its far reaching and promising contributions with lots of research publications available [2–4].

$scCO_2$ dyeing is a process which does not employ the conventional use of water as a solvent base, rather it uses carbon dioxide as supercritical fluid. Carbon dioxide is the most preferred and popularly used supercritical fluid as a result of its unique and desirable characteristics such low cost, fire-resistant nature, absence of toxicity, and lower critical conditions in comparison with other supercritical fluids. Other attributes include the similarity of its viscosity and diffusivity properties to that of gases in its supercritical state and density properties similar to that of liquid. Also, its unique dissolving powers on disperse dyes and its swelling and plasticization ability makes it ideal for dyeing of synthetic textiles [5, 6].

2. Characteristics of textile fibers

Textile fibers can be broadly divided into two categories: natural fibers and synthetic fibers. Natural fibers are obtained from nature (plants and animals) and are mostly cellulosic (e.g., cotton, flax) or proteinic (e.g., wool, silk) in nature. Examples of natural fibers are cotton, wool, silk, sisal, flax, hemp, and jute. Cotton is the most important of all natural fibers. Synthetic fibers on the other hand are man-made polymers synthesized from petroleum products. Polyester, polyamide, rayon, acetate, and acrylic are examples of synthetic fibers. Polyester is the most valuable and desirable synthetic fiber. Polyethylene terephthalate (PET) and polybutylene terephthalate (PBT) are common examples of polyester-based synthetic fibers. However, PET synthetic fiber has been widely investigated and frequently used in the textile industry due to its desirable qualities such as high glass transition temperature, slow rate of crystallization, etc. Generally, cotton and PET fibers are the two most important fibers in the textile industry [7].

On the basis of dyeing, textile fibers can be referred to as either hydrophilic or hydrophobic depending on their water-sorption properties. A material that exhibits good wettability when brought into contact with water is referred to as hydrophilic whereas a material that exhibits poor wettability when exposed to water is referred to as hydrophobic. Thus, hydrophilic fibers (cotton, wool) can be easily dyed with water-soluble dyes (acid, direct dyes) while hydrophobic fibers (synthetic textiles) require the use of almost water-insoluble disperse dyes [8]. The contact angles of $\theta \leq 90°$ and $\theta > 90°$ are usually generated for hydrophilic and hydrophobic fibers respectively when in contact with water.

3. Textile dyes

Dyes are substances with coloring abilities capable of adhering onto specific surfaces. Dyes constitute an important requirement in the textile industry with over 700,000 tons/year of synthetic dyes produced globally [1]. Textile dyes should have strong affinity, be difficult to fade, uniform in color, and cost friendly in order to produce and maintain suitable fastness properties when exposed to light, sweat, and washing conditions prior to-and-after considerable usage [9].

Commonly used disperse dyes include azo disperse dyes, benzothiazoleazo disperse dyes, anthraquinon disperse dyes, and mordant disperse dyes. Of these dyes, the dominant of them is the azo dyes which make up about 70% of global dye production [10]. There desirable fastness properties in terms of light and wetness, high structural diversity, simplicity of synthesis, and their versatility of applications to other areas beside textile industry such as paint, food, pharmaceutical, leather, and cosmetics industries make them highly sought after [10].

A distinction between dyes and pigments which can also be used for dyeing with the help of binders should be noted. Whereas dyes have strong affinity for fibers, pigments do not have affinity (physical or chemical) for the fiber [11].

4. Overview of the dyeing process

Dyeing is frequently performed in the textile industry as a batch process although it can be carried out as a batch or continuous process. For dyeing operations performed under continuous processes, the fabrics are exposed to heat and steam as they pass through several solutions of concentrated chemicals in different equipment sections. After which rinsing is performed to remove most of the chemicals used in the preparation phase. On the other hand, batch process dyeing of textiles occurs in a single closed equipment containing dilute chemicals. After each phase, water is added to the equipment and then drained. Due to the relatively long time required for batch process dyeing, it is often referred to as "exhaust dyeing" [12].

The dyeing process in terms of conventional aqueous dyeing operations, generally involves huge volumes of water for cleaning, dyeing of the fiber, and rinsing of the dyed fiber. Basically, the dyeing process can be grouped into three phases namely: the preparation phase, the dyeing phase, and the finishing phase.

Preparation phase: Here some form of pretreatment is carried out on the fiber to ensure it is free from impurities before the dyeing phase using either aqueous alkaline substances, detergents, or enzymes [12].

Dyeing phase: Here dyes and other chemicals additives such as surfactants, promoting agents, carriers, emulsifying oils, alkalis, acids, softening, and leveling agents are applied to the fiber to obtain certain color/color blends with suitable fastness properties and color uniformity [12, 13]. The transfer of dyes to the fiber is achieved through a series of diffusion and adsorption steps. Diffusion of the dye into the liquid takes place and is adsorbed onto the fiber's external surface, which subsequently diffuses into the fiber and adsorbed onto the fiber's internal surface [14].

Finishing phase: Here the quality of the dyed textile is enhanced through special treatments with chemicals. Examples of these finishing treatments include softening, stain release, waterproof, antistatic and microbial protection, and permanent press treatment [12].

5. Supercritical CO_2

Supercritical fluids are pure substances that have pressure and temperature values beyond their critical temperatures and critical pressures. They have properties similar to that of a gas and also that of a liquid which are easily changed through simple pressure adjustments [15]. This is usually achieved by increasing the temperature and pressure of the gas above their critical values. This supercritical state occurs when a closed system attains critical values such that the existing liquid and gaseous state boundaries can no longer be distinguished. At this point, subsequent increase in pressure will result in very significant increase in the system's dielectric constant, thereby bestowing dissolving abilities to a system that probably had none under normal pressure and temperature conditions [16].

Carbon dioxide exhibits both gas and liquid properties above its critical condition. When the critical temperature and pressure conditions are exceeded, CO_2 changes into a supercritical state and becomes a supercritical fluid. $scCO_2$ has been frequently used as a solvent especially in the textile industry with several advantages arising from its desirable characteristics like low viscosity, low surface tension, high diffusivity which facilitate dyeing operations as the dyes are able to better penetrate into the textiles thereby reducing the dyeing time required [17]. In terms of its density, $scCO_2$ offers a wider solvation range since it has a density that is higher than its gas phase and also lower than its liquid phase. Due to its liquid-like density, it is able to dissolve hydrophilic dyes; while its gas-like nature helps in reducing the dyeing times [5, 18]. The ease of adjusting the density through pressure control, temperature control, or both is an important characteristic of $scCO_2$. Increasing the temperature and/or pressure increases the solubility. Conversely, decreasing the temperature and/or pressure improves its capacity to facilitate separation [19]. Moreover, with $scCO_2$ the selective separation of products and residual substrates is possible at the separation stage through a combination of several depressurization steps thereby eliminating complex distillation processes [19].

$scCO_2$ is gaining popularity and increasing research interest due to its eco-friendly nature and potential to curb environmental pollution. The unique properties and dissolving power of supercritical fluids on organic molecules have prompted their use in several areas. Supercritical fluids have been used in processes such as extraction of natural products, tobacco and hop extraction, and caffeine extraction from coffee beans [16, 20–22]. In the leather industry, processes such as chromo tanning, degreasing, dyeing, de-liming, fiber separation, fat-liquoring, and finishing operations have utilized $scCO_2$ [23, 24]. Other reported applications of supercritical fluids are in separation processes [25], chromatographic operations [26], soil treatment [27–29], preparation of catalysts [30], and nanoscale production [19].

6. Application of $scCO_2$ to textile dyeing

$scCO_2$ has been extensively used for dyeing operations in the textile industry compared to any other area with several studies available in literature. Huang et al. [31] successfully used $ScCO_2$ for the continuous dyeing of zipper tapes. Very good color consistency was reported for the dyed zipper tapes. High color fastness properties with ratings of 4–5 were also reported for wetting, fading, drying, and staining. Zheng et al. [32] designed an industrial-scale equipment that incorporates both 'inner/outer' dyeing and "dynamic/static" dyeing processes for $scCO_2$ dyeing of bobbins with high levelness and reproducibility of the dye bobbins. Desirable color fastness properties of commercial standard with ratings of 5, 5–6, and 4–5, were obtained for exposure to washing, light, and rubbing. Dyeing operating conditions of 120°C temperature, 24 MPa pressure, and 60 min dyeing time were recommended to ensure low operating cost. Long et al. [33] studied the fastness of reactive dyes in $scCO_2$ on cotton fiber by means of heterogeneous and phase transfer catalytic reactions. Zheng et al. [34] examined the influence of $scCO_2$ temperature on several properties/characteristics of disperse red 167. Comparison was also carried out between the initial disperse dye and the recycled disperse red 167 dye. The results obtained using the recycled dye were as good as those obtained the starting disperse dye. It was also observed that under increased temperature conditions, melting recrystallization of disperse red 167 took place without any effect on the dyeing performance. $scCO_2$ has also been applied in the treatment and extraction of ceramides from wool fibers, alongside dyeing of cotton [35].

7. Overview of the $scCO_2$ dyeing process

The process of supercritical fluid dyeing of PET fibers in terms of process equipment design and operation conditions have benefited immensely from several pivotal research studies [5, 36–38]. These contributions have led to the development of systems with improved dyeing efficiency, lower energy consumption, and compact sizes. However, researches into several aspects of the technology are still ongoing.

Basically, a $scCO_2$ dyeing process comprises a dye bath, CO_2 tank, dyeing vessel, compression, and circulating pumps [4]. The fabric to be processed is sent to an autoclave which is then pressurized and heated to the specified process requirements. Next is the dyeing phase which is achieved using a circulating pump after which the pressure of the system is partially reduced and the dyed fabric washed with fresh $scCO_2$ to remove any excess dye. After the process is completed, CO_2 and residual dye are recovered for subsequent dyeing operations [39].

8. Advantages of $scCO_2$ in textile dyeing

Carbon dioxide is the preferred supercritical fluid especially in the textile industry because it is environmentally friendly and also safe to handle. $scCO_2$ is a widely recognized and used dyeing medium in the textile industry for dyeing of synthetic fibers with several advantages arising from its usage due to its desirable qualities such low viscosity, low surface tension, and high diffusivity which make it easy for the dyes to penetrate into the textiles thereby reducing the dyeing time required [17]. Other advantages of $scCO_2$ are low cost, fire-resistant nature, absence of toxicity, and lower critical conditions ($T_c = 31.1°C$, $P_c = 73.7$ bar) in comparison with other supercritical fluids [5]. With $scCO_2$, processes such as spinning oils extraction, dyeing of fibers, and removal of surplus dye can be accomplished in a single unit during the carbon dioxide dyeing process requiring only tuning of temperature or pressure. Unlike the conventional water-based dyeing medium which involves drying thereby increasing the operational time, energy requirements, and cost of the overall process, $scCO_2$ does not require drying since CO_2 is released as a gas after the completion of the process thereby saving cost and time. Recycling of the carbon dioxide is very easy with about 90% recovery rate after the extracted matter precipitated in a separator [5]. Problems associated with water pollution are completely avoided alongside the attendant cost and problem of waste water treatment encountered during conventional aqueous dyeing media. With $scCO_2$, the use of cost-intensive additives (surfactants and dispersing agents) required in the traditional aqueous-based dyeing process are avoided [16, 17].

9. Effect of $scCO_2$ on textile fibers

Since the dyeing medium affects the nature and properties of the fiber (shrinkage, damage, crystallinity, polymer orientation) which ultimately influences the dyeing quality of the fiber alongside the reproducibility of the fiber dye-uptake, the effect of $scCO_2$ as a dyeing medium on textile fibers will be examined.

10. Shrinkage behavior

From studies carried out, the shrinkage behavior of PET materials has been reported to be a function of the process medium used [40]. For dyeing processes for PET fibers performed with $scCO_2$, shrinkage factor of about 10% and between 0% and 2% have been reported for nonheat-set PET fibers and heat-set PET fibers, respectively [41]. Similar trends for the heat-set PET material have been reported for other synthetic materials such as polyethylene, polypropylene (PP), polyamide 6,6 at pressure conditions of 280 bar, up to a temperature of 120°C over time interval of 30–60 min in $scCO_2$ [42, 43].

11. Plasticizing effect of CO_2 on fiber polymer

The use of $scCO_2$ in dyeing permits alterations in the level of plasticization and swelling occurring in glasslike and partly crystalline polymers and also in the available volume when its density is adjusted. The plasticizing effect of CO_2 occurs due to its ability to access desirable active sites in the polymer. An explanation for this behavior is possibly the binding of CO_2 to the phenyl group or the carbonyl bond within the polymer structure [44].

12. Glass transition temperature

From previous studies available in the literature for $scCO_2$ dyeing of PET materials, $scCO_2$ has been reported to significantly influence the glass transition temperature (T_g). A reduction in the T_g by about 6–70°C was obtained for some composition of PET fibers [40, 45].

Hou et al. [46] investigated the effect of $scCO_2$ dyeing on the chemical and morphological structures of PET fabrics. The dyeing conditions used were temperature range of 80–130°C and pressure of 20 MPa for 40 min. Fourier transform infrared spectroscopy (FTIR), scanning electron microscopy (SEM), X-ray diffraction (XRD), and differential scanning calorimetry (DSC) were used for the characterization of the fabrics. From the analysis of the results, it was observed that the dyeing of PET fabrics using $scCO_2$ had no effect on the chemical structure of the material when compared with that of untreated PET fabrics. However, the crystallinity of the PET fabric dyed in $scCO_2$ decreased. Slight variations in the crystal size and T_{mp} of the $scCO_2$ dyed fibers were also reported. When compared with the untreated PET material, there was no effect in terms of cracking, pitting, or crazing after dyeing at higher temperatures in $scCO_2$, although changes in the surface morphology with respect to increased oligomer migration were reported for the PET fabric dyed in $scCO_2$.

13. Factors affecting $scCO_2$ dyeing

13.1 Uniformity of dye distribution

For the production of high-quality dyed fabrics, it is important for the dye to be evenly distributed. This is to prevent the occurrence of stains in the fabric. The nonuniformity of dye distribution may be caused by the presence of undissolved dyestuffs, variation in the supercritical fluid flow distribution, inconsistency in the pressure and temperature conditions, excess release of carbon dioxide during the depressurization phase and the poor flow rate selection during the dyeing phase for the circulating bath among others [4, 47]. These causative factors of nonuniformity of dyes are further discussed below.

The presence of undissolved dyes which led to the formation of lumps of dye or dye aggregates in the dyeing vessel occurs when the dye does not dissolve properly in the $scCO_2$. These undissolved dye particles can cause unwanted dye spots to appear and also reduce the uniformity of the dye when used in dyeing fabrics. To overcome this problem, several solutions have been proffered in the literature to ensure complete homogeneity of the dye in the $scCO_2$ such as arranging a dye container in a circulating loop with a cyclone whereby the undissolved dye particles are separated from the $scCO_2$ dye solution and returned to the dye container while the $scCO_2$ dye solution is sent to the dyeing vessel.

Variation in the supercritical fluid distribution happens when the supercritical fluid experiences resistance during its circulation due to an increase in the number of fabric layers. This is common with fabrics wound on perforated beams. This obstruction in the flow distribution may cause the fluid to pass through weak sites present in the dyeing beam which have minimal flow resistance. This can be controlled by separating the beam and the PET fabric with cotton materials and stainless steel mesh, or with beams fitted with fluid distribution baffles [4, 47].

Since the dye partitioning and sorption kinetics between the fabric and the solvent is affected by the temperature and pressure conditions, proper regulation of the temperature and pressure of the dyeing bath is necessary to achieve uniformity of dye distribution. Increasing CO_2 flow rates can also help in improving the homogeneity of the dye distribution.

The method of flow reversal which is commonly used in traditional aqueous-based dyeing processes can also be used to enhance the uniformity of the dye distribution. When longer outside to inside-and-inside to outside reversal time intervals were used, the uniformity of the dye distribution has been reported to have increased [48].

Also, increasing the time required for the depressurization phase and the stirring of the dye vessel as the $scCO_2$ is being added can help in achieving homogenous dye distribution. The process of stirring can help to eliminate the formation of dye aggregates/lumps thereby increasing the uniformity of dye distribution. On the other hand, more dye can be absorbed into the fabrics when the time required to release CO_2 is increased [47].

13.2 The use of dye mixtures for dyeing

With mixed dyes, different shades of colors can be obtained during dyeing. Dye mixtures are generally used in aqueous-based dyeing media. However, for dyeing operations carried out using $scCO_2$, the reported results in terms of coloration yields, dye uptake, and exhaustion curves are not as good as those obtained when aqueous-based dyeing media are used [49]. Therefore, current color matching systems for conventional water-based dyeing operations cannot be directly applied to supercritical fluid dyeing processes. This is because of the variation in the dyeing media used for the dye uptake and coloration yields of each dye which may be different from that obtained for a single dye as the temperature and pressure are changed for $scCO_2$ dyeing medium [50].

14. The washing step (postdyeing)

One important advantage of using a supercritical fluid dyeing medium is the replacement of the reduction clearing process with a washing process (using clean $scCO_2$) particularly for the dyeing of polyester and PET fabrics. However, the temperature of the $scCO_2$ used in the washing process should be below that of the PET glass transition temperature; hence, the need for careful selection. This is to prevent unwanted dye extraction from the polymer [6].

15. Solubility of dyes in $scCO_2$

Dye solubility in $scCO_2$ is a very important mixture property to be considered during the dye selection process. In order to develop very efficient dyeing systems, a good understanding of the solubility of dyes together with their representation is required. The choice of dye used will affect the process temperature and pressure used and also the optimal process conditions. Therefore, detailed analysis and information of the solubility mixture of the supercritical fluid is crucial for sound design and development of supercritical fluid processes [51]. To achieve this, the dependence of temperature, pressure, and invariably the density of dyestuff solubilities required to be known by studying the behavior of different dye materials in $scCO_2$ at varying conditions of pressure and temperature.

Commonly used mathematical models for modeling of dye solubility data in $scCO_2$ are the thermodynamic models such as the Equation of State (EOS) models, and the Density Based Models. The density-based models are quite simple to use when correlating the solubility of the solute. Dye properties such as critical properties, sublimation pressure, and acentric factors are not required, instead common independent variables such as pressure, temperature, and the density of the $scCO_2$ are required. The equations which usually have three or more variables can be readily obtained using the method of least squares for simple error minimization. Some of these density-based models are the

Chrastil model [52], Kumar and Johnston (K-J) model [53], Bartle's model [54], Mendez-Santiago and Teja Model [55], Garlapati-Madras empirical (GME) model [56], and Bian et al.'s model [57].

Equations of state thermodynamic models have been widely used for predicting phase-equilibrium data at high pressures. Examples of equations of state models are the Peng-Robinson (P-R) model, Soave-Redlich-Kwong (SRK) model, and Redlich-Kwong (RK) model. With thermodynamic models, the solubility behavior of the solids in compressed gaseous systems as solvents can be predicted and correlated although several exceptions exist. Of the equations of state models, the P-R model [58] is widely known and frequently used EOS model for correlating the solubility data of supercritical fluids. However, the use of P-R EOS model requires information of other physical properties of the dye that are quite difficult to determine experimentally and not readily available in the literature, resulting in the use of estimated values which increase the uncertainty and unreliability of the P-R EOS model [3]. Other limitations of the EOS models include difficulty in determining the solute-fluid interaction factor (the fine-tuning parameter, k_{ij}) due to its strong dependence on temperature and density, and the nonavailability of physicochemical data of nonvolatile solutes within certain conditions of temperature and pressure needed to calculate the vapor pressure [51]. Because of these limitations, the equations of state models are not usually used for the correlation and prediction of solubility data for supercritical fluids.

The solubilities of different dyes in $scCO_2$ have been investigated by several researchers. Lots of data are available on the effects of pressure and temperature on the solubility of disperse dyes in $scCO_2$ [59, 60]. The equilibrium solubilities of some selected dyes (disperse red 82, red 153, blue 79, yellow 119, modified yellow 119, and blue 79:1) in supercritical fluids have been studied within temperature intervals of 353.15 and 393.15 K, and pressure value of about 30 MPa [61, 62]. These studies reported improved solubilities of about 100 times when the pressure progressed from 15 to 30 MPa. Gordillo et al. [63] examined the solubility of 1,4-dimethylamino-anthraquinone (disperse blue 14) in $scCO_2$ using different group combination methods to estimate the critical properties and physical parameters. The predicted solubility obtained from the group combination methods was reported to give better results compared to the three individual equations of state models used. KhazaiePoul et al. [64] formulated a robust artificial neural network (ANN) model which had seven inputs using a multilayer perception layer network to predict the solubility of some disperse dyes (orange 3, blue 79, and solvent brown 1) in both $scCO_2$ and a cosolvent mixture comprising of 5.5 mol% of ethanol and $scCO_2$. Temperature values of 353.2, 373.2, and 393.2 K as well as pressure intervals of 15.92 and 30.11 MPa were used. From statistical analysis, their ANN models had a correlation coefficient of 0.998, discrepancy ratio of 1.053, and Nash-Sutcliffe model efficiency coefficient of 0.992; while acentric factor, molecular weight, pressure, and density (in reverse order) were reported as the most important factors from their sensitivity analysis. Sridar et al. [51] correlated the solubility data of 29 dyes in scCO_2 obtained from literature using several semiempirical models. Response surface methodology was used in developing a new model which was reported to correlate better in comparison to the other commonly used density-based models investigated. Percentage average absolute relative deviation of 12.68% was obtained for the new model developed.

Other researchers have tried correlating the solubility of carbon dioxide with the properties of the dye (molecular weight, melting point, dye structure, heat of fusion, and entropy). The dye structure to a large extent was found to be consistent for correlating the solubility of CO_2 [65].

16. Dye distribution between fiber and CO_2

The dye partitioning within the fiber and CO_2 is quite intricate and largely controlled by the dyeing time alongside other dyeing properties (temperature, pressure, density of CO_2) that affect the dye solubility, dye diffusion coefficient, and the dye affinity of the fiber [66]. To achieve maximum exhaustion rates, it is preferable to perform the scCO_2 dyeing process under operating conditions where very high values of the equilibrium partition coefficient (K_{eq}) can be obtained. The equilibrium partition coefficient which depends on the equilibrium concentration of the dye in the solution can be determined using the adsorption and desorption rate constants [66].

Disperse dyes are broadly categorized into either temperature-controlled dyes or density-controlled dyes. For the former, the solubility can be reduced by lowering the temperature while a decrease in the density of CO_2 which happens when pressure is decreased at a fixed temperature reduces the solubility of the latter category of dyes [66]. This enhances the distribution of dye between the fiber and CO_2 by causing the dye to be well partitioned in the direction of the dyed fiber which is sorbed by the fiber due to suitable dye uptake conditions during dye exhaustion.

Elaborate analysis of the behavior of the dye is required in order to design dyeing processes occurring in scCO_2, since variations in the dye conditions can lead to very high exhaustion rate of the dye from the solution beyond the sorption limit of the fiber which causes poor fastness quality due to the presence of dye precipitates on the surface of the fiber.

Also, variations in the dye yield and color shades of the fiber due to adjustments in pressure and temperature of the dyeing conditions have been reported for dyeing experiments involving combinations of dye mixtures [50, 60]. These deviations were explained to be as a result of competitions among the dyes for the available pores. Here, the molecular weights of the dyes were vital in the mobility of the dyes. The dye with the lowest molecular weight experienced increased dye uptake than those with higher molecular weights.

17. Mass transfer phenomena between fiber and CO_2

Supercritical processes have been reported to significantly increase with faster dyeing rates as observed for supercritical dyed PET than that obtained with the use of aqueous-based media [67]. The analysis of mass transfer data for the polymer and liquid phases yields the overall mass transfer rates. The polymer mass transfer rate can be explained on the basis of Fick's law of diffusion where the coefficients of diffusion (ranging from 10^{-11} to 10^{-15} m^2/s) have been reported for dyes dispersed in PET films and fibers using $scCO_2$. These values were higher than those obtained using aqueous medium for dyeing of hydrophobic textiles. Furthermore, the fluid phase mass transfer rate was affected by the fluid dynamics within the supercritical solvent and diffusion coefficients of the solute among others [67, 68]. Hou and Dai [69] reported activation energy values of 67.47 and 187.02 kJ/mol for the diffusion of disperse blue 79 in $scCO_2$ and aqueous-based medium respectively for PET.

18. Supercritical dyeing of synthetic fibers

Although much studies on supercritical fluid dyeing processes involving carbon dioxide have been centered on PET fibers, other synthetic textiles have been dyed in $scCO_2$ using disperse dyes. Synthetic fibers such as nylon, polylactic acid (PLA) fibers, and PP fabrics have been dyed using dispersive dyes in $scCO_2$ although with much difficulty. Compared to PET fibers dyed in $scCO_2$ using disperse dyes, reduced dye sorption rate and poor fastness quality have been reported for these fibers because of their nonpolar aliphatic structure and high crystalline nature [5]. The fastness quality can be enhanced by using carefully selected disperse dyes which can interact with the polymer to form covalent bonds with functional groups [5].

19. Supercritical dyeing of natural fibers

$scCO_2$ can also be used for dyeing natural fibers. However, unlike dyeing of PET which is relatively easy to perform, dyeing of natural fibers using $scCO_2$ is quite difficult. These difficulties arise because polar dyes such as reactive, direct, and acid dyes commonly used in aqueous-based dyeing processes are almost insoluble in $scCO_2$ [5]. This causes poor interaction between the natural fibers which are polar in nature and the disperse dyes [39]. This happens because of the insufficient breaking of the hydrogen bonds present in the fiber by CO_2, thereby limiting the dyes from diffusing into the internal structures of the natural fibers [16].

To overcome the challenge of dyeing natural textiles in $scCO_2$, several methods have been suggested. There are basically two methods popularly researched in the literature to overcome the challenge of dyeing natural textiles using $scCO_2$. These are the reverse micellar techniques which involve the solubilization of conventionally used polar dyes (for water dyeing process) in $scCO_2$ using water/CO_2 microemulsions, and the synthesis of novel reactive dyes which combines physical pretreatment of the fiber and the use of suitable cosolvents. Jun et al. [70] and Sawada et al. [71] have used the reverse micellar techniques to dye natural fibers in $scCO_2$.

Others methods are as follows:

(a) Use of impregnating agent

Impregnating agent like polyethylene oxide or PP glycol has been used to increase the rate of diffusion of the dye into the pores of the fibers. The impregnating agents accelerate the dissolution of disperse dyes during dyeing in the $scCO_2$ and also break down the hydrogen bonds through swelling of the fiber, thereby making it easier for the dyes to access the fiber [5, 40, 72]. Drawback to this approach includes the use of aqueous medium during the process of impregnating and removal of the substances which further requires two drying operations, poor fastness quality, and significant reduction in the color strength after washing.

(b) Use of cosolvents

Cosolvents can be used to increase the polarity and dissolving properties of $scCO_2$. Prior studies on the use of cosolvents for dyeing of natural textiles using disperse dyes in $scCO_2$ were performed by Bach et al. [5] using both alcohol and water as cosolvents. Since then several studies have been carried out on the use of cosolvents [73]. Drawback of this procedure is the problem of cleaning during the course of the process since the dye has to be changed when the dye concentration increases in the fluid phase. Moreover, drying is needed after the dyeing operation to remove the cosolvent from the fiber which is not required with conventional $scCO_2$ dyeing processes.

20. Fiber modification technologies

This involves the permanent modification of the fiber surface in order to enhance the natural textile dye uptake in $scCO_2$. This is achieved through the addition of hydrophobic functional groups that have good interactions with disperse dyes, thus improving the dye affinity to the fiber [16]. Drawback of this procedure is the requirement of large amounts of modifying agents before desirable color depths can be obtained. Also, addition of large quantities of the modifying agents negatively interferes with the properties of the fibers.

Generally, the application of these alternatives (at certain stages) requires the use of water which deviates from the concept of a waterless dyeing technique while also requires further drying operations thereby increasing the energy requirement of the dyeing process contrary to the conventional $scCO_2$ dyeing process [5].

21. Possibility for optimization

Since $scCO_2$ processes require the use of high pressures which affect their economics in terms of equipment cost and operating cost of skilled workers, the cosolvents can be used to optimize the application of supercritical fluid processes. By introducing small amounts of a cosolvent (say ethanol) into the $scCO_2$, the operating pressure can be reduced considerably thereby increasing the dissolving properties of the supercritical fluid at lower pressure conditions [74].

Also, the solubility of solutes in $scCO_2$ can be significantly increased when small amounts of cosolvents are added to the supercritical fluid. This is possible since the addition of cosolvents increases the density of the solvent while intensifying the molecular interactions taking place within the solvent, solute, and cosolvent molecules [75]. However, caution should be applied with the addition of solvents because of their effects on the environment. An approach to this is to use environmentally friendly solvents, or devising ways to reduce the amount of solvents to be used, while also finding ways to optimize solvent recovery from the dyeing process.

22. Challenges and limitations of $scCO_2$ dyeing

The use of supercritical fluid for the dyeing of natural fibers is a major challenge [76]. This is because of the nonpolar nature of $scCO_2$ which hinders its application in the dyeing of polar hydrophilic textiles. On the contrary, this low polarity of $scCO_2$ makes it an excellent solvent in the dyeing of hydrophobic fibers [39]. Natural fibers on the other hand are hydrophilic and have polar groups which result in poor color fastness when used for $scCO_2$ dyeing process due to low interactions between the hydrophobic disperse dyes and the natural fibers [77]. An approach to overcome this challenge is using modified disperse reactive dyes with desired functional groups for the $scCO_2$ dyeing of natural textiles [77, 78]. Reactive dispersed dyes with acyl fluoride functional groups have been used to dye cotton fiber in $scCO_2$ by Long et al. [78] with significant improvements in terms of dye fixation. This happened due to the formation of covalent bonds between the functional groups of the reactive disperse dyes and the cotton fibers. Color fastness leveling of 4–5, 4–5, and 4 for staining, dry, and wet respectively has been reported.

23. Future prospects

Further investigations need to be carried out to properly understand the mechanism of dye solubility in $scCO_2$ and other cosolvent processes. More ANNs in determining dye solubility in $scCO_2$ processes should be vigorously explored considering their ability to make reliable and accurate predictions without the use of often complex mathematical descriptions of the desired process [64, 79]. The tuning properties of $scCO_2$ should be further exploited since merely changing the temperature or pressure can result in the formation of a variety of dye shades without having to alter the initial quantity of dye charged into the dyeing vessel.

Other emerging technologies of interest as alternatives to the current supercritical fluid dyeing processes that are being researched are the development of a supercritical fluid spray dyeing technology and the development of a supercritical fluid dyeing technology that incorporates electrophoresis in its operation. The latter could help in the dyeing of natural fibers.

More studies need to be carried out to improve the mechanical properties and fastness quality of PP, PLA, and Aramid synthetic fibers. The use of blended fibers (synthetic and natural fabrics) for dyeing operations should be explored as an option. The development of a dyeing procedure in which blends of PET/cotton fibers can be dyed together in a single step would be a welcome innovation in the textile industry.

24. Conclusion

The generation of large volumes of effluents containing water-soluble dyes used in conventional dyeing processes has motivated the adoption of scCO_2 dyeing as a suitable alternative. scCO_2 dyeing has some important advantages over conventional dyeing processes such as environmental friendliness and ease of handling. Dye solubility is an important factor that must be given ample consideration during scCO_2 dyeing. Although most of the works in the literature have focused on the dyeing of synthetic fibers, scCO_2 dyeing can also be applied to natural fibers albeit with some difficulties. These difficulties could be overcome by the use of impregnating agents, cosolvents, solubilization of the conventionally used polar dyes in scCO_2 using water/CO_2 microemulsions, or the synthesis of entirely new dyes with desirable properties.

References

[1] C.J. Ogugbue, T. Sawidis, Bioremediation and detoxification of synthetic wastewater containing triarylmethane dyes by Aeromonas hydrophila isolated from industrial effluent, Biotechnol. Res. Int. 2011 (2011).
[2] A.I. Cooper, Polymer synthesis and processing using supercritical carbon dioxide, J. Mater. Chem. 10 (2000) 207–234.
[3] Z. Huang, Y.H. Guo, G.B. Sun, Y.C. Chiew, S. Kawi, Representing dyestuff solubility in supercritical carbon dioxide with several density-based correlations, Fluid Phase Equilib. 236 (2005) 136–145.
[4] J.J. Long, Y.Q. Ma, J.P. Zhao, Investigations on the level dyeing of fabrics in supercritical carbon dioxide, J. Supercrit. Fluids 57 (2011) 80–86.
[5] E. Bach, E. Cleve, E. Schollmeyer, Past, present and future of supercritical fluid dyeing technology–an overview, Rev. Prog. Color. Relat. Top. 32 (2002) 88–102.
[6] A. Hou, B. Chen, J. Dai, K. Zhang, Using supercritical carbon dioxide as solvent to replace water in polyethylene terephthalate (PET) fabric dyeing procedures, J. Clean. Prod. 18 (2010) 1009–1014.
[7] P. Gregory, Dyes and dye intermediates, Kirk-Othmer Encycl, Chem. Technol. (2000) 1–66.
[8] S.M. Burkinshaw, Physico-chemical Aspects of Textile Coloration, John Wiley & Sons, 2016.

[9] C.C. Guaratini, M.V.B. Zanoni, Textile dyes, Quim Nova 23 (2000) 71–78.
[10] A. Bafana, S.S. Devi, T. Chakrabarti, Azo dyes: past, present and the future, Environ. Rev. 19 (2011) 350–371, https://doi.org/10.1139/a11-018.
[11] H. Zollinger, Color Chemistry: Syntheses, Properties, and Applications of Organic Dyes and Pigments, John Wiley and Sons, Inc, 2003.
[12] S.B. Moore, L.W. Ausley, Systems thinking and green chemistry in the textile industry: concepts, technologies and benefits, J. Clean. Prod. 12 (2004) 585–601.
[13] K.R. Beck, M.A. Hubbe, W.G. O'Neal, Y.C. Sharma, Cellulosic substrates for removal of pollutants from aqueous systems: a review, Bioresources 7 (2012) 2592–2687.
[14] V. Vassileva, E. Valcheva, Z. Zheleva, The kinetic model of reactive dye fixation on cotton fibres, J. Univ. Chem. Technol. Metall. 43 (2008) 323–326.
[15] G.H. Brunner, Supercritical Fluids as Solvents and Reaction Media, Elsevier, 2004.
[16] W. Saus, D. Knittel, E. Schollmeyer, Dyeing of textiles in supercritical carbon dioxide, Text. Res. J. 63 (1993) 135–142.
[17] J. Fasihi, F. Yamini, Y. Nourmohammadian, N. Bahramifar, Investigations on the solubilities of some disperse azo dyes in supercritical carbon dioxide, Dyes Pigments 63 (2004) 161–168.
[18] J. Von Schnitzler, R. Eggers, Mass transfer in polymers in a supercritical CO2-atmosphere, J. Supercrit. Fluids 16 (1999) 81–92.
[19] K.P. Johnston, P.S. Shah, Making nanoscale materials with supercritical fluids, Science (80-.) 303 (2004) 482–483.
[20] M. Hamburger, D. Baumann, S. Adler, Supercritical carbon dioxide extraction of selected medicinal plants-effects of high pressure and added ethanol on yield of extracted substances, Phytochem. Anal. 15 (2004) 46–54.
[21] E. Reverchon, L.D. Marco, Supercritical fluid extraction and fractionation of natural matter, J. Supercrit. Fluids 46 (2006) 146–166.
[22] X. Han, M. Poliakoff, Continuous reactions in supercritical carbon dioxide: problems, solutions and possible ways forward, Chem. Soc. Rev. 41 (2012) 1428–1436.
[23] W. Deng, D. Chen, M. Huang, J. Hu, L. Chen, Carbon dioxide deliming in leather production: a literature review, J. Clean. Prod. 87 (2015) 26–38.
[24] J. Hu, W. Deng, Application of supercritical carbon dioxide for leather processing, J. Clean. Prod. 113 (2016) 931–946.
[25] G.J.K. Bonthuys, C.E. Schwarz, A.J. Burger, J.H. Knoetze, Separation of alkanes and alcohols with supercritical fluids. Part I: phase equilibria and viability study, J. Supercrit. Fluids 57 (2011) 101–111.
[26] C.J. Cowper, A.J. Derose, Supercritical Fluid Chromatography, RSC, London, 1988.
[27] G. Madras, C. Erkey, A. Akgerman, Supercritical extraction of organic contaminants from soil combined with adsorption onto activated carbon, Environ. Prog. 13 (1994) 45–50.
[28] J.R. Forero-Mendieta, H. Castro-Vargas, F. Parada-Alfonso, J.A. Guerrero-Dallos, Extraction of pesticides from soil using supercritical carbon dioxide added with methanol as co-solvent, J. Supercrit. Fluids 68 (2012) 64–70.
[29] A. Geranmayeh, A. Mowla, H. Rajaei, F. Esmaeilzadeh, J.F. Kaljahi, Extraction of hydrocarbons from the contaminated soil of Pazanan II production unit by supercritical carbon dioxide, J. Supercrit. Fluids 72 (2012) 298–304.
[30] J.D. Grunwaldt, A. Baiker, In situ spectroscopic investigation of heterogeneous catalysts and reaction media at high pressure, Phys. Chem. Chem. Phys. 7 (2005) 3526–3539.
[31] T. Huang, H. Cui, D. Yang, X. Kong, J. Lin, Continuous dyeing processes for zipper tape in supercritical carbon dioxide, J. Clean. Prod. 158 (2017) 95–100, https://doi.org/10.1016/j.jclepro.2017.04.167.
[32] H. Zheng, J. Zhang, J. Yan, L. Zheng, An industrial scale multiple supercritical carbon dioxide apparatus and its eco-friendly dyeing production, J. CO_2 Util. 16 (2016) 272–281.
[33] J.J. Long, C.-L. Cui, Y.-Q. Zhang, G.-H. Yuan, Clean fixation of dye on cotton in supercritical carbon dioxide with a heterogeneous and phase transfer catalytic reaction, Dyes Pigm. 115 (2015) 88–95.
[34] H. Zheng, Y. Zhong, Z. Mao, L. Zheng, CO2 utilization for the waterless dyeing: Characterization and properties of Disperse Red 167 in supercritical fluid, J. CO_2 Util. 24 (2018) 266–273.

[35] J.-J. Long, C.-L. Cui, L. Wang, H.-M. Xu, Z.-J. Yu, X.-P. Bi, Effect of treatment pressure on wool fiber in supercritical carbon dioxide fluid, J. Clean. Prod. 43 (2013) 52–58.
[36] W.A. Hendrix, Progress in supercritical CO_2 dyeing, J. Ind. Text. 31 (2001) 43–56.
[37] M. Van Der Kraan, M. Vanesa Fernandez Cid, G.F. Woerlee, W.J. Veugelers, G. J. Witkamp, Equilibrium study on the disperse dyeing of polyester textile in supercritical carbon dioxide, Text. Res. J. 78 (2007) 550–558.
[38] M. Banchero, S. Sicardi, A. Ferri, L. Manna, Supercritical dyeing of textiles—from the laboratory apparatus to the pilot plant, Text. Res. J. 78 (2008) 217–223.
[39] M. Banchero, Supercritical fluid dyeing of synthetic and natural textiles—a review, Color. Technol. 129 (2013) 2–17, https://doi.org/10.1111/cote.12005.
[40] P.L. Beltrame, A. Castelli, E. Selli, A. Mossa, G. Testa, A.M. Bonfatti, A. Seves, Dyeing of cotton in supercritical carbon dioxide, Dyes Pigm. 39 (1998) 335–340.
[41] E. Cleve, E. Bach, E. Schollmeyer, Untersuchungen zum Schmelz- und Strukturverhalten von Polyethylenterephthalatfasern in Luft bei 1 bar und in überkritischem CO2 bis 280 bar, Die Angew. Makromol. Chemie. 256 (1998) 39–48, https://doi.org/10.1002/(SICI)1522-9505(19980401)256:1<39::AID-APMC39>3.0.CO;2-S.
[42] S.K. Liao, P.S. Chang, Y.C. Lin, Analysis on the dyeing of polypropylene fibers in supercritical carbon dioxide, J. Polym. Res. 7 (2000) 155–159.
[43] S.K. Liao, P.S. Chang, Y.C. Lin, Dyeing of polypropylene fibers with disperse dyes by supercritical carbon dioxide, J. Text. Eng. 46 (2000) 123–129.
[44] R. Li, Z. Zhang, T. Fang, Experimental research on swelling and glass transition behavior of poly (methyl methacrylate) in supercritical carbon dioxide, J. Supercrit. Fluids (2016) 110–116.
[45] M.J. Drew, J. Eggleston, Book of Papers, in: AATCC Int. Conf. Exhib., Nashville, USA, 1996, p. 360.
[46] A. Hou, K. Xie, J. Dai, Effect of supercritical carbon dioxide dyeing conditions on the chemical and morphological changes of poly (ethylene terephthalate) fibers, J. Appl. Polym. Sci. 93 (2004) 2008–2012.
[47] S. Okubayashi, T. Suzuma, C. Zhao, K. Miyakazi, K. Hirogaki, I. Tabata, T. Hori, Supercritical dyeing of polyester fibers in a mini-plant possessing internal circulator, Sen'i Gakkaishi 67 (2011) 27–33.
[48] G. Huang, Y. Xing, J. Dai, Computer distributed control and intelligent environmental monitoring, in: Int. Conf. Chang. China, 2011, p. 1097.
[49] G. Huang, J. Dai, F. Dong, J. Wang, Y. Jia, Compatibility of a disperse dye mixture in supercritical carbon dioxide dyeing, Color. Technol. 129 (2013) 305–311.
[50] L. Tusek, V. Golob, Z. Knez, The effect of pressure and temperature on supercritical CO2 dyeing of PET-dyeing with mixtures of dyes, Int. J. Polym. Mater. 47 (2000) 657–665.
[51] R. Sridar, A. Bhowal, C. Garlapati, A new model for the solubility of dye compounds in supercritical carbon dioxide, Thermochim. Acta 561 (2013) 91–97.
[52] J. Chrastil, Solubility of solids and liquids in supercritical gases, J. Phys. Chem. 86 (1982) 3016–3021.
[53] S.K. Kumar, K.P. Johnston, Modelling the solubility of solids in supercritical fluids with density as the independent variable, J. Supercrit. Fluids 1 (1988) 15–22.
[54] K.D. Bartle, A.A. Clifford, S.A. Jafar, G.F. Shilstone, Solubilities of solids and liquids of low volatility in supercritical carbon dioxide, J. Phys. Chem. Ref. Data 20 (1991) 713–756.
[55] J. Mendez-Santiago, A.S. Teja, Solubility of solids in supercritical fluids: consistency of data and a new model for cosolvent systems, Ind. Eng. Chem. Res. 39 (2000) 4767–4771.
[56] C. Garlapati, G. Madras, New empirical expressions to correlate solubilities of solids in supercritical carbon dioxide, Thermochim. Acta 500 (2010) 123–127.
[57] X. Bian, Z. Du, Y. Tang, An improved density-based model for the solubility of some compounds in supercritical carbon dioxide, Thermochim. Acta 519 (2001) 16–21.
[58] D.Y. Peng, D.B. Robinson, A new two-constant equation of state, Ind. Eng. Chem. Res. 15 (1976) 59–64.
[59] B. Guzel, A. Akgerman, Solubility of disperse and mordant dyes in supercritical CO2, J. Chem. Eng. Data 44 (1999) 83–85.
[60] C.T. Wang, W.F. Lin, Scouring and dyeing of polyester fibers in supercritical carbon dioxide, J. Chem. Eng. Japan 34 (2001) 244–248.

[61] H.M. Lin, C.Y. Liu, C.H. Cheng, Y.T. Chen, M.J. Lee, Solubilities of disperse dyes of blue 79, red 153, and yellow 119 in supercritical carbon dioxide, J. Supercrit. Fluids 21 (2001) 1–9.
[62] H.M. Lin, C.C. Ho, M.J. Lee, Solubilities of disperse dyes of blue 79: 1, red 82 and modified yellow 119 in supercritical carbon dioxide and nitrous oxide, J. Supercrit. Fluids 32 (2004) 105–114.
[63] M.D. Gordillo, C. Pereyra, E.M. de la Ossa, Solubility estimations for Disperse Blue 14 in supercritical carbon dioxide, Dyes Pigm. 67 (2005) 167–173.
[64] A. Khazaie Poul, M. Soleimani, S. Salahi, Solubility prediction of disperse dyes in supercritical carbon dioxide and ethanol as co-solvent using neural network, Chin. J. Chem. Eng. 24 (2016) 491–498.
[65] S.L. Draper, G.A. Montero, B. Smith, K. Beck, Solubility relationships for disperse dyes in supercritical carbon dioxide, Dyes Pigm. 45 (2000) 177–183.
[66] G.A. Montero, C.B. Smith, W.A. Hendrix, D.L. Butcher, Supercritical fluid technology in textile processing: an overview, Ind. Eng. Chem. Res. 39 (2000) 4806–4812.
[67] M. Banchero, A. Ferri, Simulation of aqueous and supercritical fluid dyeing of a spool of yarn, J. Supercrit. Fluids 35 (2005) 157–166.
[68] S. Sicardi, L. Manna, M. Banchero, Comparison of dye diffusion in poly (ethylene terephthalate) films in the presence of a supercritical or aqueous solvent, Ind. Eng. Chem. Res. 39 (2000) 4707–4713.
[69] A. Hou, J. Dai, Kinetics of dyeing of polyester with CI Disperse Blue 79 in supercritical carbon dioxide, Color. Technol. 121 (2005) 18–20.
[70] J.H. Jun, K. Sawada, M. Ueda, Application of perfluoropolyether reverse micelles in supercritical CO2 to dyeing process, Dyes Pigm. 61 (2004) 17–22.
[71] K. Sawada, M. Oshima, M. Sugimoto, H. Urakawa, M. Ueda, Complexation of ionic substances with surfactants for solubilisation in supercritical carbon dioxide, Dyes Pigm. 76 (2008).
[72] B. Gebert, D. Saus, W. Knittel, H.J. Buschmann, E. Schollmeyer, Dyeing natural fibers with disperse dyes in supercritical carbon dioxide, Text. Res. J. 64 (1994) 371–374.
[73] M.V. Fernandez, J. van Spronsen, G.F. Woerlee, G.J. Witkamp, A significant approach to dye cotton in supercritical carbon dioxide with fluorotriazine reactive dyes, J. Supercrit. Fluids 40 (2007) 477–484.
[74] L.U. Huisheng, R.E.N. Miaomiao, M. Zhang, C.H.E.N. Ying, Pretreatment of corn stover using supercritical CO_2 with water-ethanol as co-solvent, Chinese, J. Chem. Eng. 2 (2013) 551–557.
[75] J.B. Ellington, K.M. Park, J.F. Brennecke, Effect of local composition enhancements on the esterification of phthalic anhydride with methanol in supercritical carbon dioxide, Ind. Eng. Chem. Res. 33 (1994) 965–974.
[76] D. Gao, H.S. Cui, T.T. Huang, D.F. Yang, J.X. Lin, Synthesis of reactive disperse dyes containing halogenated acetamide group for dyeing cotton fabric in supercritical carbon dioxide, J. Supercrit. Fluids 86 (2013) 108–114.
[77] Y.Q. Zhang, X.C. Wei, J.J. Long, Ecofriendly synthesis and application of special disperse reactive dyes in waterless coloration of wool with supercritical carbon dioxide, J. Clean. Prod. 133 (2016) 746–756.
[78] J.J. Long, G.D. Xiao, H.M. Xu, L. Wang, C.L. Cui, J. Liu, M.Y. Yang, K. Wang, C. Chen, Y.M. Ren, T. Luan, Z.F. Ding, Dyeing of cotton fabric with a reactive disperse dye in supercritical carbon dioxide, J. Supercrit. Fluids 69 (2012) 13–20.
[79] B. Vaferi, M. Karimi, M. Azizi, H. Esmaeili, Comparison between the artificial neural network, SAFT and PRSV approach in obtaining the solubility of solid aromatic compounds in supercritical carbon dioxide, J. Supercrit. Fluids 77 (2013) 44–51.

CHAPTER 14

Hydrogenation of fats and oils using supercritical carbon dioxide

Wan-Hui Wang*,†, Xin Liu*,†, Ming Bao*,†
*State Key Laboratory of Fine Chemicals, Dalian University of Technology, Ganjingzi District, Dalian China
†School of Petroleum and Chemical Engineering, Dalian University of Technology, Liaodongwan New District, Panjin, China

Fats and oils are complicated mixtures that contain mainly mono-, di-, and triglyceride, as well as free fatty acids and other components. The products of hydrogenation of fats and oils are widely employed in the cosmetic, food, oleochemical industries, and also in the development of biofuels [1, 2].

Unsaturated fatty acids in triglycerides usually contain two or more carbon-carbon double bonds. The unsaturation of the fatty acid is usually quantified by iodine value (IV), which is defined as the number of grams of I_2 consumed by 100 g of sample. Hydrogenation of fats and oils allows the modification of fatty acid and triglyceride compositions to generate fatty alcohols, saturated fatty acids, and hydrocarbons. According to the Horiuti-Polanyi mechanism of hydrogenation, hydrogen molecule is adsorbed and activated by metal catalyst. The following addition of one hydrogen atom to β-carbon of the substrate is reversible. Therefore isomerization and generation of *trans* isomers is inevitable (Scheme 1) [3]. In food industry, hydrogenated fats and oils contain plenty of *trans* fatty acids (TFA), which cause increasing health and nutrition concerns because TFA are closely related to cardiovascular disease [4, 5]. In 2003, Denmark first implemented legislation to limit the maximum level of TFA. After 3 years, United States enacted the labeling regulations to display the contents of TFA [6]. Great efforts have been devoted to reducing the generation of TFA in hydrogenation. Among the options, hydrogenation in supercritical fluids has attracted much attention. Supercritical fluids exhibit excellent properties for dissolving gases and liquids and are capable to enhance the hydrogenation rate and decrease the isomerization [4, 7].

In fat and oil processing, supercritical carbon dioxide ($scCO_2$) is a favorable alternative to conventional organic solvents [8, 9]. Different from other supercritical fluids [10, 11], CO_2 is abundant, cheap, nonflammable, and nontoxic. More importantly, it possesses low critical temperature of 31°C and moderate critical pressure of 7.4 MPa [12]. At supercritical conditions, CO_2 keeps the properties of gas and is fully miscible with H_2. Moreover, it owns high density as liquid and similar diffusivity and viscosity as gas. These properties significantly enhance the mass transfer, as well as the hydrogenation efficiency

Green Sustainable Process for Chemical and Environmental Engineering and Science
https://doi.org/10.1016/B978-0-12-817388-6.00014-3

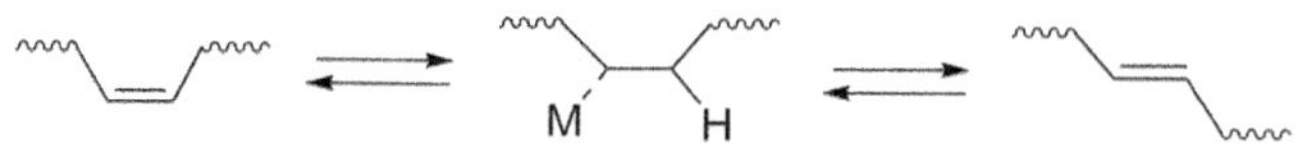

Scheme 1 Isomerization in the catalytic hydrogenation.

and allow hydrogenation to perform at markedly milder conditions, compared with conventional processes.

This chapter reviews the developments and challenges in fat and oil hydrogenation using $scCO_2$ over the previous decades. Some insights into future potential are also provided.

In 1996, Tacke et al. first studied the hydrogenation of fats with $scCO_2$ and supercritical propane (scC_3H_8) respectively in fixed bed reactors [13]. The effects of temperature (60–160°C), total pressure (8.0–16.0 MPa), and space velocity (LHSV; 5–60 h^{-1}) were studied. In the free fatty acid hardening, reaction yields using a fixed bed catalyst of 1 wt% palladium supported by DELOXAN are 6–15 times higher than that using trickle bed process with an activated carbon supported 2 wt% palladium fixed bed catalyst. Moreover, a significant decrease in undesired by-products (e.g., TFA) was achieved in the edible oil hardening in $scCO_2$ with DELOXAN supported 1 wt% platinum fixed bed catalyst.

In 2000, Blomberg et al. reported the hydrogenation of fatty acid methyl esters (FAME) of soybean oil to fatty alcohols in $scCO_2$ and scC_3H_8 with two commercially available catalysts, namely, copper chromite catalyst and chromium-free catalyst [14]. The flow reaction system is depicted in Fig. 1. Both catalysts are tested under the same reaction conditions (150–250 bar, 210–250°C, H_2 mole fraction 0.10–0.25, residence time 4–9 s) with $scCO_2$ and scC_3H_8, except the substrate flow rate (25–50 μL/min for $scCO_2$ and 50–250 μL/min for scC_3H_8). Under the optimum conditions, both catalysts exhibited similar activity and can be reused multiple times. High conversions were obtained at 250°C and at a hydrogenation mole fraction of 0.25. Large amounts of *n*-alkane (4.6%) containing C16 and C18 were generated as by-products in the H_2/scC_3H_8 system. On the contrary, no alkanes were found in the $H_2/scCO_2$. This study demonstrated that it is feasible to perform continuous hydrogenation of FAME from vegetable oil resources to produce fatty alcohols using $scCO_2$.

In 2001, King et al. investigated the hydrogenation of soybean oil with $H_2/scCO_2$ in a batch, slurry reactor using a conventional nickel catalyst in the range of 120–140°C at 2000 psi [15]. The reaction system is shown in Fig. 2. At a constant temperature, they found that the reaction rate in the mixture of $scCO_2$ and H_2 was slower than that found by using pure hydrogen. This observation is different from the reported rate increase using flow-reactor and noble-metal catalysts in $scCO_2$ [10, 13, 16]. Nonselective hydrogenation was observed at high pressures (1000 psi CO_2/1000 psi H_2). However, the oils produced were found to be with low TFA content of 6.4 wt% (120°C, 4 h).

In 2008, Larrayoz et al. reported supercritical hydrogenation of sunflower oil in a continuous hydrogenation system using 0.5% Pd/Al_2O_3, 1% vegetable oil, 4% H_2, and 95% CO_2/hexane (80/20 v/v) [17]. Lipid solubility can be improved by using

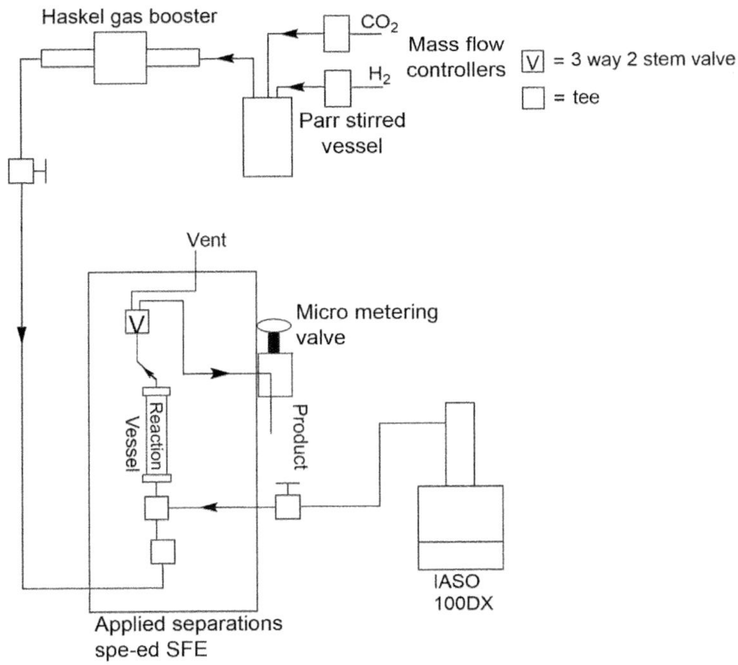

Fig. 1 Reaction system for hydrogenating FAMES to alcohols using binary fluid mixtures of $H_2/scCO_2$ or H_2/scC_3H_8. *(Reproduced with permission from M.B.O. Andersson, J.W. King, L.G. Blomberg, Synthesis of fatty alcohol mixtures from oleochemicals in supercritical fluids, Green Chem. 2 (2000) 230–234. Copyright (2000) Royal Society of Chemistry.)*

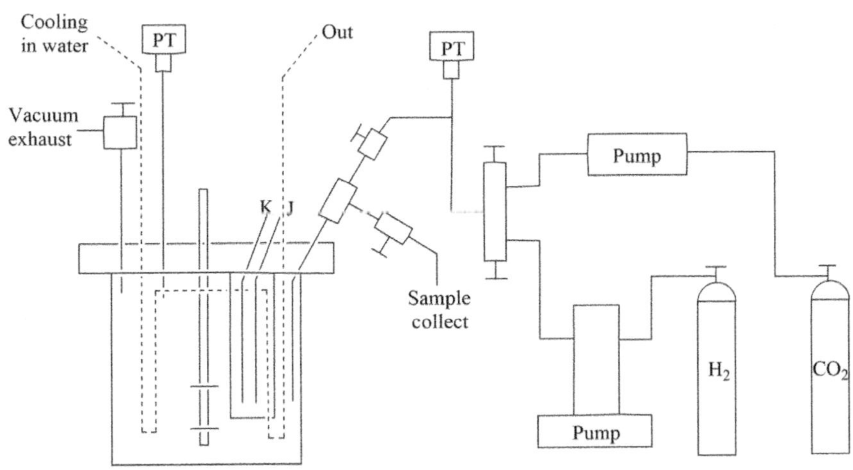

Fig. 2 High-pressure hydrogenation system for conducting conventional and binary hydrogenations on soybean oil. PT, pressure transducer; J and K, thermocouples. *(Reproduced with permission from J.W. King, R.L. Holliday, G.R. List, J.M. Snyder, Hydrogenation of vegetable oils using mixtures of supercritical carbon dioxide and hydrogen, J. Am. Oil Chem. Soc. 78 (2001) 107–113. Copyright (2001) Wiley-VCH Verlag GmbH & Co, KGaA, Weinheim.)*

hexane as a cosolvent. The study showed that the hydrogenation in $scCO_2$ and hexane presented low content of TFA (<5%) in the products. On the other hand, the use of modified $scCO_2$ under mild conditions could increase the viability of continuous hydrogenation due to the minor hazard in comparison with some flammable solvents, such as scC_3H_8 or dimethyl ether.

Wheat germ oil is rich in nutrients and has great benefits to the human body. In 2009, Wang et al. studied the hydrogenation of cold-pressed wheat germ oil with Pd/C as a catalyst (Pd content of 5%) in $scCO_2$ [18]. The effect of hydrogen pressure, reaction time, reaction temperature, stirring speed, and catalyst loading on the hydrogenation reaction was investigated by single factor and orthogonal experiments. Under the optimal conditions (80°C, 8 MPa, 40 min), the IV of the product was 72.6 g/100 g, and the content of Vitamin E was 2.453 g/kg. At the same time, Vitamin E content did not change substantially after the reaction. Thus, the nutrient loss was negligible. Moreover, hydrogenation of wheat germ oil in $scCO_2$ avoids the formation of TFA compared with conventional hydrogenation processes.

Soybean oil has acceptable lubricity, high viscosity index, and low volatility. However, soybean oil is rich in polyunsaturated fatty acids, which are easily oxidized and cannot be directly used as lubricating base oil. In 2011, Wang et al. investigated the hydrogenation of soybean oil with Pd/C catalyst in $scCO_2$ conditions to produce base oil of lubricant [19]. Under the optimal conditions (50°C, 9 MPa, 45 min), the typical indexes of the product were determined as follows: IV of 88.20 gI/100 g; viscosity value of 9.89 Pas; peroxide value of 5.6 mmol/kg; and acid value of 0.68 mg KOH/g. The produced lubricating oil is a renewable and green lubricating oil and can be applied to food industry.

In 2011, Recasens et al. used Aspen and reported a simulation study of the fatty oil hydrogenation under supercritical conditions involving two kinds of heterogeneous catalysts, namely, Pd/C and Pd/Al_2O_3 [20]. Reactor-catalyst combination and reaction conditions were both optimized. Furthermore, they investigated three supercritical solvents, namely, (i) scC_3H_8, (ii) cosolvent hexane-modified $scCO_2$, and (iii) pure liquid hexane. The process with neat CO_2 is unfavorable because the supercritical pressure is approximately 40 MPa, thereby resulting in significant increase in the equipment cost. Moreover, the solubility of oil in $scCO_2$ is still extremely low under these conditions. An addition of hexane results in remarkable decrease in the supercritical pressure. Thus, the supercritical conditions can be easily achieved. The safety differences in the three processes were evaluated by using the Dow Fire and Explosion Index (F&EI). The pure propane supercritical process reaches the highest Dow F&EI rating of 150 due to the flammability. The least hazardous process is that uses pure CO_2 (Dow F&EI < 30), but the process is impractical due to the extremely high equipment cost and operating pressure (40 MPa). Considering the safety issue, the hexane-modified CO_2 process is preferred because it has the similar risk as that of a conventional hydrogenation process.

In 2014, Bogel-Lukasik et al. conducted the hydrogenation of cattle fat in $scCO_2$ by using a commercially available heterogeneous fluid cracking catalyst (FCC) in a batch reactor [21]. The reaction was performed at 340°C with constant 5.5 bar of H_2 and 50–140 bar total pressure of CO_2/H_2 mixture for 1–6 h. The hydrogenation reaction of cattle fat results in a great variety of products. The GC analysis suggests that saturated odd-carbon hydrocarbons (defined as $2n-1$), saturated even-carbon hydrocarbons (defined as $2n$), and unsaturated hydrocarbons are the main products. In addition, the formation of small amount of benzene derivatives was detected. The typical reaction occurring in the hydrogenation is described in Scheme 2. The acid value of the product mixture is higher than that of the feedstock, suggesting that free fatty acids were generated from glycerides via hydrolysis reaction. The released glycerol from the hydrolysis was hydrogenated to obtain propane. The fatty acids of even-carbons are mainly found in the raw materials, whereas the major products of the hydrogenation reactions were found to be odd-carbon hydrocarbons. These results indicated that the hydrogenation is accompanied by decarboxylation to release CO_2 and H_2O as coproducts.

The increased total pressure caused by increased CO_2 partial pressure showed positive influences on the hydrogenation. The concentration of $2n-1$ and overall saturated hydrocarbons ($2n$ and $2n-1$) increase with the increasing total pressure (Fig. 3). The beneficial effect was attributed to the increased solubility of H_2 in the $scCO_2$, which could increase the mass transfer of gas.

The increase in CO_2 partial pressure also significantly influences the composition of the product mixture. Hydrocarbons with short alkyl chains, such as undecane, were found to be predominant when the hydrogenation was performed in the absence of CO_2 (Fig. 4). However, pentadecane generated as dominant higher aliphatic hydrocarbon in the presence of CO_2 (Fig. 4).

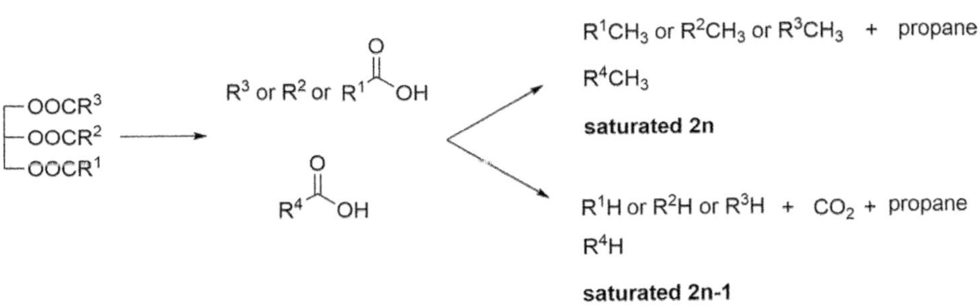

Scheme 2 Typical reactions in the hydrogenation of cattle fat catalyzed by FCC in $scCO_2$. Triglyceride represents all existing and formed glycerides in the raw material. R^4COOH represents all free fatty acids present in the feedstock. R^1, R^2, or R^3COOH symbolize the free fatty acids obtained from the hydrolysis of glycerides. The scheme presents two major routes to produce saturated $2n$ and $2n-1$ hydrocarbons and coproducts of these reactions. *(Reprinted with permission from A.R.C. Morais, A.M. da Costa Lopes, P. Costa, I. Fonseca, I.N. Nogueira, A.C. Oliveira, R. Bogel-Lukasik, Cattle fat valorisation through biofuel production by hydrogenation in supercritical carbon dioxide, RSC Adv. 4 (2014) 32081–32091. Copyright (2014) Royal Society of Chemistry.)*

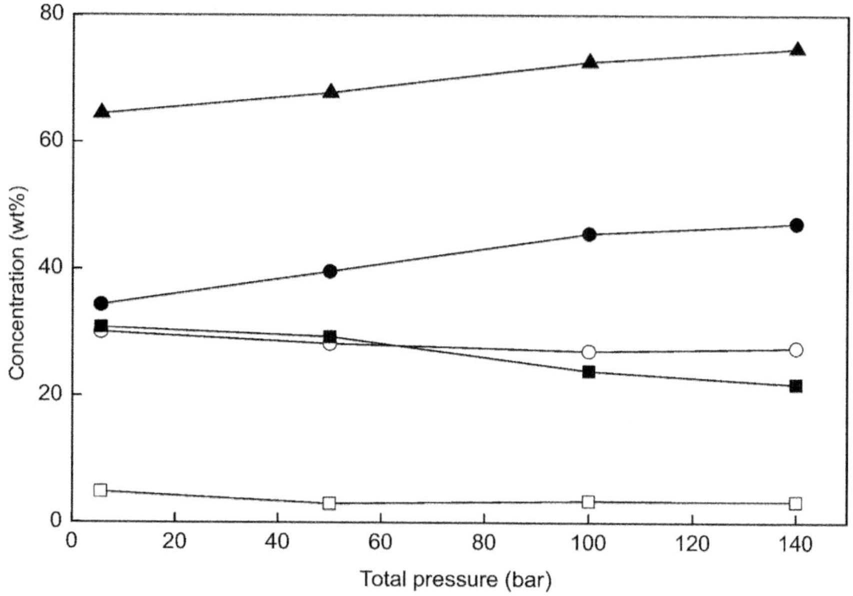

Fig. 3 Concentration of ●—saturated $2n-1$, ○—saturated $2n$, ▪—unsaturated, □—other, and ▲—total saturated hydrocarbons of hydrogenated cattle fat as a function of the total pressure used in the reaction. *(Reprinted with permission from A.R.C. Morais, A.M. da Costa Lopes, P. Costa, I. Fonseca, I.N. Nogueira, A.C. Oliveira, R. Bogel-Lukasik, Cattle fat valorisation through biofuel production by hydrogenation in supercritical carbon dioxide, RSC Adv. 4 (2014) 32081–32091. Copyright (2014) Royal Society of Chemistry.)*

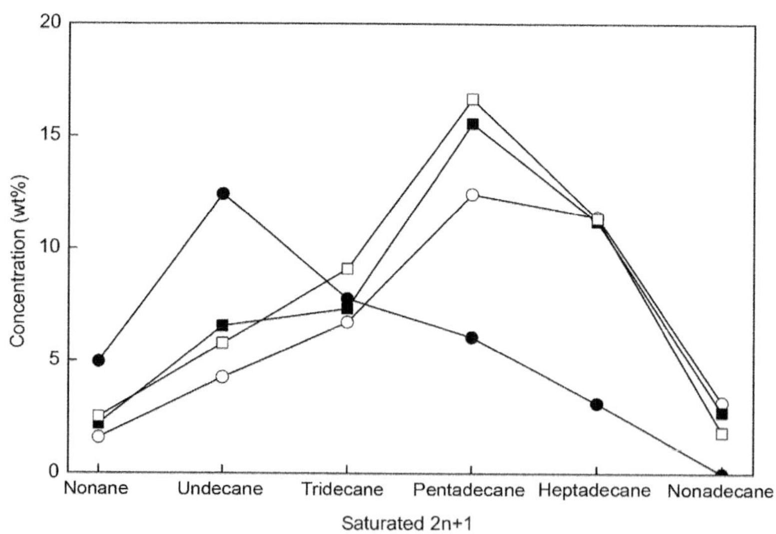

Fig. 4 Influence of the total pressure (●—5.5, ○—50, ▪—100, and □—140 bar) on the concentration of saturated $2n-1$ hydrocarbons. *(Reprinted with permission from A.R.C. Morais, A.M. da Costa Lopes, P. Costa, I. Fonseca, I.N. Nogueira, A.C. Oliveira, R. Bogel-Lukasik, Cattle fat valorisation through biofuel production by hydrogenation in supercritical carbon dioxide, RSC Adv. 4 (2014) 32081–32091. Copyright (2014) Royal Society of Chemistry.)*

Under the optimum conditions (140 bar total pressure of H_2/CO_2, 5.5 bar H_2, 340°C, 6 h), the content of saturated hydrocarbons reached 90 wt% of the total hydrocarbons. >55% of diesel-like products (C14–C20) could be obtained in the presence of CO_2, which is about two times higher than that observed in the absence of CO_2 (only 20%). Interestingly, the reused catalyst showed better selectivity for generating saturated hydrocarbons than the fresh catalyst under the same reaction conditions because CO_2 helps to orient FCC catalysts and change the catalyst morphology.

In 2018, Yu and Elfalleh et al. prepared two series of Cu- and Ni-based catalysts by co-impregnation method for the catalytic hydrogenation of soybean oil in $scCO_2$ [22]. The effect of precious metals on the catalytic activity was investigated. The isomer characterizations of hydrogenated soybean oil show low TFA content when noble metal promoters are added in the catalysts. The addition of precious metals facilitates the even dispersion of catalyst on the molecular sieve carrier, thereby increasing the catalyst surface and enhancing the rate of hydrogenation. Consequently, the chance of fatty acid isomerization and the content of TFA in products are significantly reduced. The hydrogenation procedure proposed by Albright is shown in Eq. (1).

$$\text{Linolenic (C18:3)} \rightarrow \text{Linoleic (C18:2)} \rightarrow \text{Oleic (C18:1)} \rightarrow \text{Stearic (C18:0)} \quad (1)$$

Among the catalysts examined, Cu-Ag/SBA15 showed better linolenic selectivity, while Ni-Ag/SBA15 showed superior linoleic selectivity. Therefore, a stepwise hydrogenation was performed to control the reaction selectivity. The hydrogenation was initially carried out with Cu-Ag/SBA15 for 20 min, then transferred to consequent catalytic system of Ni-Ag/SBA15 for 60 min. The analysis results showed that a high oleic acid content of 50.27% was obtained, and the content of TFA was considerably reduced to 10.43%.

In 2017, Wang et al. reported electrochemical hydrogenation of soybean oil with $scCO_2$ to decrease the production of TFA [23]. The hydrogenation was performed in a home-made reactor as shown in Fig. 5. The reaction kettle was equipped with electrodes at the top and electrolytic chambers inside (Fig. 5A). A Nafion 117 cation-exchange membrane was used to separate the cathode and anode chambers (Fig. 5B). In the cathode chamber, soybean oil, Pd/C (0.4%), and aqueous solution of HCO_2Na were added. In addition, a phase transfer catalyst ethyl dimethyl ammonium bromide (EDDAB, 0.4 g) was used to increase the solubility of oil. To the anode chamber, an aqueous solution of sulfuric acid was added. CO_2 was charged through gas inlet. When the reaction was completed, crude product was removed from the cathode chamber and converted to their corresponding FAME and analyzed by gas chromatography.

The authors studied the solubility of CO_2 in electrolytes and the corresponding electrical conductivity at different pressures. These results suggested that the solubility and conductivity increased along with increasing pressure and reached stable trend above 8 MPa. They also studied the effect of current, temperature, and stirring speed on TFA content and IV. Finally, the optimal hydrogenation parameters were obtained. Under the $scCO_2$ environment, the optimal conditions for the electrochemical

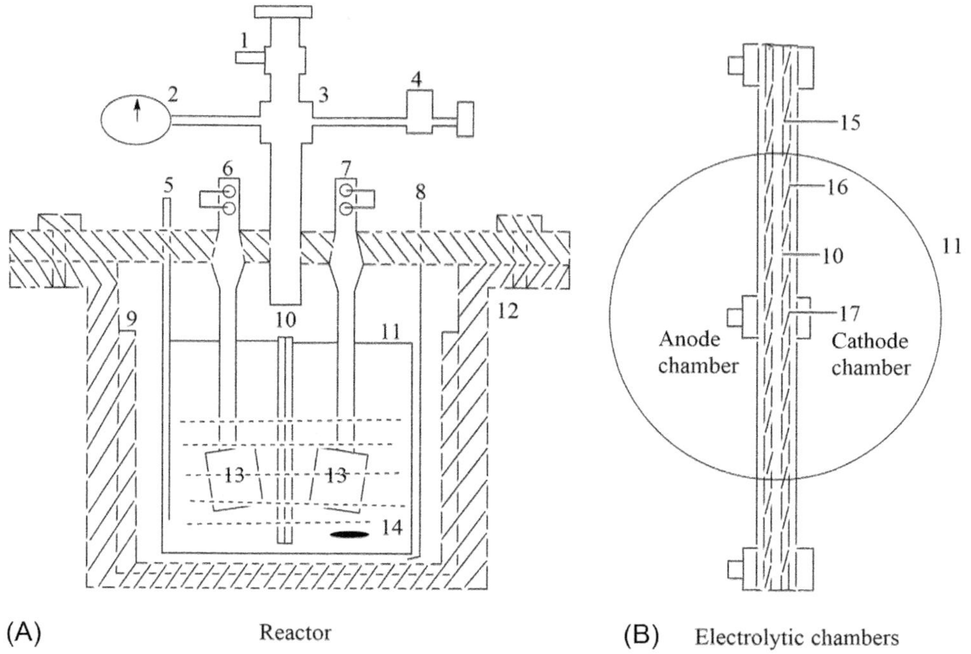

Fig. 5 Electrochemical hydrogenation reactor: (1) gas inlet, (2) pressure gauge, (3) safety valve, (4) gas outlet, (5) thermometer, (6) anode, (7) cathode, (8) sampling tube, (9) insulating sleeve, (10) cation-exchange membrane, (11) electrolytic chambers, (12) reaction kettle body, (13) platinum sheet, (14) rotor, (15) flange, (16) insulating sheet, and (17) insulating sleeve. *(Reproduced with permission from D. Yu, X. Li, Y. Wang, D. Zou, L. Hu, H. Zheng, L. Jiang, L. Wang, W. Elfalleh, The influence of supercritical carbon dioxide (SC-CO2) on electrolytes and hydrogenation of soybean oil, J. Am. Oil Chem. Soc. 94 (2017) 993–1001. Copyright (2017) Wiley-VCH Verlag GmbH & Co, KGaA, Weinheim.)*

hydrogenation are as follows: 125 mA current at 8 MPa and 48°C and 300 rpm of stirring speed for 8 h. The composition of fatty acid, TFA content, and IV of the product were tested. For comparison, the electrochemical hydrogenation was also evaluated at ambient pressure for 12 h. The results are listed in Table 1. Compared with the raw soybean oil, the IV decreased from 128.0 to 102.7 after hydrogenation at ambient pressure for 12 h, whereas hydrogenation in $scCO_2$ for 8 h lead to a low IV of 101.3. More importantly, the TFA content of hydrogenated oil obtained in $scCO_2$ was only 5.3%, which is 35.8% lower than that observed for the hydrogenated soybean oil at normal pressure.

Compared with industrial hydrogenation process at high temperature (150–225°C), the electrochemical hydrogenation of soybean oil in $scCO_2$ exhibits several merits, as follows: proceeding at low temperature, thereby considerably decreasing the TFA content in product and shortening the reaction time.

In summary, hydrogenation of different types of fats and oils in $scCO_2$ using various catalysts has been investigated with traditional approach or electrochemical method. $scCO_2$ showed different effect on the rate and composition of the product in batch

Table 1 Fatty acid profile, IV, and TFA content of the hydrogenated soybean oils at normal pressure and in $scCO_2$

Fatty acid and IV	Refined soybean oil	Hydrogenated soybean oil at normal pressure	Hydrogenated soybean oil in $scCO_2$
C18:0 (%)	4.0	11.4	9.3
C18:1 (%)	24.5	41.2	44.7
C18:2 (%)	55.1	26.1	29.4
C18:3 (%)	7.7	4.8	3.1
TFA (%)	0.7	8.1	5.2
IV	128.0	102.7	101.3

Adapted from D. Yu, X. Li, Y. Wang, D. Zou, L. Hu, H. Zheng, L. Jiang, L. Wang, W. Elfalleh, The influence of supercritical carbon dioxide (SC-CO2) on electrolytes and hydrogenation of soybean oil, J. Am. Oil Chem. Soc. 94 (2017) 993–1001.

reaction and continuous hydrogenation. Simulation study has also been performed to evaluate the process in $scCO_2$, considering the safety and cost concern. For fat and oil hydrogenation, the pure CO_2 showed low solubility for higher fatty acids [24], and high pressure is required to achieve supercritical conditions, which significantly increase the cost. Compared with scC_3H_8, $scCO_2$ exhibits number of advantages such as nontoxicity and no risk of explosion in view of green chemistry. The products of hydrogenation of fats and oils are usually employed in food industry. Considering food safety issue and the environmental protection requirement of avoiding the employment of organic solvents, the hydrogenation of fats and oils in $scCO_2$ has a bright future and deserves more investigation.

References

[1] H.B.W. Patterson, The hydrogenation reaction, in: G.R. List, J.W. King (Eds.), Hydrogenation of Fats and Oils: Theory and Practice, AOCS Press, Urbana, IL, 2011, pp. 1–32.

[2] S.K. Kim, J.Y. Han, S.-A. Hong, Y.-W. Lee, J. Kim, Supercritical CO_2-purification of waste cooking oil for high-yield diesel-like hydrocarbons via catalytic hydrodeoxygenation, Fuel 111 (2013) 510–518.

[3] C. Hu, D. Creaser, S. Siahrostami, H. Grönbeck, H. Ojagh, M. Skoglundh, Catalytic hydrogenation of C=C and C=O in unsaturated fatty acid methyl esters, Catal. Sci. Technol. 4 (2014) 2427–2444.

[4] E.S. Jang, M.Y. Jung, D.B. Min, Hydrogenation for low trans and high conjugated fatty acids, Compr. Rev. Food Sci. Food Saf. 1 (2005) 22–30.

[5] E. Jenab, F. Temelli, J.M. Curtis, Y.-Y. Zhao, Performance of two immobilized lipases for interesterification between canola oil and fully-hydrogenated canola oil under supercritical carbon dioxide, LWT-Food Sci. Technol. 58 (2014) 263–271.

[6] A. Philippaerts, P.A. Jacobs, B.F. Sels, Is there still a future for hydrogenated vegetable oils? Angew. Chem. Int. Ed. 52 (2013) 5220–5226.

[7] W.-L. Yang, K.-R. Gu, Research progress on hydrogenation of fats and oils in supercritical fluids, Cereals & Oils (2010) 1–3.

[8] E.J. Beckman, Supercritical and near-critical CO_2 in green chemical synthesis and processing, J. Supercrit. Fluids 28 (2004) 121–191.

[9] Ž. Knez, E. Markočič, M. Leitgeb, M. Primožič, M. Knez Hrnčič, M. Škerget, Industrial applications of supercritical fluids: a review, Energy 77 (2014) 235–243.
[10] M. Härröd, P. Møller, Hydrogenation of fats and oils at supercritical conditions, in: P.R. von Rohr, C. Trepp (Eds.), Process Technology Proceedings, Elsevier, 1996, pp. 43–48.
[11] S.v.d. Harka, M. Härröda, P. Møller, Hydrogenation of fatty acid methyl esters to fatty alcohols at supercritical conditions, J. Am. Oil Chem. Soc. 76 (1999) 1363–1370.
[12] R.S. Oakes, A.A. Clifford, C.M. Rayner, The use of supercritical fluids in synthetic organic chemistry, J. Chem. Soc. Perkin Trans. 1 (2001) 917–941.
[13] T. Tacke, S. Wieland, P. Panster, Hardening of fats and oils in supercritical CO_2, in: P.R. von Rohr, C. Trepp (Eds.), Process Technology Proceedings, Elsevier, 1996, pp. 17–21.
[14] M.B.O. Andersson, J.W. King, L.G. Blomberg, Synthesis of fatty alcohol mixtures from oleochemicals in supercritical fluids, Green Chem. 2 (2000) 230–234.
[15] J.W. King, R.L. Holliday, G.R. List, J.M. Snyder, Hydrogenation of vegetable oils using mixtures of supercritical carbon dioxide and hydrogen, J. Am. Oil Chem. Soc. 78 (2001) 107–113.
[16] T. Tacke, S. Wieland, P. Panster, M. Blankmann, R. Brand, H. Magerlein, U.S. Patent 5,734,070, 1998.
[17] A. Santana, A. Huguet, B. Pina, M. Larrayoz, in: AIChE Annual Meeting, Philadelphia, PA, 2008.
[18] S. Wang, Q. Jia, K. Yu, F. Hou, D. Yu, Hydrogenation of wheat germ oil in supercritical CO_2, China Oils and Fats 34 (2009) 27–30.
[19] Y. Wang, Y. Qi, H. Shao, X. Zhu, D. Yu, Preparation of lubricant base oil by hydrogenation of soybean oil in supercritical CO_2 state, China Oils and Fats 36 (2011) 61–63.
[20] E. Ramírez, M.J. Mayorga, D. Cuevas, F. Recasens, Fatty oil hydrogenation in supercritical solvents: process design and safety issues, J. Supercrit. Fluids 57 (2011) 143–154.
[21] A.R.C. Morais, A.M. da Costa Lopes, P. Costa, I. Fonseca, I.N. Nogueira, A.C. Oliveira, R. Bogel-Lukasik, Cattle fat valorisation through biofuel production by hydrogenation in supercritical carbon dioxide, RSC Adv. 4 (2014) 32081–32091.
[22] Y. Zhao, Y. Ren, R. Zhang, L. Zhang, D. Yu, L. Jiang, W. Elfalleh, Preparation of hydrogenated soybean oil of high oleic oil with supported catalysts, Food Biosci. 22 (2018) 91–98.
[23] D. Yu, X. Li, Y. Wang, D. Zou, L. Hu, H. Zheng, L. Jiang, L. Wang, W. Elfalleh, The influence of supercritical carbon dioxide (SC-CO_2) on electrolytes and hydrogenation of soybean oil, J. Am. Oil Chem. Soc. 94 (2017) 993–1001.
[24] S. Pereda, S.B. Bottini, E.A. Brignole, Phase equilibrium engineering of supercritical hydrogenation reactors, AICHE J. 48 (2002) 2635–2645.

CHAPTER 15

Extraction of bioactives from citrus

Mudasir Yaqoob[a], Poonam Aggarwal[a], Raouf Aslam[b], Jagbir Rehal[a]
[a]Department of Food Science and Technology, Punjab Agricultural University, Ludhiana, India
[b]Department of Processing and Food Engineering, Punjab Agricultural University, Ludhiana, India

Contents

1. Introduction

Citrus, belonging to family *Rutaceae*, subfamily *Aurantoideae*, and tribe *Citrae*, is one of the primordial crops, being cultivated from 2100 BC [1]. In fact, Citrus involves all species of sexually compatible genera, *Citrus*, *Fortunella*, and *Poncirus* [2]. Based on various taxonomic classification systems, it is indicated that 16–162 species are included in the genus Citrus [3], though the recent molecular and phenotypical data specify presence of only three basic taxa: *Citrus medica* (citron), *Citrus maxima* (pomelo), and *Citrus reticulata* (mandarin). The hybridization of these basic taxa in complex patterns and mutations are considered to be the origin of the rest of the species [1, 2]. Southeast Asia, mainly east India, southwest China, and north Burma is thought to be the place of origination and diversity of this genus [4].

Citrus is the most copious crop in the world, having a production of more than 124.4 million tons [5]. A huge quantity (about 50% of the originally processed fruit mass) of waste like peels, seeds, and pulps is generated by its processing by food and agro-food processing industries [6]. Citrus peel, a rich source of bioactive compounds, has attracted the interest of scientists for its utilization. These wastes can be fermented and dried to create some important products like biogas, ethanol, animal feed (for ruminants [7, 8] and pigs [9]), or volatile compounds [10]. It can also be utilized in the extraction and purification processes of bioactive compounds like carotenoids, polyphenols, essential

Green Sustainable Process for Chemical and Environmental Engineering and Science
https://doi.org/10.1016/B978-0-12-817388-6.00015-5

oils [11], additives, and colorants [12]. Hence, a range of constructive opportunities in the technological and medical aspects are exhibited by wastes and by-products from citrus fruits. The bioactive compounds like polyphenols and carotenoids are recognized to have various health benefits, mainly due to their antioxidant activity [13]. Polyphenols have substantial potential to be utilized as a raw material for the manufacture of cosmetics, functional foods, and pharmaceuticals. The polyphenols possess tremendous therapeutic and health benefits, like antiviral, antibacterial, antineoplastic, antifungal, antimalarial, antimutagenicity, antiallergenicity, anticarcinogenicity, and antiaging activity [14–16]. Owing to their great nutraceutical properties and economic importance, extraction and identification of bioactive compounds from citrus has garnered substantial research attention.

Extraction of functional ingredients from citrus waste has also attracted significant scientific interest. The sensitivity of the target molecules of interest makes the extraction process a difficult step for valorization of citrus wastes. Extraction can be done conventionally by using solvents or by utilizing progressive approaches like microwave-assisted extraction (MAE), ultrasound-assisted extraction (UAE), high-hydrostatic-pressure extraction (HHPE), and supercritical and subcritical extraction (SCE) [17, 18]. Thus, this chapter is focused on reviewing the bioactives, their nutritional composition, different extraction processes, and their utilization.

2. Composition of citrus fruits

Citrus pulps contain mostly water (85%–90%); however, there are almost 300 vital components in the edible portion depending on the cultivar, species, climatic conditions, and cultivation practices. Citrus fruits have an abundance of organic acids, sugars, and phenolic compounds, which vary according to species, climate, rootstock, soil, environmental conditions, and horticultural practices [19]. Compounds are present from various groups like flavonoids, steroids, hydroxyamines, alkanes, fatty acids, coumarins, peptides, carbohydrates, carbamates, and carotenoids [20]. The approximate composition of citrus fruits like lemon, lime, and sweet orange shows the moisture content, ash, lipid, fiber, protein, and carbohydrates in the range of 75.64%–85.1%, 0.15%–0.60%, 0.05%–1.15%, 0.05%–2.5%, 1.03%–1.56%, and 10.61%–23.05%, respectively [21]. The main carbohydrates present in citrus fruits are sucrose, glucose, and fructose, usually in the ratio of 2:1:1, and are the reason for imparting the sweetness to the juice [22]. The peel of *Citrus sinensis* has been shown to contain 10.31% moisture, 5.51% ash, 2.78% fat, 16.51% protein, and 12.47% crude fiber on dry weight basis [23]. The peels of lemons, oranges, and grapefruits have a content of total fiber, soluble fiber, and insoluble fiber of 2.47–2.49, 0.82–0.83, and 1.60–1.65 g/100 g on a fresh weight basis, respectively [24].

Owing to the typical flavor of the citrus essential oils, they find extensive applications in food and pharmaceutical industries [25]. The oil content is higher in the peels than the leaves [26]. The yield of essential oils obtained by hydro-distillation technique from peels was found to be highest in mandarin (2.70%), lemon (1.30%), orange (0.74%), and white bitter orange (0.46%) [25], whereas the oil expressed by the cold-pressing method from peels of citrus fruits gave the highest oil yield for Malta (1.21%), Eureka lemon (1.12%), Mousambi (0.98%), and grapefruit (0.73%) [27]. A total polyphenol content of 164, 154, and 135 mg/100 g is present in peeled lemons, oranges, and grapefruits, respectively, while their peels have 190, 179, and 155 mg/100 g of the same in fresh fruit. Among the polyphenols; flavonoids, phenolic acid, and coumarins are the most studied in citrus. The phenolic acids like ferulic, sinapic, and *p*-coumaric were found to be significantly higher in lemons than in oranges and grapefruits [24]. The Vitamin C content in five species of citrus ranged from 19.36 to 61.60 mg/100 g and other vitamins were also present, like thiamine (0.06–0.88 mg/100 g), riboflavin (0.01–0.11 mg/100 g), and niacin (0.03–0.43 mg/100 g) [28]. The ascorbic acid content is significantly higher in peels than in peeled fruits. It was highest in peeled fruit and peels of lemons (47.9 and 59.8 mg/100 g) and oranges (47.7 and 58.6 mg/100 g), as compared to grapefruit (35.1 and 43.8 mg/100 g) [24]. Lemon showed the highest total radical-trapping antioxidative potential (TRAP) of 4480 nmoL/mL for peeled fruits and 6720 nmoL/mL for its peels.

Pectin is a heteropolysaccharide present in citrus fruits, and can be extracted from their peel. The amount of pectin extracted from orange peels was found to be less (15.25%) than that extracted from lemon peels (20.75%) [29]. The pectin content in eight cultivars of citrus grown in Taiwan was seen to be in the range 36.0 ± 1.46 to 86.4 ± 3.36 mg/g db [30]. The elemental composition of citrus fruits like orange, lemon, and grapefruit shows that they have appreciable amounts of minerals like potassium (99.4–200.0 mg/100 g), calcium (20.0–70.0 mg/100 g) and magnesium (10.0–16.2 mg/100 g), which are required for various metabolic activities in the body [31].

3. Bioactives in citrus

Different types of citrus varieties contain different bioactive compounds. Some of them are discussed below.

3.1 Phenolic compounds

This family of compounds is huge and contains a group of compounds varying from complex polymerized compounds to simple phenols. They mainly contain hydroxybenzoic acids and hydroxycinnamic acids, based on their type of structure. They also comprise flavonoids, tannins, and coumarins [32]. In Citrus, they are synthesized by shikimate pathway, which initiates with the formation of tyrosine and phenylalanine. Phenylalanine is deaminated to form cinnamic acid in a reaction catalyzed by phenylalanine ammonia-lyase (PAL), which is the main enzyme in the whole biosynthesis. *p*-Coumaroyl coenzyme A and hydroxycinnamoyl esters are the basic structural precursors of various phenolic compounds. The methoxylation and hydroxylation leads to the formation of hydroxycinnamic acids and breakage of acetate leads to hydroxybenzoic acids. The ultimate products of the flavonoid branches remain anthocyanins, which start with the formation of chalcones, and the rest of the flavonoid compounds being formed from intermediates of anthocyanin biosynthesis [33].

3.1.1 Flavonoids

Flavonoids are a group of polyphenolic compounds that originate mainly from plant sources. They can be flavonols, flavones, flavonones, isoflavones, catechins, anthocyanidins, and chalcones, which are differentiated on the basis of their chemical structure. Citrus fruits are rich sources of these compounds [34, 35]. Flavonoids generally occur as glycosides, in which the nonsugar part is covalently bonded to a sugar moiety, while the "polymethoxylated flavones" (PMFs) are unique—they occur as the free aglycones. [36]. The citrus peels have been discovered to have greater amounts of PMFs as compared to the other parts [37]. The basic structure of flavonoid compounds consist of a pyran ring (C) and a flavan nucleus that is connecting to two aromatic rings (A and B). The flavonoids, isoflavonoids, and neoflavonoids are differentiated by the difference in the position of linkage between B-ring and C-ring [38].

Flavonoids are very diverse in nature. As many as 4000 types have been identified; most of these are present in fruits, vegetables, and some beverages. The most abundant flavonoid content in *C. sinensis* juice, in order of the amount present, in mg/100 mL, are hesperidin (28.6), narirutin (5.2), didynin (1.89), and vicenin-2 (5.72). The commercial grapefruit juices have higher concentrations of the main components, naringin (43.5), naringenin (4.20), and narirutin (9.90) mg/100 mL. The bitter taste in grapefruit juice is due to its flavonoid, naringin [39]. Furthermore, these flavonoid compounds have been stated to show miscellaneous health benefits including antiallergic, anticancer, antiinflammatory, and antimicrobial activity. Most citrus fruits are rich in quercetin, myricitin, naringin, rutin, hesperidin, and tangeritin [40].

3.1.2 Essential oils

Citrus peels consist of two layers; Flavedo and albedo (Fig. 1). Albedo is the peel's inside layer and flavedo is the outer layer [41]. Both of the layers are rich sources of phytochemicals. The flavedo part mostly consists of essential oils, which are used in the food industry for flavor and fragrance [42]. The physic-chemical properties of different citrus peel oils are different. The composition of these oils are mainly monoterpenes (90%), with some minor oxygenated compounds (5%), along with nonvolatile compounds such as pigments and wax in very minor proportions (<1%) [43]. Terpenoids are the plant-origin hydrocarbons with general formula $(C_5H_8)_n$, classified on the basis of number of C atoms into mono-, di-, tri-, and sesquiterpenoids. Citrus plants are rich sources of these terpenoids. Limonene is the major terpene present in peels of the citrus fruits [44].

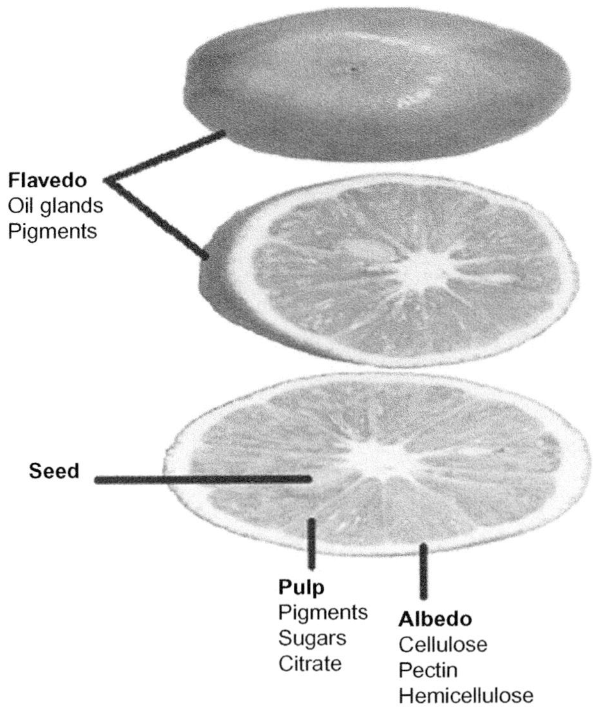

Fig. 1 Different parts of citrus fruit. *(Adapted from S. Rafiq, R. Kaul, S.A. Sofi, N. Bashir, F. Nazir, G. Ahmad, Citrus peel as a source of functional ingredient: a review, J. Saudi Soc. Agric. Sci. 17 (2018) 351–358. https://doi.org/doi:10.1016/j.jssas.2016.07.006.)*

The other components of essential oils that have been found in citrus peel are alpha-pinene, octanol, and beta mycrene.

3.1.3 Carotenoids

Carotenoids are colored pigments with yellow, orange, or red color, and can originate from both animals and plants [45]. These phytochemicals are responsible for the internal and external coloration of the fruit in citrus. Almost 110 different xanthophylls and carotenes have been reported in these fruits. Carotenoids are the main pigments, and contribute to their nutritional and antioxidant value. Mandarins accumulate carotenoid compounds mainly in the peel (94%). In Dancy mandarin, xanthophylls are the major carotenoids (68%–90%). The comparative concentration of violaxanthin differs among cultivars and tissues, and the *cis*-isomer is more abundant than the *trans*-isomer [46].

3.1.4 Limonoids

Limonoids are a family of phytochemicals found mostly in citrus fruits including lemon, orange, mandarin, grapefruit, bergamot, and lime [47]. Limonoids are highly oxygenated, secondary polycyclic metabolites and are chemically related to terpenoids [48]. Rutaceae family accumulates three forms of limonoids: monolactone, dilactone, and glucosides. Among the large number of limonoids present in citrus, nomilin, limonins, ichangin, and nomilinic acid are known to be bitter. Limonin is the major cause of citrus juice bitterness [49].

In the fresh fruit, most of the limonin is in the form of A-ring monolactone, a natural precursor of limonin, which is tasteless; it is converted to intensely bitter limonin in the juice after extraction from the fruit [50]. The seeds contain maximum limonin (9.50 mg/g), followed by peel (4.69 mg/g), and juice (0.218 mg/mL) [51]. Limonoids have been described to stimulate the activity of glutathione S-transferase (GST), which has the potential to inhibit tumor formation. These phytochemicals possess many physiological and ecological roles and are widely distributed as secondary metabolites that help the plant in various defense mechanisms.

4. Health benefits of these bioactives

Nonnutrient, plant based bioactive compounds have been attributed to reduction in the risk of major health problems worldwide. Some of the citrus bioactives and their health benefits are shown in Table 1. A wide range of chronic human diseases are considered to originate from oxidative stress caused by free radicals [73]. These free radicals may come from water, air, and food, or may originate from the metabolic processes within the cell itself. If these oxidants are overproduced, they may lead to oxidative stress in the body [74]. This stress can have detrimental effects on macromolecules such as DNA, proteins, and lipids, posing a threat to human health [73]. Fruits and vegetables are rich in a wide range of antioxidants like carotenoids and phenolics, which have antioxidant properties and can be beneficial in combatting the oxidative stress caused by free radicals.

Anticarcinogenic properties of phytochemicals present in citrus fruits were established in a study by Jacob et al. [75]. The authors reported that administration of five limonoids (limonin, isoobacunoic acid, ichangin, nomilin, and obacunone) into mice potentially induced significant proportions of an antioxidant enzyme (Glutathione-S-transferase), which is a part of the enzyme detoxification system. Moreover, at 10 mg/dose, limonin and nomilin were observed to halve the percentage of tumor occurrence. The anticarcinogenic properties of naringenin and limonoids present in citrus were further confirmed by the study of Arul and Subramaniam [76].

In addition to anticarcinogenic effects, bioactive compounds have been reported to be weakly or significantly beneficial in numerous other health problems. In a study by Arai et al. [77], intake of flavonoids like quercetin, ficetin, luteolin, kaempferol, and myricetin were observed to have an inverse impact on low-density lipoprotein (LDL) and plasma total cholesterol concentrations. Moreover, Joshipura et al. [78] observed that intake of >4 servings/day of vegetables and fruits decreased the risk of coronary artery disease.

Phytochemical class	Compound	Health benefits	Reference (s)
Flavonoids	Naringenin	Beneficial in gastric carcinogenesis	[52, 53]
	Hesperetin	Reduces activity of bacterial activity in Colon Cancer	[54, 55]
	Apigenin	Cell death of pancreatic cancer cells	[56]
	Poncirin	Inhibition of human gastric cancer cells	[57]
	Tangeretin	Growth inhibition in the incubation of colon adenocarcinoma cells.	[58]
	Quercetin	Suppressive effect in colon carcinogenesis and human cervical cancer cells	[59]
	Tangeretin and Nobiletin	Lowers blood cholesterol and triglyceride concentrations	[60]
	Nobiletin	Reduction in hepatic and intestinal triglyceride accumulation. Attenuation of very low-density lipoprotein triglyceride secretion Normalization of insulin sensitivity	[61]
	Hesperidin, naringin, and nobiletin	Antidiabetic effects	[62]
	Naringenin and hesperetin	Antiatherogenic effects	[63]
	Naringin, hesperidin, neohesperidin, and Nobiletin	Neuroprotective effects	[64]
	Nobiletin	Improves the cognition of patients with Alzheimer's disease.	[65]
	Heptamethoxyflavone	Induces brain-derived neurotrophic factor production in astrocytes and enhance neurogenesis after brain Ischemia	[66]
	Hesperidin	Balance in bone metabolism on bone health	[67]
Limonoid	Methyl nomilinate, isoobacunoic acid, isolimonexic acid, and limonexic acid	Protective effects in colon cancer Cyotoxic effects in estrogen-induced breast cancers.	[68]
Coumarins	5-geranyloxy-7-methoxycoumarin, limettin, and isopimpinellin	Inhibition of human colon cancer cell proliferation	[69]
	Auraptene and 7-isopentenyloxycoumarin	Protective effects in excitatory neurotoxicity in mixed cortical cell cultures	[70]
	Auraptene	Inhibits microglia activation	[71]
Carotenoids	β-Cryptoxanthin	Inhibition of skin tumorigenesis and colon carcinogenesis in mice.	[72]

Phytochemicals associated with citrus fruits have shown promising health benefits. A number of studies related to citrus fruits have revealed that these phytochemicals have cardioprotective, antihyperglycemic, and several other benefits [79]. In one study, Odhner et al. [80] induced diabetes into rats with streptozotocin, which was demonstrated by a high insulin level. Ethanolic extracts from *C. medica* were then observed to reduce glucose levels in blood plasma of these rats, hence showing antihyperglycemic effects. Similar effects were observed by Hetta et al. [81] where the authors found that 70% MeOH leaf extract exhibited a significant antihyperglycemic effect in mice, possibly due to the presence of flavonoids. These authors also suggested consideration of the product for diabetes management in humans after clinical trials.

In a recent study with mice, flavonoids in *C. medica* were reported to have anticardiotoxicity effects [82]. The ethanolic extracts from *C. medica* brought the isoproterenol induced high levels of cholesterol, triglycerides, low-density lipoproteins, and very low-density lipoproteins to normal levels. Moreover, essential oils found in citrus have been regarded as beneficial in treatment of gallstones, headaches, and stomach disorders [79]. Hence, the nonnutritious bioactive compounds in plants are beneficial against several human degenerative pathologies.

5. Extraction of bioactive compounds from citrus

5.1 Conventional methods

5.1.1 Maceration

In earlier days, this process was used for the preparation of homemade tonics. This technique is also used in wine making and now it has been adopted as an inexpensive technique for the extraction of different phenolic compounds as well. Maceration basically involves soaking the sample with solvent with frequent agitation at room temperature for 3 days. The objective of the process is softening and breaking of the plant cell wall so that phytochemicals can escape. The principle of diffusion (Fick's law) is used for the extraction process, so to increase maceration speed, system heating is important [83].

Safdar et al. [34] related the effectiveness of maceration and ultrasound on the extraction and quantification of polyphenols from kinnow (*Citrus reticulate* L.) It was found that UAE was a superior technique in polyphenol extraction. The use of solvent ethanol at a concentration level of 80% extracted the highest yield. When 80% ethyl acetate was used in the process of maceration, minimum phenolics of kinnow were obtained. The analysis of the extracted phenolics was done by HPLC, which showed that six flavonoids and five phenolic acids were present. The most common of the polyphenols were hesperidin and ferulic acid.

Al-juhaimi [84] used peel and pulp of lemon, mandarin, and orange, using 80% ethanol with maceration technique for extraction of phenolic compounds. The Folin-Ciocalteau reagent assay was used for the evaluation of the same compounds. They found that orange, mandarin, and lemon peels contained 178.90 mg GAE/100 g, 169.54 mg GAE/100 g, and 61.22 mg GAE/100 g total phenolics respectively, which were higher in amount than the pulp extracted phenolic compounds.

5.1.2 Refluxing

The process of refluxing is more efficient as compared to maceration, with less solvent and extraction time required. The only demerit of this technique is that it cannot be used for heat labile plant products. Aqueous or nonaqueous solvents are used at heated temperature for refluxing extraction of phenols. Solute/solvent ratio, temperature, time, and extraction cycles are the important parameters that determine the high yield of the phytochemicals; thus making for a rather laborious process in understanding the best and most efficient solvent for extraction. Most of the studies have shown that using methanol produced high yield of citrus phytochemicals. However, the toxic nature of this solvent to our environment does not support its use in food applications. More nontoxic and greener solvents such as isopropanol, ethanol, and petroleum ether can be used as a replacement to methanol [17], with ethanol being the most recommended among them.

5.2 Nonconventional methods

The health-related benefits of certain phenolic compounds in the citrus fruits make it important to develop techniques for their extraction and quantification minimizing reduction in potency and retaining their functional value. The waste generated by the citrus processing industry is a potential goldmine of components of immense value and applications in food, flavor, fragrance, and pharmaceutical industries. Solvents are used conventionally to extract these. Conventional methods utilized maceration, refluxing, solvent extraction, and hydro-distillation method. Magwaza et al. [85] developed a rapid method by extraction of phenolic compounds from freeze dried rind powder with methanol, water, and HCl (70:29.5:0.5 v/v/v) for 30 min for optimum yield. The factors that are relevant during the extraction process are the plant part, solvent used, temperature, time, and pressure combination employed [86]. Extraction by solvents is time consuming, causes possible degradation of thermo-sensitive compounds, and requires large quantities of solvent. Moreover, energy requirements for separation of solute as well as co-extraction of undesirable components has to be dealt with [85].

A number of recent techniques have been described to extract bioactive compounds from plant sources; these include MAE, UAE, high pressure-assisted extraction (HPE), enzyme-assisted extraction (EAE), and supercritical CO_2 extraction (SFE).

5.2.1 Ultrasound assisted extraction

The discharge of phytochemicals from plant matrices can be enhanced by the breaking of cell walls induced by cavitational forces of ultrasound, which make it a nonconventional method for the extraction of bioactive compounds from citrus [87]. Ultrasounds can be applied as low frequency/high intensity or high frequency/low intensity; the latter does not change the properties of the substrate, whereas the high intensity generates high pressure and temperature gradient leading to physical and chemical effects [88]. Typical advantages of ultrasound-assisted extraction are short extraction times, decrease of solvent consumption, increased extraction rate, better quality of extracts, and better quality of thermo-sensitive compounds.

Alissandrakis et al. [89] used ultrasound assisted extraction to extract semi- and fully volatile compounds from flowers and honey of citrus using *n*-pentane: diethyl ether solvent for 10 min, and their identification to authenticate the origin of honey [89]. Ultrasound assisted extraction using the temperature of 25°C for 55 min using ethanol concentration 40 g/100 g gave the best extraction yields of flavonoids from grapefruit solid wastes [90].

It was reported that choice of solvent, frequency of ultrasound, and processing temperature are the most critical factors for optimizing yields. [63] formed the optimum ultrasonic conditions of frequency of 60 kHz for 60 min at 40°C utilizing methanol as solvent for optimizing hesperidin yield from the peels of *C. reticulata* [63].

Ultrasound assisted extraction with the use of food grade solvents for extraction purposes was necessitated due to the recommendation by FDA. Khan et al. [91] undertook the extraction of phenolic antioxidants with ethanol-water mixtures and, in comparison to the conventional solvent extraction, it was found that the duration of ultrasound assisted were reduced drastically. The optimized conditions were using 4 ethanol:1 water as solvent and a sonication power of 150 W at 40°C [91].

5.2.2 Microwave assisted extraction

The combined dipole rotation and ionic conduction are responsible for heating of dielectric material or solvents under microwave, which causes the heating up of moisture in the cell, its subsequent evaporation, resulting in tremendous pressure and cell wall rupturing, and releasing the active components leading to increased yields [92]. The parameters having a strong influence on the extract yield are temperature of extraction, power applied, solvent composition, extraction time, solid to feed ratio, and material characteristics (moisture and size) [93]. Solvent is regarded as the major factor as it influences the solubility of the desired biomolecules and the microwave energy absorption is given by the dissipation factor of the solvent [94].

Solvent free microwave extraction (SFME) is another development where water contained in the substrate is used as an extractor solvent, so fresh or moistened samples with at least 70% moisture content are required [93]. Essential oils, when extracted from orange peels utilizing microwave heating at atmospheric pressure and without using any solvent, resulted in better yields of oil in less time as compared to hydro-distillation [95].

D-limonene, a monoterpene present in citrus fruits, has a wide application in food, flavor, and pharmaceutical industries. The D-limonene, extracted from orange wastes utilizing MAE, reported shortened extraction duration and twice the yield of D-limonene when compared to conventional heating. The optimal conditions for extraction of essential oils from *Citrus limon* leaves with 60% hexane using the solvent:substrate ratio of 20:1 and a microwave power of 500 W for 10 min yielded a recovery of 2.2% oil [96, 97].

Hesperidin, a flavanone having many functional applications, can be extracted from citrus peels for valorization of citrus produce. Inoue et al. [98] suspended *Citrus unshiu* peel powder in 70% ethanol, homogenized them, followed by microwave irradiation at 62–180°C for 2–12 min in a closed system and obtained an extraction of hesperidin and narirutin as 58.6 mg/100 g and 13.1 mg/100 g, respectively. Prakash Maran et al. [99] validated the extraction of pectin from orange peels using MAE, and reported that using a microwave power of 422 W, pH of 1.4, irradiation time of 169 s and solid-liquid ratio of 1:16.9 g/mL resulted in maximum pectin extract yield of 19.24%. Hayat et al. [100] also optimized the conditions for extraction of phenolic acids from citrus mandarin peels by employing the microwave power of 152 W for 44 s time and methanol concentration of 66% [100].

5.2.3 Enzyme-assisted extraction

Due to the various limitations attached to the conventional methods, nonconventional methods are being exploited. EAE is specific and efficient approach of extraction and helps in preserving the potency of the desired product in an eco-friendly way. The various groups of enzymes employed include hydrolyzing enzymes, isomerizing enzymes, oxidation-reduction enzymes, etc., which act by increasing the cell wall permeability, resulting in better extraction yields of the desired bioactive component [101]. A majority of phenolic compounds are encased within the polysaccharides of the cell wall, like pectin, cellulose, and hemicellulose, hence enzymes such as pectinase, cellulose, and hemicellulases are employed to weaken the cell wall and release the entrapped biomolecules. Cheng et al. [102] reported better extraction ease using cell wall degrading enzyme, cellulase, which weakened the integrity of the cell wall of the five citrus peels (Yen Ben lemon, Meyer lemon, grapefruit, mandarin, and orange) under study and facilitated the extraction of phenolics [102].

The factors to be kept under control during enzyme assisted extraction include time of reaction, temperature, pH, concentration, and the particle size of the substrate [17]. The EAE is especially useful for thermo-sensitive molecules, is a "green" technology, results in reduction of environmental pollution, and helps in preserving the efficacy of the biomolecule [103]. At the same time, there is a limited industrial application of EAE due to the cost of the enzyme involved, and the specific and rigid action and conditions required for the activity of the enzymes. The use of EAE in tandem with other nonconventional techniques, such as ultrasound, microwave, supercritical fluid, and high pressure, will help to increase the extraction yield of desired compounds.

5.2.4 Supercritical CO_2 extraction

Supercritical CO_2 extraction (SCFE) is a nonconventional technique, which is applied to extract many bioactive compounds. It is considered to be an environmental friendly and a clean extraction method. This technique extracts the bioactive compounds in absence of oxygen and light along with low temperature, which prevents their thermal degradation. The efficiency of the process is affected mostly by parameters like temperature, pressure, supercritical CO_2 flux, presence of co-solvent, and extraction time [17, 104]. Mira et al. [105] extracted the essential oils from orange peel. Limonene was the main component that was extracted at 12.5 MPa and 308 K which was almost 99.5%. It was also found that Linalool was also found at higher concentration at 80 bar and 35°C.

Suetsugu et al. [106] carried out the experiment for the extraction of flavonoids by SFE, from *Citrus junos*. The ratio of ethanol to water was used as a cosolvent because naringin and hesperidin have glycoside group and can be easily extracted by these solvents. An increase in ethanol concentration tended to increase the yield of tangeretin.

The temperature and pressure at which the liquid and gaseous phases of a substance become indistinguishable is referred to as critical point on the phase diagram of that substance. When both the pressure and temperature of a fluid increase above the critical point values, it is said to be in a supercritical state. However, solid-solid, solid-liquid, and solid-gas equilibrium lines do not end at the critical point. A substance is said to be a supercritical fluid when it is exposed to its particular critical temperature and pressure. The critical pressure and temperature are the two most important parameters that affect the selectivity of an appropriate solvent for supercritical fluid extraction. For example, the extraction process will be more capital intensive if fluids of high critical pressure are used. Such pressures will require special equipment; moreover, the maintenance of high pressure will also involve greater expenditure. In a similar way, extraction with fluids having higher critical temperature will pose a threat to heat-sensitive components. In addition to this, chemical stabilities of the solvent with process equipment and with the food components is another important criterion to arrive on the most appropriate solvent. The ideal solvent should be readily available, legally approved, and should not be toxic, corrosive, explosive, or flammable. Carbon dioxide best fulfills the requirements of an ideal solvent and should ideally be used. It again becomes necessary to explain these properties that determine the state of a substance and efficiency of extraction. However, in addition to critical constants, knowledge of certain important properties of the solvent and other compounds of interest, such as density, compressibility, thermal conductivity, viscosity, diffusivity, dielectric constant, and heat capacity is necessary to maximize the output of the process.

The addition of a co-solvent, often referred to as a modifier or entrainer, to a mixture containing a supercritical component can enhance the extraction selectivity and yield. Liquid entrainers, which have higher solubility parameters than the SCF component, are generally used to enhance the solvent power of SCF. Ethanol is the preferred co-solvent because of its nontoxic nature and GRAS approved status [107]. The extraction of bioactives is specific at different critical temperatures and pressures. Some of them are given in Table 2.

6. Utilization of citrus bioactives

Currently, the food processing industries are focused on changing the trend of using synthetic antioxidants to using natural ones. Scientific studies have shown that the incorporation of natural and synthetic additives does not cause any change in the finished product. In biscuits, there was no significant change in color or in nutritional value, when compared with control samples. Both synthetic and natural additives showed similar antioxidant activity in the bakery products. Therefore, natural additives can be exploited for consumers who prefer foods free from synthetic additives [115].

Zia-ur-Rehman [116] conducted a study in 2005 to find the practicability of natural sources of antioxidants in citrus peel extract. The antioxidant activity of the compounds extracted from citrus peel was evaluated on refined corn oil for a period of 6 months. A refluxing method was used to collect the extract. Ethanol, methanol, dichloromethane, hexane, acetone, and diethyl ether were used as solvents to be mixed together with dried ground peel. Zia-ur-Rehman found that methanol was the most efficient solvent for extraction of these bioactives. Corn oil having citrus peel extract in concentrations of 1600 and 2000 ppm showed higher iodine values, and lower free fatty acid (1.5% and 1.0%) and peroxide values (8.38 and 7.0 meq kg) [101, 116]. The values signified that citrus peel extract has a strong antioxidant activity that was approximately of the same quantity to that of BHA and BHT.

Ojha et al. [117] incorporated mandarin peel powder in biscuits (6%). They found that the fortified biscuits were comparable with the control biscuits. The fiber, ash, ascorbic acid, carotenoids, polyphenolic compounds, and antioxidant activity was found to be significantly higher than the control biscuits. Kohajdova et al. [118] prepared biscuits with high citrus fiber. The study indicated that citrus dietary fiber preparations had low fat and protein, but high crude fiber. Sensory evaluation revealed that the 5% peel concentration revealed the most acceptable biscuits. Singh and Immanuel [119] utilized peels of orange, lemon, and pomegranate as sources of natural antioxidants. Lemon extract contained the maximum of total phenolic activity (0.9 mg/g). Peel extracts were added to paneer samples and it was found that 2% was the most acceptable level, which did not hinder the sensory quality and had greater ability to prevent peroxide formation. Hence, the shelf lives of food products containing fat and oil content can be enhanced by the addition of these natural antioxidants, by their action of preventing rancidity.

Table 2 SFE extraction of bioactives at different pressures, temperatures, and with different co-solvents

Bioactive compound	Botanical material	*T* (K)	*P* (MPa)	Co-solvent	Yield	References
Essential oils	*Orange*	308	12.5	–	–	[105]
	Citrus reticulate Blanco	318–323	10		2.26 (%)	[108]
Lemonoids (limonin)	*Citrus paradisi* Macf.	323	48.3	–	6.3 (mg/g seeds)	[109]
(limonin-17-β-D-glucopyranoside)		333	41.4	30% (ethanol)	0.62 (mg/g seeds)	
Flavonoids	*Citrus grandis (L.) Osbeck*	80	39	85% (ethanol)	2.37%	[110]
(Naringin)	*Citrus paradise*	332	9.5	15 (ethanol)	14.4 (g/kg)	[111]
(β-Carotene)	Tomato	338.15	30	5% ethanol	171.7–899.97 (μg/g feed)	[112]
Lycopene		328	30	5% ethanol	53.93%	
Tocopherols	Dried pomace of Tomato (*Lycopersicon esculentum* Mill.)	313–353	30–46	–	984–5920	[113]
	Grape seeds from several varieties	338	37	–	153.2–569	[114]
Tocotrienols	Dried pomace of Tomato (*L. esculentum* Mill.)	313–353	30–46	–	0–42	[113]

7. Conclusion

Citrus fruits are important sources of bioactive compounds like phenolics, flavonoids, carotenoids, lemonoids, and essential oils. These phytochemicals are very important in terms of their natural antioxidant activity. Conventional processes such as refluxing and maceration are used, which have disadvantages like thermal degradation of valuable compounds, using toxic solvents, and long extraction time. Therefore, novel methods of extraction like UAE, MAE, EAE, and SFE have been developed, which utilize less solvent, require less time, and deliver efficient yields.

References

[1] G.A. Moore, Oranges and lemons: clues to the taxonomy of Citrus from molecular markers, Trends Genet, 17 (2001) 536–540, https://doi.org/10.1016/S0168-9525(01)02442-8.

[2] F.R. Tadeo, M. Cercós, J.M. Colmenero-Flores, D.J. Iglesias, M.A. Naranjo, G. Ríos, E. Carrera, O. Ruiz-Rivero, I. Lliso, R. Morillon, P. Ollitrault, M. Talon, Molecular physiology of development and quality of Citrus, Adv, Bot. Res. 47 (2008) 147–223, https://doi.org/10.1016/S0065-2296(08)00004-9.

[3] W.T. Swingle, P.C. Reece, The botany of citrus and its wild relatives, in: W. Reuther, H.J. Webber, L.D. Batchelor (Eds.), The Citrus Industry, vol. I, University of California, Division of Agriculture and Natural Resources, U.S.A., 1967, pp. 190–430.

[4] D.J. Mabberley, Citrus (Rutaceae): a review of recent advances in etymology, systematics and medical applications, Blumea Biodivers. Evol. Biogeogr. Plants 49 (2004) 481–498, https://doi.org/10.3767/000651904X484432.

[5] Y.S. Huang, S.C. Ho, Polymethoxy flavones are responsible for the anti-inflammatory activity of citrus fruit peel, Food Chem. 119 (2010) 868–873, https://doi.org/10.1016/j.foodchem.2009.09.092.

[6] F. Anwar, R. Naseer, M.I. Bhanger, S. Ashraf, F.N. Talpur, F.A. Aladedunye, Physico-chemical characteristics of citrus seeds and seed oils from Pakistan, J. Am. Oil Chem. Soc. 85 (2008) 321–330, https://doi.org/10.1007/s11746-008-1204-3.

[7] A. Nazok, M. Rezaei, H. Sayyahzadeh, Effect of different levels of dried citrus pulp on performance, egg quality, and blood parameters of laying hens in early phase of production, Trop. Anim. Health Prod. 42 (2010) 737–742, https://doi.org/10.1007/s11250-009-9481-x.

[8] M. Lanza, A. Priolo, L. Biondi, M. Bella, H. Ben Salem, Replacement of cereal grains by orange pulp and carob pulp in faba bean-based diets fed to lambs: effects on growth performance and meat quality, Anim. Res. 50 (2001) 21–30, https://doi.org/10.1051/animres:2001101.

[9] T.C. O'Sullivan, P.B. Lynch, P.A. Morrissey, J.F. O'Grady, Evaluation of citrus pulp in diets for sows and growing pigs, Irish J. Agric. Food Res. 42 (n.d.) 243–253. https://doi.org/10.2307/25562493.

[10] S. Djilas, J. Čanadanović-Brunet, G. Ćetković, By-products of fruits processing as a source of phytochemicals, Chem. Ind. Chem. Eng. Q. 15 (2009) 191–202, https://doi.org/10.2298/CICEQ0904191D.

[11] R. Casquete, S.M. Castro, M.C. Villalobos, M.J. Serradilla, R.P. Queirós, J.A. Saraiva, M.G. Córdoba, P. Teixeira, High pressure extraction of phenolic compounds from citrus peels†, High Pressure Res. 34 (2014) 447–451, https://doi.org/10.1080/08957959.2014.986474.

[12] K. Sharma, N. Mahato, M.H. Cho, Y.R. Lee, Converting citrus wastes into value-added products: economic and environmently friendly approaches, Nutrition 34 (2017) 29–46, https://doi.org/10.1016/J.NUT.2016.09.006.

[13] V. Dragović-Uzelac, D. Bursać Kovačević, N. Krasnići, Poljoprivredna znanstvena smotra = Agriculturae Conspectus Scientificus: ACS, Faculty of Agriculture, University of Zagreb, 2009. https://hrcak.srce.hr/47391. Accessed 30 December 2018.

[14] P. Putnik, D. Bursać Kovačević, A. Režek Jambrak, F. Barba, G. Cravotto, A. Binello, J. Lorenzo, A. Shpigelman, Innovative "green" and novel strategies for the extraction of bioactive added value compounds from citrus wastes—a review, Molecules 22 (2017) 680, https://doi.org/10.3390/molecules22050680.
[15] D. Bursać Kovačević, P. Putnik, V. Dragović-Uzelac, S. Pedisić, A. Režek Jambrak, Z. Herceg, Effects of cold atmospheric gas phase plasma on anthocyanins and color in pomegranate juice, Food Chem. 190 (2016) 317–323, https://doi.org/10.1016/J.FOODCHEM.2015.05.099.
[16] T.Y. Forbes-Hernández, F. Giampieri, M. Gasparrini, L. Mazzoni, J.L. Quiles, J.M. Alvarez-Suarez, M. Battino, The effects of bioactive compounds from plant foods on mitochondrial function: a focus on apoptotic mechanisms, Food Chem. Toxicol. 68 (2014) 154–182, https://doi.org/10.1016/j.fct.2014.03.017.
[17] N. M'hiri, I. Ioannou, M. Ghoul, N.M. Boudhrioua, Extraction methods of citrus Peel phenolic compounds, Food Rev. Int. 30 (2014) 265–290, https://doi.org/10.1080/87559129.2014.924139.
[18] A.R. Fakhari, P. Salehi, R. Heydari, S.N. Ebrahimi, P.R. Haddad, Hydrodistillation-headspace solvent microextraction, a new method for analysis of the essential oil components of *Lavandula angustifolia* mill, J. Chromatogr. A 1098 (2005) 14–18, https://doi.org/10.1016/j.chroma.2005.08.054.
[19] A. Bermejo, A. Cano, Analysis of Nutritional Constituents in Twenty Citrus Cultivars from the Mediterranean Area at Different Stages of Ripening, vol. 2012, 2012, pp. 639–650.
[20] J.M.J. Favela-hernández, O. González-santiago, M.A. Ramírez-cabrera, P.C. Esquivel-ferriño, M.R. Camacho-corona, Chemistry and Pharmacology of *Citrus sinensis*, Molecules 6 (2016), https://doi.org/10.3390/molecules21020247.
[21] E. Chuku, N. Akani, Determination of proximate composition and microbial contamination of fresh juice from three citrus species, Int. J. Biol. Med. Res 1 (2015) 1–8.
[22] Z. Liu, A. Quek, S. Kent Hoekman, R. Balasubramanian, Production of solid biochar fuel from waste biomass by hydrothermal carbonization, Fuel 103 (2013) 943–949, https://doi.org/10.1016/j.fuel.2012.07.069.
[23] E. Adewole, D.F. Adewumi, J. Jonathan, Phytochemical Constituents and Proximate Analysis of Orange Peel (Citrus Fruit) Phytochemical Constituents and Proximate Analysis of Orange Peel (Citrus Fruit), 2014, pp. 3–5, https://doi.org/10.15297/JABZ.V1I3.02.CITATIONS.
[24] S. Gorinstein, O. Martı, Y. Park, R. Haruenkit, A. Lojek, C. Milan, Comparison of some biochemical characteristics of different citrus fruits, Food Chem. 74 (2001) 309–315.
[25] S. Bourgou, F.Z. Rahali, I. Ourghemmi, M. Sa, Changes of peel essential oil composition of four tunisian citrus during fruit maturation, Sci. World J. 2012 (2012), https://doi.org/10.1100/2012/528593.
[26] M. Mushtaq, A. Et, Genetic variability to essential oil composition in four citrus fruit species, Pak. J. Bot. 38 (2006) 319–324.
[27] A. Aslin Sanofer, Role of Citrus fruits in health, J. Pharm. Sci. Res. 6 (2014) 121–123.
[28] D.E. Okwu, Citrus fruits: a rich source of phytochemicals and their roles in human health, Int. J. Chem. Sci. 6 (2008) 451–471.
[29] A.M.E. Sulieman, K.M.Y. Khodari, Z.A. Salih, Extraction of pectin from lemon and orange fruits peels and its utilization in jam making, Int. J. Food Sci. Nutr. Eng. 3 (2013) 81–84, https://doi.org/10.5923/j.food.20130305.01.
[30] Y.-C. Wang, Y.-C. Chuang, H.-W. Hsu, The flavonoid, carotenoid and pectin content in peels of citrus cultivated in Taiwan, Food Chem. 106 (2008) 277–284, https://doi.org/10.1016/J.FOODCHEM.2007.05.086.
[31] D.K. Paul, R.K. Shaha, Nutrients, vitamins and minerals content in common Citrus fruits in the northern region of Bangladesh, Pak. J. Biol. Sci. 7 (2004) 238–242, https://doi.org/10.3923/pjbs.2004.238.242.
[32] N. G Hounsome, B. Hounsome, D. Tomos, G. Edwards-Jones, Plant metabolites and nutritional quality of vegetables, J. Food Sci. 73 (2008) 48–65, https://doi.org/10.1111/j.1750-3841.2008.00716.x.
[33] E.G.W.M. Schijlen, C.H.R. De Vos, A.J. Van Tunen, A.G. Bovy, Modification of flavonoid biosynthesis in crop plants, Phytochemistry 65 (2004) 2631–2648, https://doi.org/10.1016/j.phytochem.2004.07.028.

[34] M.N. Safdar, T. Kausar, S. Jabbar, A. Mumtaz, K. Ahad, A.A. Saddozai, Science direct extraction and quantification of polyphenols from kinnow (Citrus reticulate L.) peel using ultrasound and maceration techniques, J. Food Drug Anal. 5 (2016), https://doi.org/10.1016/j.jfda.2016.07.010.
[35] M. Senevirathne, Y. Jeon, J. Ha, S. Kim, Effective drying of citrus by-product by high speed drying: a novel drying technique and their antioxidant activity, J. Food Eng. 92 (2009) 157–163, https://doi.org/10.1016/j.jfoodeng.2008.10.033.
[36] C. Sturt, P.O. Box, W. Wagga, Critical review analytical chemistry of fruit bioflavonoids a review, Analyst 122 (1997) 11–34.
[37] L. Wang, J. Wang, L. Fang, Z. Zheng, D. Zhi, S. Wang, S. Li, C. Ho, H. Zhao, Anticancer activities of citrus peel polymethoxyflavones related to angiogenesis and others, Biomed. Res. Int. (2014) 1–10.
[38] J.P.J. Marais, B. Deavours, R.A. Dixon, D. Ferreira, The stereochemistry of flavonoids, in: E. Grotewold (Ed.), The Science of Flavonoids, Springer, New York, NY, 2006, pp. 1–46, https://doi.org/10.1007/978-0-387-28822-2_1.
[39] G. Gattuso, D. Barreca, C. Gargiulli, U. Leuzzi, C. Caristi, C. Organica, U. Messina, S. Sperone, Flavonoid composition of citrus juices, Molecules 12 (2007) 1641–1673.
[40] D.E. Okwu, I.N. Emenikie, Evaluation of the phytonutrients and vitamins content of Citrus fruits, Int. J. Mol. Med. Adv. Sci. 2 (2006) 1–6. (Accessed 29 December 2018). http://www.medwelljournals.com/abstract/?doi=ijmmas.2006.1.6.
[41] S. Rafiq, R. Kaul, S.A. Sofi, N. Bashir, F. Nazir, G. Ahmad, Citrus peel as a source of functional ingredient: a review, J. Saudi Soc. Agric. Sci. 17 (2018) 351–358, https://doi.org/10.1016/j.jssas.2016.07.006.
[42] A.K. Bejar, N.B. Mihoubi, N. Kechaou, Moisture sorption isotherms—experimental and mathematical investigations of orange (Citrus sinensis) peel and leaves, Food Chem. 132 (2012) 1728–1735, https://doi.org/10.1016/j.foodchem.2011.06.059.
[43] M. Kondo, M. Goto, A. Kodama, T. Hirose, Fractional extraction by supercritical carbon dioxide for the deterpenation of bergamot oil, Ind. Eng. Chem. Res. (2000), https://doi.org/10.1021/IE000224Q.
[44] D.C. Sikdar, R. Menon, K. Duseja, P. Kumar, P. Swami, Extraction of citrus oil from orange (*Citrus sinensis*) peels by steam distillation and its characterizations, Int. J. Tech. Res. Appl. 4 (2016) 341–346.
[45] M. Butnariu, Methods of analysis (extraction, separation, identification and quantification) of carotenoids from natural products, J. Ecosyst. Ecography (2016), https://doi.org/10.4172/2157-7625.1000193.
[46] X. Hou, W. Peng, L. Xiong, C. Huang, X. Chen, X. Chen, W. Zhang, Engineering *Clostridium acetobutylicum* for alcohol production, J. Biotechnol. 166 (2013) 25–33, https://doi.org/10.1016/j.jbiotec.2013.04.013.
[47] M. Russo, I. Bonaccorsi, V. Inferrera, P. Dugo, L. Mondello, Underestimated sources of flavonoids, limonoids and dietary fiber: availability in orange's by-products, J. Funct. Foods 12 (2015) 150–157, https://doi.org/10.1016/j.jff.2014.11.008.
[48] H. Bilal, P. Health, J. Amanah, W. Akram, S.A. Hassan, H.S. Academy, S. Sahar, Determination of limonin and nomilin contents in different citrus cultivars using high performance liquid chromatography, Pak. J. Sci. Ind. Res. 56 (2013) 36–40.
[49] R. Gualdani, M.M. Cavalluzzi, G. Lentini, S. Habtemariam, The chemistry and pharmacology of *citrus limon*oids, Molecules (2016), https://doi.org/10.3390/molecules21111530.
[50] V.P. Maier, D.A. Margileth, Limonoic acid a-ring lactone, a new limonin derivative in Citrus, Phytochemistry 8 (1969) 243–248.
[51] B. Premi, B. Lal, V. Joshi, Debittering of kinnow juice with amberlite XAD-16 resin, Indian Food Pack. (1995).
[52] G. Ekambaram, P. Rajendran, V. Magesh, D. Sakthisekaran, Naringenin reduces tumor size and weight lost in N-methyl-N′-nitro-N-nitrosoguanidine-induced gastric carcinogenesis in rats, Nutr. Res. 28 (2008) 106–112, https://doi.org/10.1016/j.nutres.2007.12.002.
[53] J. Renugadevi, S.M. Prabu, Cadmium-induced hepatotoxicity in rats and the protective effect of naringenin, Exp. Toxicol. Pathol. 62 (2010) 171–181, https://doi.org/10.1016/j.etp.2009.03.010.
[54] S. Aranganathan, J.P. Selvam, N. Nalini, Effect of hesperetin, a citrus flavonoid, on bacterial enzymes and carcinogen-induced aberrant crypt foci in colon cancer rats: a dose-dependent study, J. Pharm. Pharmacol. 60 (2008) 1385–1392, https://doi.org/10.1211/jpp/60.10.0015.

[55] S. Aranganathan, N. Nalini, Antiproliferative efficacy of hesperetin (citrus flavanoid) in 1,2-dimethylhydrazine-induced colon cancer, Phytother Res. 27 (2012) 999–1005.
[56] J.L. Johnson, E.G. De Mejia, Flavonoid apigenin modified gene expression associated with inflammation and cancer and induced apoptosis in human pancreatic cancer cells through inhibition of GSK-3β/NF-κB signaling cascade, Mol. Nutr. Food Res. 57 (2013) 2112–2127, https://doi.org/10.1002/mnfr.201300307.
[57] X. Zhu, F. Luo, Y. Zheng, J. Zhang, J. Huang, C. Sun, X. Li, K. Chen, Characterization, purification of poncirin from edible citrus ougan (Citrus reticulate cv. Suavissima) and its growth inhibitory effect on human gastric cancer cells SGC-7901, Int. J. Mol. Sci. 14 (2013) 8684–8697, https://doi.org/10.3390/ijms14058684.
[58] M.-H. Pan, Tangeretin induces cell-cycle G1 arrest through inhibiting cyclin-dependent kinases 2 and 4 activities as well as elevating Cdk inhibitors p 21 and p 27 in human colorectal carcinoma cells, Carcinogenesis 23 (2002) 1677–1684, https://doi.org/10.1093/carcin/23.10.1677.
[59] V.E. Steele, R.C. Moon, R.A. Lubet, C.J. Grubbs, B.S. Reddy, M. Wargovich, D. L. McCormick, M.A. Pereira, J.A. Crowell, D. Bagheri, C.C. Sigman, C.W. Boone, G. J. Kelloff, Preclinical efficacy evaluation of potential chemopreventive agents in animal carcinogenesis models: methods and results from the NCI chemoprevention drug development program, J. Cell. Biochem. 56 (1994) 32–54, https://doi.org/10.1002/jcb.240560905.
[60] I. Demonty, Y. Lin, Y.E.M.P. Zebregs, M.A. Vermeer, H.C.M. van der Knaap, M. Jakel, E. A. Trautwein, The Citrus flavonoids hesperidin and Naringin do not affect serum cholesterol in moderately Hypercholesterolemic men and women, J. Nutr. 140 (2010) 1615–1620, https://doi.org/10.3945/jn.110.124735.
[61] E.E. Mulvihill, J.M. Assini, B.G. Sutherland, A.S. Dimattia, M. Khami, J.B. Koppes, C.G. Sawyez, S. C. Whitman, M.W. Huff, Naringenin decreases progression of atherosclerosis by improving dyslipidemia in high-fat-fed low-density lipoprotein receptor-null mice, Arterioscler. Thromb. Vasc. Biol. 30 (2010) 742–748, https://doi.org/10.1161/ATVBAHA.109.201095.
[62] S. Akiyama, S. Katsumata, K. Suzuki, Y. Ishimi, J. Wu, Dietary hesperidin exerts in streptozotocin-induced marginal type 1 diabetic rats, J. Clin. Biochem. Nutr. (2010) 87–92.
[63] Y. Ma, X. Ye, Y. Hao, G. Xu, G. Xu, D. Liu, Ultrasound-assisted extraction of hesperidin from Penggan (Citrus reticulata) peel, Ultrason. Sonochem. 15 (2008) 227–232, https://doi.org/10.1016/j.ultsonch.2007.03.006.
[64] M. Wu, H. Zhang, C. Zhou, H. Jia, Z. Ma, Z. Zou, Identification of the chemical constituents in aqueous extract of Zhi-Qiao and evaluation of its antidepressant effect, Molecules 20 (2015) 6925–6940, https://doi.org/10.3390/molecules20046925.
[65] T. Seki, T. Kamiya, K. Furukawa, M. Azumi, S. Ishizuka, S. Takayama, S. Nagase, H. Arai, T. Yamakuni, N. Yaegashi, Nobiletin-rich Citrus reticulata peels, a kampo medicine for Alzheimer's disease: a case series, Geriatr Gerontol Int 13 (2013) 236–238, https://doi.org/10.1111/j.1447-0594.2012.00892.x.
[66] S. Okuyama, N. Shimada, M. Kaji, M. Morita, K. Miyoshi, S. Minami, Y. Amakura, M. Yoshimura, T. Yoshida, S. Watanabe, M. Nakajima, Y. Furukawa, Heptamethoxyflavone, a citrus flavonoid, enhances brain-derived neurotrophic factor production and neurogenesis in the hippocampus following cerebral global ischemia in mice, Neurosci. Lett. 528 (2012) 190–195, https://doi.org/10.1016/j.neulet.2012.08.079.
[67] S.M. Sacco, M.N. Horcajada, E. Offord, Phytonutrients for bone health during ageing, Br. J. Clin. Pharmacol. 75 (2013) 697–707, https://doi.org/10.1111/bcp.12033.
[68] J. Kim, G.K. Jayaprakasha, B.S. Patil, Limonoids and their anti-proliferative and anti-aromatase properties in human breast cancer cells, Food Funct. 4 (2013) 258–261, https://doi.org/10.1039/c2fo30209h.
[69] J.R. Patil, G.K. Jayaprakasha, J. Kim, K.N.C. Murthy, M.B. Chetti, S.Y. Nam, B.S. Patil, 5-Geranyloxy-7-methoxycoumarin inhibits colon cancer (SW480) cells growth by inducing apoptosis, Planta Med. 79 (2013) 219–226, https://doi.org/10.1055/s-0032-1328130.
[70] F. Epifano, G. Molinaro, S. Genovese, R.T. Ngomba, F. Nicoletti, M. Curini, Neuroprotective effect of prenyloxycoumarins from edible vegetables, Neurosci. Lett. 443 (2008) 57–60, https://doi.org/10.1161/CIRCRESAHA.109.210864.

[71] S. Okuyama, S. Minami, N. Shimada, N. Makihata, M. Nakajima, Y. Furukawa, Anti-inflammatory and neuroprotective effects of auraptene, a citrus coumarin, following cerebral global ischemia in mice, Eur. J. Pharmacol. 699 (2013) 118–123, https://doi.org/10.1016/j.ejphar.2012.11.043.

[72] T. Narisawa, Y. Fukaura, S. Oshima, T. Inakuma, M. Yano, H. Nishino, Chemoprevention by the oxygenated carotenoid β-cryptoxanthin of N-methylnitrosourea-induced colon carcinogenesis in F334 rats, Jpn. J. Cancer Res. 90 (1999) 1061–1065, https://doi.org/10.1111/j.1349-7006.1999.tb00677.x.

[73] B.N. Ames, L.S. Gold, Endogenous mutagens and the causes of aging and cancer, Mutat. Res. Fundam. Mol. Mech. Mutagen. 250 (1991) 3–16.

[74] R.H. Liu, J.H. Hotchkiss, Potential genotoxicity of chronically elevated nitric oxide: a review, Mutat. Res. 339 (1995) 73–89.

[75] R. Jacob, S. Hasegawa, G. Manners, The Potential of *citrus limon*oids as Anticancer Agents, Perish. Handl. Q. 102 (2000) 6–8.

[76] D. Arul, P. Subramanian, Naringenin (citrus flavonone) induces growth inhibition, cell cycle arrest and apoptosis in human hepatocellular carcinoma cells, Pathol. Oncol. Res. (2013), https://doi.org/10.1007/s12253-013-9641-1.

[77] Y. Arai, S. Watanabe, M. Kimira, K. Shimoi, R. Mochizuki, N. Kinae, Dietary intakes of flavonols, flavones and isoflavones by Japanese women and the inverse correlation between quercetin intake and plasma LDL cholesterol concentration, J. Nutr. (2000) 2243–2250.

[78] K.J. Joshipura, F.B. Hu, J.E. Manson, M.J. Stampfer, E.B. Rimm, F.E. Speizer, G. Colditz, A. Ascherio, B. Rosner, D. Spiegelman, W.C. Willett, The effect of fruit and vegetable intake on risk for coronary heart disease, Ann. Intern. Med. 134 (2001) 1106–1114. http://www.ncbi.nlm.nih.gov/pubmed/11412050. Accessed 29 December 2018.

[79] N. Chhikara, H.R. Devi, S. Jaglan, P. Sharma, P. Gupta, Bioactive compounds, food applications and health benefits of Parkia speciosa (stinky beans): a review, Agric. Food Secur. (2018) 1–9, https://doi.org/10.1186/s40066-018-0197-x.

[80] P.B. Odhner, I.S. Horvath, M.M. Kabir, A. Schabbauer, Biogas From Lignocellulosic Biomass, Avfall Sverige, 2012, pp. 1–55. http://www.avfallsverige.se/fileadmin/uploads/Rapporter/U2012-07.pdf.

[81] M.H. Hetta, N.Z. Yassin, E.M. Kadry, Phytochemical and Antihyperglycemic Studies on *Citrus medica* L. Leaves (Etrog) Growing in Egypt, Int. J. Pharmacogn. Phytochem. Res. 5 (2013) 271–277.

[82] M.A. Al-yahya, R.A. Mothana, M.S. Al-said, K.E. El-tahir, M. Al-sohaibani, S. Rafatullah, *Citrus medica* "Otroj": attenuates oxidative stress and cardiac dysrhythmia in isoproterenol-induced cardiomyopathy in rats, Nutrients (2013), https://doi.org/10.3390/nu5114269.

[83] D. Naviglio, Laboratory production of lemon liqueur (limoncello) by conventional maceration and a two-syringe system to illustrate rapid solid–liquid dynamic extraction, J. Chem. Educ. (2015), https://doi.org/10.1021/ed400379g.

[84] F.Y. Al-juhaimi, Citrus fruits by-products as sources of bioactive compounds with antioxidant potential, Pak. J. Bot. 46 (2014) 1459–1462.

[85] L.S. Magwaza, U.L. Opara, P.J.R. Cronje, S. Landahl, J.O. Ortiz, L.A. Terry, Rapid methods for extracting and quantifying phenolic compounds in citrus rinds, Food Sci. Nutr. 4 (2016) 4–10, https://doi.org/10.1002/fsn3.210.

[86] G.D. Saratale, S. Der Chen, Y.C. Lo, R.G. Saratale, J.S. Chang, Outlook of biohydrogen production from lignocellulosic feedstock using dark fermentation—a review, J. Sci. Ind. Res. (India) 67 (2008) 962–979.

[87] C.M. Galanakis, Recovery of high added-value components from food wastes: conventional, emerging technologies and commercialized applications, Trends Food Sci. Technol. 26 (2012) 68–87, https://doi.org/10.1016/J.TIFS.2012.03.003.

[88] T.J. Mason, L. Paniwnyk, J.P. Lorimer, The uses of ultrasound in food technology, Ultrason. Sonochem. 3 (1996), https://doi.org/10.1016/S1350-4177(96)00034-X.

[89] E. Alissandrakis, D. Daferera, P. Tarantilis, M. Polissiou, P. Harizanis, Ultrasound-assisted extraction of volatile compounds from citrus flowers and citrus honey, Food Chem. 82 (2003) 575–582, https://doi.org/10.1016/S0308-8146(03)00013-X.

[90] M. Boluda-Aguilar, L. García-Vidal, F. d P. González-Castañeda, A. López-Gómez, Mandarin peel wastes pretreatment with steam explosion for bioethanol production, Bioresour. Technol. 101 (2010) 3506–3513. https://doi.org/10.1016/j.biortech.2009.12.063.

[91] M.K. Khan, M. Abert-Vian, A.-S. Fabiano-Tixier, O. Dangles, F. Chemat, Ultrasound-assisted extraction of polyphenols (flavanone glycosides) from orange (Citrus sinensis L.) peel, Food Chem. 119 (2010) 851–858, https://doi.org/10.1016/J.FOODCHEM.2009.08.046.
[92] P. Tatke, Y. Jaiswal, An overview of microwave assisted extraction and its applications in herbal drug research, Res. J. Med. Plants 5 (2011) 21–31, https://doi.org/10.3923/rjmp.2011.21.31.
[93] E. Destandau, T. Michel, C. Elfakir, Chapter 4. Microwave-assisted Extraction, 2013. https://doi.org/10.1039/9781849737579-00113.
[94] X. Lu, B. Xi, Y. Zhang, I. Angelidaki, Microwave pretreatment of rape straw for bioethanol production: focus on energy efficiency, Bioresour. Technol. 102 (2011) 7937–7940, https://doi.org/10.1016/j.biortech.2011.06.065.
[95] M.A. Ferhat, B.Y. Meklati, J. Smadja, F. Chemat, An improved microwave Clevenger apparatus for distillation of essential oils from orange peel, J. Chromatogr. A 1112 (2006) 121–126, https://doi.org/10.1016/j.chroma.2005.12.030.
[96] T.M. Attard, B. Watterson, V.L. Budarin, J.H. Clark, A.J. Hunt, Microwave assisted extraction as an important technology for valorising orange waste, New J. Chem. 38 (2014) 2278–2283, https://doi.org/10.1039/C4NJ00043A.
[97] A.S. Bale, N.H. Shinde, Research article microwave assisted extraction of essential oil from lemon leaves, Int. J. Recent Sci. Res. 4 (2013) 1414–1417, ISSN: 0976-3031.
[98] T. Inoue, S. Tsubaki, K. Ogawa, K. Onishi, J. ichi Azuma, Isolation of hesperidin from peels of thinned Citrus unshiu fruits by microwave-assisted extraction, Food Chem. 123 (2010) 542–547, https://doi.org/10.1016/j.foodchem.2010.04.051.
[99] J. Prakash Maran, V. Sivakumar, K. Thirugnanasambandham, R. Sridhar, Optimization of microwave assisted extraction of pectin from orange peel, Carbohydr. Polym. 97 (2013) 703–709, https://doi.org/10.1016/J.CARBPOL.2013.05.052.
[100] K. Hayat, X. Zhang, H. Chen, S. Xia, C. Jia, F. Zhong, Liberation and separation of phenolic compounds from citrus mandarin peels by microwave heating and its effect on antioxidant activity, Sep. Purif. Technol. 73 (2010) 371–376, https://doi.org/10.1016/J.SEPPUR.2010.04.026.
[101] X. Cheng, L. Bi, Z. Zhao, Y. Chen, Advances in enzyme assisted extraction of natural products, in: Proc. 3rd Int. Conf. Mater. Mech. Manuf. Eng., Atlantis Press, Paris, France, 2015. https://doi.org/10.2991/ic3me-15.2015.72.
[102] B.B. Li, B. Smith, M.M. Hossain, Extraction of phenolics from citrus peels: II. Enzyme-assisted extraction method, Sep. Purif. Technol. 48 (2006) 189–196, https://doi.org/10.1016/J.SEPPUR.2005.07.019.
[103] S.S. Nadar, P. Rao, V.K. Rathod, Enzyme assisted extraction of biomolecules as an approach to novel extraction technology: a review, Food Res. Int. 108 (2018) 309–330, https://doi.org/10.1016/J.FOODRES.2018.03.006.
[104] B. Díaz-Reinoso, A. Moure, H. Domínguez, J.C. Parajó, Supercritical CO_2 extraction and purification of compounds with antioxidant activity, J. Agric. Food Chem. 54 (2006) 2441–2469, https://doi.org/10.1021/jf052858j.
[105] B. Mira, M. Blasco, A. Berna, S. Subirats, Supercritical CO2 extraction of essential oil from orange peel. Effect of operation conditions on the extract composition, J. Supercrit. Fluids 14 (1999) 95–104.
[106] T. Suetsugu, H. Iwai, M. Tanaka, M. Hoshino, A. Quitain, M. Sasaki, M. Goto, Extraction of citrus flavonoids from peel of citrus junos using supercritical carbon dioxide with polar solvent, Chem. Eng. Sci. 1 (2013) 87–90, https://doi.org/10.12691/ces-1-4-7.
[107] N.T. Dunford, J.A. Teelb, J.W. Kingb, A continuous countercurrent supercritical fluid deacidification process for phytosterol ester fortification in rice bran oil, Foodserv. Res. Int. 36 (2003) 175–181. https://doi.org/10.1016/S0963-9969(02)00134-5.
[108] L. Ram, D. Kumar, S. Kumar, Supercritical extraction of aroma/flavor and essential oil from nagpur mandarin (*citrus reticulata* blanco) peel, J. Biol. Chem. Res. 30 (2013) 522–528.
[109] J. Yu, D.V. Dandekar, R.T. Toledo, R.K. Singh, B.S. Patil, Supercritical fluid extraction of limonoids and naringin from grapefruit (Citrus paradisi Macf.) seeds, Food Chem. 105 (2007) 1026–1031, https://doi.org/10.1016/J.FOODCHEM.2007.04.062.

[110] J. He, P. Shao, J. Liu, Q. Ru, Supercritical carbon dioxide extraction of flavonoids from pomelo (*citrus grandis* (L.) osbeck) peel and their antioxidant activity, Int. J. Mol. Sci. (2012) 13065–13078, https://doi.org/10.3390/ijms131013065.

[111] A.N. Giannuzzo, H.J. Boggetti, M.A. Nazareno, H.T. Mishima, Supercritical fluid extraction of naringin from the peel of Citrus paradisi, Phytochem. Anal. 14 (2003) 221–223, https://doi.org/10.1002/pca.706.

[112] S. Machmudah, S. Zakaria, M. Winardi, M. Sasaki, N. Goto, K.H. Kusumoto, Lycopene extraction from tomato peel by-product containing tomato seed using supercritical carbon dioxide, J. Food Eng. 108 (2012) 290–296, https://doi.org/10.1016/J.JFOODENG.2011.08.012.

[113] E. Vági, B. Simándi, K.P. Vásárhelyiné, H. Daood, Á. Kéry, F. Doleschall, B. Nagy, Supercritical carbon dioxide extraction of carotenoids, tocopherols and sitosterols from industrial tomato by-products, J. Supercrit. Fluids 40 (2007) 218–226, https://doi.org/10.1016/J.SUPFLU.2006.05.009.

[114] T.H.J. Beveridge, B. Girard, T. Kopp, J.C.G. Drover, Yield and composition of grape seed oils extracted by supercritical carbon dioxide and petroleum ether: varietal effects, J. Agric. Food Chem. 53 (2005) 1799–1804, https://doi.org/10.1021/jf040295q.

[115] C. Caleja, L. Barros, A.L. Antonio, M.B.P.P. Oliveira, I.C.F.R. Ferreira, A comparative study between natural and synthetic antioxidants: evaluation of their performance after incorporation into biscuits, Food Chem. 216 (2017) 342–346, https://doi.org/10.1016/j.foodchem.2016.08.075.

[116] Zia-ur-Rehman, Citrus peel extract—a natural source of antioxidant, Food Chem. 99 (2006) 450–454, https://doi.org/10.1016/J.FOODCHEM.2005.07.054.

[117] P. Ojha, S. Thapa, Quality evaluation of biscuit incorporated with mandarin peel powder, Chem. Chem. Eng. Biotechnol. Food Ind. (2017). http://pubs.ub.ro/dwnl.php?id=CSCC6201701V01S01A0003. Accessed 30 December 2018.

[118] Z. Kohajdova, J. Karovicova, M. Jurasova, K. Kukurova, Application of citrus dietary fibre preparations in biscuit production, J. Food Nutr. Res. 50 (2011) 182–190.

[119] S. Singh, G. Immanuel, Extraction of antioxidants from fruit peels and its utilization in paneer, J. Food Process. Technol. (2014). http://agris.fao.org/agris-search/search.do?recordID=US2016B00446. Accessed 30 December 2018.

CHAPTER 16

Solubility of organic compounds in $scCO_2$

Noor U Din Reshi[a], Masood Ahmad Rizvi[a], Syed Kazim Moosvi[b], Mudasir Ahmad[c], Adil Gani[c]
[a]Department of Chemistry, University of Kashmir, Srinagar, India
[b]School Education Department, J&K Government, Srinagar, India
[c]Department of Food Science and Technology, University of Kashmir, Srinagar, India

Contents

Green Sustainable Process for Chemical and Environmental Engineering and Science
https://doi.org/10.1016/B978-0-12-817388-6.00016-7

Conventional methods of organic transformations involve the use of volatile organic compounds (VOCs) as solvents. The problems associated with the use of these solvents include toxicity, flammability, and environmental hazards [1–3]. Besides, recycling of solvents is not only energy demanding but often complicated and uneconomical. Consequently, they end up as contaminants in the environment. Moreover, purity concerns, solvent removal difficulties in case of heat sensitive products, tedious recyclability, and disposal challenges limit the use of conventional solvents in major industries (pharmaceutical, food processing, and polymer) [4].

The economic and environmental concerns have led to the search for nonconventional solvents and several replacements like water, "renewable" solvents derived from biomass, ionic liquids, fluorous solvents, and supercritical fluids (SCFs) were attempted to make chemical processes eco-friendly and safe [4]. These nonconventional solvents are not in competition but rather complement each other, with their individual advantages and limitations. For example, removal of water can be energy demanding. Similarly, ionic liquids are expensive to generate and difficult to purify and recycle [5].

SCFs exist in a supercritical state, the state where temperature and pressure are above a critical point. Among the nonconventional solvents, use of SCFs as solvents to replace VOCs is an attractive choice. SCFs are desirable as they combine the properties of the liquid and gaseous state. The supercritical state offers diffusion rates close to gases and liquid-like densities, hence allowing application as a medium for reactions and separations. The use of SCFs is environment-friendly, and easy recovery of solvent ensures zero levels of product contamination with no solvent loss or environmental load [3].

In the case of SCFs, the critical point offers control center for modulation, a minor variation in temperature or pressure can change properties such as density, solubility, partition coefficient, and dielectric constant drastically. Thus, the solvation power of a SCF can be modulated through changes in temperature and pressure which can have widespread and potential applications. Hence, solutes dissolved in SCFs are conveniently precipitated by reducing the pressure of the fluid phase and the gas can be recycled for repeated use [6].

The use of SCFs for applications such as reaction medium, extracting bioactive molecules from plants etc., dairy product processing, oil and metal ion recovery, polymer and nanomaterial synthesis, textile dyeing, and leather processing represents an efficient and green alternative to conventional application of organic solvents for these processes. Besides, SCFs have become important for the production of micro- and nano-size particles [7].

1. Suitability of CO_2 as solvent for SCF

SCFs tailor their solvency with pressure. The solvation power of a SCF is determined by its temperature and pressure. Table 1 presents the critical temperature (T_c), critical pressure (P_c) of some common solvents [8]. Under critical conditions, the kinetic energy of molecules is counterbalanced by attractive molecular forces in SCF. Generally, more polar solvents have elevated T_c primarily because high amount of kinetic energy is required to counter the attractive interactions. It is clear from Table 1 that CO_2 is the most suitable solvent for SCF application on account of its moderate critical constants. In addition to this, CO_2 has many other desirability [9] as depicted in Fig. 1.

Moreover, it is safe for the ozone layer unlike chlorofluorocarbons (CFCs) that were used extensively in polymer manufacturing before being barred. $scCO_2$ is an attractive solvent for processes such as separation of biomolecules, processing of pharmaceuticals

Table 1 Critical constants for some solvents [8]

Entry	Solvent	Critical temperature (°C)	Critical pressure (bar)
1	Ammonia	132.5	112.8
2	Benzene	289.0	48.9
3	Carbon dioxide	32.1	73.8
4	Cyclohexane	280.3	40.7
5	Ethane	32.2	48.8
6	Ethylene	9.3	50.4
7	Isopropanol	235.2	47.6
8	Propane	96.7	42.5
9	Propylene	91.9	46.2
10	Toluene	318.6	41.1

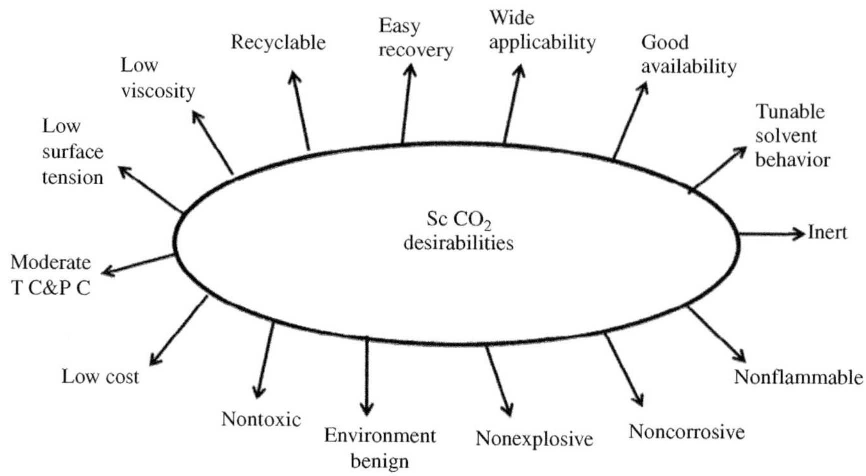

Fig. 1 Desirability of CO_2 as an SCF solvent.

applications, and other heat sensitive systems due to its moderate critical temperature, nontoxicity, and easy removal [10].

The rate of molecular diffusion and heat transfer are high in $scCO_2$ which shall significantly improve the rates of reactions limited by diffusion rates in a liquid state. Furthermore, in $scCO_2$, reactions such as hydrogenation, oxidation, and hydroformylation which operate under bi- or triphasic conditions using organic liquids as solvents, become single phasic due to high miscibility of gaseous ingredients (H_2, O_2, etc.) in $scCO_2$ [9]. Similarly, low viscosity and high diffusion rates of $scCO_2$ allows for the better design of reactors, where lifetime of catalysts can be enhanced. Since the solvent strength of $scCO_2$ changes with temperature and pressure, variation in temperature and pressure of the system can be used to fine-tune rate or product distribution of a reaction and catalyst activity. Unlike conventional organic solvents, $scCO_2$ as the solvent can induce altered reactivity or selectivity in certain chemical reactions for preferred applications like asymmetric catalysis.

A thorough understanding of solvency and solvation phenomenon of CO_2 requires a molecular level observation. CO_2 with a zero dipole moment, low molecular polarizability, and dielectric constant is expected to dissolve nonpolar systems efficiently. Surprisingly, nonpolar CO_2 is more miscible in H_2O than polar carbon monoxide (CO). Besides, many molecules with hydrocarbon chains are insoluble in CO_2 [1]. Similarly, sugar acetates which are rather polar are quite soluble in $scCO_2$. Anomalous results such as this get simplified on realizing CO_2 as a charge-separated polar molecule with significant nonzero bond dipole moment. Due to the separation of charge, CO_2 molecule can be thought to possess quadrupole moment similar to other nondipolar solvents such as dioxane. Besides being a quadrupolar solvent, CO_2 is capable of dissolving solutes by virtue of its feeble Lewis acidity [11] and Lewis basicity [8]. Many experimental and computational investigations have indicated that CO_2 can interact with various proton donors via hydrogen bonds (both red-shifting and blue-shifting) [12–17]. Hence, CO_2 can dissolve various dipolar molecules through these interactions. Molecular dynamics studies have predicted a hydrogen bond between CO_2 and H_2O (CO_2...H_2O). It has also been established computationally that CO_2 and H_2O interact significantly via H_2O...CO_2 interaction. Fluorocarbons are fairly soluble in $scCO_2$ and many explanations have been provided to explain their solubility. For example, Ab initio calculations have indicated that C-H bonds in fluorinated hydrocarbons can form C-H...O type hydrogen bonding interaction with oxygen atoms of CO_2. These calculations further indicated that interactions are stronger with a more polar fluorinated hydrocarbon [18]. Similarly, spectroscopic studies using NMR, IR, and Raman spectroscopic techniques have indicated a weak C—H...O type hydrogen bond between various carbonyls and CO_2 molecule. Hence many carbonyl-containing molecules are also very soluble in $scCO_2$ [17, 19]. The different types of interactions possible in CO_2 solvent are shown in Fig. 2.

2. Enhancing solubility in $scCO_2$

The practical application of $scCO_2$ as the solvent is often limited by the poor solubility of many classes of compounds in $scCO_2$. This necessitates a high operating pressure of CO_2 required to solubilize even small amounts of these compounds. However, to broaden the use of $scCO_2$ to a variety of chemical processes "CO_2-philes" which are some molecular fragments to enhance the solubility of otherwise poorly soluble molecular systems in $scCO_2$ are employed. With the use of "CO_2-philes" we can increase the effective concentration by efficiently transporting insoluble compounds into $scCO_2$. Although fluorocarbons are the most effective CO_2-philes, poly(ether-carbonate) copolymers and

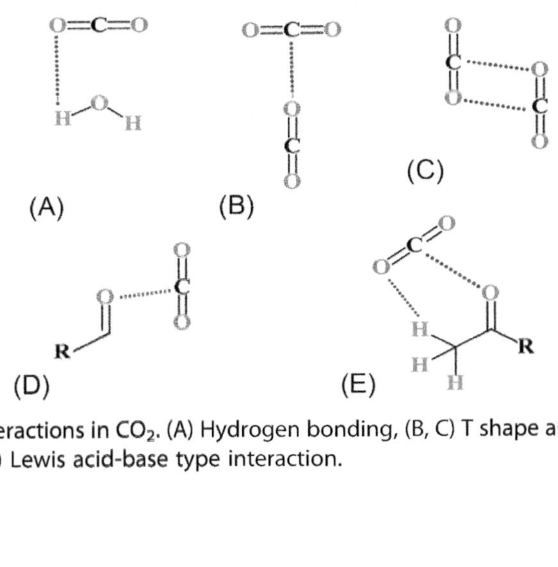

Fig. 2 Site-specific interactions in CO_2. (A) Hydrogen bonding, (B, C) T shape and slipped parallel self-interactions, and (D, E) Lewis acid-base type interaction.

sugar acetates are more economical alternatives [17, 19]. Carbonyl group in these systems interacts with C of CO_2 via Lewis acid-base interactions, thus aiding the solubilization of these compounds. Ab initio calculations indicated that the carbonyl group in these systems interacts with C of CO_2 via Lewis acid-Lewis base (LA-LB) interactions. The calculations also indicated the presence of a weak synergistic hydrogen bonding interaction involving methyl group hydrogen atom and CO_2 oxygen (see Fig. 2) [16]. These interactions result in high solubility, melting point decrease, and liquid swelling. In the case of hydrocarbons containing acetamide group, carbonyl group forms a weak complex with CO_2 resulting in enhanced solubility in $scCO_2$ [20].

Other approaches to improve the solubility of compounds in $scCO_2$ include use of a cosolvent [9, 21], complexing agent [22], or even an inverse micelle [4]. The addition of a cosolvent can be also helpful for improving the selectivity of a transformation. For example, the carbonylation of norbornene in $scCO_2$ using a Pd(II) + $CuCl_2$ catalytic system resulted in the formation of three products [21]. The addition of an optimum amount of MeOH in the reaction mixture improved the dissolution of catalyst. Thus, the efficiency and selectivity of the process improved for the major product (diester). Another important approach involves the use of complexing agents for enhancement of metal ion solubility in $scCO_2$ (elaborate discussion in the section for metal extraction).

3. Use of models to correlate experimental solubility data

The experimental data for solubility of compounds in $scCO_2$ is not always available due to limitations of immensely large number of compounds. So many methods including the equation of state, semiempirical methods, etc. are investigated to correlate, predict, and estimate the solubility of a compound in $scCO_2$ [23–28]. Although equation of state is the method of choice to estimate the solubility of a compound in SCFs, its use is limited due to unavailability of thermo-physical data particularly critical properties, sublimation pressure, and acentric factors which are necessary for its application. Hence, these parameters are predicted using methods such as contribution methods which increase the uncertainty of the results obtained [29]. This has led to the use of many semiempirical methods such as density-based equations; as use of most of these methods does not require any thermophysical data. Hence these models have become simple predictive tools. The four mainly used correlations based on density are those proposed by Mendez-Santiago and Teja (MST) [25], Bartle et al. [30], Chrastil [27], and Kumar and Johnston (K-J) [28]. The experimental solubility data is used to calculate fitting parameters for these models using simple data fitting or mean square error approaches. These models can be used to extrapolate solubility data and to correlate solubilities under various temperatures and pressures. Average absolute relative deviation (AARD%) can be calculated and is indicative of how good the experimental results correlate with the calculated values.

4. Case studies of solubility of organic compounds in $scCO_2$

$scCO_2$ has been used for extracting and purifying natural products, for recrystallizing, solubilizing, and forming particles (micronization) of drugs and as a nonhazardous solvent for the synthesis of pharmaceuticals [31]. One possible application of the fact that $scCO_2$ causes swelling of many polymers due to significant mass transport inside polymer can be to dissolve drugs in $scCO_2$ and to load these on biodegradable polymers which can be used as efficient drug delivery system [32]. The experimental determination of solubility is vital for the development of these newer technologies employing $scCO_2$. The following is a concise summary of solubility data of various organic compounds reported in recent literature.

5. Amoxicillin

Amoxicillin belongs to a class of antibiotics known as penicillins and is used in the treatment of a variety of bacterial infections. Esmaeilzadeh group investigated the solubility of amoxicillin in $scCO_2$ at pressure16–40 MPa and temperature 308.15–338.15 K using a gravimetric method [31]. They reported the solubility of amoxicillin in the range of 1.08×10^{-5} and 7.23×10^{-3} under the temperature and pressure used in terms of mole fraction of amoxicillin. Furthermore, the solubility increased with an increase in pressure and this effect was more prominent at higher temperatures. The crossover pressure was found to be 16–18 MPa for this system. The experimental solubility correlated satisfactorily with solubility calculated using MST, Bartle et al., Chrastil, and K-J models. The fitting parameters for models were calculated using a simple data fitting which led to the calculation of solubility using Bartle et al., MST, K-J, and Chrastil methods with AARD % of 5.26%, 6.26%, 7.05%, and 7.53%, respectively. Interestingly, the self-consistency analysis indicated the Bartle et al. model is most suitable for extrapolation of solubility.

6. Antiinflammatory drugs

Kikic et al. reported the solubilities of Ketoprofen, Piroxicam, and Nimesulide which are inflammation inhibitors, in the range 4×10^{-6}–15×10^{-4} mole fraction in $scCO_2$, in the pressure range of 100–220 bar and at 312.5 and 331.5 K [33]. The solubility obtained experimentally showed a trend which is representative of the solubility of non VOCs in $scCO_2$. Moreover, the solubility depended on the solvent density and could be precisely correlated using K-J model.

7. Artemisinin

Artemisinin (Qinghaosu), an endoperoxide sesquiterpene lactone, holds promise for its use against chloroquine-resistant strains of Plasmodium falciparum to cure cerebral malaria [34]. This compound is extracted from plant Artemisia annua L. using solvents such as toluene, hexane, or petroleum ether. To examine the possibility of SFC, Ren et al. investigated the solubility of artemisinin in $scCO_2$ over the temperature range 310.1–338.1 K and pressure from 10 to 27 MPa. The solubility of artemisinin in $scCO_2$ was observed to be $(0.098–2.66) \times 10^{-3}$. It was found that in pressure range 10–19 MPa, the solubility increases with increase in pressure and decreasing temperature but pressure above 19 MPa, the solubility increases with increase in the temperature or pressure. The measured solubilities agreed well with the calculated solubilities using, four-parameter semiempirical model.

8. Anesthetics

Weinstein and coworkers measured the solubility of benzocaine, lidocaine, and procaine which have similar structures and are used as local anesthetics in liquid and $scCO_2$ [35]. The solubilities were obtained at three temperatures (298, 308, and 318) K and in the pressure range 70–280 bar. Mole fraction solubility of benzocaine, procaine, and lidocaine were reported to be in the range of $(1.75–21) \times 10^{-5}$, $(5.82–17) \times 10^{-5}$, and $(1.55–28) \times 10^{-5}$, respectively. Hydrochlorides of procaine and lidocaine were almost insoluble over the conditions investigated. Moreover for lidocaine and benzocaine solubility was dependent on the density whereas for procaine solubility was dependent on density as well as temperature.

The solubilities could not be predicted by the Peng-Robinson equation of state. However, the solubilities were efficiently predicted using an enhancement factor dependent on density and vapor pressure.

9. Cholesterol

Foster group determined the solubility of cholesterol in $scCO_2$ at three temperatures (40°C, 50°C, 60°C) and reported a typical trend of a solid/SCF system with a crossover pressure of approximately 160 bar [36]. They have reported the mole fraction solubility of cholesterol to be $0.23–15 \times 10^{-5}$. The solubility data could be adequately correlated by K-J Model.

10. Dexamethasone

Sousa and coworkers determined the solubility of solid dexamethasone, a glucocorticoid which can be used to cure chronic immune and inflammatory disorders, in $scCO_2$ [37]. The solubility was determined at 308.2, 318.2, and 328.2 K in the pressure range15.1–35.7 MPa. The measured solubility as mole fraction varied between 1.25×10^{-6} at 308.2 K and 15.1 MPa and 2.81×10^{-6} at 328.2 K and 34.8 MPa. The best results for solubility data correlation were obtained using Chrastil model with AARD lower than 13%.

11. Flurbiprofen

Flurbiprofen is a nonsteroidal antiinflammatory drug. Duarte et al. measured solubility of flurbiprofen in $scCO_2$ by a static analytical method in the pressure range 8.0–25.0 MPa, at three temperatures (303.0, 313.0, and 323.0) K [18]. The reported solubility is (1–20) × 10^{-5}, (AARD of 9%) in terms of mole fraction of flurbiprofen. The solubility increased with increase in pressure under isothermal conditions. Below 12.0 MPa, solubility decreases on increasing temperature, whereas above 12.0 MPa, solubility increases with increase in temperature. Furthermore, the authors also studied cosolvent effect of EtOH on the solubility of flurbiprofen in $scCO_2$ at 18.0 MPa and 313.0 K using EtOH (2.5, 5.0, 7.5, and 10.0 mol%). The data was correlated using Chrastil's model (with a maximum AARD of 10%).

12. Isoniazid

In the context of developing improved technologies for processing of Isoniazid (pyridine-4-carbohydrazide) which is the main drug to cure tuberculosis (TB), Hasan et al. measured the solubility of isoniazid in $scCO_2$ at three temperatures 308, 311, and 313 K and in the pressure range 13.0–18.5 MPa [38]. The study indicated that isoniazid solubility decreases with the decrease of pressure under isothermal conditions and it decreases with the decrease of temperature. The experimental data could be adequately correlated using the Chrastil and MST models.

13. 2,2′-Bipyridine and 4,4′-dimethyl-2,2′-bipyridine

Guo et al. reported low solubility of 2,2′-bipyridine and 4,4′-dimethyl-2,2′-bipyridine in $scCO_2$ in the temperature and pressure range 308–323 K and 8.1–25.9 MPa respectively [39]. The mole fraction solubility of bipyridine and dimethyl bipyridine was observed to be (1–15) × 10^{-3} and (5–12) × 10^{-4}, respectively. It was concluded that the solubilities of both the compounds in $scCO_2$ are very low, hence, not good for extraction of metal ions. The solubility of 2,2′-Bipyridine was observed to be higher than 4,4′-Dimethyl-2,2′-bipyridine. The experimental results were in agreement with the results obtained using a semiempirical method with a difference of 4.6% and 34.4% between experimental and calculated values.

14. Amide compounds

Li and coworkers measured the solubilities of three amides in $scCO_2$ in the temperature range of 313–333 K and pressure ranging from 8.7 to 16.4 MPa [20]. The reported solubilities of amide compounds in $scCO_2$ under different conditions of temperatures are tabulated in Table 2. The solubilities decreased with decreasing pressure under isothermal

Table 2 Mole fraction solubility of some amide compounds

S. no.	Name of compound	Mole fraction solubility
1.	2,2′-oxybis(N,N-diethylacetamide)	$(2\text{–}12) \times 10^{-3}$
2.	2,2′-oxybis(N,N-dibutylacetamide)	$(1.5\text{–}9) \times 10^{-3}$
3.	2,2′-oxybis(N,N-dihexylacetamide)	$(1\text{–}5) \times 10^{-3}$

conditions and increased with decreasing temperature and molecular weight under isobaric conditions in the pressure range of 8.2–16.4 MPa, and in temperature range of 313–333 K in $scCO_2$. The observed solubilities were correlated using a semiempirical model. The calculated results adequately agreed with measured solubilities.

15. Polynuclear aromatic hydrocarbons

Experimental data for solubility of polycyclic aromatic hydrocarbons in $scCO_2$ is scarce due to their very low solubility. Barna et al. developed a method for measuring low solubilities, and determined solubilities of three polycyclic aromatic hydrocarbons in the temperature range 308.15–328.15 K and pressure ranging from 84 to 251 bar [40]. Solubility of fluoranthene, chrysene, and triphenylene came out to be 9.13×10^{-6}–9.3×10^{-4}, 7.29×10^{-7}–8.84×10^{-6}, and 1.01×10^{-6}–4.22×10^{-5}, respectively. The results thus indicate very low solubility of polycyclic aromatic hydrocarbons in $scCO_2$. There was a satisfactory correlation between the experimental results and the excess function equation of state model by optimization an interaction parameter.

16. Maleic acid

Ghaziaskar et al. measured solubility of maleic acid in $scCO_2$ at three temperatures 318.15, 333.15, and 348.15 K in the pressure range from 7 to 30 MP using a static apparatus [41]. The authors reported the solubility from 1.30×10^{-5} at 318.15 K and 7 MPa to 59.17×10^{-5} at 348.15 K and 30 MPa. The experimental solubility data was correlated using Chrastil and MST models.

17. Menthol

Solubility of menthol ($C_{10}H_{20}O$) in temperature and pressure range of (33–55)°C and (6.7–11.6) MPa, respectively, were measured via flow-type apparatus and were observed to be $(0.2\text{–}13) \times 10^{-3}$ in terms of mole fraction [42]. Adachi-Lu equation was used to correlate with the density of pure solvent. This approach yielded deviations of 4.4%.

18. Methyl salicylate

Ismadji measured the solubilities of methyl salicylate in $scCO_2$ in static mode at 343.15, 363.15, 383.15, 403.15, and 423.15 K, at pressures up to 31 MPa [43]. The mole fraction solubility of methyl salicylate at different temperatures (343.15–423.15) and pressure from 9 to 31 MPa was investigated by S. Ismadj and was observed to be $(0.51–75.33) \times 10^{-3}$. Moreover, Chrastil model could satisfactorily correlate the measured solubilities. Critical constants of methyl salicylate were predicted using models proposed by Joback and Lee-Kesler.

19. Phenacetin

Madras et al. used a saturation method to determine the solubility of the drug, N-(4-ethoxyphenyl) ethanamide (phenacetin), in $scCO_2$ at three temperatures 308, 318, and 328 K and pressure range 9–19 MPa [44]. The authors reported solubilities (as mole fraction) of phenacetin in the range of 1.29×10^{-5}–2.88×10^{-5}, 1.13×10^{-5}–3.65×10^{-5}, and 0.91×10^{-5}–4.28×10^{-5} at 308, 318, and 328 K, respectively. The solubility data was correlated by using the equation of state (Peng-Robinson) and MST model.

20. Oxymatrine

Oxymatrine is a quinolizidine alkaloid extracted mainly from three medicinal plants Sophora japonica, Sophora subprostrata, and Sophora alopecuroides. Oxymatrine has antiarrhythmic and antitumor activities. In the context of developing SFE method, Su and coworkers used a continuous flow-type apparatus to determine the solubility of oxymatrine in $scCO_2$ at three temperatures (308.15, 318.15, and 328.15) K and the pressure 11–21 MPa [45]. Solubility of oxymatrine in terms of mole fraction was found to be 6.8×10^{-7}–24.6×10^{-7} in $scCO_2$. The correlation was done using Bartle method with AARD% of 12.7%.

21. Polyacrylamide

Jin et al. used a dynamic method to measure the solubility of polyacrylamide (PAM) with different molecular weight (5×10^6, 7×10^6, 14×10^6 g mol^{-1}) in $scCO_2$ at three temperatures (313, 323, and 333 K) and pressure 9.0–18.0 MPa [46]. The solubility of PAM with a molecular weight of 5×10^6 g mol^{-1}, 7×10^6 g mol^{-1}, and 14×10^6 g mol^{-1} in $scCO_2$ was determined to be respectively $2.34–7.72 \times 10^{-10}$, $1.55–5.24 \times 10^{-10}$, and $0.59–2.29 \times 10^{-10}$ (as mole fraction). The correlation of experimental solubility was done using Chrastil, M-T, A-L, S-S, and Bartle methods. A-L model gave the best-correlated results. The self-consistency of solubility of PAM was established by M-T model. Furthermore, Chrastil and Bartle models were used to estimate the enthalpy values of PAM.

22. 1,4-Dimethoxybenzene

1,4-Dimethoxybenzene is used in cosmetics, flavoring agents, as a fixative in perfumes, dyes, and resins, and as a weathering agent in paints and plastics. Lee et al. determined the solubility of 1,4-Dimethoxybenzene in $scCO_2$ in temperature and pressure range 313.15–338.15 K and 200–280 bar respectively [47]. It was found to be 4.7×10^{-5}–19×10^{-5} in terms of mole fraction. The experimental solubilities were modeled using Chrastil model.

23. Troeger's base

Troeger's base has drawn considerable attention for its use as a chiral solvating agent, as DNA-binding ligand, and for the construction of biomimetic molecular receptors and clathrate hosts. Ren et al. measured the solubility of Troeger's base in $scCO_2$ at 308, 318, and 328 K and in the pressure range of 8–19 MPa [48]. Mole Fraction Solubility of Troeger's Base under different conditions of temperatures and pressures was observed to be 0.3×10^{-4}–8×10^{-4} in $scCO_2$.

24. Palmitic acid + capsaicin

Elizalde-Solis et al. measured the solubilities of a solid mixture of palmitic acid and capsaicin in $scCO_2$ at 308.15 and 328.15 K, and at pressure 10–35 MPa [49]. The self-consistency of experimental solubilities was examined using the MST method. Moreover, the solubility of the binary mixture did not change significantly as compared to that of the solubility of pure palmitic acid and pure capsaicin in $scCO_2$.

25. Phenol and pyrocatechol

Rodríguez et al. determined the solubilities of phenol and pyrocatechol (1,2-dihydroxybenzene) in $scCO_2$ [50]. Measurements were carried out using a recirculating equilibrium apparatus and the results were correlated using Bartle model. Mole fraction solubility of phenol and pyrocatechol at a temperature range of 333.15–363.15 K and pressure range of 100–350 bar were found to be $(1.14–91)\times10^{-3}$ and $(1.22–45)\times10^{-4}$, respectively. The higher solubility of phenol (≈20-folds more) than pyrocatechol at higher pressure was also observed in the findings.

26. 2H-chromene derivatives

Yamini and coworkers measured the solubilities of 2H-1-benzopyran (2H-chromene) derivatives in $scCO_2$ [51]. The measurements were performed in temperature and pressure range of 308–348 K and 12.2–35.5 MPa respectively. The calculated results using a semiempirical model showed good correlation with experimental solubilities. The mole fraction solubilities of the four reported chromenes derivatives are shown in Table 3. The order of solubilities of chromene derivatives was observed as $A_1 > A_2 > A_3$. The solubility of A_3 was found comparable to A_4. Solubility was found to increase with an increase in pressure at an isothermal condition owing to the increase in solvent density. However, the solubility of chromene derivatives in $scCO_2$ was found to decrease with temperature under isobaric conditions.

27. Fat-soluble vitamins A, D, E, and K

Johannsen and Brunner measured the solubility data of eight variants of the fat-soluble vitamins in $scCO_2$ [52]. The solubilities were measured at 313, 333, and 353 K and pressure 20–35 MPa. The solubilities were found to be nearly 10 g/kg, except provitamin A, which has solubility less by three orders of magnitude. The solubility increased with decreasing molecular mass of the vitamin. The solubility increased with increased density under isothermal conditions for all solutes investigated. Furthermore, at constant density, solubility increased with the increase in temperature.

The solubilities of fat-soluble vitamins (Retinol, α-Tocopherol, δ-Tocopherol, Carotene, Vitamin D_3, Vitamin D_2, Vitamin K_1, and Vitamin K_3) were studied by M. Johannsen and others in $scCO_2$ by making use of an analytical method [28], under different conditions of temperatures (313, 333, and 353) K and pressures (20–35) MPa. The solubility of the vitamins is shown in Table 4. Solubility of all fat-soluble vitamins except β-carotene was observed to be around 10 g/Kg. The solubility of β-carotene was reported to be three orders of magnitude less than other fat-soluble vitamins. The solubility of vitamins in supercritical CO_2 was also observed to decrease with molecular mass. At isothermal conditions, the solubility of all substances was found to increase with

Table 3 Mole fraction solubilities of some chromene derivatives

S. no	Chromene derivatives	Mole fraction solubility
1	A_1 ($C_{13}H_{11}O_5$)	$(5\text{–}28) \times 10^{-3}$
2	A_2 ($C_{23}H_{26}O_5$)	$(2\text{–}100) \times 10^{-4}$
3	A_3($C_{13}H_{11}BrO_5$)	$(0.4\text{–}25) \times 10^{-4}$
4	A_4($C_{17}H_{14}O_5$)	$(1.3\text{–}17.5) \times 10^{-4}$

Table 4 Solubility of some fat-soluble vitamins

S. no.	Fat-soluble vitamin	Solubility (g/kg)	Mole fraction solubility
1	Retinol	9.2–28	1.4×10^{-3}–4.3×10^{-3}
2	α-Tocopherol	6.2–34	0.64×10^{-3}–3.6×10^{-3}
3	δ-Tocopherol	4.9–40.2	0.54×10^{-3}–3.7×10^{-3}
4	Carotene	0.11–4	0.09×10^{-6}–3.24×10^{-6}
5	Vitamin D_3	4.0–44	0.46×10^{-3}–5.21×10^{-3}
6	Vitamin D_2	2.6–9.3	0.29×10^{-3}–1.04×10^{-3}
7	Vitamin K_1	3.0–35.3	0.29×10^{-3}–3.6×10^{-3}
8	Vitamin K_3	23–26	6×10^{-3}–7×10^{-3}

density. Moreover, at constant density, solubility was observed to increase with temperature elevation, attributed to the increasing vapor pressure of the solid.

28. Disperse dyes

J.W. Lee by making use of flow-type equilibrium cell measured the solubilities of two disperse dyes (CI disperse red 60 and CI disperse orange 3) at temperature and pressure range of (323.7–413.7) K and (100–300) bar respectively. Mole fraction solubility range was found to be (0.138–38) $\times 10^{-6}$ in case of CI disperse red 60. The mole fraction solubility range was reported to be (0.094–92.4) $\times 10^{-6}$ in case of CI disperse orange 3 [53]. The solubility data showed satisfactory agreement with an expanded liquid model as well as the empirical method.

29. Anthracene, phenanthrene, and carbazole mixture

The solubilities of anthracene, phenanthrene, and carbazole in $scCO_2$ at different temperatures and pressures have been determined by various people [54]. Goodarznia and Esmaeilzadeh measured the solubility data of a mixture of anthracene, phenanthrene, and carbazole in $scCO_2$ at 308.2 K and pressure 111–201 bar. Solubilities of anthracene, carbazole, and phenanthrene in $scCO_2$ were reported to be (3.5–11.65) $\times 10^{-5}$, (1.2–3) $\times 10^{-5}$, and (77–144) $\times 10^{-5}$ respectively.

30. 6-Caprolactam

Antero Laitinen and coworkers. reported the solubility of 6-caprolactam in $scCO_2$ in temperature and pressure range of 307–324 K and 100–230 bar, respectively using a static variable volume view cell apparatus. 6-Caprolactam was observed to depict a relatively high solubility (0.02–0.07 in terms of mole fraction) [55].

31. Solubility of energetic materials in SCFs

Due to lower solubilities of energetic materials in the $scCO_2$, comprehensive experimental solubility data for these compounds in $scCO_2$ is scarce. One exception to this is TNT, which is soluble in $scCO_2$ (up to 2 wt%) [18]. However, solubility can be increased by changing the polarity of fluid using additives such as MeOH [7]. Various theoretical investigations have been used to predict the solubility of these compounds in $scCO_2$ and mixture of CO_2 and modifiers. Isothermal-isobaric MC simulations and the Widom insertion method have been used to study the enhancement of solubility of 1,3,5-trinitro-s-triazine in $scCO_2$ by polar additives such as DMSO, CH_3CN, or CH_3OH [56]. The results showed that solubility depends on factors such as dipole moment, additive concentration, and interaction between RDX and additive. The solute-solvent interactions were described using potentials between functional groups on the explosives and CO_2 and calibrated using two experimentally determined solubility values. The comparison with experimental results showed a good correlation. The authors also used simple pair additive interaction potentials to generate the results. The results obtained using single or multisite interactions did not differ significantly at low-densities. However, at high densities, multisite interaction potentials were important to rationalize the predicted solubility. Similar investigations have been carried out to model the solubility of 1,3,5-trinitro-s-triazine and trinitrotoluene in $scCO_2$ [57]. The predicted solubilities matched satisfactorily with experimental values.

32. Extraction of metal ions

Extraction using SCFs particularly $scCO_2$ also referred to as supercritical fluid extraction (SFE) has drawn wide attention as environmentally benign separations in the industry [22]. One such application is to remove metals from industrial effluents, metal contaminated wastes, and extraction of precious metals. Since metal ions have poor solubility in $scCO_2$ due to weak solute-solvent interactions, chelating ligands are used to transport and concentrate metal ions in a single step which ensures that no subsequent concentration steps are required. Some of the ligands which have been used for chelation and selective extraction of metal ions in $scCO_2$ include dithiocarbamates, β-diketones, organophosphorus ligands, and macrocycles [58]. Wai and coworkers used fluorinated dithiocarbamate ligand, bis(trifluoroethyl)dithiocarbamate in $scCO_2$ to extract copper ions from solid materials, such as celite [59]. A less soluble metal chelate was formed when a non-fluorinated ligand, diethyldithiocarbamate, was used in $scCO_2$. One important example is the use of tributyl phosphate (TBP) in $scCO_2$ for extracting uranyl ions from HNO_3 solutions [60]. This observation can lead to the development of novel and green technology for processing the nuclear waste. The current method for the processing of nuclear waste involves use of harsh nitric acid solution (3–6 M) and subsequent extraction using kerosene in the presence of TBP. This method produces a lot of organic and acid waste contamination in the nuclear waste discharge. Sievers et al. used 2,2,7-trimethyl-3,5-octanedionate, (tod), 2,2,6,6-tetramethyl-3,5-octanedionate, (thd), and 2,4-pentanedionate, (acac) in $scCO_2$ to chelate Ce(IV), Tb(III), and Fe(III) [61]. The solubilities of the resultant metal chelates were determined in $scCO_2$ at 323 and 333 K and pressure 10–35 MPa using near-infrared spectroscopy with a high-pressure optical cell. The solubility was found to strongly depend on the nature of the ligand and varied as tod > thd > acac for all metal chelates. The higher solubility of tod chelates was rationalized in terms of higher shielding of the metal ion due to higher flexibility of the isobutyl group on tod ligand. The experimental solubilities were satisfactorily correlated with Chrastil model (AARD of 6.2%–20.0%).

33. Catalysis in $scCO_2$

Catalytic processes producing products with absolute selectivity under milder operating conditions are vital from an economic and environmental point of view. These processes are made desirable, sustainable, and safer on replacing VOCs with $scCO_2$ [4]. An enzyme-catalyzed process when performed in $scCO_2$ profits from the easy separation of the products from the reaction medium [62]. The use of $scCO_2$ also allows for the recycling of catalysts from the reaction medium by modulating pressure and temperature condition, at the reactor output. Besides, in many cases, improved selectivity (chemo-, regio-, or stereo) has been obtained by carrying out catalytic reactions in $scCO_2$ [4, 60]. Furthermore, modulation of density, temperature, and pressure conditions of $scCO_2$ strongly affects chemical equilibrium and reaction rate which can be utilized to tune reactivity and selectivity of catalytic reactions. One such example of selectivity control is metathesis of alkenes catalyzed by ruthenium catalysts, where there is a competition between ring closing metathesis and acyclic diene metathesis [63]. At a high density of $scCO_2$ mole fraction of the reactants is reduced in reaction medium resulting in effective dilution favoring intramolecular reaction [63]. The low density of $scCO_2$, increases concentration and suppresses the intramolecular reaction [63]. The synergistic effect of $scCO_2$ and cosolvents in improving the selectivity of a chemical transformation is also known. For example, norbornene when carbonylated in $scCO_2$, produces cis-exo-diester, cis-exo-β-chloride ester, and exo-chloronorbornane [21]. Adding an optimum amount of MeOH as cosolvent enhances dissolution of catalyst, consequently improving efficiency and selectivity toward the major product (diester). However, increasing temperature or pressure lead to a decrease in the selectivity for diester product [21]. The chemo- and regioselectivity in the hydroboration of vinyl arenes using a rhodium catalyst got significantly improved with $scCO_2$ as a solvent [64]. Compared to the mixture of four products formed in THF solvent, only a single isomer was formed in $scCO_2$. Similarly using $scCO_2$ solvent, favorable changes were observed in the regioselectivity of

Mizoroki-Heck reaction catalyzed by palladium-nanoparticles [65]. When S-oxidation of cysteine derivatives was carried in $scCO_2$, diastereoselectivity was shown to be a function of density [66].

A key factor responsible for change of reaction rates in $scCO_2$ is physicochemical behavior of the supercritical state which ensures higher mass transfer rate in case of heterogeneous systems. Similarly, increased rates for homogeneous gas reactions such as CO_2 hydrogenation and hydroformylation are due to higher reactant (gas) availability in $scCO_2$. Replacing organic solvents with $scCO_2$ can alter reaction profile such as a change of rate-limiting step [4]. For example, the rate of hydrogenation of imines increases noticeably in $scCO_2$ which cannot be rationalized simply in terms of increased gas availability [67]. In $scCO_2$ the substrate binds quantitatively and rate of binding is very fast than the reaction with H_2 as indicated by zero-order kinetics.

The inert nature of $scCO_2$ is a major attraction for its application in transformations such as C-H functionalization or oxidations where organic solvents cannot be used because they can react with highly active intermediates [4]. For example, $scCO_2$ has been used in the photochemical functionalization of methane under monophasic conditions [68].

The reactivity of CO_2 with some functional groups can also be explored for carrying out chemical transformations more conveniently. For example, cyclization of substrates containing unprotected NH group cannot be carried out as such in organic solvents because of irreversible deactivation of ruthenium catalysts used for RCM by NH groups. Hence, cyclization of such substrates in conventional solvents requires additional protection/deprotection steps for NH group. In $scCO_2$ NH group reversibly forms carbamic acids or carbamates which can be converted back to free amine upon pressure release, thus eliminating the need for protection and deprotection of NH group [69]. Besides protecting NH group, $scCO_2$ also affects the reaction profile for hydroaminomethylation of allylic amines. Whereas in conventional solvents cyclic amides are exclusively formed, saturated heterocyclic are selectively obtained in $scCO_2$.

The use of catalysts immobilized on solid supports has also been used as an effective approach to develop reactors for continuous operation [70–72].

Some selected examples of catalytic reactions that have been carried out in $scCO_2$ are:

34. Palladium-catalyzed C-C coupling reactions

C-C coupling reactions catalyzed by palladium are important for synthesis of fine chemicals such as pharmaceuticals. These reactions have been carried out with high rates and selectivity in $scCO_2$. For example, Crooks et al. used palladium nanoparticles decorated with fluorinated CO_2-philic dendrimers for catalyzing C—C bond forming reactions under homogeneous conditions in $scCO_2$, yielding a 2-phenylacrylate product with high selectivity [65]. It is well known that some polymeric compounds plasticize and swell in the presence of $scCO_2$ which allows for faster mass transfer between polymeric and supercritical phase [4]. This effect has been used in many coupling and polymerization processes. For example, McConvey and coworkers carried out coupling reactions on substrates attached to solid-phase using $Pd(OAc)_2(P^tBu_3)$ which is soluble in $scCO_2$ [73]. Products bound to the polymer are obtained by washing excess reagents and catalysts with $scCO_2$ eliminating steps for drying or purification. This technique can be used for parallel or high-throughput synthesis of desired molecules such as pharmaceuticals.

35. Hydrogenation and hydroformylation

Since $scCO_2$ is nonexplosive and nonflammable its use as a solvent reduces potential hazards in many reactions particularly where very reactive intermediates or pressurized conditions are involved. Homogeneous reaction conditions eliminate mass transfer restrictions and increase the effective concentration of reactants for reactions involving gaseous reactants [4].

The reactions that utilize CO_2 can be carried out in $scCO_2$ by taking advantage of the inherent high concentrations of CO_2 in $scCO_2$ [2, 4, 9]. One such reaction is hydrogenation of CO_2 which can be efficiently carried out in $scCO_2$. The catalytic hydrogenation of CO_2 to formic acid is conventionally carried out in an organic solvent in the presence of a base. Leitner et al. carried out hydrogenation of CO_2 to produce HCOOH in $scCO_2$ using $[Ru\ (P(CH_3)_4(Cl_2)]$ as a precatalyst [73]. The rate of the reaction in $scCO_2$ increased by an order of magnitude as compared to reaction carried out in conventional organic solvents due to higher concentration of CO_2 in $scCO_2$ [74]. This method has also been used to carry out the tandem processes by coupling the hydrogenation reaction with subsequent steps [75, 76]. For example, the use of methanol led to in situ esterification of the formic acid to form methyl formate. The same process happened in subcritical systems albeit with decreased rate by two orders of magnitude. Similarly, the hydrogenation of $scCO_2$ at 100°C produced N-substituted methanamides in the presence of amines.

Hydroformylation in $scCO_2$ can be followed by a subsequent SFE workup step in which high-boiling long-chain aldehydes or side products could be removed to regenerate precious catalysts insoluble in $scCO_2$ [77, 78]. If catalysts are soluble in $scCO_2$ soluble, changes in pressure or temperature allow partitioning organic components into two phases (supercritical phase and liquid phase) whereas organometallic catalyst remains exclusively in the supercritical phase [79]. This technique has been used for asymmetric hydroformylation of styrene using Rh/3-H2F6-BINAPHOS with high recyclability [80].

36. Asymmetric hydrogenation (AH)

Many studies suggest the potential of $scCO_2$ as a novel solvent for asymmetric catalysis. Chiral rhodium complexes (Rh/DuPhos) are efficient catalysts for the AH of α-enamide esters to α-amino acids in conventional solvents. Interestingly, higher ee was obtained in the AH of disubstituted enamides in $scCO_2$ than conventional solvents (MeOH or hexane). The authors also ruled out a simple high-pressure effect responsible for higher ee [81].

When the AH of itaconates was carried in hexane, a decrease in ee was observed with a decrease in H_2 pressure but no such effect was observed in $scCO_2$. A study of mechanism indicated that there is no significant difference in the mechanism of the transformation in two solvents. However, the difference in response to H_2 pressure in two solvents could be rationalized by the increased availability of H_2 in $scCO_2$ [82, 83].

37. Polymerization solvent

37.1 Homogeneous polymerization

Fluoropolymers are very important for their wide applications but being insoluble in solvents except ozone-depleting CFCs, their use is limited [84]. However, fluoropolymers are highly soluble in $scCO_2$. Hence, $scCO_2$ provides an eco-friendly alternative to CFCs as solvent for synthesis of fluoropolymers and telomeres. Many examples have demonstrated that $scCO_2$ can be used in the synthesis of perfluoropolymers with benefits like improved initiator life and efficiency. DeSimone and coworkers carried out the homogeneous solution free-radical polymerization of 1,1-dihydroperfluorooctyl acrylate (FOA) in $scCO_2$ using AIBN [2,2′-azobis(isobutyronitrile)] as an initiator to synthesize fluoropolymers [84]. Copolymerization of FOA with methyl 2-methylprop-2-enoate, vinylbenzene, butyl-2-propenoate, and ethene was also reported [9]. Some other processes that have been efficiently carried out in $scCO_2$ include cationic polymerization of fluorinated vinyl ethers and epoxides, radical polymerization of 1,1-difluoroethylene and tetrafluoroethylene, and copolymerization of tetrafluoroethylene and other alkenes [85].

37.2 Heterogeneous polymerization

Only a few polymers have sufficient solubility in $scCO_2$ to carry out homogeneous polymerization in $scCO_2$. However, many polymers swell and plasticize by exposure to $scCO_2$ which allows for high rates of mass transport between polymeric and bulk phases which can be useful in metal-catalyzed polymerizations in $scCO_2$ [86, 87]. Some examples where $scCO_2$ has been successfully used are ring-opening metathesis polymerization (Ru) [63, 88], olefin copolymerization (Pd) [89], olefin/CO copolymerization (Ni) [90], and formation of poly(phenylacetylene) (Rh) [91]. Moreover, significant changes in stereoregularity of PPA obtained with rhodium catalysts were observed [91]. In addition to these polymerizations, synthesis of hydrocarbon polymers can also be accomplished in $scCO_2$ by the use of an appropriate stabilizer [9].

37.3 Oxidation

As is well known, the use of $scCO_2$ eliminates mass transfer restrictions and increases the effective concentration of reactants for reactions involving gaseous reactants. This is particularly interesting for oxidations where oxygen/substrate ratio can be considerably increased without compromising on explosion limits. Besides, $scCO_2$ has excellent heat transport capacity, thus enabling improved heat control in highly exothermic reactions. Interestingly, the inertness of $scCO_2$ ensures that no side products are formed by the oxidation of the solvent [4]. The direct synthesis of H_2O_2 from H_2 and O_2 using Pd catalyst under homogeneous conditions in $scCO_2$ can be a good example [92].

One approach to initiate oxidation pathways in $scCO_2$ is to use an aldehyde, which is sacrificially oxidized [93–95]. In $scCO_2$, high pressure and stainless steel of reactor initiate the formation of peroxy acyl radicals from a mixture of aldehyde and O_2. These radicals generate peracids or metal-oxo species (in presence of metal catalysts). All these species can initiate oxidation pathways of substrates.

37.4 Diels-Alder reaction

The Diels-Alder reaction involves the addition of an alkene or alkyne to the 1,4 positions of a diene to make six-membered rings. Cycloaddition of cyclopentadiene with methyl acrylate takes place efficiently at ambient temperature in $scCO_2$ similar to organic solvent (toluene) [96]. Similarly, the cycloaddition of maleic anhydride and isoprene to form methyl cyclohexene dicarboxylic anhydride is also known in $scCO_2$ [97]. The cycloaddition reaction of isoprene and methyl propenoate using $AlCl_3$ as a catalyst in $scCO_2$ was studied by Ikushima et al. [98].

37.5 Free-radical reactions

Radical addition reactions do not involve charged species such as nucleophiles or electrophiles. Hence, these reactions can be performed in $scCO_2$ which is a nonpolar solvent [9, 99, 100]. McHugh et al. studied the oxidation of isopropylbenzene in $scCO_2$ [101]. Later on, Johnston et al. performed the dimerization of isoacetophorone and dimerization of benzyl radical [101]. Tanko and Blackert studied bromination of the side chain of methyl and ethylbenzene in $scCO_2$ [102]. They found that selectivity does not change significantly in $scCO_2$ as compared to organic solvents. Their results also showed that brominations using N-bromosuccinimide which are conventionally carried in CCl_4 can be performed in $scCO_2$.

In $scCO_2$, Pd-catalyzed radical addition reaction of CO and CCl_4 to 1-octene was successfully carried out with selectivity different from the case when the reaction was carried out in EtOH [99, 100]. The yield of carbonylation product increased in $scCO_2$ probably due to higher solubility of CO in $scCO_2$. Moreover, in $scCO_2$ temperature and pressure of the reaction mixture influences the reaction selectivity and conversion.

37.6 Use of ionic liquids (ILs) or H_2O along with $scCO_2$ for catalysis

A combination of $scCO_2$ and ionic liquids can be used to isolate products and recycle catalysts. A hydroformylation catalyst (Rh/TPPDS) was in effect immobilized in ILs for continuous flow setup [103]. Occasionally molecular interactions are possible between the catalyst and catalyst-support (ionic liquid). For example, Ni-based precatalyst for the hydrovinylation reaction was dissolved in an ionic liquid which besides a solvent activates the catalyst by abstracting chloride from it in the presence of $scCO_2$ [103, 104].

CO_2 can be used to dissolve nonpolar reactants and products whereas catalysts remain in aqueous medium if these are soluble in H_2O or using colloids [104, 105]. Besides easy product isolation and catalyst recycling, use of emulsions enhances reaction rate by improving the contact between reactants and catalysts [106]. Since reactants and products remain in H_2O, no decrease in pressure is required to obtain the products, which decreases the cost of this method. Moreover, the product is obtained as its aqueous solution free of any catalyst and organic solvent impurity. This is desirable for the processing of chemicals, which are required as aqueous solution for their applications.

37.7 Enzymes

Enzymes on exposure to $scCO_2$ can produce a different effect on their stability and activity depending on the amount of water present, pressure, and temperature of the reaction system [32, 62]. Extreme conditions of temperature and pressure render enzymes denatured and inactive due to changes in a three-dimensional structure [107]. This effect can lead to the inactivation of microbes on $scCO_2$ treatment. Hence, $scCO_2$ treatment has been used as a safe and effective method for sterilization in the processing of foods and bioactive products at a convenient temperature [108–124].

However, if the conditions are modest, $scCO_2$ treatment of enzyme may not change protein structure drastically. The small changes can result as positive regulators of enzyme activity, substrate specificity, selectivity, and attenuated side reactions of the native enzyme. Thus, the increase in enzyme activity in the $scCO_2$ medium cannot be rare [123]. Due to the specific advantage of $scCO_2$ as a convenient nonaqueous solvent for enzyme-catalyzed reactions, its use as an enzyme solubilizing solvent can be interesting for enzymologists for a plethora of biotechnological applications. For example, some enzymes seen suitable for catalyzing reactions in $scCO_2$ include lipases, several phosphatases, dehydrogenases, oxidases, and amylases [62]. These enzymes have been used in various hydrolytic reactions and synthesis of biomolecules. Examples of some industrially important enzyme-catalyzed processes in $scCO_2$ are shown in Table 5 with emphasis on hydrolysis to produce soaps, long-chain alcohols, and monoglycerides [152–154].

Table 5 Selected enzyme catalyzed industrial processes in $scCO_2$ solvent

Entry	Enzyme	Reaction	Reference
1	Lipase (from Candida Antarctica)	Hydrolysis of triglycerol	[125]
2	Lipozyme	Hydrolysis of canola oil	[126, 127]
3	Lipozyme	Hydrolysis of blackcurrant oil	[126, 127]
4	Lipase	Hydrolysis of sunflower oil	[128–130]
5	α-Amylase and glucoamylase	Hydrolysis of starch	[131]
6	Cellulase	Hydrolysis of cellulose	[132]
7	Lipase	Amidation	[132]
8	Pyrrole-2-carboxylate decarboxylase	Carboxylation	[133–137]
9	Lipase	Esterification	[138–144]
10	Lipase	Acetylation	[143]
11	Cutinase	Transesterification	[145, 146]
12	Novozym 435	Esterification	[138, 147]
13	Alcohol dehydrogenase	Reduction	[148–150]
14	Sucrose phosphorylase	Transglycosylation	[151]

37.8 Other reactions in $scCO_2$

The synthesis of $Cp^*Ir(CO)(H)_2$ and $Cp^*Ir(CO)(H)C_2H_5)$ has been carried out in $scCO_2$ by exposing $Cp^*Ir(CO)_2$ to UV radiation in the presence of H_2 or C_2H_6 [155]. Similarly, synthesis of nanoparticles of spherical titanium dioxide particles in $scCO_2$ was carried out by reacting titanium alkoxide with water in the presence of anionic surfactant Zony FSJ $(F(CF_2CF_2)_zCH_2CH_2O)_xP(O)(ONH_4)_y$, where $x=1$ or 2, $y=2$ or 1, and $z=1$–7 [156].

37.9 Extraction

Extractions using organic solvents are common in industry and laboratory for separations. Since these solvents are often difficult to recycle, these methods produce a large amount of solvent waste which is a threat to the environment. Moreover, these methods are expensive, labor intensive, and time-consuming. Hence, there is a serious need for efficient and selective techniques of extraction without involving toxic organic solvents.

SFE is one such efficient, economical, and eco-friendly extraction process. SFE produces extracts with improved purity, with no solvent contamination, and in high yields [157–159]. SFE has been used to extract caffeine [160], essential oils [161–163], cholesterol, pesticides, alkaloids, steroids, taxanes, vitamins, metabolites [7, 164–167]. SFE has also been used for purifying polymers and monomers [7].

37.10 Supercritical fluid nucleation

The use of technologies employing SCFs is exciting for applications where manufacturing very fine particle is desirable such as energetic materials (explosives and propellants) [7, 168, 169]. The conventional methods for these processes are spray drying, grinding and crystallization, and liquid crystallization which suffers from limitations such as requirement of high temperature to remove the solvent, making it unsuitable for thermally labile systems, inefficient control on size of particles particularly when very small size is desirable, and residual solvent in the solid [7, 159, 170–172]. This method has been demonstrated for a wide variety of materials including polymers, dyes, and energetic materials [159, 172].

37.11 Dyeing

Dyeing using classical methods wastes a large amount of fresh water and it is very difficult to treat wastewater contaminated with coloring, dispersing agents, and surfactants. Hence, dyeing using SCFs has been investigated [173–175]. Dyeing of fabrics using SCFs has many rewards as it does not require water, surfactants, dispersing agents, and drying. Some investigators have also reported the solubility data for dyes in $scCO_2$ [176–180].

38. Summary

In replacement of VOCs with environmentally benign solvents, $scCO_2$ is being widely used for carrying out chemical extractions and reactions. The unique physicochemical properties of $scCO_2$ allow it to combine the advantages of both liquid and gaseous state. The solvent desirabilities like nontoxicity, lower viscosity, and surface tension, variable density, nonflammability, high natural abundance, and easy separation make $scCO_2$ an environmentally benign solvent. CO_2 although a nondipolar solvent has a significant quadrupole moment, giving it a "polar attribute." CO_2 also has a significant Lewis acid and base characteristic. The diverse ways by which a solute can interact with $scCO_2$ enable it to dissolve solutes with considerable polarity in addition to nonpolar and gaseous systems. Although $scCO_2$ is now used in commercial methods, however, with modulation of parameters, there can be enormous scope for its use in separation and extraction as a reaction medium for newer systems. A large number of target reactions for desired molecules have been carried in $scCO_2$ with similar or improved results to that of organic solvents in terms of easy catalyst recyclability, enhanced solubility and stability of catalyst, increased reactivity and efficiency, stereoselectivity, eliminating the need for protection/deprotection steps, inducing novel reactivity, removing toxic wastes, and side products. The practical application of $scCO_2$ on the industrial scale is complicated by the multicomponent nature of a catalytic reaction, its progression with time, and high pressure and temperature conditions. The challenges in the use of $scCO_2$ need to be addressed by the coordinated efforts of synthetic, analytical chemists, and chemical engineers. It is anticipated that such efforts would allow for the development of efficient and green methods that are also economically viable. The use of $scCO_2$ as a reaction medium for performing industrially important enzyme-catalyzed processes is going to become a vital, economic, sustainable, and environment-friendly synthetic protocol.

Acknowledgment

Mr. Mudasir Ahmad is thankful to Indian Council of Medical Research, Government of India for the award of Senior Research Fellowship (3/1/2/81/2018-nut).

References

[1] P. Raveendran, Y. Ikushima, S.L. Wallen, Polar attributes of supercritical carbon dioxide, Acc. Chem. Res. 38 (2005) 478–485, https://doi.org/10.1021/ar040082m.
[2] D. Ballivet-Tkatchenko, S. Camy, J.S. Condoret, Carbon dioxide, a solvent and synthon for green chemistry, Environ. Chem. Green Chem. Pollut. Ecosyst. (2005) 541–552, https://doi.org/10.1007/3-540-26531-7_49.
[3] H. Black, Supercritical carbon dioxide: the "Greener" solvent, Environ. Sci. Technol. 30 (1996) 124A–127A, https://doi.org/10.1021/es962136w.
[4] W. Leitner, Supercritical carbon dioxide as a green reaction medium for catalysis, Acc. Chem. Res. 35 (2002) 746–756, https://doi.org/10.1021/ar010070q.
[5] R. Skouta, Selective chemical reactions in supercritical carbon dioxide, water, and ionic liquids, Green Chem. Lett. Rev. 2 (2009) 121–156, https://doi.org/10.1080/17518250903230001.
[6] S. Mayadevi, Reactions in supercritical carbon dioxide, Indian J. Chem. Sect. A Inorg. Phys. Theor. Anal. Chem. 51 (2012) 1298–1305, https://doi.org/10.1002/anie.199414521.
[7] S.M. Pourmortazavi, S.S. Hajimirsadeghi, Application of supercritical carbon dioxide in energetic materials processes: a review, Ind. Eng. Chem. Res. 44 (2005) 6523–6533, https://doi.org/10.1021/ie0503242.
[8] S. Ag, Supercritical CO_2: a green solvent, Chem. Eng. (2010) 1–9.https://www.chemengonline.com/supercriticalco2agreensolvent/?printmode=1.
[9] Z. Gang, J. Huan-feng, C. Ming-cai, Chemical reactions in supercritical carbon dioxide, Arkivoc 2003 (2003) 191–198.
[10] X. Han, M. Poliakoff, Continuous reactions in supercritical carbon dioxide: problems, solutions and possible ways forward, Chem. Soc. Rev. 41 (2012) 1428–1436, https://doi.org/10.1039/c2cs15314a.
[11] H. Sato, N. Matubayasi, M. Nakahara, F. Hirata, Which carbon oxide is more soluble? Ab initio study on carbon monoxide and dioxide in aqueous solution, Chem. Phys. Lett. 323 (2000) 257–262, https://doi.org/10.1016/S0009-2614(00)00508-X.
[12] M.A. Blatchford, P. Raveendran, S.L. Wallen, Spectroscopic studies of model carbonyl compounds in CO2: evidence for cooperative C-H···O interactions, J. Phys. Chem. A 107 (2003) 10311–10323, https://doi.org/10.1021/jp027208m.
[13] A. Fujii, T. Ebata, N. Mikami, Direct observation of weak hydrogen bonds in microsolvated phenol: infrared of OH stretching vibrations of phenol-CO and -CO2 in S0 and D0, J. Phys. Chem. A 106 (2002) 10124–10129, https://doi.org/10.1021/jp0212601.
[14] M.A. Blatchford, P. Raveendran, S.L. Wallen, Raman spectroscopic evidence for cooperative C-H···O interactions in the acetaldehyde-CO2complex, J. Am. Chem. Soc. 124 (2002) 14818–14819, https://doi.org/10.1021/ja027337g.

[15] A. Iwasaki, A. Fujii, T. Watanabe, T. Ebata, N. Mikami, Infrared spectroscopy of hydrogen-bonded phenol—amine clusters in supersonic jets, J. Phys. Chem. 100 (1996) 16053–16057, https://doi.org/10.1021/jp960711c.

[16] P. Raveendran, S.L. Wallen, Cooperative C-H0 ··· O hydrogen bonding in CO2-Lewis base complexes: Implications for solvation in supercritical CO2, J. Am. Chem. Soc. 124 (2002) 12590–12599, https://doi.org/10.1021/ja0174635.

[17] P. Raveendran, S.L. Wallen, Sugar acetates as novel, renewable CO2-philes, J. Am. Chem. Soc. 124 (2002) 7274–7275, https://doi.org/10.1021/ja025508b.

[18] A.R.C. Duarte, P. Coimbra, H.C. De Sousa, C.M.M. Duarte, Solubility of flurbiprofen in supercritical carbon dioxide, J. Chem. Eng. Data 49 (2004) 449–452, https://doi.org/10.1021/je034099b.

[19] T. Sarbu, T. Styranec, E.J. Beckman, Non-fluorous polymers with very high solubility in supercritical CO2down to low pressures, Nature 405 (2000) 165–168, https://doi.org/10.1038/35012040.

[20] J.F. Liu, H.J. Yang, W. Wang, Z. Li, Solubilities of amide compounds in supercritical carbon dioxide, J. Chem. Eng. Data 53 (2008) 2189–2192, https://doi.org/10.1021/je800434j.

[21] L. Jia, H. Jiang, J. Li, Selective carbonylation of norbornene in scCO2, Green Chem. 1 (1999) 91–93, https://doi.org/10.1039/a901396b.

[22] W. Cross, A. Akgerman, C. Erkey, Determination of metal-chelate complex solubilities in supercritical carbon dioxide, Ind. Eng. Chem. Res. 35 (1996) 1765–1770, https://doi.org/10.1021/ie950515c.

[23] Q. Li, C. Zhong, Z. Zhang, Y. Liu, Q. Zhou, An equilibrium model for the correlation of the solubility of solids in supercritical fluids with cosolvent, Sep. Sci. Technol. 38 (2003) 1705–1719, https://doi.org/10.1081/SS-120019404.

[24] Z.B. loubna Nasri, Solubility of Solids in Supercritical Fluids: The Mendez-Santiago Teja Model Revisited, Adv. Chem. Eng. Sci. 5 (2015) 413–423, https://doi.org/10.4236/aces.2015.53042.

[25] J. Méndez-Santiago, A.S. Teja, The solubility of solids in supercritical fluids, Fluid Phase Equilib. 158–160 (1999) 501–510, https://doi.org/10.1016/S0378-3812(99)00154-5.

[26] I. Ashour, R. Almehaideb, S.E. Fateen, G. Aly, Representation of solid-supercritical fluid phase equilibria using cubic equations of state, Fluid Phase Equilib. 167 (2000) 41–61, https://doi.org/10.1016/S0378-3812(99)00314-3.

[27] D.J. Vezzetti, Solubility of solids in supercritical gases, J. Chem. Phys. 77 (1982) 1512–1516, https://doi.org/10.1063/1.443931.

[28] S.K. Kumar, K.P. Johnston, Modelling the solubility of solids in supercritical fluids with density as the independent variable, J. Supercrit. Fluids 1 (1988) 15–22, https://doi.org/10.1016/0896-8446(88)90005-8.

[29] Q. Lan, C. Liu, F. Yang, S. Liu, J. Xu, D. Sun, Synthesis of bilayer oleic acid-coated Fe3O4 nanoparticles and their application in pH-responsive Pickering emulsions, J. Colloid Interface Sci. 310 (2007) 260–269, https://doi.org/10.1080/00986448708960487.

[30] K.D. Bartle, A.A. Clifford, S.A. Jafar, G.F. Shilstone, Solubilities of Solids and Liquids of Low Volatility in Supercritical Carbon Dioxide, J. Phys. Chem. Ref. Data 20 (1991) 713–756, https://doi.org/10.1063/1.555893.

[31] M. Ahmadi Sabegh, H. Rajaei, A. Zeinolabedini Hezave, F. Esmaeilzadeh, Amoxicillin solubility and supercritical carbon dioxide, J. Chem. Eng. Data 57 (2012) 2750–2755, https://doi.org/10.1021/je3006826.

[32] N. Budisa, D. Schulze-Makuch, Supercritical carbon dioxide and its potential as a life-sustaining solvent in a planetary environment, Life (Basel) 4 (2014) 331–340, https://doi.org/10.3390/life4030331.

[33] S.J. Macnaughton, I. Kikic, N.R. Foster, P. Alessi, A. Cortesi, I. Colombo, Solubility of anti-inflammatory drugs in supercritical carbon dioxide, J. Chem. Eng. Data 41 (1996) 1083–1086, https://doi.org/10.1021/je960103q.

[34] H. Xing, Y. Yang, B. Su, M. Huang, Q. Ren, Solubility of artemisinin in supercritical carbon dioxide, J. Chem. Eng. Data 48 (2003) 330–332, https://doi.org/10.1021/je025575l.

[35] R.D. Weinstein, K.R. Muske, J. Moriarty, E.K. Schmidt, The solubility of benzocaine, lidocaine, and procaine in liquid and supercritical carbon dioxide, J. Chem. Eng. Data 49 (2004) 547–552, https://doi.org/10.1021/je034163p.

[36] S.L.J. Yun, K.K. Liong, G.S. Gurdial, N.R. Foster, Solubility of cholesterol in supercritical carbon dioxide, Ind. Eng. Chem. Res. 30 (1991) 2476–2482, https://doi.org/10.1021/ie00059a018.
[37] R.B. Chim, M.B.C. De Matos, M.E.M. Braga, A.M.A. Dias, H.C. De Sousa, Solubility of dexamethasone in supercritical carbon dioxide, J. Chem. Eng. Data 57 (2012) 3756–3760, https://doi.org/10.1021/je301065f.
[38] R. Heryanto, E.C. Abdullah, M. Hasan, Solubility of isoniazid in supercritical carbon dioxide, J. Chem. Eng. Data 55 (2010) 2306–2309, https://doi.org/10.1021/je8001656.
[39] Y. Bai, H.J. Yang, C. Quan, C.Y. Guo, Solubilities of 2,2′-bipyridine and 4,4′-dimethyl-2,2′- bipyridine in supercritical carbon dioxide, J. Chem. Eng. Data 52 (2007) 2074–2076, https://doi.org/10.1021/je700269m.
[40] L. Barna, J.M. Blanchard, E. Rauzy, C. Berro, Solubility of flouranthene, chrysene, and triphenylene in supercritical carbon dioxide, J. Chem. Eng. Data 41 (1996) 1466–1469, https://doi.org/10.1021/je960189n.
[41] M. Sahihi, H.S. Ghaziaskar, M. Hajebrahimi, Solubility of maleic acid in supercritical carbon dioxide, J. Chem. Eng. Data 55 (2010) 2596–2599, https://doi.org/10.1021/je9008393.
[42] H. Sovová, J. Jež, Solubility of menthol in supercritical carbon dioxide, J. Chem. Eng. Data 39 (1994) 840–841, https://doi.org/10.1021/je00016a045.
[43] S. Ismadji, Solubility of methyl salicylate in supercritical carbon dioxide at several temperatures, J. Chem. Eng. Data 53 (2008) 2207–2210, https://doi.org/10.1021/je800476n.
[44] R. Ch, C. Garlapati, G. Madras, E. Section, Solubility of n-(4-ethoxyphenyl)ethanamide in supercritical carbon dioxide, J. Chem. Eng. Data 55 (2010) 1437–1440.
[45] B. Su, H. Xing, Y. Yang, Q. Ren, Solubility of oxymatrine in supercritical carbon dioxide, J. Chem. Eng. Data 53 (2008) 1672–1674, https://doi.org/10.1021/je800179c.
[46] H.L. Wang, J.R. Sang, L.T. Guo, J. Zhu, J.S. Jin, Solubility of polyacrylamide in supercritical carbon dioxide, J. Chem. Eng. Data 62 (2017) 341–347, https://doi.org/10.1021/acs.jced.6b00677.
[47] L.S. Lee, J.H. Fu, H.L. Hsu, Solubility of solid 1,4-dimethoxybenzene in supercritical carbon dioxide, J. Chem. Eng. Data 45 (2000) 358–361, https://doi.org/10.1021/je9902591.
[48] Q. Ren, B. Su, M. Huang, P. Wu, Solubility of Troeger's base in supercritical carbon dioxide, J. Chem. Eng. Data 45 (2000) 464–466, https://doi.org/10.1021/je9902903.
[49] M.G. Arenas-Quevedo, O. Elizalde-Solis, A. Zúñiga-Moreno, R. Macías-Salinas, F. García-Sánchez, Solubilities of palmitic acid + capsaicin in supercritical carbon dioxide, J. Chem. Eng. Data 62 (2017) 3861–3871, https://doi.org/10.1021/acs.jced.7b00576.
[50] J. García-González, M.J. Molina, F. Rodríguez, F. Mirada, Solubilities of phenol and pyrocatechol in supercritical carbon dioxide, J. Chem. Eng. Data 46 (2001) 918–921, https://doi.org/10.1021/je0003795.
[51] N. Bahramifar, Y. Yamini, A. Ramazani, N. Noshiranzadeh, Solubilities of some 2H-chromene derivatives in supercritical carbon dioxide, J. Chem. Eng. Data 48 (2003) 1104–1108, https://doi.org/10.1021/je0201875.
[52] M. Johannsen, G. Brunner, Solubilities of the fat-soluble vitamins A, D, E, and K in supercritical carbon dioxide, J. Chem. Eng. Data 42 (1997) 106–111, https://doi.org/10.1021/je960219m.
[53] J.W. Lee, J.M. Min, H.K. Bae, Solubility measurement of disperse dyes in supercritical carbon dioxide, J. Chem. Eng. Data 44 (1999) 684–687, https://doi.org/10.1021/je9802930.
[54] I. Goodarznia, F. Esmaeilzadeh, Solubility of an anthracene, phenanthrene, and carbazole mixture in supercritical carbon dioxide, J. Chem. Eng. Data 47 (2002) 333–338, https://doi.org/10.1021/je010093f.
[55] A. Laitinen, M. Jäntti, Solubility of 6-caprolactam in supercritical carbon dioxide, J. Chem. Eng. Data 41 (1996) 1418–1420, https://doi.org/10.1021/je9600313.
[56] P.M. Agrawal, B.M. Rice, D.C. Sorescu, D.L. Thompson, NPT-MC simulations of enhanced solubility of RDX in polar-modified supercritical CO2, Fluid Phase Equilib. 166 (1999) 1–19, https://doi.org/10.1016/S0378-3812(99)00292-7.
[57] P.M. Agrawal, B.M. Rice, D.C. Sorescu, D.L. Thompson, Models for predicting solubilities of 2,4,6-trinitrotoluene (TNT) and 1,3,5-trinitro-1,3,5-s-triazine (RDX) in supercritical CO2: Isothermal-

isobaric Monte Carlo simulations, Fluid Phase Equilib. 187–188 (2001) 139–153, https://doi.org/10.1016/S0378-3812(01)00533-7.
[58] Y. Lin, N.G. Smart, C.M. Wai, Supercritical fluid extraction and chromatography of metal chelates and organometallic compounds, Trends Anal. Chem. 14 (1995) 123–133, https://doi.org/10.1016/0165-9936(95)94045-G.
[59] K.E. Laintz, C.M. Wai, C.R. Yonker, R.D. Smith, Extraction of metal ions from liquid and solid materials by supercritical carbon dioxide, Anal. Chem. 64 (1992) 2875–2878, https://doi.org/10.1021/ac00046a039.
[60] Y. Lin, N.G. Smart, C.M. Wai, Supercritical fluid extraction of uranium and thorium from nitric acid solutions with organophosphorus reagents, Environ. Sci. Technol. 29 (1995) 2706–2708, https://doi.org/10.1021/es00010a036.
[61] W.C. Andersen, R.E. Sievers, A.F. Lagalante, T.J. Bruno, Solubilities of cerium(IV), terbium(III), and iron(III) ??-diketonates in supercritical carbon dioxide, J. Chem. Eng. Data 46 (2001) 1045–1049, https://doi.org/10.1021/je000257c.
[62] Z. Wimmer, M. Zarevúcka, A review on the effects of supercritical carbon dioxide on enzyme activity, Int. J. Mol. Sci. 11 (2010) 233–253, https://doi.org/10.3390/ijms11010233.
[63] A. Fürstner, L. Ackermann, K. Beck, H. Hori, D. Koch, K. Langemann, M. Liebl, C. Six, W. Leitner, Olefin metathesis in supercritical carbon dioxide, J. Am. Chem. Soc. 123 (2001) 9000–9006, https://doi.org/10.1021/ja010952k.
[64] C.A.G. Carter, R.T. Baker, S.P. Nolan, W. Tumas, Enhanced regioselectivity of rhodium-catalysed alkene hydroboration in supercritical carbon dioxide, Chem. Commun. (2000) 347–348, https://doi.org/10.1039/a909636a.
[65] L.K. Yeung, C. Ted Lee Jr., K.P. Johnston, R.M. Crooks, Catalysis in supercritical CO2 using dendrimer-encapsulated palladium nanoparticles, Chem. Commun. (2001) 2290–2291, https://doi.org/10.1039/b106594g.
[66] R. Scott Oakes, A.A. Clifford, K.D. Bartle, M.T. Pett, C.M. Rayner, Sulfur oxidation in supercritical carbon dioxide: dramatic pressure dependant enhancement of diastereoselectivity for sulfoxidation of cysteine derivatives, Chem. Commun. (1999) 247–248, https://doi.org/10.1039/a809434i.
[67] S. Kainz, A. Brinkmann, W. Leitner, A. Pfaltz, Iridium-catalyzed enantioselective hydrogenation of imines in supercritical carbon dioxide, J. Am. Chem. Soc. 121 (1999) 6421–6429, https://doi.org/10.1021/ja984309i.
[68] J.C. Choi, Y. Kobayashi, T. Sakakura, C-H bond activation in dense carbon dioxide: Rhodium-catalyzed methane carbonylation and alkane dehydrogenation, J. Organomet. Chem. 66 (2001) 5262–5263, https://doi.org/10.1021/jo0102967.
[69] K. Wittmann, W. Wisniewski, R. Mynott, W. Leitner, C.L. Kranemann, T. Rische, P. Eilbracht, S. Kluwer, J.M. Ernsting, C.J. Elsevier, Supercritical carbon dioxide as solvent and temporary protecting group for rhodium-catalyzed hydroaminomethylation, Chem. A Eur. J. 7 (2001) 4584–4589, https://doi.org/10.1002/1521-3765(20011105)7:21<4584::AID-CHEM4584>3.0.CO;2-P.
[70] N.J. Meehan, A.J. Sandee, J.N.H. Reek, P.C.J. Kamer, P.W.N.M. Van Leeuwen, M. Poliakoff, Continuous, selective hydroformylation in supercritical carbon dioxide using an immobilised homogeneous catalyst, Chem. Commun. (2000) 1497–1498, https://doi.org/10.1039/b002526g.
[71] R.J. Sowden, M.F. Sellin, N. De Blasio, D.J. Cole-Hamilton, Carbonylation of methanol in supercritical CO2 catalysed by a supported rhodium complex, Chem. Commun. (1999) 2511–2512, https://doi.org/10.1039/a906653e.
[72] L. Schmid, M. Rohr, A. Baiker, A mesoporous ruthenium silica hybrid aerogel with outstanding catalytic properties in the synthesis of N,N-diethylformamide from CO2, H2and diethylamine, Chem. Commun. (1999) 2303–2304, https://doi.org/10.1039/a906956i.
[73] D.K. Morita, D.R. Pesiri, S.A. David, W.H. Glaze, W. Tumas, Palladium-catalyzed cross-coupling reactions in supercritical carbon dioxide, Chem. Commun. (1998) 1397–1398, https://doi.org/10.1039/a802621a.
[74] P.G. Jessop, T. Ikariya, R. Noyori, Homogeneous catalytic hydrogenation of supercritical carbon dioxide, Nature 368 (1994) 231–233, https://doi.org/10.1038/368231a0.

[75] W. Leitner, Carbon dioxide as a raw material: the synthesis of formic acid and its derivatives from CO2, Angew. Chem. Int. Ed. Eng. 34 (1995) 2207–2221, https://doi.org/10.1002/anie.199522071.
[76] P.G. Jessop, Y. Hsiao, T. Ikariya, R. Noyori, Homogeneous Catalysis in Supercritical Fluids: Hydrogenation of Supercritical Carbon Dioxide to Formic Acid, Alkyl Formates, and Formamides, J. Am. Chem. Soc. 118 (1996) 344–355, https://doi.org/10.1021/ja953097b.
[77] M.F. Sellin, D.J. Cole-Hamilton, Hydroformylation reactions in supercritical carbon dioxide using insoluble metal complexes, J. Chem. Soc. Dalton Trans. (2000) 1681–1683, https://doi.org/10.1039/b001317j.
[78] K. Schwirten, R. Kummer, W. Richter US Patent 4,568,653 1986, Working up of hydroformylation or carbonylation reaction mixtures, Google Patents. (n.d.). https://patents.google.com/patent/US4568653A/en, Accessed 26 December 2018.
[79] D. Koch, W. Leitner, Rhodium-catalyzed hydroformylation in supercritical carbon dioxide, J. Am. Chem. Soc. 120 (1998) 13398–13404, https://doi.org/10.1021/ja980729w.
[80] G. Franciò, K. Wittmann, W. Leitner, Highly efficient enantioselective catalysis in supercritical carbon dioxide using the perfluoroalkyl-substituted ligand (R,S)-3-H2F6-BINAPHOS, J. Organomet. Chem. 621 (2001) 130–142, https://doi.org/10.1016/S0022-328X(00)00778-6.
[81] M.J. Burk, M.F. Gross, S. Feng, W. Tumas, Asymmetric catalytic hydrogenation reactions in supercritical carbon dioxide, J. Am. Chem. Soc. 117 (1995) 8277–8278, https://doi.org/10.1021/ja00136a029.
[82] S. Lange, W. Leitner, Mechanistic aspects of hydrogen addition during enantioselective rhodium-catalysed reduction of C=C double bonds with formic acid/triethylamine or molecular hydrogen, in: Dalt. Trans., 2002, pp. 752–758, https://doi.org/10.1039/b108774f.
[83] S. Lange, A. Brinkmann, P. Trautner, K. Woelk, J. Bargon, W. Leitner, Mechanistic aspects of dihydrogen activation and transfer during asymmetric hydrogenation in supercritical carbon dioxide, Chirality 12 (2000) 450–457, https://doi.org/10.1002/(SICI)1520-636X(2000)12:5/6<450::AID-CHIR26>3.0.CO;2-H.
[84] S. Solomon, R.R. Garcia, F.S. Rowland, D.J. Wuebbles, On the depletion of the antartic ozone, Nature 321 (1986) 755. https://www.nature.com/articles/321755a0. Accessed 26 December 2018.
[85] M.R. Clark, J.M. DeSimone, Cationic polymerization of vinyl and cyclic ethers in supercritical and liquid carbon dioxide, Macromolecules 28 (1995) 3002–3004, https://doi.org/10.1021/ma00112a059.
[86] J.L. Kendall, D.A. Canelas, J.L. Young, J.M. DeSimone, Dispersion polymerizations in supercritical carbon dioxide, Science (80-.) 265 (1994) 356–359, https://doi.org/10.1021/cr9700336.
[87] S.P. Nalawade, F. Picchioni, L.P.B.M. Janssen, Supercritical carbon dioxide as a green solvent for processing polymer melts: processing aspects and applications, Prog. Polym. Sci. 31 (2006) 19–43, https://doi.org/10.1016/j.progpolymsci.2005.08.002.
[88] J.G. Hamilton, J.J. Rooney, J.M. DeSimone, C. Mistele, Stereochemistry of ring-opened metathesis polymers prepared in liquid CO2 at high pressure using Ru(H 2 O) 6 (Tos) 2 as catalyst, Macromolecules 31 (1998) 4387–4389, https://doi.org/10.1021/ma9715536.
[89] T.J. de Vries, M.A.G. Vorstman, J.T.F. Keurentjes, R. Duchateau, Polymerization of olefins catalyzed by a palladium complex in supercritical carbon dioxide, Chem. Commun. (2000) 263–264, https://doi.org/10.1039/a908966g.
[90] W. Kläui, J. Bongards, G.J. Reiß, Novel nickel(II) complexes for the catalytic copolymerization of ethylene and carbon monoxide: polyketone synthesis in supercritical carbon dioxide, Angew. Chem. Int. Ed. 39 (2000) 3894–3896, https://doi.org/10.1002/1521-3773(20001103)39:21<3894::AID-ANIE3894>3.0.CO;2-K.
[91] H. Hori, C. Six, W. Leitner, Rhodium-catalyzed phenylacetylene polymerization in compressed carbon dioxide, Macromolecules 32 (1999) 3178–3182, https://doi.org/10.1021/ma981723u.
[92] D. Hâncu, E.J. Beckman, Generation of hydrogen peroxide directly from H2 and O2 using CO2 as the solvent, Green Chem. 3 (2001) 80–86, https://doi.org/10.1039/b008515o.
[93] F. Loeker, W. Leitner, Steel-promoted oxidation of olefins in supercritical carbon dioxide using dioxygen in the presence of aldehydes, Chem. A Eur. J. 6 (2000) 2011–2015, https://doi.org/10.1002/1521-3765(20000602)6:11<2011::AID-CHEM2011>3.0.CO;2-L.

[94] N. Theyssen, W. Leitner, Selective oxidation of cyclooctane to cyclootanone with molecular oxygen in the presence of compressed carbon dioxide, Chem. Commun. 2 (2002) 410–411, https://doi.org/10.1039/b111212k.

[95] C. Bolm, C. Palazzi, G. Franciò, W. Leitner, Baeyer–Villiger oxidation in compressed CO2Safety warning: the use of compressed gases and especially oxygen in the presence of organic substrates requires appropriate safety precautions and must be carried out using suitable equipment only, Chem. Commun. (2002) 1588–1589, https://doi.org/10.1039/b202727e.

[96] J.A. Hyatt, Liquid and supercritical carbon dioxide as organic solvents, J. Organomet. Chem. 49 (1984) 5097–5101, https://doi.org/10.1021/jo00200a016.

[97] M.E. Paulaitis, G.C. Alexander, Reactions in supercritical fluids. A case study of the thermodynamic solvent effects on a Diels-Alder reaction in supercritical carbon dioxide, Pure Appl. Chem. 59 (1987) 61–68, https://doi.org/10.1351/pac198759010061.

[98] Y. Ikushima, N. Saito, M. Arai, Supercritical carbon dioxide as reaction medium: examination of its solvent effects in the near-critical region, J. Phys. Chem. 96 (1992) 2293–2297, https://doi.org/10.1021/j100184a050.

[99] J.-H. Sun, H.-F. Jiang, Palladium-catalyzed addition reaction of 1-Octene, carbon tetrachloride and carbon monooxide in alcohols, Chin. J. Chem. 19 (2010) 518–521, https://doi.org/10.1002/cjoc.20010190516.

[100] S. Peach, S. Liping, S. Qiaozhen, Y. Liu, X. Majian, T. Meng, Y.S. Zhixing, Preparation of nano-TiO2 catalyst and its photocatalytic activity, Lanzhou Univ. Nat. Sci. Ed. 36 (2000) 75–78. http://www.cqvip.com/qk/95222x/200006/8576333.html. Accessed 26 December 2018.

[101] G.J. Suppes, R.N. Occhiogrosso, M.A. McHugh, Oxidation of Cumene in Supercritical Reaction Media, Ind. Eng. Chem. Res. 28 (1989) 1152–1156, https://doi.org/10.1021/ie00092a006.

[102] J.M. Tanko, J.F. Blackert, Free-radical side-chain bromination of alkylaromatics in supercritical carbon dioxide, Science (80-.) 263 (1994) 203–205, https://doi.org/10.1126/science.263.5144.203.

[103] C. Bianchini, G. Giambastiani, Continuous flow homogeneous catalysis: hydroformylation of alkenes in supercritical fluid-ionic liquid biphasic mixtures: supported ionic liquid catalysis—a new concept for homogeneous hydroformylation catalysis, Chemtracts 16 (2003). http://pubs.rsc.org/en/content/articlehtml/2001/cc/b101046h. Accessed 26 December 2018.

[104] A. Wegner, W. Leitner, Nickel-catalysed enantioselective hydrovinylation of styrenes in liquid or supercritical carbon dioxide, Chem. Commun. (1999) 1583–1584, https://doi.org/10.1039/a904799i.

[105] A. Bösmann, G. Franci, E. Janssen, M. Solinas, W. Leitner, P. Wasserscheid, Activation, tuning, and immobilization of homogeneous catalysts in an ionic liquid/compressed CO2 continuous-flow system, Angew. Chem. Int. Ed. 40 (2001) 2697–2699, https://doi.org/10.1002/1521-3773(20010716)40:14<2697::AID-ANIE2697>3.0.CO;2-W.

[106] G.B. Jacobson, C.T. Lee Jr., K.P. Johnston, W. Tumas, Enhanced catalyst reactivity and separations using water/carbon dioxide emulsions, J. Am. Chem. Soc. 121 (1999) 11902–11903, https://doi.org/10.1021/ja993208x.

[107] J.A. m De Bont, Solvent-tolerant bacteria in biocatalysis, Trends Biotechnol. 16 (1998) 493–499, https://doi.org/10.1016/S0167-7799(98)01234-7.

[108] S. Damar, M.O. Balaban, Review of dense phase CO2 technology: microbial and enzyme inactivation, and effects on food quality, J. Food Sci. 71 (2006) (2006). https://onlinelibrary.wiley.com/doi/abs/10.1111/j.1365-2621.2006.tb12397.x. Accessed 26 December 2018.

[109] L. Garcia-Gonzalez, A. Geeraerd, High pressure carbon dioxide inactivation of microorganisms in foods, Int. J. Food Microbiol. 117 (2007) 1–28. https://www.sciencedirect.com/science/article/pii/S0168160507001584. Accessed 26 December 2018.

[110] O. Erkmen, Inactivation of Salmonella typhimurium by high pressure carbon dioxide, Food Microbiol. 17 (2000) 225–232, https://doi.org/10.1006/fmic.1999.0308.

[111] S.R. Kim, M.S. Rhee, B.C. Kim, H. Lee, K.H. Kim, Modeling of the inactivation of Salmonella typhimurium by supercritical carbon dioxide in physiological saline and phosphate-buffered saline, J. Microbiol. Methods 70 (2007) 132–141, https://doi.org/10.1016/j.mimet.2007.04.003.

[112] O. Erkmen, Antimicrobial effects of pressurised carbon dioxide on Brochothrix thermosphacta in broth and foods, J. Sci. Food Agric. 80 (2000) 1365–1370, https://doi.org/10.1002/1097-0010(200007)80:9<1365::AID-JSFA652>3.0.CO;2-P.

[113] O. Erkmen, Antimicrobial effect of pressurized carbon dioxide on Enterococcus faecalis in physiological saline and foods, J. Sci. Food Agric. 80 (2000) 465–470, https://doi.org/10.1002/(SICI)1097-0010(200003)80:4<465::AID-JSFA550>3.0.CO;2-E.

[114] O. Erkmen, Effect of carbon dioxide pressure on Listeria monocytogenes in physiological saline and foods, Food Microbiol. 17 (2000) 589–596, https://doi.org/10.1006/fmic.2000.0361.

[115] O. Erkmen, Effects of high-pressure carbon dioxide on Escherichia coli in nutrient broth and milk, Int. J. Food Microbiol. 65 (2001) 131–135, https://doi.org/10.1016/S0168-1605(00)00499-2.

[116] J.C. Liao, E.N. Lightfoot, Extending the quasi-steady state concept to analysis of metabolic networks, J. Theor. Biol. 126 (1987) 253–273, https://doi.org/10.1016/S0022-5193(87)80234-5.

[117] D. Kincal, W.S. Hill, M.O. Balaban, K.M. Portier, C.I. Wei, M.R. Marshall, A continuous high pressure carbon dioxide system for microbial reduction in orange juice, J. Food Sci. 70 (2005) M249–M254, https://doi.org/10.1111/j.1365-2621.2005.tb09979.x.

[118] J. Zhang, T.A. Davis, M.A. Matthews, M.J. Drews, M. LaBerge, Y.H. An, Sterilization using high-pressure carbon dioxide, J. Supercrit. Fluids 38 (2006) 354–372, https://doi.org/10.1016/j.supflu.2005.05.005.

[119] S.-I. Hong, W.-S. Park, Y.-R. Pyun, Inactivation of Lactobacillussp. from Kimchiby high pressure carbon dioxide, LWT Food Sci. Technol. 30 (1997) 681–685, https://doi.org/10.1006/fstl.1997.0250.

[120] A. Isenschmid, I.W. Marison, U. von Stockar, The influence of pressure and temperature of compressed CO_2on the survival of yeast cells, J. Biotechnol. 39 (1995) 229–237, https://doi.org/10.1016/0168-1656(95)00018-L.

[121] S. Hong, Y.-R. Pyun, Inactivation Kinetics of Lactobacillus plantarum by high pressure carbon dioxide, J. Food Sci. 64 (1999) (1999) 728–733.https://onlinelibrary.wiley.com/doi/abs/10.1111/j.1365-2621.1999.tb15120.x. Accessed 26 December 2018.

[122] W.Y. Jung, Y.M. Choi, M.S. Rhee, Potential use of supercritical carbon dioxide to decontaminate Escherichia coli O157:H7, Listeria monocytogenes, and Salmonella typhimurium in alfalfa sprouted seeds, Int. J. Food Microbiol. 136 (2009) 66–70, https://doi.org/10.1016/j.ijfoodmicro.2009.08.014.

[123] K.A.M.I.H.I.R.A. Masamichi, T.A.N.I.G.U.C.H.I. Masayuki, K.O.B.A.Y.A.S.H.I. Takeshi, Sterilization of microorganisms with supercritical carbon dioxide (food & nutrition), Agric. Biol. Chem. 51 (1987) 407–412. https://www.jstage.jst.go.jp/article/bbb1961/51/2/51_2_407/_article/-char/ja/. Accessed 26 December 2018.

[124] O. Erkmen, H. Karaman, Kinetic studies on the high pressure carbon dioxide inactivation of Salmonella typhimurium, J. Food Eng. 50 (2001) 25–28, https://doi.org/10.1016/S0260-8774(00)00191-6.

[125] J.W. Hampson, T.A. Foglia, Effect of moisture content on immobilized lipase-catalyzed triacylglycerol hydrolysis under supercritical carbon dioxide flow in a tubular fixed-bed reactor, JAOCS, J. Am. Oil Chem. Soc. 76 (1999) 777–781, https://doi.org/10.1007/s11746-999-0065-8.

[126] K. Rezaei, F. Temelli, On-line extraction-reaction of canola oil using immobilized lipase in supercritical CO_2, J. Supercrit. Fluids 19 (2001) 263–274, https://doi.org/10.1016/S0896-8446(00)00099-1.

[127] K. Rezaei, F. Temelli, Lipase-catalyzed hydrolysis of canola oil in SC-CO_2, J. Am. Oil Chem. Soc. 77 (2000) 903–909.https://link.springer.com/article/10.1007/s11746-000-0143-0.

[128] M. Primožič, M. Habulin, Ž. Knez, Modeling of kinetics for the enzymatic hydrolysis of sunflower oil in a high-pressure reactor, J. Am. Oil Chem. Soc. 82 (2005) 543–547, https://doi.org/10.1007/s11746-005-1107-0.

[129] M. Primožič, M. Habulin, Ž. Knez, Parameter optimization for the enzymatic hydrolysis of sunflower oil in high-pressure reactors, J. Am. Oil Chem. Soc. 80 (2003) 643–646, https://doi.org/10.1007/s11746-003-0752-7.

[130] M. Primožič, M. Habulin, Ž. Knez, Thermodynamic properties of the enzymatic hydrolysis of sunflower oil in high-pressure reactors, J. Am. Oil Chem. Soc. 80 (2003) 785–788, https://doi.org/10.1007/s11746-003-0772-3.

[131] H.S. Lee, W.G. Lee, S.W. Park, H. Lee, H.N. Chang, Starch hydrolysis using enzyme in supercritical carbon dioxide, Biotechnol. Tech. 7 (1993) 267–270, https://doi.org/10.1007/BF00150896.
[132] M. Paljevac, M. Primožič, M. Habulin, Z. Novak, Ž. Knez, Hydrolysis of carboxymethyl cellulose catalyzed by cellulase immobilized on silica gels at low and high pressures, J. Supercrit. Fluids 43 (2007) 74–80, https://doi.org/10.1016/j.supflu.2007.05.006.
[133] I. Sharma, D. Paul, Prevalence of community acquired urinary tract infections in silchar medical college, Assam, India and its antimicrobial susceptibility profile, Indian J. Med. Sci. 66 (2012) 273, https://doi.org/10.1016/S1381-1177(00)00038-2.
[134] M. Wieser, T. Yoshida, T. Nagasawa, B. Assoc Italiana, S.C.I.W. Bioseparazioni, B. Party Appl, Carbon dioxide fixation by reversible pyrrole-2-carboxylate decarboxylase and its application, in: 4th Int. Symp. Biocatal. Biotransformations, 1999, pp. 179–184. https://www.sciencedirect.com/science/article/pii/S1381117700000382. Accessed 26 December 2018.
[135] M. Wieser, N. Fujii, T. Yoshida, T. Nagasawa, Carbon dioxide fixation by reversible pyrrole-2-carboxylate decarboxylase from Bacillus megaterium PYR2910, Eur. J. Biochem. 257 (1998) 495–499, https://doi.org/10.1046/j.1432-1327.1998.2570495.x.
[136] T. Matsuda, Y. Ohashi, T. Harada, R. Yanagihara, T. Nagasawa, K. Nakamura, Conversion of pyrrole to pyrrole-2-carboxylate by cells of Bacillus megaterium in supercritical CO2, Chem. Commun. 21 (2001) 2194–2195, https://doi.org/10.1039/b105137g.
[137] M. Wieser, T. Yoshida, T. Nagasawa, Microbial synthesis of pyrrole-2-carboxylate by Bacillus megaterium PYR2910, Tetrahedron Lett. 39 (1998) 4309–4310, https://doi.org/10.1016/S0040-4039(98)00718-7.
[138] C. Peres, M.D.R. Gomes Da Silva, S. Barreiros, Water activity effects on geranyl acetate synthesis catalyzed by novozym in supercritical ethane and in supercritical carbon dioxide, J. Agric. Food Chem. 51 (2003) 1884–1888, https://doi.org/10.1021/jf026071u.
[139] A. Overmeyer, S. Schrader-Lippelt, V. Kasche, G. Brunner, Lipase-catalysed kinetic resolution of racemates at temperatures from 40°C to 160°C in supercritical CO2, Biotechnol. Lett. 21 (1999) 65–69, https://doi.org/10.1023/A:1005422524509.
[140] M.C. Almeida, R. Ruivo, C. Maia, L. Freire, T. Corrêa De Sampaio, S. Barreiros, Novozym 435 activity in compressed gases. Water activity and temperature effects, Enzym. Microb. Technol. 22 (1998) 494–499, https://doi.org/10.1016/S0141-0229(97)00241-X.
[141] T. Olsen, F. Kerton, R. Marriott, G. Grogan, Biocatalytic esterification of lavandulol in supercritical carbon dioxide using acetic acid as the acyl donor, Enzym. Microb. Technol. 39 (2006) 621–625, https://doi.org/10.1016/j.enzmictec.2005.11.025.
[142] Z.J. Dijkstra, R. Merchant, J.T.F. Keurentjes, Stability and activity of enzyme aggregates of Calb in supercritical CO2, J. Supercrit. Fluids 41 (2007) 102–108, https://doi.org/10.1016/j.supflu.2006.08.013.
[143] H. Michor, R. Marr, T. Gamse, T. Schilling, E. Klingsbichel, H. Schwab, Enzymatic catalysis in supercritical carbon dioxide: comparison of different lipases and a novel esterase, Biotechnol. Lett. 18 (1996) 79–84, https://doi.org/10.1007/BF00137815.
[144] J.F. Martins, I.B. de Carvalho, T.C. de Sampaio, S. Barreiros, Lipase-catalyzed enantioselective esterification of glycidol in supercritical carbon dioxide, Enzym. Microb. Technol. 16 (1994) 785–790, https://doi.org/10.1016/0141-0229(94)90036-1.
[145] P. Vidinha, V. Augusto, M. Almeida, I. Fonseca, A. Fidalgo, L. Ilharco, J. M. Joaquim, S. Barreiros, Sol-gel encapsulation: an efficient and versatile immobilization technique for cutinase in non-aqueous media, J. Biotechnol. 121 (2006) 23–33, https://doi.org/10.1016/j.jbiotec.2005.06.018.
[146] S. Garcia, P. Vidinha, H. Arvana, M.D.R. Gomes Da Silva, M.O. Ferreira, J.M.S. Cabral, E. A. MacEdo, N. Harper, S. Barreiros, Cutinase activity in supercritical and organic media: water activity, solvation and acid-base effects, J. Supercrit. Fluids 35 (2005) 62–69, https://doi.org/10.1016/j.supflu.2004.10.011.
[147] M. Romero, L. Calvo, C. Alba, A. Daneshfar, H. Ghaziaskar, Enzymatic synthesis of isoamyl acetate with immobilized lipase in -hexane, Enzym. Microb. Technol. 37 (2005) 42–48, https://doi.org/10.1016/j.enzmictec.2004.12.033.

[148] T. Matsuda, K. Watanabe, T. Kamitanaka, T. Harada, K. Nakamura, Biocatalytic reduction of ketones by a semi-continuous flow process using supercritical carbon dioxide, Chem. Commun. 3 (2003) 1198–1199, https://doi.org/10.1039/b301452e.

[149] T. Matsuda, T. Harada, K. Nakamura, Alcohol dehydrogenase is active in supercritical carbon dioxide, Chem. Commun. (2000) 1367–1368, https://doi.org/10.1039/b004069j.

[150] J.L. Panza, A.J. Russell, E.J. Beckman, Synthesis of fluorinated NAD as a soluble coenzyme for enzymatic chemistry in fluorous solvents and carbon dioxide, Tetrahedron 58 (2002) 4091–4104, https://doi.org/10.1016/S0040-4020(02)00259-4.

[151] M.H. Shin, N.Y. Cheong, J.H. Lee, K.H. Kim, Transglucosylation of caffeic acid by a recombinant sucrose phosphorylase in aqueous buffer and aqueous-supercritical CO2media, Food Chem. 115 (2009) 1028–1033, https://doi.org/10.1016/j.foodchem.2009.01.013.

[152] H. Baumann, M. Bühler, H. Fochem, F. Hirsinger, H. Zoebelein, J. Falbe, Natural fats and oils—renewable raw materials for the chemical industry, Angew. Chem. Int. Ed. Eng. 27 (1988) 41–62, https://doi.org/10.1002/anie.198800411.

[153] R.D. Schmid, R. Verger, Lipases: interfacial enzymes with attractive applications, Angew. Chem. Int. Ed. 37 (1998) 1608–1633, https://doi.org/10.1002/(SICI)1521-3773(19980703)37:12<1608::AID-ANIE1608>3.0.CO;2-V.

[154] P.H.L. Moquin, F. Temelli, Production of monoolein from oleic acid and glycerol in supercritical carbon dioxide media: a kinetic approach, J. Supercrit. Fluids 44 (2008) 40–47, https://doi.org/10.1016/j.supflu.2007.08.011.

[155] M. Jobling, S.M. Howdle, M. Poliakoff, Supercritical fluids in organometallic chemistry: IR evidence for the photochemical C–H activation of high density polyethylene by [(η^5-C_5 Me_5)Ir$(CO)_2$], J. Chem. Soc. Chem. Commun. 0 (1990) 1762–1764, https://doi.org/10.1039/C39900001762.

[156] M.E. Tadros, C.L.J. Adkins, E.M. Russick, M.P. Youngman, Synthesis of titanium dioxide particles in supercritical CO2, J. Supercrit. Fluids 9 (1996) 172–176, https://doi.org/10.1016/S0896-8446(96)90029-7.

[157] M. McHugh, V.K. Process, Supercritical fluid extraction, n.d. https://books.google.co.in/books?hl=en&lr=&id=VTIvBQAAQBAJ&oi=fnd&pg=PP1&dq=Supercritical+fluid+extraction&ots=QIacS-YLxR&sig=2UTjsMUtEiiv_0vA5ZPQGQ2as4w#v=onepage&q=Supercriticalfluidextraction&f=false, Accessed 26 December 2018.

[158] M.H. Mark, K. Val, Supercritical Fluid Extraction: Principles and Practice, https://books.google.co.in/books?hl=en&lr=&id=VTIvBQAAQBAJ&oi=fnd&pg=PP1&dq=Mark+McHugh+and+V.+Krukonis,+Supercritical+fluid+Extraction+Principle+and+Practice,+Butterworth-Hernemann,+1994.&ots=QIacSVSItK&sig=sVj8CpAHZOUnhw34CvXc6i3WR20, 1994. Accessed 26 December 2018.

[159] M. Paulaitis, V.J. Krukonis, R.T. Kurnik, R.C. Reid, Supercritical fluid extraction, Rev. Chem. Eng. 1 (1983) 179–250.https://jhu.pure.elsevier.com/en/publications/supercritical-fluid-extraction.

[160] O. Vitzthum, P. Hubert, Method for the Manufacture of Caffeine Free Black Tea, https://patents.google.com/patent/US4167589A/en, 1979. Accessed 27 December 2018.

[161] S.P. Weaver, C. Passmore, B. Collins, E. Fung, Vitamin D, sunlight exposure, and bone density in elderly African American females of low socioeconomic status, Fam. Med. 42 (2010) 47–51, https://doi.org/10.1021/jf0341619.

[162] S.M. Pourmortazavi, P. Baghaee, M.A. Mirhosseini, Extraction of volatile compounds from Juniperus communis L. leaves with supercritical fluid carbon dioxide: Comparison with hydrodistillation, Flavour Fragr. J. 19 (2004) 417–420, https://doi.org/10.1002/ffj.1327.

[163] S.M. Pourmortazavi, M. Ghadiri, S.S. Hajimirsadeghi, Supercritical fluid extraction of volatile components from Bunium persicum Boiss. (black cumin) and Mespilus germanica L. (medlar) seeds, J. Food Compos. Anal. 18 (2005) 439–446, https://doi.org/10.1177/1468797614550957.

[164] Y. Ikushima, M. Arai, K. Hatakeda, S. Ito, N. Saito, T. Goto, Selective extraction of a mixture of stearic, oleic, linoleic, and linolenic acid methyl esters with supercritical carbon dioxide using a gas-flow method, J. Chem. Eng. Japan 21 (1988) 439–441, https://doi.org/10.1252/jcej.21.439.

[165] M.B. King, T.R. Bott, Extraction of Natural Products Using Near-Critical Solvents, Springer Netherlands, Dordrecht, 1993. https://doi.org/10.1007/978-94-011-2138-5_1.

[166] C.L. Phelps, N.G. Smart, C.M. Wai, Past, present, and possible future applications of supercritical fluid extraction technology, J. Chem. Educ. 73 (1996) 1163, https://doi.org/10.1021/ed073p1163.
[167] L.A. Blanchard, D. Hancu, E.J. Beckman, J.F. Brennecke, Green processing using ionic liquids and CO2, Nature 399 (1999) 28–29, https://doi.org/10.1038/19887.
[168] R. Gupta, P. Chattopadhyay, Method of forming nanoparticles and microparticles of controllable size using supercritical fluids and ultrasound, https://patents.google.com/patent/US6620351B2/en, 2001. Accessed 27 December 2018.
[169] J. Jung, M. Perrut, Particle design using supercritical fluids: literature and patent survey, J. Supercrit. Fluids 20 (2001) 179–219, https://doi.org/10.1016/S0896-8446(01)00064-X.
[170] B. Subramaniam, S. Saim, R.A. Rajewski, V. Stella, Methods for particle micronization and nanonization by recrystallization from organic solutions sprayed into a compressed antisolvent, 2011 https://doi.org/10.1016/j.powtec.2011.06.024.
[171] U. Teipel, U. Förter-Barth, P. Gerber, H.H. Krause, Formation of particles of explosives with supercritical fluids, Propellants Explos. Pyrotech. 22 (1997) 165–169, https://doi.org/10.1002/prep.19970220313.
[172] U. Teipel, H. Kröber, H.H. Krause, Formation of energetic materials using supercritical fluids, Propellants, Explos. Pyrotech. 26 (2001) 168–173, https://doi.org/10.1002/1521-4087(200110)26:4<168::AID-PREP168>3.0.CO;2-X.
[173] W. Saus, D. Knittel, E. Schollmeyer, Dyeing of textiles in supercritical carbon dioxide, Text. Res. J. 63 (1993) 135–142, https://doi.org/10.1177/004051759306300302.
[174] B. Gebert, W. Saus, D. Knittel, H. Buschmann, E. Schollmeyer, Dyeing natural fibers with disperse dyes in supercritical carbon dioxide, Text. Res. J. 64 (1994) 371–374, https://doi.org/10.1177/004051759406400701.
[175] K.H. Chang, H.K. Bae, J.J. Shim, Dyeing of pet textile fibers and films in supercritical carbon dioxide, Korean J. Chem. Eng. 13 (1996) 310–316, https://doi.org/10.1007/BF02705955.
[176] U. Haarhaus, P. Swidersky, G.M. Schneider, High-pressure investigations on the solubility of dispersion dyestuffs in supercritical gases by VIS/NIR-spectroscopy. Part I — 1,4-Bis-(octadecylamino)-9,10-anthraquinone and disperse orange in CO2 and N2O Up to 180 MPa, J. Supercrit. Fluids 8 (1995) 100–106.https://www.sciencedirect.com/science/article/pii/0896844695900225.
[177] P. Swidersky, D. Tuma, G.M. Schneider, High-pressure investigations on the solubility of anthraquinone dyestuffs in supercritical gases by VIS-spectroscopy. Part II - 1,4-bis-(n-alkylamino)-9,10-anthraquinones and Disperse Red 11 in CO2, N2O, and CHF3up to 180 MPa, J. Supercrit. Fluids 9 (1996) 12–18, https://doi.org/10.1016/S0896-8446(96)90039-X.
[178] A.S. Ozcan, A.A. Clifford, K.D. Bartle, D.M. Lewis, Solubility of disperse dyes in supercritical carbon dioxide, J. Chem. Eng. Data 42 (1997) 590–592, https://doi.org/10.1021/je960337+.
[179] S.N. Joung, K.-P. Yoo, Solubility of disperse anthraquinone and azo dyes in supercritical carbon dioxide at 313.15 to 393.15 K and from 10 to 25 MPa, J. Chem. Eng. Data 43 (1998) 9–12, https://doi.org/10.1021/je9701285.
[180] S.N. Joung, H.Y. Shin, Y.H. Park, K.P. Yoo, Measurement and correlation of solubility of disperse anthraquinone and azo dyes in supercritical carbon dioxide, Korean J. Chem. Eng. 15 (1998) 78–84, https://doi.org/10.1007/BF02705309.

CHAPTER 17

Supercritical fluid based extraction of marigold principles

Muhammad Mushtaq[a], Sumia Akram[b]
[a]Department of Chemistry, GC University, Lahore, Pakistan
[b]University of Education, Bank Road Campus, Lahore, Pakistan

Contents

1. Introduction

The word marigold has been assigned to the flowers of two strongly scented genera of the Asteraceae family, i.e., *Tagetes*, which includes African and American Marigolds and *Celandula*, covering Pot Marigolds. The well-known variants include *Tagetes erecta* L. (African or Aztec marigold, *Tagetes patula* (French marigold or Safari), *Tagetes erecta* × *patula* (triploid marigolds), *Tagetes tenuifolia* (Signet marigold), Mule Marigolds (A sterile hybrid of African and French Marigolds), and *Calendula officinalis*. The flowers or florets of marigold are hardy, attractive, bear good fragrance, and are available in a variety of colors. The plant may reach a height of 150 cm and is widely grown for ornamental purposes in countries with tropical or subtropical climates. The flowers of the cultivated marigold have become a staple crop in the counties of Central America, Asia, and Europe for its ability to bloom brightly during winter and continue all summer long.

In addition to ornamental uses, the flowers are recognized as a source for natural colorants, essential oils, and thiophenes. The distillates or extracts of marigold flowers, especially of genus *Celandula*, have often been used as a herbal decoction to treat burns, sores, and several other skin ailments, rheumatism, malaria, ear-aches, cuts and wounds, eye diseases, ulcers, cold, and bronchitis [1]. In many parts of the world, flowers of *Celandula* species have been used as food additives or colorants. For example, marigold

Green Sustainable Process for Chemical and Environmental Engineering and Science
https://doi.org/10.1016/B978-0-12-817388-6.00017-9

flowers are often mixed with chicken feed to make egg yolks a darker yellow. The marigold seeds are half black and half white and house inside the petals. It often takes a couple of days when the bundle dries, which can be easily snapped off the stem and stored in a dry, cool space for the germination in the subsequent season. Like other plant parts, marigold seeds have been used for the preparation of folk remedies, soups, teas, and other beverages for human consumption.

Marigold flowers, leaves, and seeds have been extensively investigated for volatile and non-volatile oleaginous compounds for their characteristic and penetrating fragrance. It has been observed that marigold flowers, bearing a wide range of striking colors, mainly compose of carotenoids and essential oils produced during photosynthesis (except those present in bacteria and fungi). Many of these compounds cannot be synthesized by the human or animal body but their presence in diets generally retards aging and protects against certain types of stress-oriented diseases and cancers. For their established health benefits and uses in poultry feed, pharmaceutics, food, and cosmetic industries, analytical chemists and food scientists are attempting to devise an efficient, selective, economical method to recover carotenoids and other bioactives from marigold.

In general, marigold flowers have been declared to be a good source of carotenoids; particularly lycopene, lutein, and xanthophylls pigments, phenolic acids (syringic, chlorogenic, rosmarinic, caffeic dihydroxybenzonic, gallic, and dihydroxy-5-methoxy-benzoic acids), flavonoids, phenylpropanoids (coumarin, scopoletine), thiophenes and its derivatives, benzofunan derivatives, terpenoids, and steroids. The selection of an appropriate extraction/isolation technique solely depends upon the class of bioactive compounds targetted and the nature of the biological matrix from where isolation is needful. Moreover, careful optimization of extraction conditions and solvent polarity is equally critical for good quality and quantity of envisioned bioactive compounds. It would be more practical to have an idea about the solubility of target analytes before considering the selection of extraction solvent or techniques. In supercritical fluid-based extraction technologies, the solvent applied, i.e., SC-CO_2 has surface tension about 100 times smaller than liquid (water), and its density is very much tunable around the critical pressure (Fig. 1) and solubility of analytes increases with the increase in density, irrespective to their nature. Therefore, depressurization, in any case, will crystalize out the analytes solubilized at higher pressure.

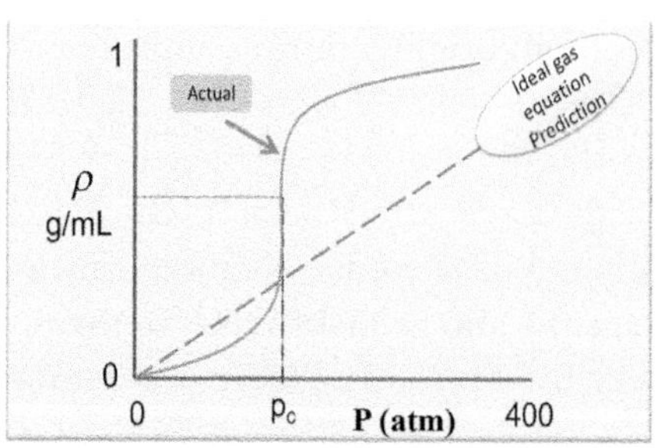

Fig. 1 Variation of density (ρ) of Supercritical carbon dioxide around critical pressure.

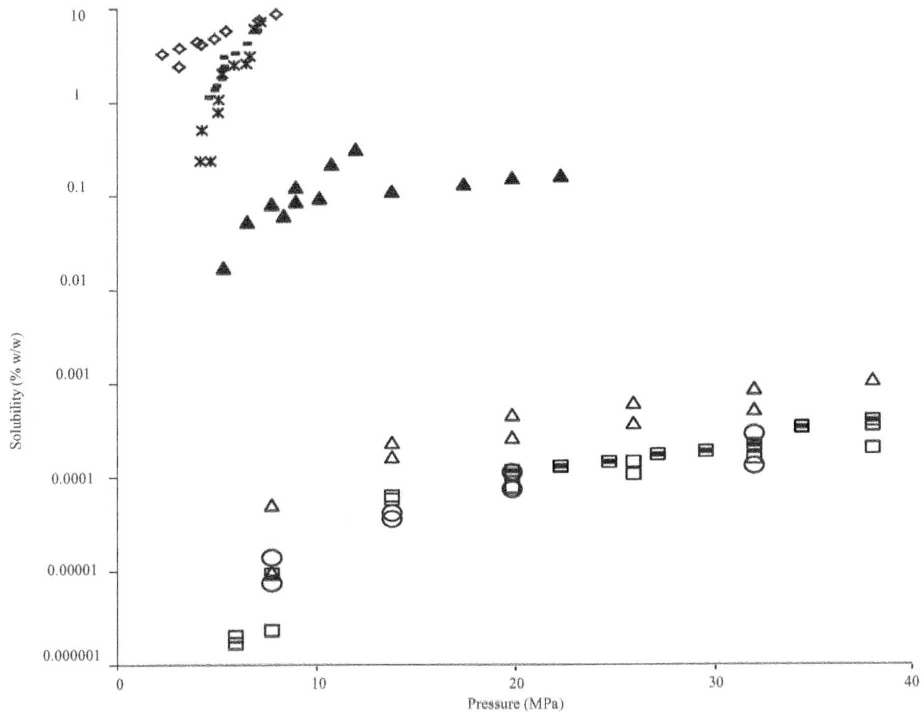

Fig. 2 Solubility of carotenoids (▪), Phenolic compounds (○: protocatechuic acid, □: *p*-coumaric acid phenolic acids, △: methyl gallate), waxes (▲), and Essential oil constituents (◊: linalool, –: α-pinene and ×: limonene) in SC-CO_2 at 35–50°C [2].

However, the applicable range of pressure would increase with the increase in polarity and molecular weight of the target analytes. The similar behavior of well-known classes of bioactive compounds can be established in Fig. 2. A look at the plotted data indicates that an increase in pressure increases the solubility of all the bioactives; however, a suitable range of pressure varies in each case, which points out the need for optimization of extraction conditions.

We are pretty sure that the working principle and experimental setup of supercritical fluids based extraction technologies have been comprehensively discussed in other parts of this monograph. The present chapter will explore the potential and benefits of supercritical fluids based extraction (SFE) technologies towards the recovery and isolation of marigold principle phytoconstituents.

2. Marigold carotenoids

Over the last couple of decades, a large number of researchers across the globe have investigated what exactly present in various parts of the marigold and how to isolate/harmonize major phytochemicals for human consumption/utilization. A wide range of bioactives has been identified and isolated from the root, stem, flowers, and seeds of marigold, whose general composition and anticipated health benefits vary with agro-climate conditions and plant ecotype. The most fascinating class of bioactives attributed to marigold includes carotenoids or other pigments present in flowers bearing a variety of striking colors. These bioactive were initially focused on as natural food colorants, but current technical data reveals their exceptional antioxidant and anticancer behaviors. In general, the presence of impurities may reduce the potential health benefits of marigold extracts or decoctions, or often make them unfit for direct consumption. Moreover, the availability of certain bioactive compounds like lutein decreases substantially in the presence of impurities. The supercritical fluids based extraction methods only have the ability to produce extracts free of organic solvents and other impurities. The carotenoids fall among the most abundant class of marigold bioactives. Although the majority of these compounds are non-polar in nature their tendency to interact with lipophilic solvents would vary with the structural variation taking place along the backbone side chains, which might be long unsaturated aliphatic like fatty acids. Moreover, structural diversity among carotenoids, their thermal and photolabile behavior, extensive conjugation reject or at least challenge the majority of conventional extraction methods.

In addition to the ornamental uses, flowers of marigold or their extracts have been investigated from a medicinal and functional point of view, and many researchers in this respect have labeled marigold as medicinal plants. Currently, extracts of marigold flower petals are directly used as dietary supplements, marketed in the forms of teas, beverages, or purified products. The first report of marigold photochemistry appeared in 1950, which discloses that flowers of marigold contain substantial amounts of carotenoids. The species bearing striking color flowers have been declared an outstanding source of carotenoids, especially lutein and lycopene (Fig. 3).

Valadon and Mummery [4] observed that the orange flowers of *Tagetes erecta* contain 3.23 mg/g carotenoids, including phytoene, phytofluene, carotenes, and their epoxy derivatives, lycopene, mutatochrome, flavochrome, gazaniaxanthine, cryptoxanthin, zeaxanthin, flavoxanthin, neoxanthin, lutein, and lutein epoxides. Later on, Alam et al. [5] separated the pigments of marigold flower petals by thin layer chromatography. The authors refluxed petal powder with a solvent comprising *n*-hexane, methanol, and potassium hydroxide at 60°C for 20 min and diluted the aliquot with a sufficient amount of water. The pigments from the digested solution were initially extracted using di-ethyl ether, which was further fractionated using two solvents of counter polarity.

The phytochemical analysis indicates that flower petals of *Tagetes erecta* may contain carotenoids pigments up to 5 g/kg of petal powder. The majority of these carotenoids are xanthophylls (98%), containing 64% lutein, 31% antheraxanthin, 3% α-cryptoxanthin, and traces of γ-carotene, phytofluene, and carotenes. Luteins, the principle bioactive pigments in marigold flower petals, are mainly present in the form of fatty acid esters of lauric, myristic, or palmitic acids. Many authors have labeled lutein as carotenoids vitamin, or

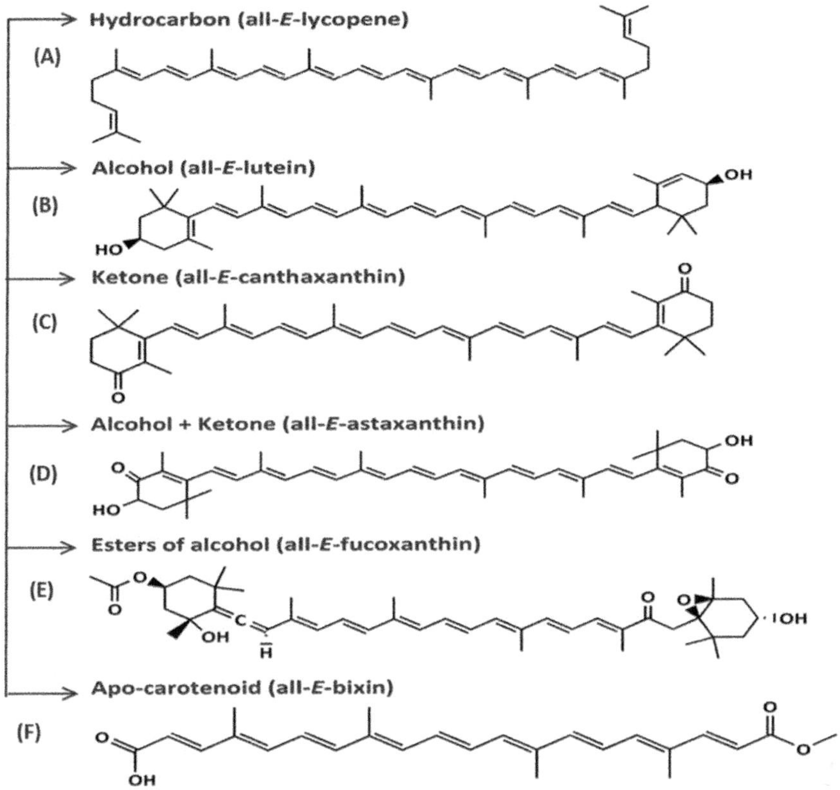

Fig. 3 Generic structures of various carotenoid compounds [3].

eye-vitamin, supplied to the retina by blood through macula lutea. Its uses in poultry feed to improve the color of egg yolk and chicken and dietary supplements for eye health require free and pure lutein. Nevertheless, lutein in marigold flowers petals mainly presents in the form of their fatty acid esters, which are ultimately converted into free lutein via saponification. These processes involve multi-step crystallization, isolation, and purification that may contaminate or abate the biological potential of lutein. Supercritical fluid based technologies have potential to saponify marigold extracts directly and produce good quality lutein for its applications in food industries.

3. Extraction of Marigold carotenoids

The behavior of carotenoids towards various environmental factors has been keenly investigated and must be considered before the analytical or preparative type separation of these pigments. These compounds are mostly unstable and may oxidize simultaneously into various products when initiated by light, temperature, and oxygen. In addition, the composition of processing/extraction media also affects the degradation of carotenoids, i.e., β-carotene has a half-life of 9.62 days in water, which decreases to 3.5 days when stored in cyclohexane. Generally, non-polar solvents, like *n*-hexane, chloroform, dichloromethane, dimethylsulphoxide, petroleum ether, and tetrahydrofuran, or their mixtures are used for the extraction carotenes or esterified xanthophylls (non-polar) by orbital shaker (mechanical stirring) or soxhlet type extractors (refluxing). The application of solvents such as acetone, acetonitrile, and ethanol bearing an intermediate polarity is preferred for the extraction of polar carotenoids. Recently, Abdel-Aal and Rabalski [6] used different solvents, including methyl tert-butyl ether, water-saturated *n*-butanol, and 2-propanol for the recovery of lutein from marigold flowers, lutein dietary supplements, and herbal teas of marigold. The authors observed that certain varieties of French marigold may contain lutein up to 5 mg/g of fresh flower petals (85% moisture). Moreover, methyl tert-butyl ether was found to be a more efficient solvent for the recovery of lutein esters from marigold. Acetone and Tetrahydrofuran (THF) also work for the extraction of carotenoids from plant sources, but care needs to be practiced for THF, which can initiate autoxidation (peroxidation) of carotenoids very quickly. A possible solution for this is the use of synthetic antioxidants such as butylated hydroxytoluene to prevent the formation of peroxides.

The incorporation of electromagnetic radiation based pre-treatment to break cell structure can obviously increase the mass transfer rate but these pre-treatments may deteriorate the antioxidant and functional character of liberated carotenoids. Similarly, the traditional extraction methods commonly involve at least one or more extractions with organic solvents, whose traces in food and pharmaceutical materials may impose a diversity of complications. Similarly, during the processing and drying of final extracts, a reasonable quantity of these toxic solvents is discharged into the environment. Consequently, faster and benign extraction/processing methods that are compatible with food production are extremely important.

The use of hydrolytic enzymes to macerate cell structure seems more productive [7–10], but it calls for complete knowledge of the biological matrix and careful investigation of incubation conditions. The costs of hydrolytic enzymes also limit the commercialization of enzyme-assisted extraction methodologies. The carotenoids are more susceptible to acids or alkaline hydrolysis, so none of these pre-treatments is advisable to enhance the liberation rate. Some other maceration techniques involve the use of least hazardous surfactants [11] are capable to disrupt cell mass structure but none of these have been practiced for the extraction of carotenoids at industrial scale.

4. SFE of Marigold carotenoids

Supercritical fluids (SCFs) have emerged as benign extraction solvents, particularly for non-polar macromolecules. The most characteristic features of SCFs are their ability to produce relatively cleaner extracts (selectivity) and tuneable polarity, i.e., polarity of SCFs, sharply changes with the change in thermodynamic conditions, which facilitates the selective precipitation of analytes. Moreover, the use of co-solvents has further extended the application range of SCFs. Although there is a wide range of SCFs, carbon dioxide based non-toxic SCF (SC-CO_2) has been frequently used for the extraction of biomolecules. The most reasonable evidence for the non-toxic character of SC-CO_2 has been linked with its presence in human and animal bodies (digestive and respiration systems).

Although SCFs falls among green and safe solvent choices for the recovery of thermally liable and oxidation prone biomolecules like carotenoids, the solubility of these compounds in SC-CO_2 is relatively low. Moreover, the capital investment and operational cost of supercritical based technologies become another matter of concern, particularly when the concentration of the analyte in the target matrix becomes too low. One evidence for such concern is the number of research reports available regarding extraction of lycopene. Interestingly, most of the researchers have tried extraction of lycopene from tomatoes because other plants or vegetable may not offer economically favorable extraction yield. In order to address the solubility problem, the flexibility in nature of SCFs with the change in phase conditions has been successfully utilized, while installation and operation cost of SCFs can only be compensated with increased clarity, quality, and precipitation of carotenoids.

The carotenoids based colorants are practically insoluble in water, which limits their application in food products. Their dissolution rate is often increased by increasing temperature, changing non-polar chain length, or encapsulating carotenoids inside hydrophilic materials. Most of these approaches are associated with precipitation and SCFs are found to be more intelligent precipitating agents for thermally labile compounds like carotenoids than organic solvents (solvent effect), temperature, and salts (salting out). The pressure and temperature dependent solubility of carotenoids allows the use of SC-CO_2 as an anti-solvent for their precipitation (Fig. 2). There are many ways in which SCFs can be used to carry out precipitation and the most suitable, i.e., gas anti-solvent process, involves the addition of SC-CO_2 into a saturated solution of carotenoids in some organic solvent. The addition of SC-CO_2 decreases the solubility of carotenoids, which nucleate and can be easily precipitated out. The higher mixing velocities of SC-CO_2 and miscibility with a wide range of organic solvents have broadened its use as an antisolvent for carotenoids extraction/processing.

The hydrocarbon carotenoids (Class A in Fig. 3) containing α and β-carotene, lycopene, and phytophene contribute about 1%–2% of marigold flower pigments. For the extraction of non-polar carotenoids (class A), many researchers have applied organic solvents and SC-CO_2 along with organic modifiers like methanol, ethanol, acetone, and olive oils. Actually, the non-polar class of carotenoid compounds, like lycopene, α, β, and ζ-carotene fall under minor constituents of marigold, so no data is available regarding their extraction from these plants or their flowers. Nevertheless, the recovery of these carotenoids from tomato and carrot has been extensively investigated for organic and supercritical fluids based solvents. The recovery of hydrocarbon carotenoids; for many reasons is influenced by the nature of the matrix or sample, sample preparation, moisture level, modifiers, and extraction conditions. Most of the researchers have investigated the effect of extraction conditions, i.e., temperature, pressure, and SC-CO_2 flow rate while keeping the moisture level and matrix particle size at the lowest possible ranges.

Cadoni et al. [12] investigated the effect of temperature on the SC-CO_2 based extraction of lycopene and β-carotene from tomato while keeping the pressure at 4000 Psi and flow rate of SC-CO_2 500 mL/min. Although the authors could recover a very small amount of lycopene and β-carotene (64.41 ± 0.12 and 34.88 ± 0.42 mg/100 g, respectively) at an elevated temperature of 80°C, the yield of both hydrocarbon carotenoids increased with the increase in temperature. The smaller extraction yields might be due to very high flow rates. Rozzi et al. [13] also applied SC-CO_2 for the extraction of hydrocarbon carotenoids from tomato and observed that recovery of this class of phytochemical increases with the increase in temperature and pressure applied. The authors recovered almost 38.01% of lycopene from tomato peel at 86.0°C and 34.47 MPa pressure. In contrast, an increase in SC-CO_2 flow rate reduced the recovery of lycopene. Finally, it was recommended that there is no need for modifiers for the extraction of class A carotenoids.

The above hypothesis was also endorsed by the study of Reverchon and De Marco [14], who could recover relatively small amounts of class A oleoresins at low temperature and high pressure while applying ethanol as a modifier (Table 1). Yi et al. [15] observed that an increase in temperature beyond 70°C caused transformation of trans-lycopene into cis-isomers. The antioxidant capacity of extracts increased up to 80°C but further increase in temperature adversely affected the antioxidant potential of lycopene extracts. It was observed that lycopene yield continuously increased within the range of temperature (40–100°C) and pressure (20–40 MPa) tested. However, extraction of these oleoresins above 70°C adversely affected the antioxidant character might be due to detrimental changes in lycopene isomers. An increase in flow rate could not enhance the recovery of lycopene at parallel rates.

The penultimate abundant class of marigold phytochemicals is polar carotenoids (xanthophylls) mainly attributed to the flowers of marigold (Class B in Fig. 3). The most interesting and abundant members of this family are Lutein and zeaxanthin, which only differ in the position of double bonds in their cyclic structures attached to polyisoprenoids (40 carbon atoms). The chromatographic characterization of marigold xanthophylls revealed the abundance (85%–95%) of trans-lutein, flowed by zeaxanthins (2.5%–8.0%) and other geometrical isomers (0.2%–1.5%). The xanthophylls are the singular class of plant bioactives occurring in the macular region of the human retina and were found to check the age-related degradation of macula and cataracts. Lutein, perhaps the most abundant marigold carotenoid (about 2 g/kg), works as a blue light filter and potent antioxidant.

Table 1 A comparison of extraction conditions followed for Marigold Carotenoids

Carotenoid class	Source	SFE conditions: Pressure (MPa)	Temperature (°C)	Flow rate (mL/min)	Pretreatment/modifier (%)	Yield (%)	Reference
A	–	25	80	500		0.64	[12]
A	–	4.47	86	2.5		38.0	[13]
A	–	44.6	62	3.5	Ethanol (14)	33.2	[14]
A	–	40	100	1.5		31.25 μg/g	[15]
B	*Tagetes erecta* L	35	55	10 kg/h	Ultra sound	690 mg/100 g)	[16]
B	Tagetes erecta L	46.8 MPa	65.9°C	10 kg/h	*Co*-solvent (1.5% of low chain glyceride)	1263.62 mg/100	[17]
B	Tagetes erecta L	35.5	58.7	19.9 L/h	co-solvent (7.0% soybean oil)	1039.7 mg lutein /100 g	[18]
B	Tagetes erecta L	40 MPa	60	–	*Co*-solvent (10% palm oil)	87.2 ± 4.4% total xanthophylls	[19]
C & D	*Haematococcus pluvialis*	40	70	3 mL/min	co-solvent (10% soybean or olive oil)	40%	[20]
C & D	Do	22–55	80	2–4 mL/min	co-solvent (7.5% Ethanol)	80.2%	[21]
D	*Capsicum annuum* L.	19–32	313–333	–	co-solvent (4.10 mmol/mol triolein)	5.27 μmol/mol	[22]
E		20	100	2.5 mL/min	co-solvents methanol, acetonitrile, ethyl-acetate	1–2 μg/100 g	[23]
F		20–50	40–55	–	–	0.26 mg/g CO_2	[24]

Lutein is under consideration as an essential nutrient because (i) the human metabolism cannot synthesize it and it must be part of the diet; (ii) its deficiency or the constant consumption of lutein-deprived diets may cause blindness; (iii) lutein related dysfunctions can be reverted by its supplementation before the ailments become irreversible. Finally, there are more than 170 million individuals suffering visual disabilities due to age-related macular degeneration (AMD), which can be prevented before it becomes irreversible by prescribing about 5 mg lutein supplements daily. At present, the United States of America alone consumes lutein worth of 150 million US dollars in food colorants and additives. The concentrated extracts of lutein are available in the market under various trade names like Xangold (Cognis) and FloraGLO (Kemin Foods) [25].

Several researchers have analyzed marigold flowers for the incidence of free and bound xanthophylls. The majority of methods applied for lutein and zeaxanthin isolation involve extraction of their esters of epoxides, which are subsequently saponified for free lutein and zeaxanthin. This practice not only uses bulk of toxic solvents, but also faces limited recovery rates or often produces extracts of poor quality. Moreover, drying of marigold flowers may result in degradation of carotenoids based pigments; still, drying has been chosen during several extractions and food processing technologies. Drying often involves freeze drying, hot air drying, sun drying or far infrared drying, and combined far-infrared radiation with hot air convection. It has been observed that freeze-drying or drying by hot air may deplete colored pigments of marigold, whereas use of hot air in combination with infrared radiation based convection causes least depletion of colors and phenolic compounds, lutein, and lycopene [26].

It is important to mention here that Naranjo-Modad et al. [27] investigated the solubility of lutein esters in supercritical carbon dioxide (SC-CO_2) for various thermodynamic conditions. It was observed that an increase in temperature improves the mass transfer and ultimate solubility of lutein esters in supercritical fluids. Nevertheless, the thermally labile nature of oxycarotenoids restricts processing of marigold flowers above ambient conditions. Subsequent to temperature, increase in pressure improves the solvation of power of SC-CO_2. The pure lutein esters deposited on nonporous glass beads were solubilized in SC-CO_2 in an automatic batch extractor set to the conditions of flow rate 4 mL/min, total CO_2 volume 7 mL, temperatures 40, 45, 50, or 55°C, and holding time was varied from 0 to 60 min. It was observed for a particular density range of solvent (0.7, 0.8, or 0.9 g/mL), an increase in temperature or pressure enhanced the solubility of esters in SC-CO_2. The solubility of lutein diesters in SC-CO_2 was also increased in the presence of modifiers tested (methanol, ethanol, acetone, acetonitrile, *n*-hexane, and chloroform) and in this context, chloroform can increase the solubility of lutein esters by 2.8 fold.

Although, xanthophyll content of marigold flower may vary for species to species, flower maturity, and agro-climate conditions. In general, fresh or properly preserved flowers of *Tagetes erecta* L. may contain up to 105.19 g xanthophyll/kg of petals, which may decline to 58.87 g/kg, if the flowers are not properly preserved. The SFE based extraction of xanthophyll depends on extraction conditions (temperature, pressure, SC-CO_2 flow rate), sample morphology (particle size and structure), and moisture level. The sample morphology and ultra-structure are very crucial to optimize in addition to the suitable selection of extraction conditions, solvent system, and modifiers. It is easy to understand that distribution of extraction solvent solely depends upon sample morphology, whereas its ability to solubilize xanthophylls changes with the composition of the matrix. Ma et al. [18] dried marigold flower petals at 50°C in a vacuum drying chamber to a moisture level of 7.35%, ground to 0.15 to 0.30 mm particle size and processed under various extraction conditions while using soybean oil as a co-solvent. The authors of this study observed that temperature, pressure, solvent flow rate, and their interactions can significantly affect the liberation of lutein esters. Overall, the extraction conditions observing pressure 35.50 MPa, the SC-CO_2 flow rate of 19.90 L/h, temperature 58.70°C, and 6.90% of soybean oil as co-solvent offered 1039.7 mg lutein/100 g marigold petal powder.

Gao et al. [16] applied ultrasound radiation prior to the SFE extraction of lutein esters from marigold flowers. The authors observed that application of ultrasound radiation based maceration improved the mass transfer coefficient in the solid phase 1.38 fold. Moreover, the matrix particle size, temperature, SC-CO_2, and pressure affected the recovery of lutein esters. This group recovered 690 mg of lutein ester/100 g petal powder of particle size 0.245–0.350 mm using SC-CO_2 set at 10 kg/h, 32.5 MPa, and 55.0°C. The ultrasound maceration was carried out at 25 kHz, ultrasonic power of 400 W, and irradiation time/interval of 6/9 s. Overall, increase in temperature, pressure, SC-CO_2 flow rate, and ultrasound power increased the lutein ester yield up to 55°C, 32.5 MPa, 10 kg/h, and 400 W, respectively, whereas an increase in ultrasound frequency beyond 25 kHz and irradiation time above 9 s reduced the recovery rates.

The same group [17] also used medium-chain triglycerides like rapeseed oil, sunflower seed oil, soybean oil, and *n*-hexane as co-solvent to enhance the recovery of lutein from marigold (*Tagetes erecta* L.) flowers during SFE. The results showed that extraction conditions observing 46.8 MPa pressure, 65.9°C temperature, 10 kg/h CO_2 flow rate, and 1.5% w/w of CO_2 addition of medium chain triglycerides offered 1263.62 mg lutein/100 g marigold during an extraction time of 3 h. Palumpitag et al. [19] also reported similar trends during the SFE of lutein esters from marigold flowers. In the absence of co-solvent, the recovery of xanthophylls increased to $74.4 \pm 0.9\%$ with the increase in temperature and pressure up to 60°C and 40 MPa, respectively. Under the same conditions addition of palm oil as co-solvent 10% (w/w) enhanced xanthophylls recovery to $87.2 \pm 4.4\%$ (16% increase). Finally, lutein esters were saponified into free lutein when treated with 40% (*w*/*v*) KOH solution at the oleoresin to solution ratio of 1:2 (w/v) for 3 h at 75°C.

Another study regarding the extraction of marigold lutein fatty acid esters was reported by Palumpitag et al. [19], who investigated the use of vegetable oils (palm, soybean, and olive oil) and ethanol as co-solvents and post-extraction saponification. In the absence of co-solvents, the optimum conditions, i.e., 60°C and 40 MPa pressure offered lutein fatty acid esters (total xanthophylls) up to 74.4 ± 0.9% of marigold flower powder of 0.19–0.22 mm particle size. However, incorporation of 10% palm oil (w/w) as co-solvent along with SC-CO_2 enhanced the total xanthophylls recovery to 87.2 ± 4.4% (overall 16% increase). The saponification of extracted lutein esters with 40% w/v KOH solution at extract to solution ratio of 1:2 (mass/volume) for 3 h at 75°C produced the maximum amount of free lutein.

A closer look at the experimental conditions (Table 1) indicates that recovery of class A and class B carotenoids superbly increased with the increase in temperature and pressure of supercritical fluids. The key reason behind this kind of behavior is limited solubility of these bioactive compounds in supercritical carbon dioxide. On the other hand, low solubility of carotenoids in SC-CO_2 eases quick precipitation of carotenoids (anti-solvent effect). In general pressure as high as 50–60 MPa and temperature higher than 50°C offered comparable yields of oxycarotenoids from marigold flowers. Although not investigated by any researcher dealing with supercritical fluid-based extraction of oxycarotenoids, bed geometry, including bed height, bed diameter, and thickness, and likewise, matrix particle size or bed porosity would also affect the kinetics of the extraction process. The extraction time is another important parameter, which depends upon solvent flow rate, matrix property, and nature of analytes of interest.

The ketocarotenoids (class C and D) like canthaxanthin and astaxanthin are usually present in traces in the flower petals of marigold but these carotenoids possess antioxidant characteristics about 10 times stronger than the most abundant carotenoids like β-carotene, lutein, and zeaxanthin, and a hundred times stronger than vitamin E. Although no record could be found regarding extraction of this class from marigold flower, Bustamante et al. [28] presented their research regarding solubility of these carotenoids in SC-CO_2 under different thermodynamic conditions. The data plotted in Fig. 4 indicates that for low values of extraction parameters the recovery yields were too low, but an increase in temperature and pressure and the incorporation of entrainer enhanced the recovery rates of Astaxanthin.

Krichnavaruk et al. [20] extracted ketocarotenoids from *Haematococcus pluvialis* using SFE with soybean and olive oils as continuous co-solvents. In the absence of co-solvents, the recovery rates were 25.40 ± 0.79%, with SC-CO_2 operated at a flow rate of 3 mL/min, 70°C, 40 MPa, and 5 h extraction time. The incorporation of soybean or olive oil at 10% (v/v) as a co-solvent enhanced the oil yield up to 36.36 ± 0.79%. Previously, Machmudah et al. [21] recovered almost 80% astaxanthin from the same matrix at 80°C and ethanol as co-solvent (7.5). Araus et al. [22] carefully investigated the solubility of capsanthin isolated from red pepper under different thermodynamic conditions in the presence and absence of co-solvents. The authors observed that solubility of natural capsanthin (purity about 86%) in pure supercritical carbon dioxide increases with the increase in pressure in the range below 20 MPa to 33 MPa, which might be due to increasing in the

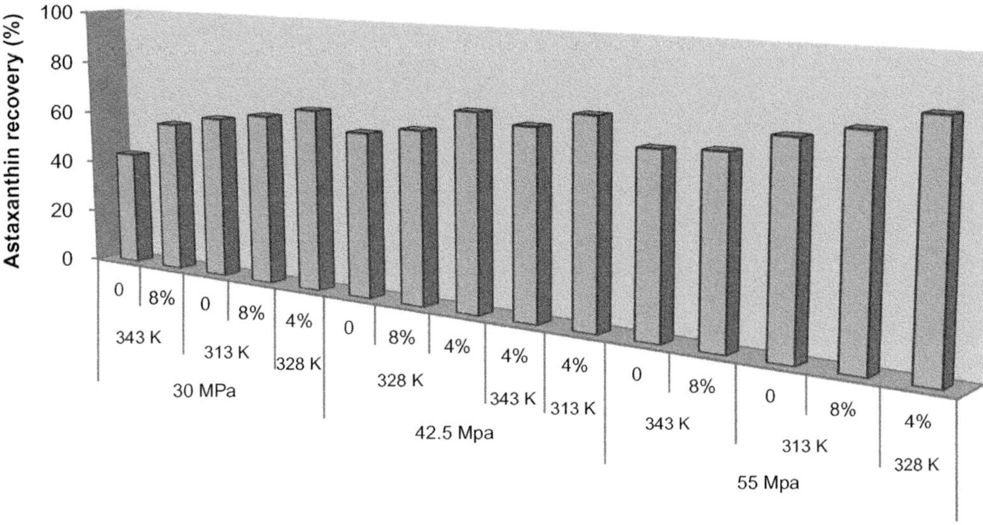

Fig. 4 Effect of pressure (MPa), temperature (K) and entrainer (ethanol) levels on ketocarotenoids data presented by Bustamante et al. [28].

density and solvation power of SC-CO_2. An increase in temperature while keeping the SC-CO_2 close to its critical pressure (7.38 MPa), could not increase the solubility of capsanthin because the density and vapor pressure compensate each other. The pressure corresponding to these conditions is called "crossover pressure" and close to this pressure increase in temperature may reduce the solubility of analytes in supercritical fluid systems. The crossover pressure for SC-CO_2 + capsanthin and SC-CO_2 + triolein was found to be 16.7 and 31.8 MPa. Overall, change in thermodynamic conditions, i.e., temperature (beyond crossover pressure) and pressure, can enhance the solubility of capsanthin in pure SC-CO_2 by 3 folds and presence of triolein as co-solvent can do the same without changing the thermodynamic conditions.

The carotenoids of class E and F, like fucoxanthin and bixin, are often present in traces but are equally valuable for pharmaceutical potential and antioxidant character of marigold flowers. Like other carotenoids, the solubility of fucoxanthin and bixin increases with the increase in temperature, pressure, SC-CO_2 density, and presence of co-solvents. Degnan et al. [24] observed an increase in solubility of pure bixin with the increase in temperature of 55°C and pressure 6000 psi.

5. SFE of faradiol esters from Marigold

The cosmetic and other topical applications of marigold, especially those attributed to flower heads of *Calendula officinalis* L. are due to the presence of faradiol esters or alcohols (Fig. 5). These bioactives have established anti-oedematous and anti-inflammatory

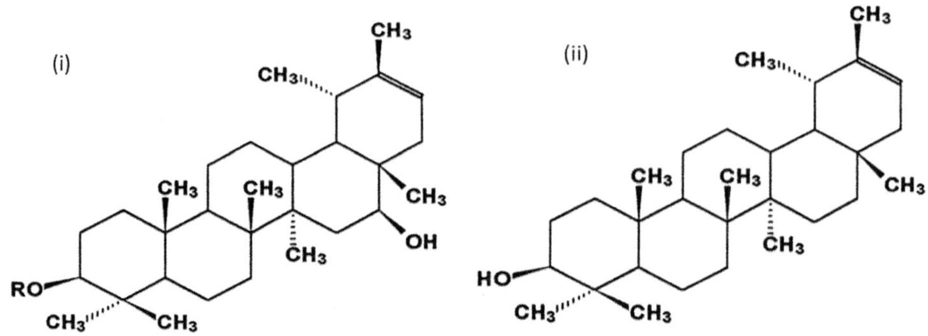

Fig. 5 Typical Structure of faradiol (i) esters and (ii) alcohols [29].

properties responsible for the pronounced antiphlogistic activity of lipophilic extracts, as well as free faradiol and ψ-taraxasterol. The lipophilic extracts of *Calendula officinalis* L flower petals or heads prepared in peanut oil, wool fat, or lard is supposed to contain >20 percent of these bioactives and offer a wide range of medicinal benefits against dermatological diseases like burns, ulcers, wounds, eczemas, contusions, eruptions, varicose veins, and hemorrhoids. The SC-CO_2 being non-polar provides cleaner technology for the recovery of faradiol esters [29].

The SFE of marigold (*Calendula officinalis* L.) with SC-CO_2 often furnishes dark orange color extract rich in carotenoids and lycopene. These bioactives have antioxidant properties and are believed to stimulate the granulation and formation of the skin's cell tissue in the wound healing process. In addition to carotenoids, SC-CO_2 extracts of *Calendula officinalis* contain >20% of esters of pentacyclic triterpene monols and diols like faradiol monoester (98%) and faradiol diester (2%) as the diester of mainly myristic (51%), palmitic (28%), and lauric (8%) acids. Faradiol esters are being increasingly focused on, due to their high activity and their quantitative prevalence of about 20% in the CO_2-extract.

Hamburger et al. [30] attempted preparative scale purification of faradiol esters; the major anti-inflammatory triterpenoid esters of the medicinal marigold *Calendula officinalis*. The authors successfully purified faradiol esters of lauric, palmitic, and myristic acids from the flower heads of marigold by the combination of SFE and High-Performance Liquid Chromatography. The ambient dried flower heads (1.0 kg) of 350 mm particle size were processed in large-scale vessels each of 6.5 and 4.0-liter capacity observing conditions: extraction temperature 50°C, pressure 500 bar, time 3 h, and SC-CO_2 flow rate of 35 kg/h. The extraction yields were up to 5% when calculated on a dry mass basis and approximately 85% of total faradiol esters in the flower heads of a medicinal plant were extracted. However, the addition of 20% adsorbent from the bottom of the extraction vessel removed 95% of the total pigments and the yield of faradiol esters was observed to be 75%. Overall, dried marigold flower buds may contain 18%–20% faradiol esters.

6. SFE of Marigold oleoresin

In addition to carotenoids, marigold flowers and other plant parts contain handful amounts of fatty acids, their esters, and other liposoluble of high value. The majority of these lipids have been extracted using soxhlet type extractors. Danielski et al. [31] applied SC-CO_2 to extract lipids for marigold flowers. The marigold flower powder (0.62 mm) commercially available in Brazil was processed under mild conditions of temperature (20–40°C), pressure (12–20 MPa), and SC-CO_2 flow rate (0.79–1.67 g/min). The solubility of marigold oleoresins in SC-CO_2 under the above-mentioned conditions varied: 4.74×10^{-4} to 17.04×10^{-4} g/g of SC-CO_2. Fig. 6 helps us to understand the effect of experimental conditions (temperature, pressure, solvent density, and flow rate) on the solubility (determined in terms of constant extraction rate (CER) and represented by Y_{CTE}) of marigold oleoresin in supercritical carbon dioxide. The SC-CO_2 flow rate (g/min) results indicate that optimum values of solubility (Y_{CTE}) for each set of temperature and pressure were observed close to the SC-CO_2 flow rate of near 1.10 g/min. At this flow rate, it can be assumed that the oleoresins saturate solvent, and that switching to lower or higher flow rates may cause a decreased solubility due to increase in intraparticle solid diffusion or decrease in contact time. The solubility of marigold oleoresins increased with the increase in solvent density because, at elevated pressures, SC-CO_2 presents liquid-like density, which enhances the number of soluble components, but may decrease selectivity. It is evident from plotted data that maximum solubility of marigold oleoresin has been observed with SC-CO_2 of 841.5 g/m^3 density set at a flow rate of 0.94 g at 40°C, 200 bar pressure.

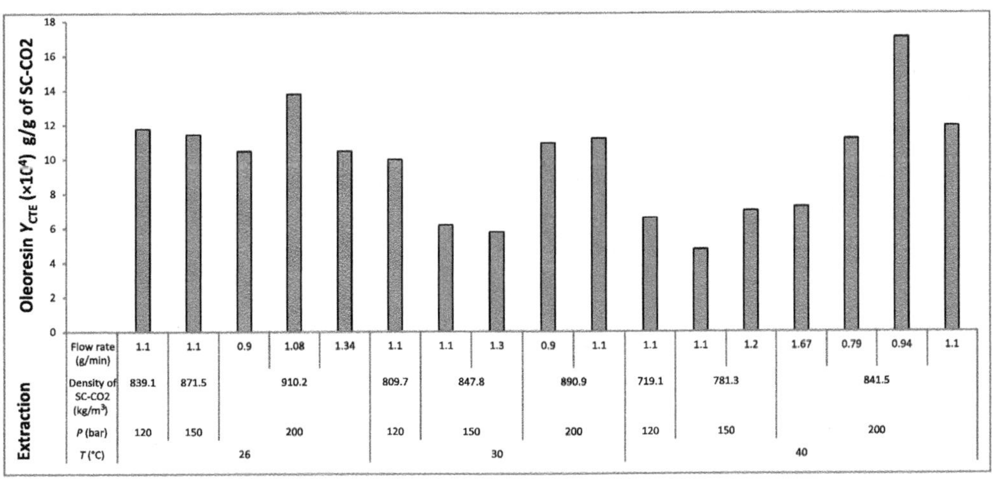

Fig. 6 Solubility of marigold oleoresins in supercritical carbon dioxide under different temperature, pressure, solvent density, and flow rates plotted from data observed by Danielski et al. [31].

It was interesting to note that the composition of SFE extract changes with the change in extraction solvent and experimental conditions. Marigold oleoresins obtained at 40°C and 188 bar pressure consisted of major constituents (relative peak areas) Acetyl eugenol (7.44), Phenol-4-octyl (1.07), Guaiol (5.07), Octadecane (4.14), Docosane (4.88), Tetracosane (10.47), Hexacosane (18.23), Octacosane (9.95), and Taraxasterol (1.19).

Gong et al. [32] also investigated the effect of thermodynamic conditions on the recovery of lipids from marigold flower powders containing particles of diameter 0.4–0.8 mm (40%), 0.2–0.4 mm (25%), 0.1–0.2 mm (10%), and < 0.1 mm (25%). In each experiment, 150 g of marigold powder, having a moisture level up to 5.65%, were charged to the extraction vessel of 1.0 L capacity. The samples were held there for 5 h under various conditions of temperature and pressure while keeping the SC-CO_2 flow rate of 15 kg/h. It was observed that supercritical temperature and pressure affected the recovery, composition, and antioxidant character of lipids. In the present case, a decline in lipid recovery was observed for the rise in temperature up to 40°C might be due to the reduction in density of the extraction solvent; however, a further increase in temperature improved the yield of oleoresins, which could be attributed to increases in vapor pressure or solubility of lipids. In general, an increase in temperature of SC-CO_2 charged at 30 MPa increased the recovery of oleoresin whereas an increase in pressure while operating the SFE at 50°C enhanced the extraction yield up to 30 MPa and a further increase in pressure could not improve the recovery of oleoresins.

The data plotted in Fig. 7 indicates the variation in the fatty acid profile of marigold lipids produced under various conditions of temperature and pressure. The

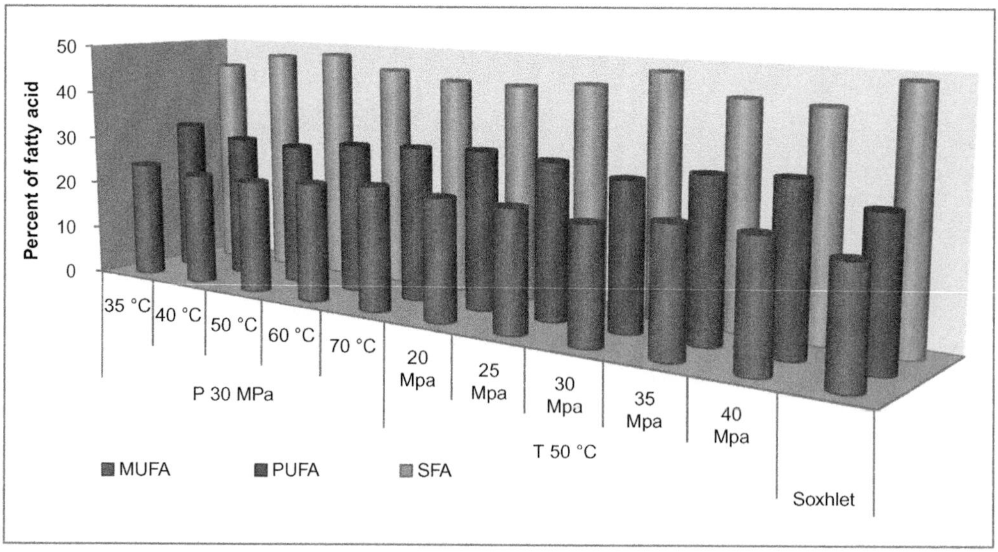

Fig. 7 Fatty acid profile of marigold lipids produced under various conditions of temperature and pressure plotted from results observed by Gong et al. [32].

chromatographic analysis of lipids produced via SFE and soxhlet type extraction by *n*-hexane revealed the presence of fourteen fatty acids, with relative percentages of linoleic (26%), palmitic (24%), oleic (20%), stearic (8%), and myristic acid (7%). The relative content of unsaturated and saturated fatty acids in SC-CO_2 and soxhlet based extracts were different. The unsaturated fatty acids in SC-CO_2 extracts at 30 MPa increased with the increase in temperature from 35 to 40°C, but further increase in temperature between 40 and 50°C could not cause a significant increase in unsaturated fatty acids. It was interesting to note that an increase in temperature from 60 to 70°C while keeping the pressure at 30 MPa again increased the unsaturated fatty acid content of extracts. Likewise, an increase or decrease in pressure around the 30 MPa while keeping the extraction temperature at 50°C decreased the level of unsaturated fatty acids of extracts. Interestingly, the oleoresin produced via soxhlet based extraction offered relatively better antioxidant characteristics in free radical based in-vitro antioxidant activities.

7. SFE of Marigold phenolic bioactives

In addition to carotenoids, and fatty acids, the flowers of marigold (*Tagetes erecta* L.) also contain esters/ethers of kaempferol, myricetin, quercetin, quercetagetin, quercetagetrin, patulitrin, patuletin, and rhamnetin. The presence of these bioactive compounds also offers a variety of functional and sensory characteristics to the flowers of marigold; however, supercritical fluid technology has not been applied for their recovery from this plant or its parts. On the other hand, SFE has been frequently used for the extraction of phenolic compounds from other agricultural resources. Most of the phenolic compounds, as their name indicates, contain phenol ring or simply hydroxyl group, so the direct application of supercritical fluid may not extract these polar bioactives. However, an addition of some polar but low-density solvents like methanol and ethanol as co-solvent render SFE an intelligent choice for the production of cleaner extracts. Previously, Li et al. [33] optimized SFE conditions for the recovery of kaempferol glycosides from tea seed cake. It was observed that pressure, temperature, solvent flow rate and polarity, and withholding time control the recovery rates. The highest yield 11.4 ± 0.4 mg kaempferol glycosides/g tea seed cake was observed within an extraction time of 150 min at a pressure of 20 MPa, temperature 80°C, with 60% aqueous ethanol (v/v) as co-solvent. Recently, we observed that an enzymatically macerated pomegranate peel and black tea leftovers were processed with SC-CO_2 set at 2 g/min at 300 bar and 55°C offered the extract yield comparable with the non-macerated sample within 60 min of extraction time. Overall, it was observed that incorporation of an enzymatic maceration prior to SFE can render it greener and suitable for the recovery of a non-extractable phenolic antioxidant [8, 34].

8. SFE of Marigold essential oil

According to the International Standard Organization on Essential oils (ISO 9235; 2013), an essential oil covers the products obtained from plant-based raw material via steam, hydro, dry distillation, or cold pressing in the case of citrus fruits. By virtue of weight, the essential oil covers a small fraction of marigold but actually, it confers a wide range of benefits attributed to marigold. The essential oil may include millions of compounds starting with simple oxygenated hydrocarbons (oxides, alcohols, ethers, aldehydes, ketones, acetates, and esters), to complex hydrocarbon of oxygenated terpenes and sesquiterpene. Although distillation via hot water is a simple process, it is associated with the transformation of many active aroma constituents like cis-sabinene and its derivatives, linalyl acetate, and other fragrance molecules.

Supercritical fluids based extraction (SFE) technologies are considered to be a more benign and more intelligent choice for the extraction of essential oils (volatiles and aroma compounds). The extracts or volatile fractions collected via SFE are cleaner and superior from a quality point of view as compared to those obtained by conventional extraction methods like hydro-distillation, steam distillation, or liquid–solid extraction. SFE is normally performed with pure CO_2 but the application of some co-solvents has been also reported. Like the extraction of other bioactives, SFE of essential oil is unambiguously affected by extraction conditions like temperature, pressure, SC-CO_2 flow rate, extraction time, matrix/sample characteristics, and composition. Similarly, the understanding regarding structure and solubility of essential oil constituents in SC-CO_2 is also necessary to establish favorable extraction conditions. In this respect, the non-polar nature of SC-CO_2 favors rapid dissolution of essential oil constituents yet matrix effects vary from situation to situation. The similarity in nature advises SFE under mild conditions of pressure (9–12 MPa) and temperature (35–50°C) can solubilize aroma compounds (Fig. 2). Nevertheless, higher pressure can be applied to tune the selectivity of SC-CO_2 and solvation power up to a certain extent. As essential oils consist of a large number of families of volatiles biomolecules, a decrease in temperature or pressure can be followed to fractionate these families. The modern SFE instruments provide the facility of on-line fractionation of extracts produced under same SFE conditions. This is usually accomplished by using several separators in series set at different pressure or temperatures, which can precipitate volatiles of corresponding molecular mass as a function of their saturation temperature or pressure.

Marigold flowers mainly contain menthol, methyl chavicol, α-humulene acetaldehyde, toluene, acetone, acyclic monoterpene ketones, piperitenone, myrecene, limonene, sabinene, ocimenes, thujones, 3-methylbutyl acetate, linalyl acetates, terpinolene, aromadendrene, phenylacetaldehyde, salicylaldehyde, ethyl-2- methylbutyrate, caryophyllene, linalool, tagetone and n-nonyl aldehyde, phenylethyl alcohol, 2-hexen-1-al, eudesmol, p-cymene, eugenol, d-carvone, α-thujene, β-pinene, camphene, α-phellandrene, α-terpinene, dihydrotagetone, geraniol, thymolhydroquinone dimethyl-ether, isopiperitenone, nerolidol, thymol, terpinolene, and myroxide [35]. The marigold essential oil is usually collected by hydrodistillation or using Clevenger type extractors. Fornari et al. [2] fractionated marigold flower bioactives when processed at 30 MPa and 40°C in the absence of cosolvent and observed that fraction 2 collected at smaller pressure

(5 MPa) comprised 100 percent of essential oil. The authors recovered about 2.20 g of essential for 100 g of flower matrix. The other fraction of essential oil collected at high pressure (10 MPa) did not contained any aroma compounds, wherease fraction of marigold flower collected at 5 MPa contained 1,8-Cineole, Sabinene hydrate (cis & trans), Linalool, Camphor, 1-Terpinene-4-ol, α-Terpineol, Eugenol, Cinamic acid methyl ester, Caryophyllene, α-bBergamatone, γ-Cadinene, and δ- Cadinene.

Normally, an increase in either temperature above 50°C or pressure beyond 50 MPa may cause detrimental changes in the essential oil produced. The SC-CO_2 flow rate is a crucial parameter regarding solvent distribution and mass transfer rates, but this variable again can be tuned over a range. Maceration prior to SFE extraction can be a good alternative to improve the solvent distribution and control resistance to mass transfer. Grinding normally increases the extraction yield, but very small particles can reduce the bed porosity and cause resistance to mass transfer. Moreover, an attempt to reduce the particle size generally via mill grinding may transfer energy sufficient for the escape of aroma compounds. Several authors have advised cryo-grinding or milling of chilled raw material to minimize the evaporation of volatile biomolecules. The last but not least variable that must be considered prior to sample charge into extraction vessel is its moisture level: the higher the moisture level the greater the possibility for the formation of ice. Generally, moisture level below 15% by weight of matrix is acceptable for the SFE of an agricultural product; but in the present case, drying will obviously reduce the essential oil yield.

9. Conclusion and future prospectus

Supercritical CO_2 extraction of marigold phytochemicals represents a cleaner choice as compared to organic solvents like *n*-hexane and alcohols, many of which may cause certain environmental and health complications in workers and consumers. The investor and consumers must be aware that marigold extract produced via supercritical fluid technology would not only be free of impurities and contain relatively higher amounts or activities of target chemopreventive agents and functional foods. Moreover, supercritical fluids, being more sensitive to thermodynamic conditions, particularly temperature and pressure, offer more flexible experimental conditions. Recently, López-Padilla et al. [36] evaluated the scaleability of supercritical fluid based extraction model using semi-empirical engineering scaling criteria. The authors have suggested Broken and Intact Cell (BIC) and constant solvent residence time (CSRT) models regarding the SFE of marigold bioactives. According to these models, if the residential time of SC-CO_2 is kept constant (CSRT) in extraction vessels of variable sizes, the kinetic behavior should be reproducible. Unfortunately, the results observed were not in accordance with those predicted by the above models. According to this group of researchers, the CSRT model only justify the experimental data over a scaling ratio of about 4.9, however, constant SC-CO_2

flow rate model can be applied up to scaling factor of 19. In order to commercialize SFE of marigold principles a more comprehensive model is required, covering sample particle size, porosity, extraction cell shape and geometry, and Schmidt number (a mass transfer analogy).

References

[1] N. Maity, N.K. Nema, M.K. Abedy, B.K. Sarkar, P.K. Mukherjee, Exploring Tagetes erecta Linn flower for the elastase, hyaluronidase and MMP-1 inhibitory activity, J. Ethnopharmacol. 137 (2011) 1300–1305.

[2] T. Fornari, G. Vicente, E. Vázquez, M.R. García-Risco, G. Reglero, Isolation of essential oil from different plants and herbs by supercritical fluid extraction, J. Chromatogr. A 1250 (2012) 34–48.

[3] R.K. Saini, Y.-S. Keum, Carotenoid extraction methods: A review of recent developments, Food Chem. 240 (2018) 90–103.

[4] L.R.G. Valadon, R.S. Mummery, Carotenoids of certain compositae flowers, Phytochemistry 6 (1967) 983–988.

[5] A.U. Alam, J.R. Couch, C.R. Creger, The carotenoids of the marigold, Tagetes erecta, Can. J. Bot. 46 (1968) 1539–1541.

[6] E.-S.M. Abdel-Aal, I. Rabalski, Composition of lutein Ester Regioisomers in Marigold flower, dietary supplement, and herbal tea, J. Agric. Food Chem. 63 (2015) 9740–9746.

[7] M. Mushtaq, B. Sultana, S. Akram, F. Anwar, A. Adnan, S.S. Rizvi, Enzyme-assisted supercritical fluid extraction: An alternative and green technology for non-extractable polyphenols, Anal. Bioanal. Chem. 409 (2017) 3645–3655.

[8] M. Mushtaq, B. Sultana, F. Anwar, A. Adnan, S.S. Rizvi, Enzyme-assisted supercritical fluid extraction of phenolic antioxidants from pomegranate peel, J. Supercrit. Fluids 104 (2015) 122–131.

[9] M. Mushtaq, B. Sultana, H.N. Bhatti, M. Asghar, RSM based optimized enzyme-assisted extraction of antioxidant phenolics from underutilized watermelon (Citrullus lanatus Thunb.) rind, J. Food Sci. Technol. 52 (2015) 5048–5056.

[10] M. Mushtaq, B. Sultana, H.N. Bhatti, M. Asgher, Optimization of enzyme-assisted revalorization of sweet lime (Citrus limetta Risso) peel into phenolic antioxidants, Bioresources 9 (2014) 6153–6165.

[11] B. Mukhtar, M. Mushtaq, S. Akram, A. Adnan, Maceration mediated liquid–liquid extraction of conjugated phenolics from spent black tea leaves extraction of non-extractable phenolics, Anal. Methods (2018).

[12] E. Cadoni, M. Rita De Giorgi, E. Medda, G. Poma, Supercritical CO2 extraction of lycopene and β-carotene from ripe tomatoes, Dyes Pigments 44 (1999) 27–32.

[13] N.L. Rozzi, R.K. Singh, R.A. Vierling, B.A. Watkins, Supercritical fluid extraction of lycopene from tomato processing byproducts, J. Agric. Food Chem. 50 (2002) 2638–2643.

[14] E. Reverchon, I. De Marco, Supercritical fluid extraction and fractionation of natural matter, J. Supercrit. Fluids 38 (2006) 146–166.

[15] C. Yi, J. Shi, S.J. Xue, Y. Jiang, D. Li, Effects of supercritical fluid extraction parameters on lycopene yield and antioxidant activity, Food Chem. 113 (2009) 1088–1094.

[16] Y. Gao, B. Nagy, X. Liu, B. Simándi, Q. Wang, Supercritical CO2 extraction of lutein esters from marigold (Tagetes erecta L.) enhanced by ultrasound, J. Supercrit. Fluids 49 (2009) 345–350.

[17] Y. Gao, X. Liu, H. Xu, J. Zhao, Q. Wang, G. Liu, Q. Hao, Optimization of supercritical carbon dioxide extraction of lutein esters from marigold (Tagetes erecta L.) with vegetable oils as continuous co-solvents, Sep. Purif. Technol. 71 (2010) 214–219.

[18] Q. Ma, X. Xu, Y. Gao, Q. Wang, J. Zhao, Optimisation of supercritical carbon dioxide extraction of lutein esters from marigold (Tagetes erect L.) with soybean oil as a co-solvent, Int. J. Food Sci. Technol. 43 (2008) 1763–1769.

[19] W. Palumpitag, P. Prasitchoke, M. Goto, A. Shotipruk, Supercritical carbon dioxide extraction of Marigold lutein fatty acid esters: effects of Cosolvents and saponification conditions, Sep. Sci. Technol. 46 (2011) 605–610.
[20] S. Krichnavaruk, A. Shotipruk, M. Goto, P. Pavasant, Supercritical carbon dioxide extraction of astaxanthin from Haematococcus pluvialis with vegetable oils as co-solvent, Bioresour. Technol. 99 (2008) 5556–5560.
[21] S. Machmudah, A. Shotipruk, M. Goto, M. Sasaki, T. Hirose, Extraction of Astaxanthin from Haematococcus pluvialis using supercritical CO2 and ethanol as Entrainer, Ind. Eng. Chem. Res. 45 (2006) 3652–3657.
[22] K.A. Araus, J.M. del Valle, P.S. Robert, J.C. de la Fuente, Effect of triolein addition on the solubility of capsanthin in supercritical carbon dioxide, J. Chem. Thermodyn. 51 (2012) 190–194.
[23] M.-K. Roh, M.S. Uddin, B.-S. Chun, Extraction of fucoxanthin and polyphenol from Undaria pinnatifida using supercritical carbon dioxide with co-solvent, Biotechnol. Bioprocess Eng. 13 (2008) 724–729.
[24] A.J. DEGNAN, J.H. VON ELBE, R.W. HARTEL, Extraction of annatto seed pigment by supercritical carbon dioxide, J. Food Sci. 56 (1991) 1655–1659.
[25] J.M. Fernández-Sevilla, F.G. Acién Fernández, E. Molina Grima, Biotechnological production of lutein and its applications, Appl. Microbiol. Biotechnol. 86 (2010) 27–40.
[26] S. Siriamornpun, O. Kaisoon, N. Meeso, Changes in colour, antioxidant activities and carotenoids (lycopene, β-carotene, lutein) of marigold flower (Tagetes erecta L.) resulting from different drying processes, J. Funct. Foods 4 (2012) 757–766.
[27] S. Naranjo-Modad, A. López-Munguía, G. Vilarem, A. Gaset, E. Bárzana, Solubility of purified lutein Diesters obtained from Tagetes erecta in supercritical CO2 and the effect of solvent modifiers, J. Agric. Food Chem. 48 (2000) 5640–5642.
[28] Bustamante, A., Roberts, P., Aravena, R., Del Valle, J. 2011. Supercritical extraction of astaxanthin from *H. pluvialis* using ethanol-modified CO2. Experiments and modeling. In: Proceedings of the 11th International Conference of Eng Food, Athens, Greece, pp. 22–26.
[29] K. Zitterl-Eglseer, S. Sosa, J. Jurenitsch, M. Schubert-Zsilavecz, R. Della Loggia, A. Tubaro, M. Bertoldi, C. Franz, Anti-oedematous activities of the main triterpendiol esters of marigold (Calendula officinalis L.), J. Ethnopharmacol. 57 (1997) 139–144.
[30] M. Hamburger, S. Adler, D. Baumann, A. Förg, B. Weinreich, Preparative purification of the major anti-inflammatory triterpenoid esters from Marigold (Calendula officinalis), Fitoterapia 74 (2003) 328–338.
[31] L. Danielski, L.M.A.S. Campos, L.F.V. Bresciani, H. Hense, R.A. Yunes, S.R.S. Ferreira, Marigold (Calendula officinalis L.) oleoresin: Solubility in SC-CO2 and composition profile, Chem. Eng. Process. Process Intensif. 46 (2007) 99–106.
[32] Y. Gong, S. Plander, H. Xu, B. Simandi, Y. Gao, Supercritical CO2 extraction of oleoresin from marigold (Tagetes erecta L.) flowers and determination of its antioxidant components with online HPLC-ABTS•+ assay, J. Sci. Food Agric. 91 (2011) 2875–2881.
[33] B. Li, Y. Xu, Y.-X. Jin, Y.-Y. Wu, Y.-Y. Tu, Response surface optimization of supercritical fluid extraction of kaempferol glycosides from tea seed cake, Ind. Crop Prod. 32 (2010) 123–128.
[34] M. Mushtaq, B. Sultana, S. Akram, F. Anwar, A. Adnan, S.S. Rizvi, Enzyme-assisted supercritical fluid extraction: An alternative and green technology for non-extractable polyphenols, Anal. Bioanal. Chem. (2017) 1–11.
[35] P. Gupta, N. Vasudeva, A potential ornamental plant drug, Hamdard Medicus. 55 (2012).
[36] A. López-Padilla, A. Ruiz-Rodriguez, G. Reglero, T. Fornari, Supercritical carbon dioxide extraction of Calendula officinalis: Kinetic modeling and scaling up study, J. Supercrit. Fluids 130 (2017) 292–300.

CHAPTER 18

Industrial polymer synthesis using supercritical carbon dioxide

Natascha Cheikhyoussef[a], Ahmad Cheikhyoussef[b]
[a]Ministry of Higher Education, Training and Innovation, Windhoek, Namibia
[b]University of Namibia, Windhoek, Namibia

Contents

1. Introduction

Polymers are macromolecules with subunits of monomers that react to form long chains or three-dimensional structures [1]. The extensive developments in the fields of polymer science are based on the work and discoveries of Hermann Staudinger, for which, in 1953, he received the Nobel Prize for Chemistry [2]. About 20 years ago the first polymerization process using supercritical carbon dioxide was reported. Stricter emissions standards [3] have contributed to the increased use of more environmentally friendly solvents. Carbon dioxide that is maintained beyond supercritical conditions at 31.1°C and 7.4 MPa is termed supercritical carbon dioxide ($scCO_2$) [4]. The application of $scCO_2$ in the synthesis of polymers has been widely reviewed by several groups, including the different methodologies such as step-growth and chain-growth polymerizations (Table 1). Synthesis reactions involving step growth and chain growth polymerizations have efficiently utilized $scCO_2$ as the inert solvent [8]. The additional development and use of surfactants have also made the application of heterogeneous polymerizations possible in $scCO_2$ [8].

Green Sustainable Process for Chemical and Environmental Engineering and Science
https://doi.org/10.1016/B978-0-12-817388-6.00018-0

Table 1 Reviews in the field of polymer synthesis using supercritical carbon dioxide

Title of the review	Reference
Polymerization reactions at high pressure and supercritical conditions	Scholsky [5]
Chain polymerizations in inert near- and supercritical fluids	Shaffer and DeSimone [6]
Polymer synthesis and characterization in liquid/supercritical carbon dioxide	Cooper and DeSimone [7]
Polymerizations in liquid and supercritical carbon dioxide	Canelas and DeSimone [8]
Carbon dioxide as a continuous phase for polymer synthesis	Canelas et al. [9]
Polymer synthesis and processing using supercritical carbon dioxide	Cooper [10]
Polymerizations in supercritical carbon dioxide	Kendall et al. [11]
Generation of microcellular biodegradable polymers using supercritical carbon dioxide	Sparacio and Beckman [12]
Frontiers in green chemistry utilizing carbon dioxide for polymer synthesis and applications	Young and DeSimone [13]
Polymerizations and polymer modification reactions in near and supercritical fluids	Kiran et al. [14]
Synthesis of fluoropolymers in liquid and supercritical carbon dioxide solvent systems	DeYoung et al. [15]
Polymer synthesis in supercritical carbon dioxide	Filardo et al. [16]
Synthesis and characterization of polymers: from polymeric micelles to step-growth polymerizations	Young [17]
Supercritical fluid applications in polymer nanocomposites	Tomasko et al. [18]
Supercritical carbon dioxide and polymers	Said-Galiyev et al. [19]
Green synthesis of polymers using supercritical carbon dioxide	Wood et al. [20]
Production of fluoropolymers in supercritical carbon dioxide	Wood et al. [21]
Heterogeneous polymerization in supercritical carbon dioxide	Mueller et al. [22]
Synthesis of advanced materials using supercritical fluids	Cooper [23]
Supercritical carbon dioxide: in polymer reaction engineering	Kemmere and Meyer [24]
Supercritical carbon dioxide: A reactive medium for chemical processes involving fluoropolymers	Nikitin et al. [25]
Fluoropolymer synthesis in supercritical carbon dioxide	Du et al. [26]
Synthesis of polymer-inorganic filler nanocomposites in supercritical CO_2	Haldorai et al. [27]
Supercritical carbon dioxide and polymers: An interplay of science and technology	Picchioni [28]
Input of supercritical carbon dioxide to polymer synthesis: An overview	Boyere et al. [29]
Applications of supercritical carbon dioxide in materials processing and synthesis	Zhang et al. [4]
Supercritical fluids and polymers—The year in review—2014	Kiran [30]
Synthesis of polymer nanocomposites in supercritical carbon dioxide	Haldorai and Shim [31]
Dispersion polymerization in environmentally benign solvents via reversible deactivation radical polymerization	Wang et al. [32]

The first polymers synthesized using $scCO_2$ were semi-fluorinated acrylates [33]. Mingotaud et al. [34] reported the first cationic ring-opening polymerization of heterocycles, such as octamethylcyclosiloxane and phenyloxazoline, and the anionic and pseudo-anionic ring-opening polymerization of ε-caprolactone in $scCO_2$. Some of the polymers that have been synthesized with the use of $scCO_2$, and that are described in this chapter, include biopolymers, conducting polymers, polyamides, fluoropolymers, core-shell polymers, and polymer nanocomposites. The use of enzymes in $scCO_2$ for ring-opening polymerization reactions produces clean biopolymer products that are used in food and pharmaceutical applications [35]. The importance of the area of polymerizations in $scCO_2$ is highlighted by several informative reviews published in the past 30 years [5, 9, 10, 13, 16, 19, 20, 23, 24]. Kiran et al. [30] published a summary in 2014 encapsulating important information appearing in earlier reported publications on the synthesis and processing of polymers. The number of publications for the year 2014 on polymerizations in $scCO_2$ were reported to be 52, whilst nanocomposites had 15 publications [30].

2. Supercritical carbon dioxide as the polymerization solvent

A fluid is considered to be supercritical when it is above its critical temperature (Tc) and pressure (pc) [36]. Termed as a "green" solvent, $scCO_2$ is increasingly being researched and industrially applied for polymer synthesis, including polymer processing [4, 28, 29, 32]. The fact that $scCO_2$ is inert, non-toxic, non-flammable, with the recycling processes being environmentally friendly [29], allows it to be termed a "green" solvent. Apart from being an environmentally useful solvent, there are economic benefits to using $scCO_2$, since it is an inexpensive solvent with a recovery processes requiring less energy consumption [29]. Once polymerization reactions are completed, simple depressurization or venting off of $scCO_2$ leads to a dry polymer product or powder, allowing for a cost-effective method to remove the polymerization solvent [11]. A clean polymer product is also obtained that has no traces or residues of monomer or organic solvent [37]. The so-called "closed-loop polymer process" as described by Kemmere et al. [38] allows for reagents and $scCO_2$ to be recycled.

The properties of $scCO_2$ are two-fold: gas-like diffusivity combined with liquid-like density [7, 11]. The properties such as the solvent densities [11] of $scCO_2$ can be adjusted by minor changes in temperature and pressure [29] within the reaction system. There is a direct relation between the solvation effect and pressure, by which particle size and distribution are controlled by high pressures at around 200 bar [39]. Changes in the reaction parameters cause $scCO_2$ to have an impact on porosity, pore size, and morphology of the polymer [40, 41]. Being inexpensive, non-toxic, and non-flammable [31], including the unique properties of $scCO_2$, such as gas-like diffusivity and viscosity, and liquid-like density [36, 42], make it a useful solvent for synthesis at a large-scale [11].

$ScCO_2$ presents selective solubility properties towards compounds. Non-polar molecules of low molecular weight, vinyl monomers (polar molecules), polysiloxanes, silicones, and amorphous fluoropolymers are easily soluble in $scCO_2$ [11, 29]. The affinity of polymers for carbon dioxide gives such polymers the terminology of being "CO_2-philic" [7, 43]. Polymers that are termed as "CO_2-phobic" are the lipophilic and hydrophilic polymers that are less soluble in CO_2 [7].

A number of reviews (Table 1) have extensively described the different polymerization methods developed [6, 11, 14, 17, 22], including the use of $scCO_2$ as the polymerization solvent. Wang et al. [32] reviewed the reversible deactivation radical polymerization-mediated dispersion polymerization in $scCO_2$ and other environmentally friendly solvents. Fig. 1 provides an overview of polymerization processes used for synthesis of polymers using $scCO_2$. Heterogeneous polymerization methods that include dispersion, precipitation, suspension, and emulsion polymerization, with dispersion polymerization being the most widely investigated, have been used to synthesize a number of different polymers [44]. The development and use of surfactants as stabilizers, in particular for dispersion and emulsion polymerizations, widened the scope of polymerizations to be used and the polymers to be produced [11]. The first dispersion polymerization method described by DeSimone et al. [45] was used to synthesize methyl methacrylate in $scCO_2$ with excellent yield and superior properties. In particular, the availability of micelle-forming surfactants has created opportunities for polymer synthesis via heterogeneous polymerizations [8].

The effect of plasticization during heterogeneous polymerizations allows for the diffusion of monomer and initiator into the polymer phase to be increased [46]. The production of polypyrrole, polyphenoxides, silica gels, polyamides, polyesters, and polycarbonates [11, 29, 47] has involved the use of step-growth polymerizations. Chain-growth polymerizations include the cationic, anionic, metal-catalyzed, and free-radical reactions, of which the most focus has been on free-radical reactions in the application of $scCO_2$ [11, 29]. The use of controlled radical polymerizations in $scCO_2$, such as atom transfer radical polymerization (ATRP), reversible addition-fragmentation chain transfer polymerization (RAFT), nitroxide-mediated radical polymerization (NMP), reversible chain transfer catalyzed polymerization (RTCP), and cobalt-mediated radical polymerization (CMRP) have been widely reported [29]. Jaramillo-Soto et al. [48] have simulated the reversible addition-fragmentation chain transfer (RAFT) radical dispersion polymerization of vinyl monomers in $scCO_2$. Grignard et al. [49] reported a novel approach towards monitoring the kinetics of homogeneous and heterogeneous reactions with the application of Raman spectroscopy under high pressure via online mode. Haven and Junkers [50] reviewed current methods of characterization for use in the online monitoring of polymerizations.

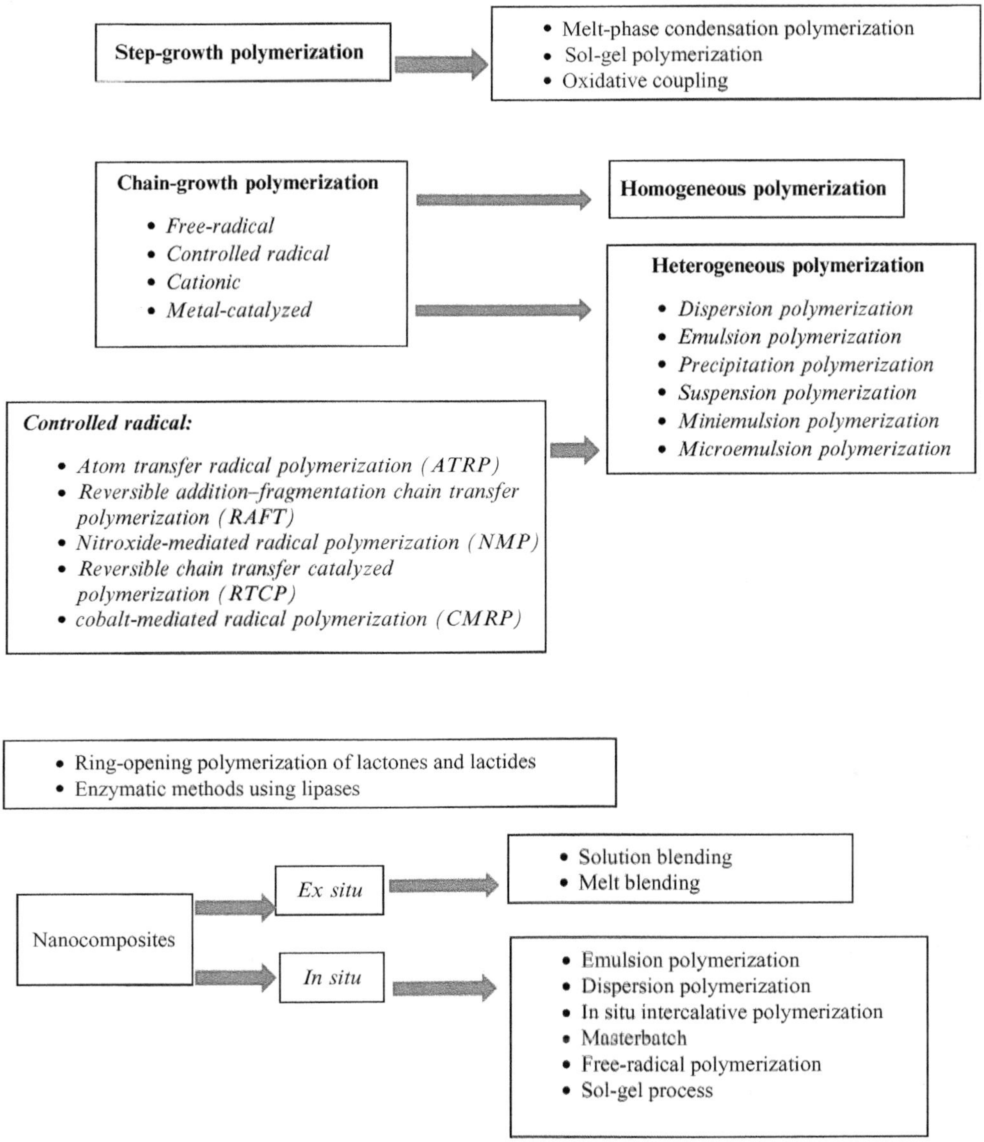

Fig. 1 Overview of synthesis processes for polymers using supercritical carbon dioxide [11, 29, 31, 32].

3. Polymers synthesized using supercritical carbon dioxide

3.1 Biopolymers

In the synthesis of biopolymers such as poly(lactic-*co*-glycolic acid [51], poly(glycolic acid) and poly(lactic acid), naturally occurring monomers such as lactic acid and glycolic acid are used [52]. Some of the most important biopolymers are polylactic acid and the polyhydroxyalkanoates [53]. By virtue of being biodegradable, these polymers are also non-toxic and biocompatible, making them useful for applications in bio-medicine [52, 54] and as bioplastics [55]. Gupta et al. [56] provided an overview of poly(lactic acid) describing its chemistry, properties, and applications, in particular in the area of tissue engineering. The use of scCO_2 has been applied in the synthesis of a number of biopolymers, including poly(lactic-*co*-glycolic acid) [12].

The use of enzymes in scCO_2 for the ring-opening polymerization of lactones [35, 57–59] has become more common as compared to the use of the Lewis acid catalyst in organic solvent [58]. This is due to the fact that the process is non-toxic—a clean extraction of monomer residues can be achieved and the enzyme can be recycled [35, 58, 60]. The kinetics of this reaction was investigated for the first time by Thurecht et al. [61]. *Candida antarctica* lipase B (CALB) catalyzed the reaction, which was found to be first order with respect to monomer (80% conversion) producing poly(ε-caprolactone) up to 50 kDa and polydispersities (range of 2) [61]. Polycaprolactone is a biodegradable, biocompatible, and hydrophobic semi-crystalline polyester [62]. Biodegradable aliphatic polyesters have been synthesized using scCO_2 via the enzymatic catalyzed (*Candida antarctica* lipase) ring-opening polymerization of lactones (ε-caprolactone) as well as the enzymatic poly-condensation of divinyl adipate and 1,4-butanediol [57].

The effects of enzyme and monomer concentrations and pressure of the reaction system were also investigated [57]. Lipase-catalyzed polymerization at pressure (145 bar), temperature (65°C), agitation (200 rpm), and lipase concentration (10 wt%) to synthesize poly(ε-caprolactone) in a batch reactor resulted in the chain size of the polymer to be 7419.6 g/gmol [59]. Poly(ε-caprolactone) (M_n = 12,000–37,000 g/mol) was synthesized via a lipase-catalyzed ring-opening polymerization [58]. The enzyme catalyst used was lipase B from *Candida antarctica* supported on macroporous beads (Novozym-435) [58]. The optimum reaction conditions to synthesize poly(ε-caprolactone) (7419.6 g/gmol) by lipase-catalyzed reactions in a batch reactor as reported by Santos et al. [59] were pressure (145 bar), temperature (65°C), agitation (200 rpm), and enzyme concentration (10 wt%). Enzymatic ring-opening polymerization in a variable-volume view reactor was used to synthesize poly(ε-caprolactone) (90 wt%) with a number-average molecular weight (13,700 Da), weight-average molecular weight (22,200 Da), and a polydispersity index (1.2–1.7) [35].

In 2017, the first enzyme (Novozym 435 lipase) catalyzed ring-opening in scCO_2 of the cyclic macrolactone, ω -pentadecalactone, was applied in the synthesis of poly (ω-pentadecalactone), resulting in 60 wt% yield and 20,700–33,000 g/mol as molecular weight [60]. The polymer has potential applications in biomedicine fields such as cartilage and bone tissue regeneration [60]. The dibutyltin dimethoxide initiation of the ring-opening polymerization of ε-caprolactone at temperature (40°C) and pressure (210–215 bar) was reported by Stassin et al. [63]. Initiators such as metal alkoxides ($Y(OiPr)_3$, $La(OiPr)_3$, and $Al(OiPr)_3$) have been used in the anionic ring-opening polymerization of ε-caprolactone in $scCO_2$ [37]. Differences in reactivity were observed in terms of metal and temperature [37]. Non-toxic hydrocarbon stabilizers have been developed for ring opening polymerization reaction in $scCO_2$ for use in the one-step synthesis of poly(glycolide) [64]. Biopolymeric particles for potential application in drug delivery systems were prepared using fluorine- or siloxane-based dispersants for the free radical dispersion copolymerization of N-vinyl-2-pyrrolidone and 2-methylene-1,3-dioxepane in $scCO_2$ with 2,2′-azobisisobutyronitrile as the initiator [44].

The effect on particle size and copolymer distribution by dispersant type and concentration, initiator concentration, ratios of co-monomers, and temperature was investigated [44]. Schmidt et al. [54] were the first to report the synthesis of polyglycolide (31,200 g/mol) using stabilizer-free $scCO_2$ via ring-opening polymerization of diglycolide at 120°C, 530 bar, in 5 h. The catalysts and initiator used were tin(II) ethyl hexanoate and 1-dodecanol, respectively. The resultant polymer was a white powder, free of solvent or additive residues [54]. Ring-opening dispersion polymerization of L-lactide to prepare poly(L-lactide) in $scCO_2$ has also been described [65]. Synthesis of cyclic L- and D-polylactides of high stereochemical purity was recently achieved using thiourea additives and an organocatalyst in $scCO_2$ under CO_2 plasticizing polymerization [66].

3.2 Electrochemical synthesis of conducting polymers

Conducting polymers have been successfully applied in light emitting devices, electrochemical devices, and chemical and optical sensors [67]. Traditional methods of synthesis include chemical or electrochemical oxidation via the application of a charge transfer agent. These methods involve the use of hazardous reagents, while the use of supercritical fluids can provide for a non-hazardous reaction environment [67]. DeSimone and Yizeng [68] have been granted a patent on the synthesis of an electrically conductive polymer via the use of a charge transfer agent in $scCO_2$. Anderson et al. [69] were the first to have successfully synthesized conducting polymers (polypyrrole and polyaniline) electrochemically using $scCO_2$ as the solvent. A patent was granted to Mabrouk [67] on his work on the electrochemical synthesis of electrically conductive polymers in $scCO_2$. Park et al. [70] have also got a patent on the use of $scCO_2$ and other supercritical fluids for the electrochemical synthesis of conductive polymers, a metallized nanotube polymer composite, with applications in shielding of radiation and electromagnetic effects.

3.3 Synthesis of polyamides

Polyamides (R-CO-NH-R′) can be synthesized from the polymerization of one lactam monomer in CO_2 and from the condensation of a nylon salt and CO_2 via a ring-opening polymerization procedure [71]. The properties (strength and toughness) exhibited by polyamides make them suitable candidates in applications such as gears, bearings, fibers, adhesives, and coatings [71]. DeSimone et al. [71] have obtained a patent on the synthesis of polyamides in $scCO_2$. Zhang et al. [72] have synthesized high molecular weight nylon 46 via a one-step, three-stage (pre-condensation, post-condensation, and vacuum) polymerization method in $scCO_2$. The salt made from the monomer 1,6-adipic acid and 1,4-diaminobutane was used for the synthesis. The raw material, urea, has been used in the synthesis of nylon 1 in $scCO_2$, by which CO_2 also adsorbed the by-product, NH_3 [73].

3.4 Synthesis of polycarbonates

Polycarbonates are a class of thermoplastic polymers containing carbonate units (−O−(C=O)−O−) in their chemical structure [74]. The polymers of this group are tough, electrically resistant, have heat distortion resistance, and are made up of aliphatic and aromatic polycarbonates [75]. Aliphatic polycarbonates have promising applications in drug delivery systems, since they show biocompatibility and result in degradation products that are non-toxic [76]. The research groups of Odell [77] and DeSimone [75] were the first to get a patent on the synthesis of polycarbonates in $scCO_2$.

Tryznowski et al. [76] have synthesized, characterized, and investigated the solubility of aliphatic hyperbranched polycarbonates in $scCO_2$, whereby synthesis involved the synthesized monomer, 5-(4-hydroxybutyl)-1,3-dioxan-2-one, in the ring-opening polymerization. The transesterification reaction in $scCO_2$ between diphenyl carbonate and bisphenol-A to synthesize polycarbonate was investigated by Xiao et al. [78] and found the optimal conditions to be rotational speed (800 rpm), temperature (120°C) for a period of 50 h.

High molecular weight aromatic polycarbonate nanoparticles (30–140 nm, 3.1105 g/mol) have been synthesized in $scCO_2$ via the transesterification between bisphenol-A and diphenyl carbonate [79]. The stabilizer used for stable dispersions was the poly(propylene oxide)-poly(ethylene oxide)-poly(propylene oxide) tri-block copolymer with a CO_2-phobic anchor and CO_2-philic tail group [79]. The benefits of using $scCO_2$ were seen in the simple removal of the by-product and the phenol from the reaction system, as a result of excellent solubility in $scCO_2$ [80]. The extraction of the final polymer product through venting of the $scCO_2$ was also possible [79].

An aromatic polycarbonate (up to $4.28 \times 10(4)$ g/mol) was synthesized via a novel process involving the transesterification between bisphenol-A (BPA) and diphenyl carbonate based on the $scCO_2$ impinging stream technique [80]. $ScCO_2$ is an excellent reaction medium since heat and mass transfer between phases are enhanced due to the impinging stream [80]. Solid-state polymerization for the synthesis of poly(bisphenol A carbonate) with $scCO_2$ was studied by Gross et al. [81], who hypothesized that higher polymerization rates in $scCO_2$ were achieved due to the fact that the amorphous regions of the polymer were plasticized increasing the chain mobility and the rate at which the by-product, phenol, diffused out from synthesized polymer. In addition, lower reaction temperature suppressed the formation of side reactions and unwanted color developments.

3.5 Synthesis of fluoropolymers

The synthesis of fluoropolymers has been reviewed previously by DeYoung et al. [15], Wood et al. [21], and Du et al. [26]. DeSimone et al. [33] reported the synthesis of fluoropolymers via free-radical homogeneous polymerization in scCO_2 using acrylic monomers that are highly fluorinated. Liu et al. [82] developed a new mechanism using scCO_2 for the free radical homo-polymerization of 2,2,2-trifluoroethyl methacrylate and the copolymerization of 2,2,2-trifluoroethyl methacrylate and *N*-vinylpyrrolidone. Beuermann and Imram ul haq [83] published the first report on the polymerization of vinylidene fluoride (stabilizer-free) in homogenous phase in scCO_2 with di-*tert* butyl peroxide as the initiator.

Du et al. [26] have done pioneering work on both the homogeneous and heterogeneous polymerization of fluorinated monomers in scCO_2. Two categories have been reviewed by Du et al. [26]. Firstly, the side-chain polymers, which include the acrylate-based polymers, styrene-based polymers and the poly(vinyl ether)s and poly(oxetane)s. The second category includes the main-chain fluoropolymers, involving the tetrafluoroethylene (TFE)-based fluoropolymers, vinylidene fluoride (VDF) fluoropolymers, perfluoropolyethers, and fluorinated poly(olefin sulfone)s [26]. Nikitin et al. [25] have reviewed the use of scCO_2 in the development of superhydrophobic surfaces through the deposition of fluoropolymers onto substrates and core-shell micro-particles. Solubility of fluoropolymer in scCO_2 is beneficial as these are insoluble in organic solvents [25]. Xu et al. [84] synthesized stable poly(dodecafluoroheptyl methacrylate)-b-poly(methyl methacrylate) diblock copolymer nanoparticles via the RAFT dispersion polymerization of methyl methacrylate. Fluorinated polymeric nanoparticles have generated an interest for further research into their different possible applications [84].

3.6 Synthesis of core-shell polymers

Core-shell polymers (CSPs) have been used in biomaterials [39], drug delivery applications, catalysis, pollution control, surface coatings, and printing, in addition to use as impact modifiers [85–89]. Ramli et al. [89] have reviewed some of the fundamental principles of CSPs and these polymers have been defined as structured composite particles composed of the core (as either solid, liquid, or gas) and the shell (usually a solid) of the particles [87, 90]. The categories of CSPs have been classified as hydrogels, non-hydrogels, nano-, and micro-gels [89].

Chambon et al. [39] used polydimethylsiloxane (PDMS)-based reactive stabilizers to synthesize "core-shell" polyurethane-polydimethylsiloxane particles in $scCO_2$ through dispersion step-growth polymerization at 60°C and above 200 bar in a stainless steel reactor. The polyurethane-polydimethylsiloxane core-shell has potential biomaterials applications. Hydrogels and micro-gels have been the part of the category of smart materials (materials that with changes in pH, temperature, and ionic strength can adjust their properties) [91]. Micro-gels such as the core-shell poly(N-isopropylacrylamide) (PNIPAM)/carboxymethyl starch (CMS) micro-particles have been produced via the graft copolymerization of N-isopropylacrylamide (NIPAM) from CMS at 70°C in $scCO_2$. Stabilization was achieved through the use of the CMS and crosslinker [92]. Abdul Haq et al. [91] have reviewed the mechanical properties of PNIPAM-based hydrogels and have discussed three potential applications: soft robotics, injectable hydrogels, and shape memory hydrogels.

The development of hydrogels using nanoparticles due to their improved mechanical strength is becoming a research area of great interest [91]. Various applications in the areas of medical biotechnology, molecular bio-imaging, cancer therapy, and drug delivery can be achieved through the use of core-shell nanoparticles [93]. Pérez-Salinas et al. [94] synthesized hydrogels from 2-hydroxyethylene methacrylate and ethylene glycol dimethacrylate via reversible addition-fragmentation radical transfer (RAFT) and free radical copolymerization in $scCO_2$. Krytox 157 FSL was used as the dispersion agent [Perez]. Significant differences in the two hydrogels were observed with the RAFT synthesized hydrogel being less heterogeneous [94].

In 1997, the synthesis of aerogels (monolithic silica, silsesquioxane gels) was reported for the first time by Loy et al. [95] via sol-gel polymerizations of alkoxysilane monomers in $scCO_2$. Sui et al. [40] have used hexamer building blocks to synthesize TiO_2 aerogels in $scCO_2$. Increase in the R value (4.0 to 5.5) affected the microstructures from being precipitate-like to micron-size rods to nanofibers, whilst pore size was related to the R value, pressure, and temperature [40]. Robust polysiloxane aerogels synthesized via a novel one-pot mechanism in $scCO_2$, were super-hydrophobic and flexible with high oil absorption ability [96]. This one-pot mechanism is less time consuming as compared to the traditional sol-gel mechanisms, with a clean product being obtained. This type of aerogel makes it suitable for applications in clean-up of oil spills and water purification [96]. Elmanovich et al. [97] further applied the one-pot hydrosilation mechanism for aerogel formation in $scCO_2$. The length of the divinyl part and the density of CO_2 have an effect on the aerogel properties [97]. The sol-gel method has been applied in the synthesis of zirconia (ZrO_2) particles (200 nm) for incorporation into poly(methyl methacrylate) (PMMA) via an in situ radical dispersion polymerization to prepare core-shell ZrO_2/PMMA composites [18]. Poly(dimethylsiloxane)-g-pyrrolidone carboxylic acid (Monosil PCA) was used as the stabilizer for the reaction. The novel method, described by Haldorai et al. [18], has also been applied to the development of polymer composites containing other inorganic oxides [18].

3.7 Synthesis of polymer nanocomposites

Polymer nanocomposites, or "nanoscale reinforced polymeric materials" [98], are hybrid materials [31] that are matrices (or the so-called polymeric host), having either inorganic or organic fillers with nanoscale distribution (10–100 nm). The fillers can be fibers, clusters, particles, or layered materials, carbon materials, metal oxides, clays, or semiconductor metallic nanocrystals [27, 99]. Carbon nanotubes, metals, semiconductors, and magnetic nanoparticles have been incorporated into polymers with the use of $scCO_2$ to produce polymeric nanocomposites [31]. Their unique properties allow for nanocomposites to be used in the manufacture of electronics, aeronautics, medical equipment [27], components for solar and fuel cells, electrolytes for batteries, thin-film capacitors, and biomaterials [99]. Nanoparticles (1–100 nm) in pure or in composite form have been suitable candidates for drug delivery systems, with a focus on controlled release of pharmaceuticals or biomolecules, due to their unique structure and properties, by which molecules can be entrapped or adsorbed [62, 100]. The use of $scCO_2$ in the synthesis of nanomaterials allows the development of unique materials, as the high surface areas and nano- and pore-structures are well-maintained [40]. In addition, the use of $scCO_2$ provides a cleaner process for synthesis reactions without use of toxic organic solvents [101].

Haldorai and Shim [31] reviewed the synthesis of polymeric nanocomposites in $scCO_2$ by describing the ex situ (solution blending and melt blending) and in situ methods. In the ex situ method, fillers are directly incorporated into the polymer via solution blending or melt blending [31]. The in situ method makes use of the polymerization process of monomers, by which fillers are then dispersed throughout the mixture [31]. Supercritical fluid technology is one of the methods by which polymer nanoparticles can be prepared, either by way of the polymerization of monomers or the dispersion of polymers [102]. Rao and Geckeler [102] have reviewed the various preparation methods and parameters for size-control of polymer nanoparticles. The two processes, namely "rapid expansion of supercritical solution" and "rapid expansion of supercritical solution into liquid solvent," for developing nanoparticles using supercritical fluids have been described in the review [102]. A review by Haldorai et al. [99] provides insight into the synthesis of polymer-inorganic filler nanocomposites in $scCO_2$. More recently, a review published by Crucho and Barros [103] discusses the several methods available to prepare polymeric nanoparticles with a section on the use of supercritical fluid technology.

Jerome et al. [104] have got a patent on the preparation of biodegradable and biocompatible nanocomposite aliphatic polyesters of high purity and enhanced thermomechanical properties in $scCO_2$. Other nanocomposites, such as montmorillonite/polystyrene nanocomposites [105], polystyrene/montmorillonite nanocomposites (free-radical precipitation polymerization) [106], poly(methyl methacrylate) and polystyrene/clay nanocomposites (pseudo-dispersion polymerizations) [107], polylactide/clay nanocomposites (in situ intercalative polymerization) [108] single-wall carbon nanotube/poly(methyl methacrylate) nanocomposite (in-situ polymerization) [109], silica/poly(divinylbenzene) composite particles (dispersion polymerization) [110], and silver-polystyrene nanocomposite particles (ex situ dispersion of silver nanoparticles within the monomers and emulsion polymerization) [111], have been synthesized in $scCO_2$.

The synthesis of polymer/nanocomposites in particular is gaining greater attention, since these have been shown to possess antimicrobial activities and are useful candidates in catalysis reactions [111, 112]. The method developed by Kamrupt et al. [111] to synthesize silver/polystyrene nanocomposite particles can be applied in the synthesis of a number of other metal-polymer nanocomposites. Yang et al. [113] have also reported a novel method on the synthesis of metal-polymer nanocomposites by in situ suspension and bulk polymerizations in $scCO_2$. Gold nanoparticles (3.7 to 6.6 nm) have successfully been incorporated into polymers such as polyamide, polypropylene, and poly(tetrafluoroethylene) [114].

Dong et al. [105] have synthesized montmorillonite/polystyrene nanocomposites by incorporation of styrene and *N,N′*-azobis(isobutyronitrile) (initsiator) into montmorillonite galleries at pressure (12 MPa) and temperature (65°C). Free-radical solid-state polymerization of macroscopic crystalline monomers (monomers, acryloyl β-alanine, and methacryloyl β-alanine) in $scCO_2$ has been applied by Ganapathy et al. [101] to produce polymer nanoparticles, poly(acryloyl β-alanine) (94 nm), and poly(methacryloyl β-alanine, Mw = 3.8 × 105 g/mol, yield ±96%) (102 nm). The reaction was conducted at temperature (65°C) and pressure (34.5 MPa) for 24 h and initiated by azobisisobutyronitrile [101], with CO_2 being vented off at the end to obtain the polymeric powder [101].

The N-isopropyl acrylamide monomer has been polymerized in the presence of ferrite nanoparticles via the use of $scCO_2$ [115]. The resultant polymer nanocomposite had 21%–77% of drug release activity, which was found to be pH-dependent, and 81%–93% of cell viability (breast cancer cells) [115]. Hua and Hua [116] reported a new method regarding the use of the bovine serum albumin-poly(methyl methacrylate) conjugate via precipitation in $scCO_2$ to produce polymer nanoparticles. Camptothecin, an antitumor drug, was loaded onto the nanoparticles. The resultant drug-encapsulated nanoparticle was shown to have greater ability in terminating cancer cells as compared to the free drug [116]. Ajiboye et al. [62] have investigated the effects of polymer and surfactant concentrations as well as polymer-to-surfactant weight ratio on the particle size and surface morphology of polycaprolactone nanoparticles. Supercritical fluid extraction (CO_2) of emulsions was used to produce the polycaprolactone (PCL) nanoparticles (monodisperse and spherical) in a short reaction period of one hour in batch mode [62]. The $scCO_2$ can be used in the particle formation method of "supercritical fluid extraction emulsion" to extract solvent or oil phase, resulting in the precipitation of the solute and finally the nanoparticles suspension [62, 102].

A novel technique based on the "CO_2-expanded liquid concept" to directly synthesize polymeric nanoparticles (100 nm) through the polymerization of methyl methacrylate has been reported by Pu et al. [100]. At the stage of the homogeneous expansion limit, an emulsion is induced due to the anti-solvent effect of dense CO_2 [100]. The polymerization utilized a macroazo initiator in CO_2-expanded ethanol at pressure (10 MPa) and temperature (70 °C) [100]. The utilization of CO_2 has been shown to produce more uniform nanoparticles with smaller size, as compared to reactions carried out without CO_2 [100].

The development of polymer-clay nanocomposites in $scCO_2$ is a research area of significant interest. Polymer-clay nanocomposites play vital roles in tensile strength, barrier properties, and thermal stability [117]. The fire resistance, mechanical characteristics, and gas permeability are improved via the addition of a small amount of the nanometric fillers [118]. A one-pot process using $scCO_2$ has been used for the synthesis of poly(styrene-*co*-acrylonitrile) (SAN)/clay nanocomposites [119]. PCL/clay masterbatches have been synthesized in situ in $scCO_2$, resulting in a ready-to-use powder with a high clay loading [118]. An in situ ring-opening polymerization of D,L-lactide in the presence of Cloisite 20A (CL20A) and Cloisite 30B (CL30B) clays has been used for the synthesis of PLA/clay nanocomposites in $scCO_2$ [108]. Kamrupi and Dolui [120] have applied emulsion polymerization to synthesize polystyrene/bentonite clay nanocomposites, for which ultrasonication was used to disperse the clay within the monomer followed by emulsion polymerization.

4. Conclusions

The use of $scCO_2$ has proven to be an effective alternate polymerization solvent to the organic solvents, not only in terms of cost savings mechanisms and environmental considerations, but also the unique effects on the properties of the polymer product. One-pot methods in batch and variable-volume reactor setups, including faster polymerization rates have been achieved with the use and optimization of the $scCO_2$ reaction process. Research activities have focused on developing continuous phase reaction systems that incorporate $scCO_2$ as the solvent for more effective polymerizations. The use of $scCO_2$ in large-scale applications has made it suitable for synthesis of polymers in industry. The development of polymer nanocomposites, in particular the core-shell nanocomposites, has provided new avenues for engineering applications of $scCO_2$.

References

[1] A. Roy, KPL International, August 18, KPL Blog, 2015, http://www.kplintl.com/blog/top-5-industrial-polymers/. Accessed 18 September 2018.

[2] American Chemical Society, Hermann Staudinger and the Foundation of Polymer Science, International Historic Chemical Landmark, 2018. https://www.acs.org/content/acs/en/education/whatischemistry/landmarks/staudingerpolymerscience.html. Accessed 22 September 2018.

[3] National Research Council, Polymer Science and Engineering: The Shifting Research Frontiers, The National Academies Press, Washington DC, 1994. https://doi.org/10.17226/2307.
[4] X. Zhang, S. Heinonen, E. Levänen, Applications of supercritical carbon dioxide in materials processing and synthesis, RSC Adv. 4 (2014) 61137–61152.
[5] K.M. Scholsky, Polymerization reactions at high pressure and supercritical conditions, J. Supercrit. Fluids 6 (1993) 103–127.
[6] K.A. Shaffer, J.M. DeSimone, Chain polymerizations in inert near- and supercritical fluids, Trends Polym. Sci. 3 (1995) 146–153.
[7] A.I. Cooper, J.M. DeSimone, Polymer synthesis and characterization in liquid/supercritical carbon dioxide, Curr. Opin. Solid St. M. 1 (1996) 761–768.
[8] D.A. Canelas, J.M. DeSimone, Polymerizations in liquid and supercritical carbon dioxide, in: D.A. Canelas, J.M. DeSimone, A. Harada, E. Ihara, K. Mashima, A. Nakamura, Y. Nakayama, H. Yasuda (Eds.), Metal Complex Catalysts Supercritical Fluid Polymerization Supramolecular Architecture. Advances in Polymer Science, Vol. 133, Springer, Berlin, Heidelberg, 1997, pp. 103–140.
[9] D.A. Canelas, A.L.C. Burke, J.M. DeSimone, Carbon dioxide as a continuous phase for polymer synthesis, Plastics Eng. 53 (1997) 37–40.
[10] A.I. Cooper, Polymer synthesis and processing using supercritical carbon dioxide, J. Mater. Chem. 10 (2000) 207–234.
[11] J.L. Kendall, D.A. Canelas, J.L. Young, J.M. DeSimone, Polymerizations in supercritical carbon dioxide, Chem. Rev. 99 (1999) 543–564.
[12] D. Sparacio, E.J. Beckman, Generation of microcellular biodegradable polymers using supercritical carbon dioxide, in: T.E. Long, M.O. Hunt (Eds.), Solvent-Free Polymerizations and Processes, American Chemical Society, Washington, DC, 1998, pp. 181–193.
[13] J.L. Young, J.M. DeSimone, Frontiers in green chemistry utilizing carbon dioxide for polymer synthesis and applications, Pure Appl. Chem. 72 (2000) 1357–1363.
[14] E. Kiran, Polymerizations and polymer modification reactions in near and supercritical fluids, in: E. Kiran, P.G. Debenedetti, C.J. Peters (Eds.), Supercritical Fluids: Fundamentals and Applications, Kluwer Academic Publishers, 2000, pp. 277–289.
[15] J.P. DeYoung, T.J. Romack, J.M. DeSimone, Synthesis of fluoropolymers in liquid and supercritical carbon dioxide solvent systems, in: G. Hougham, P.E. Cassidy, K. Johns, T. Davidson (Eds.), Fluoropolymers 1: Synthesis. Topics in Applied Chemistry, Springer, Boston, MA, 2002, pp. 191–205.
[16] G. Filardo, A. Galia, A. Giaconia, Polymer synthesis in supercritical carbon dioxide, in: M. Aresta (Ed.), Carbon Dioxide Recovery and Utilization, Kluwer Academic Publishers, 2003, pp. 181–195.
[17] J.L. Young, J.M. DeSimone, Synthesis and characterization of polymers: from polymeric micelles to step-growth polymerizations, in: J.M. DeSimone (Ed.), Green Chemistry Using Liquid and Supercritical Carbon Dioxide, Oxford University Press, Cary, 2003, p. 149.
[18] Y. Haldorai, T. Zong, J.J. Shim, Core-shell ZrO_2/PMMA composites via dispersion polymerization in supercritical fluid: Synthesis, characterization and mechanism, J. Appl. Polym. Sci. 123 (2012) 1176–1183.
[19] E.E. Said-Galiyev, I.V. Potostkaya, Y.S. Vygodskii, Supercritical carbon dioxide and polymers, Polym. Sci. Ser. C 46 (2004) 1–13.
[20] C.D. Wood, A.I. Cooper, J.M. DeSimone, Green synthesis of polymers using supercritical carbon dioxide, Curr. Opin. Solid St. Mat. Sci. 8 (2004) 325–331.
[21] C.D. Wood, J.C. Yarbrough, G.W. Roberts, J.M. DeSimone, Production of fluoropolymers in supercritical carbon dioxide, in: M.F. Kemmere, T. Meyer (Eds.), Supercritical Carbon Dioxide in Polymer Reaction Engineering, Wiley-VCH Verlag GmbH & Co. KGaA, Weinheim, Germany, 2005, pp. 189–204.
[22] P.A. Mueller, B. Bonavoglia, G. Storti, M. Morbidelli, Heterogeneous polymerization in supercritical carbon dioxide, in: M.F. Kemmere, T. Meyer (Eds.), Supercritical Carbon Dioxide in Polymer Reaction Engineering, Wiley-VCH Verlag GmbH & Co. KGaA, Weinheim, Germany, 2005, pp. 105–138.
[23] A.I. Cooper, Synthesis of advanced materials using supercritical fluids, in: M.F. Kemmere, T. Meyer (Eds.), Supercritical Carbon Dioxide in Polymer Reaction Engineering, Wiley-VCH Verlag GmbH & Co. KGaA, Weinheim, Germany, 2005, pp. 239–254.

[24] M.F. Kemmere, T. Meyer, Supercritical Carbon Dioxide: In Polymer Reaction Engineering, WILEY-VCH Verlag GmbH & Co. KGaA, Weinheim, 2005.
[25] L.N. Nikitin, M.O. Gallyamov, E.E. Said-Galiev, A.R. Khokhlov, V.M. Buznik, Supercritical carbon dioxide: a reactive medium for chemical processes involving fluoropolymers, Russ. J. Gen. Chem. 79 (2009) 578–588.
[26] L. Du, J.,.Y. Kelly, G.W. Roberts, J.M. DeSimone, Fluoropolymer synthesis in supercritical carbon dioxide, J. Supercrit. Fluids 47 (2009) 447–457.
[27] T.-C. Huang, J.-M. Yeh, C.-Y. Lai, Polymer nanocomposite coatings, in: F. Gao (Ed.), Advances in Polymer Nanocomposites Types and Applications, Woodhead Publishing Limited, 2012, pp. 605–638.
[28] F. Picchioni, Supercritical carbon dioxide and polymers: an interplay of science and technology, Polym. Int. 63 (2014) 1394–1399.
[29] C. Boyère, C. Jérôme, A. Debuigne, Input of supercritical carbon dioxide to polymer synthesis: an overview, Eur. Polym. J. 61 (2014) 45–63.
[30] E. Kiran, Supercritical fluids and polymers-the year in review-2014, J. Supercrit. Fluids 110 (2016) 126–153.
[31] Y. Haldorai, J.-J. Shim, Synthesis of polymer nanocomposites in supercritical carbon dioxide, in: V. Mittal (Ed.), Synthesis Techniques for Polymer Nanocomposites, Wiley-VCH Verlag GmbH & Co. KGaA, 2015, pp. 267–290.
[32] X. Wang, L. Shen, Z. An, Dispersion polymerization in environmentally benign solvents via reversible deactivation radical polymerization, Prog. Polym. Sci. 83 (2018) 1–27.
[33] J.M. DeSimone, Z. Guan, C.S. Elsbernd, Synthesis of fluoropolymer in supercritical carbon dioxide, Science 257 (1992) 945–947.
[34] A.-F. Mingotaud, F. Cansell, N. Gilbert, A. Soum, Cationic and anionic ring-opening polymerization in supercritical CO2. Preliminary results, Polym. J. 31 (1999) 406–410.
[35] S.R.C. Rosso, E. Bianchin, D. De Oliveira, J.V. Oliveira, S.R.S. Ferreira, Enzymatic synthesis of poly (ε-caprolactone) in supercritical carbon dioxide medium by means of a variable-volume view reactor, J. Supercrit. Fluids 79 (2013) 133–141.
[36] A.R.C. Morais, A.M. da Costa Lopes, R. Bogel-Łukasik, Carbon dioxide in biomass processing: Contributions to the green biorefinery concept, Chem. Rev. 115 (2015) 3–27.
[37] V. Bergeot, T. Tassaing, M. Besnard, F. Cansell, A.F. Mingotaud, Anionic ring opening polymerization of ε-caprolactone in supercritical carbon dioxide: parameters influencing the reactivity, J. Supercrit. Fluids 28 (2004) 249–261.
[38] M.F. Kemmere, Supercritical carbon dioxide for sustainable polymer processes, in: M.F. Kemmere, T. Meyer (Eds.), Supercritical Carbon Dioxide: In Polymer Reaction Engineering, WILEY-VCH Verlag GmbH & Co. KGaA, Weinheim, 2005, pp. 1–14.
[39] P. Chambon, E. Cloutet, H. Cramail, T. Tassaing, M. Besnard, Synthesis of core-shell polyurethane–polydimethylsiloxane particles in cyclohexane and in supercritical carbon dioxide used as dispersant media: a comparative investigation, Polymer 46 (2005) 1057–1066.
[40] R. Sui, A. Rizkalla, P.A. Charpentier, Experimental study on the morphology and porosity of TiO_2 aerogels synthesized in supercritical carbon dioxide, Microp. Mesop. Mat. 142 (2011) 688–695.
[41] X. Liao, H. Zhang, T. He, Preparation of porous biodegradable polymer and its nanocomposites by supercritical CO_2 foaming for tissue engineering, J. Nanomat. Article ID 836394. (2012). https://doi.org/10.1155/2012/836394.
[42] J.W. King, K.J. Srinivas, Multiple unit processing using sub- and supercritical fluids, Supercrit. Fluids. 47 (2009) 598–610.
[43] E.E. Maury, H.J. Batten, S.K. Killian, Y.Z. Menceloglu, J.R. Combes, J.M. DeSimone, Graft copolymer surfactants for supercritical carbon dioxide applications, ACS Polym. Prepr. 34: (1993) 664–665.
[44] S. Kwon, K. Lee, H. Kim, Y.-W. Lee, W. Bae, Synthesis of biocompatible and biodegradable polymer particles in supercritical carbon dioxide, Colloid Polym. Sci. 286 (2008) 1181–1191.
[45] J.M. DeSimone, E.E. Maury, Y.Z. Menceloglu, J.B. McClain, T.J. Romack, J. R. Combes, Dispersion polymerizations in supercritical carbon dioxide, Science 265 (1994) 356–359.
[46] Applied Separations Inc., Polymers: Uses of supercritical fluids, http://www.appliedseparations.com/polymers.html. Accessed 4 December 2018.

[47] S.P. Nalawade, F. Picchioni, L.P.M.B. Janssen, Supercritical carbon dioxide as a green solvent for processing polymer melts: processing aspects and applications, Prog. Polym. Sci. 31 (2006) 19–43.
[48] G. Jaramillo-Soto, M.L. Castellanos-Cárdenas, P.R. García-Morán, E. Vivaldo-Lima, G. Luna-Bárcenas, A. Penlidis, Simulation of RAFT dispersion polymerization in supercritical carbon dioxide, Macromol. Theory Simul. 17 (2008) 280–289.
[49] B. Grignard, B. Gilbert, C. Malherbe, C. Jerome, C. Detrembleur, C, online monitoring of heterogeneous polymerizations in supercritical carbon dioxide by Raman spectroscopy, Chem. Phys. Chem. 13 (2012) 2666–2670.
[50] J.J. Haven, T. Junkers, Online monitoring of polymerizations: current status, Eur. J. Org. Chem. (2017) 6474–6482.
[51] J. Enderle, S. Blanchard, J. Bronzino, Biomaterials: Properties, Types and Applications. Introduction to Biomedical Engineering, Academic Press, San Diego, CA, 2005.
[52] K. Pal, A.T. Paulson, D. Rousseau, Biopolymers in controlled-release delivery systems, in: S. Kasapis, I.T. Norton, J.B. Ubbink (Eds.), Modern Biopolymer Science Bridging the Divide between Fundamental Treatise and Industrial Application, Elsevier Inc, 2009, pp. 519–557.
[53] L. Jiang, J. Zhang, Biodegradable polymers and polymer blends, in: S. Ebnesajjad (Ed.), Handbook of Biopolymers and Biodegradable Plastics, Elsevier Inc, 2013, pp. 109–128.
[54] C. Schmidt, M. Behl, A. Lendlein, S. Beuermann, Synthesis of high molecular weight polyglycolide in supercritical carbon dioxide, RSC Adv. 4 (2014) 35099–35105.
[55] H. Ltd, Production Process for Polylactic Acid (PLA), https://www.hitachi.com/businesses/infrastructure/product_site/ip/process/pla.html, 2018. Accessed 12 November 2018.
[56] B. Gupta, N. Revagade, J. Hilborn, Poly(lactic acid) fiber: an overview, Prog. Polym. Sci. 32 (2007) 455–482.
[57] T. Takamoto, H. Uyama, S. Kobayashi, Lipase-Catalyzed Synthesis of Aliphatic Polyesters in Supercritical Carbon Dioxide, e-Polymers (2001) (no. 004).
[58] F.C. Loeker, C.J. Duxbury, R. Kumar, W. Gao, R.A. Gross, S.M. Howdle, Enzyme-catalyzed ring-opening polymerization of ε-caprolactone in supercritical carbon dioxide, Macromolecules 37 (2004) 2450–2453.
[59] R.D. Santos, S.R. Rosso Comim, D. de Oliveira, H. Treichel, M. Di Luccio, S.R.S. Ferreira, J. Vladimir Oliveira, Lipase-catalyzed synthesis of poly(e-caprolactone) in supercritical carbon dioxide, Biocatal. Agric. Biotechnol. 1 (2012) 280–283.
[60] A.E. Polloni, J.G. Veneral, E.A. Rebelatto, D. de Oliveira, J.V. Oliveira, P.H.H. Araújo, C. Sayer, Enzymatic ring opening polymerization of ω-pentadecalactone using supercritical carbon dioxide, J. Supercrit. Fluids 119 (2017) 221–228.
[61] K.J. Thurecht, A. Heise, M. Degeus, S. Villarroya, J. Zhou, M.F. Wyatt, S.M. Howdle, Kinetics of enzymatic ring-opening polymerization of caprolactone in supercritical carbon dioxide, Macromolecules 39 (2006) 7967–7972.
[62] A.L. Ajiboye, V. Trivedi, J.C. Mitchell, Preparation of polycaprolactone nanoparticles via supercritical carbon dioxide extraction of emulsions, Drug Deliv. Transl. Res. 8 (2018) 1790–1796.
[63] F. Stassin, O. Halleux, R.R. Jérome, Ring-opening polymerization of ε-caprolactone in supercritical carbon dioxide, Macromolecules 34 (2001) 775–781.
[64] D. Bratton, M. Brown, S.M. Howdle, Synthesis of poly(glycolide) in supercritical carbon dioxide in the presence of a hydrocarbon stabiliser, Chem. Commun. (2004) 808–809.
[65] D. Jinjun, Z. Shiping, W. Jingchang, L. Zhijun, Z. Li, Ring-opening dispersion polymerization of L-lactide with polylactide stabilizer in supercritical carbon dioxide, Austral. J. Chem. 70 (2016) 712–717.
[66] N. Mase, Moniruzzaman, S. Yamamoto, Y. Nakaya, K. Sato, T. Narumi, Organocatalytic stereoselective cyclic polylactide synthesis in supercritical carbon dioxide under plasticizing conditions, Polymers 10 (2018) 713, https://doi.org/10.3390/polym10070713.
[67] P.A. Mabrouk, Electrochemical Synthesis and Processing of Conducting Polymers in Supercritical Media, US Patent No. 7247230 B2, Jul. 24, 2007.
[68] J.M. DeSimone, Y. Ni, Synthesis of Conductive Polymers in Liquid and Supercritical CO_2, United States Patent 5855819, Jan 05, 1999.

[69] P.E. Anderson, R.N. Badlani, J. Mayer, P.A. Mabrouk, Electrochemical synthesis and characterization of conducting polymers in supercritical carbon dioxide, J. Am. Chem. Soc. 124 (2002) 10284–10285.
[70] C. Park, J.S. Harrison, N. Nazem, L.T. Taylor, J.H. Kang, J.-W. Kim, G. Sauti, P.T. Lillehei, S.E. Lowther, Metallized nanotube polymer composite (MNPC) and methods for making same, US patent 9550870 B2, Jan. 24, 2017.
[71] J.M. DeSimone, R. Givens, Y. Ni, Synthesis of Polyamides in Liquid and Supercritical CO2, Patent US006025459A. Feb. 15, 2000.
[72] Z. Zhang, K. Huang, Z. Liu, Synthesis of high molecular weight nylon 46 in supercritical carbon dioxide, Macromolecules 44 (2011) 820–825.
[73] D. Yuan, J. Bao, Y. Ren, W. Li, L. Huang, X. Cai, Synthesis of nylon 1 in supercritical carbon dioxide and its crystallization behavior effect on nylon 11, Cryst. Eng. Comm. 20 (2018) 4676–4684.
[74] Y. Wang, D.J. Darensbourg, Carbon dioxide-based functional polycarbonates: metal catalysed copolymerization of CO2 and epoxides, Coord. Chem. Rev. 372 (2018) 85–100.
[75] J.M. DeSimone, R. Givens, M. Jikei, J.D. Cohen, Synthesis of Polycarbonates Using CO2, US Patent No. 6288202 B1, Sep. 11, 2001.
[76] M. Tryznowski, K. Tomczyk, Z. Fraś, J. Gregorowicz, G. Rokicki, E. Wawrzyńska, P.G. Parzuchowski, Aliphatic hyperbranched polycarbonates: synthesis, characterization, and solubility in supercritical carbon dioxide, Macromolecules 45 (2012) 6819–6829.
[77] P.G. Odell, Polycarbonate Process with Supercritical Carbon Dioxide, US Patent No. 5698665, Dec. 16, 1997.
[78] Y. Xiao, Y. Wu, C. Wang, W. Ying, Synthesis of polycarbonate in supercritical CO_2, Huagong Xuebao/J. Chem. Indust. Engin. (China). 58 (2007) 2403–2407.
[79] J.-Y. Lee, C.-H. Song, J.-I. Kim, J.-H. Kim, Preparation of aromatic polycarbonate nanoparticles using supercritical carbon dioxide, J. Nanopart. Res. 4 (2002) 53–59.
[80] Y. Xiao, A. Zhang, Preparation and characterization of bisphenol a polycarbonate facilitated by supercritical impinging stream, Adv. Mat. Res. 875–877 (2014) 101–105.
[81] S.M. Gross, G.W. Roberts, D.J. Kiserow, J.M. DeSimone, Crystallization and solid-state polymerization of poly(bisphenol a carbonate) facilitated by supercritical CO_2, Macromolecules 37 (2000) 40–45.
[82] Z.-T. Liu, J.-G. Chen, Z.-W. Liu, J. Lu, New process for synthesizing fluorinated polymers in supercritical carbon dioxide, Macromolecules 41 (2008) 6987–6992.
[83] S. Beuermann, M. Imran ul-haq, Homogeneous phase polymerization of vinylidene fluoride in supercritical carbon dioxide, J. Polym. Sci, Part A, Polym. Chem. 45 (2007) 5626–5635.
[84] Q.L. Xu, Z. Huo, J. Ma, B. Geng, U. Azhar, L. Zhang, S. Zhang, Synthesis of fluorinated nanoparticles via RAFT dispersion polymerization-induced self-assembly using fluorinated macro-RAFT agents in supercritical carbon dioxide, RSC Adv. 7 (2017) 51612–51620.
[85] N. Sasa, T. Yamaoka, Surface-activated photopolymer microgels, Adv. Mater. 6 (1994) 417–421.
[86] W.-H. Li, H.D.H. Stöver, Monodisperse cross-linked Core-Shell polymer microspheres by precipitation polymerization, Macromolecules 33 (2000) 4354–4360.
[87] C.D. Jones, L.A. Lyon, Shell-restricted swelling and core compression in poly(N-isopropylacrylamide) core–shell microgels, Macromolecules 36 (2003) 1988–1993.
[88] A.K. Khan, B.C. Ray, S.K. Dolui, Preparation of core–shell emulsion polymer and optimization of shell composition with respect to opacity of paint film, Prog. Org. Coat. 62 (2008) 65–70.
[89] R.A. Ramli, W.A. Laftah, S. Hashim, Core–shell polymers: a review, RSC Adv. 3 (2013) 15543–15565.
[90] J.-W. Ha, I.J. Park, S.-B. Lee, D.-K. Kim, Preparation and characterization of core-shell particles containing perfluoroalkyl acrylate in the shell, Macromolecules 35 (2002) 6811–6818.
[91] M. Abdul Haq, Y. Su, D. Wang, Mechanical properties of PNIPAM based hydrogels: A review, Mater. Sci. Eng. C 70 (2017) 842–855.
[92] L. Cao, L. Chen, X. Chen, L. Zuo, Z. Li, Synthesis of smart core–shell polymer in supercritical carbon dioxide, Polymer 47 (2006) 4588–4595.

[93] K.S. Kumar, V.B. Kumar, P. Paik, Recent advancement in functional core-shell nanoparticles of polymers: synthesis, physical properties, and applications in medical biotechnology, J. Nanopart. 2013 (2013), 24 pages. https://doi.org/10.1155/2013/672059.
[94] P. Pérez-Salinas, G. Jaramillo-Soto, A. Rosas-Aburto, H. Vázquez-Torres, M.J. Bernad-Bernad, Á. Licea-Claverie, E. Vivaldo-Lima, Comparison of polymer networks synthesized by conventional free radical and raft copolymerization processes in supercritical carbon dioxide, Processes 5 (2017) 1–26.
[95] D.A. Loy, E.M. Russick, S.A. Yamanaka, B.M. Baugher, K.J. Shea, Direct formation of aerogels by sol-gel polymerizations of alkoxysilanes in supercritical carbon dioxide, Chem. Mater. 9 (1997) 2264–2268.
[96] F. Zou, L. Peng, W. Fu, J. Zhang, Z. Li, Flexible superhydrophobic polysiloxane aerogels for oil/water separation via one-pot synthesis in supercritical CO_2, RSC Adv. 5 (2015) 76346–76351.
[97] I.V. Elmanovich, T.A. Pryakhina, V.G. Vasil'ev, M.O. Gallyamov, A.M. Muzafarov, A study of the hydrosilylation approach to a one-pot synthesis of silicone aerogels in supercritical CO_2, J. Supercrit. Fluids 133 (2018) 512–518.
[98] D.L. Tomasko, X. Han, D. Liu, W. Gao, Supercritical fluid applications in polymer nanocomposites, Curr. Opin. Solid State Mater. Sci. 7 (2003) 407–412.
[99] Y. Haldorai, J.-J. Shim, K.T. Lim, Synthesis of polymer-inorganic filler nanocomposites in supercritical CO_2, J. Supercrit. Fluids 71 (2012) 45–63.
[100] D.W. Pu, M. Soh, P.B. Zetterlund, F.P. Lucien, Exploiting the homogeneous expansion limit of CO_2-expanded media for the synthesis of polymeric nanoparticles, J. Supercrit. Fluids 78 (2013) 89–94.
[101] H.S. Ganapathy, S.Y. Park, W.-K. Lee, J.M. Park, K.T. Lim, Polymeric nanoparticles from macroscopic crystalline monomers by facile solid-state polymerization in supercritical CO_2, J. Supercrit. Fluids 51 (2009) 264–269.
[102] J.P. Rao, K.E. Geckeler, Polymer nanoparticles: preparation techniques and size-control parameters, Prog. Polym. Sci. 36 (2011) 887–913.
[103] C.I.C. Crucho, M.T. Barros, Polymeric nanoparticles: a study on the preparation variables and characterization methods, Mater. Sci. Eng. C 80 (2017) 771–784.
[104] R. Jerome, C. Calberg, F. Stassin, O. Halleux, P. Dubois, N. Panoustier, M. Alexandre, B. Lepoittevin, Nanocomposite Polyester Preparation Method, US Patent 2004/0106720 A1, Jun. 3, 2004.
[105] Z. Dong, Z. Liu, J. Zhang, B. Han, D. Sun, Y. Wang, Y. Huang, Synthesis of montmorillonite/polystyrene nanocomposites in supercritical carbon dioxide, J. Appl. Polym. Sci. 94 (2004) 1194–1197.
[106] C. Yan, L. Ma, J. Yang, J. Preparation of polystyrene/montmorillonite nanocomposites in supercritical carbon dioxide, J. Appl. Polym. Sci. 98 (2005) 22–28.
[107] Q. Zhao, E.T. Samulski, A comparative study of poly(methyl methacrylate) and polystyrene/clay nanocomposites prepared in supercritical carbon dioxide, Polymer 47 (2006) 663–671.
[108] L. Urbanczyk, F. Ngoundjo, M. Alexandre, C. Jérôme, C. Detrembleur, C. Calberg, Synthesis of polylactide/clay nanocomposites by *in situ* intercalative polymerization in supercritical carbon dioxide, Eur. Polym. J. 45 (2009) 643–648.
[109] B. Yue, Y. Wang, C.Y. Huang, R. Pfeffer, Z. Iqbal, Polymeric nanocomposites of functionalized carbon nanotubes synthesized in supercritical CO_2, J. Nanosci. Nanotechnol. 7 (2007) 994–1000.
[110] M.H. Woo, K.M. Do, M.J. Moon, H.G. Kim, Y.S. Gal, K.T. Lim, Preparation of silica/poly(divinylbenzene) composite particles by dispersion polymerization in supercritical carbon dioxide, Compos. Interf. 17 (2010) 495–503.
[111] I.R. Kamrupi, P. Phukon, B.K. Konwer, S.K. Dolui, Synthesis of silver-polystyrene nanocomposite particles using water in supercritical carbon dioxide medium and its antimicrobial activity, J. Supercrit. Fluids 55 (2011) 1089–1094.
[112] V.V. Kopeikin, E.F. Panarin, Water-soluble nanocomposites of zerovalent metallic silver with enhanced antimicrobial activity, Doklady Chem. 380 (2001) 277–279.
[113] J. Yang, T. Hasell, W. Wang, S.M. Howdle, A novel synthetic route to metal-polymer nanocomposites by in situ suspension and bulk polymerizations, Eur. Polym. J. 44 (2008) 1331–1336.

[114] B. Wong, S. Yoda, S.M. Howdle, The preparation of gold nanoparticle composites using supercritical carbon dioxide, J. Supercrit. Fluids 42 (2007) 282–287.
[115] G. Bisht, M.G.H. Zaidi, S. Rayamajhi, Supercritical carbon dioxide-assisted synthesis of stimuli-responsive magnetic poly(N-isopropylacrylamide)-ferrite biocompatible nanocomposites for targeted and controlled drug delivery, Int. J. Polym. Mater. Polym. Biomater. 66 (2017) 708–716.
[116] M. Hua, X. Hua, Polymer nanoparticles prepared by supercritical carbon dioxide for in vivo anti-cancer drug delivery, Nano-Micro Lett. 6 (2014) 20–23.
[117] M. Manitiu, R.J. Bellair, S. Horsch, E. Gulari, R.M. Kannan, Supercritical carbon dioxide-processed dispersed polystyrene-clay nanocomposites, Macromolecules 41 (2008) 8038–8046.
[118] L. Urbanczyk, C. Calberg, F. Stassin, M. Alexandre, R. Jérôme, C. Jérôme, C. Detrembleur, Synthesis of PCL/clay masterbatches in supercritical carbon dioxide, Polymer 49 (2008) 3979–3986.
[119] L. Urbanczyk, C. Calberg, S. Benali, S. Bourbigot, E. Espuche, F. Gouanvé, P. Dubois, A. Germain, C. Jérôme, C. Detrembleur, M. Alexandre, Poly(caprolactone)/clay masterbatches prepared in supercritical CO_2 as efficient clay delamination promoters in poly(styrene-CO-acrylonitrile), J. Mater. Chem. 18 (2008) 4623–4630.
[120] I.R. Kamrupi, S.K. Dolui, Synthesis of polystyrene/bentonite clay nanocomposites by emulsion polymerization in aqueous and in supercritical carbon dioxide medium, J. Thermoplast. Compos. Mater. 45 (2011) 2437–2446.

CHAPTER 19

Organometallic compounds solubility in supercritical carbon dioxide ($SCCO_2$): Measurement techniques, variables affecting solubility, recent developments, and thermodynamic modeling

Shahryar Jafarinejad, Nader Vahdat
Department of Chemical Engineering, College of Engineering, Tuskegee University, Tuskegee, AL, United States

Contents

1. Introduction

In recent years, increasing attention has been focused on the application of supercritical fluids (SCFs) in a wide variety industries including chemical, biochemical, food, petroleum, extraction, separation, water and wastewater treatment, materials and particle preparation, polymer, and pharmaceutical industries [1–9]. In other words, supercritical fluid technology (SFT) is extremely important in a variety of green chemistry processing cases [10, 11].

Study of the chemistry and structure of compounds, including metal-carbon (M-C) bonds or composition, in which an organic molecule is bonded to a metal atom via an

Green Sustainable Process for Chemical and Environmental Engineering and Science
https://doi.org/10.1016/B978-0-12-817388-6.00019-2

oxygen or nitrogen atom, is commonly called organometallic chemistry [12]. Organometallic compounds can widely be applied as reagents in the production and processing of new and advanced inorganic substances, as catalysts or precursors in preparation of different chemicals, and as chemotherapy drugs or in the design of novel medical therapeutics [10, 12, 13].

Solubility data of the organometallic compounds in supercritical fluids (SCFs) are required for examining their feasibility, establishing optimum operating conditions, and designing of SCF processes [3, 10, 14–16]. In addition, better knowledge of the dissolution phenomenon can be provided by mathematical modeling of organometallic compounds solubility in SCFs. Furthermore, solubility modeling can provide an approach to estimate solubility under different operating conditions with a minimum of experimental data. All this can enhance the development and progress rate of an SCF process [3, 17]. In this chapter, types of organometallic compounds, their solubility in supercritical carbon dioxide ($SCCO_2$), measurement techniques, variables affecting solubility, recent developments, and thermodynamic modeling are discussed.

2. Organometallic compounds

Compounds with metal-carbon (M-C) bond in their structure are called organometallic compounds. Metals can be alkali, alkali earth, transition, main group metals, and lanthanides or actinides. Ligands can include organics such as alkenes, alkynes, and alkyls, as well as organic-like ligands such as phosphines, hydride, and CO [12, 18]. In organometallic compounds, metal and ligand orbitals interactions can be based on sigma and pi interactions, as illustrated in Fig. 1. Ligands may be categorized as sigma donor (e.g., NH_3, H_2O, H^-), pi donor (e.g., OH^-, Cl^-, RCO_2^-), and pi acceptor (e.g., CO, CN^-, PR_3). There is an electron pair in sigma donor ligands that can be donated to the empty orbital of a metal atom. Electrons donation can occur in pi donor ligands in a pi fashion applying a filled p orbital on a ligand. In other cases, electron density may also be accepted by the ligands from a metal using proper acceptor orbitals. For instance, the carbonyl ligand has empty pi* orbitals that may receive electron density from the metal. Rather strong bonding may be formed by this donation and acceptance [18].

Some of the supercritical fluid related studies on organometallic compounds are:

- Organotransition metal compounds application as catalysts for reactions in supercritical fluids media and in biphasic systems (e.g., homogeneously catalyzed propylene hydroformylation utilizing $Co_2(CO)_8$ precatalyst in $SCCO_2$) [19];
- Copper films deposition onto planar and etched silicon substrates by the hydrogen-assisted reduction of copper(I)(1,1,1,5,5,5,-hexafluoro-2,4-acetylacetonate)(1,5-cyclooctadiene) for advanced interconnecting structures [20];
- Organometallic compounds impregnation into polyethylene films using $SCCO_2$ for property enhancement [21];

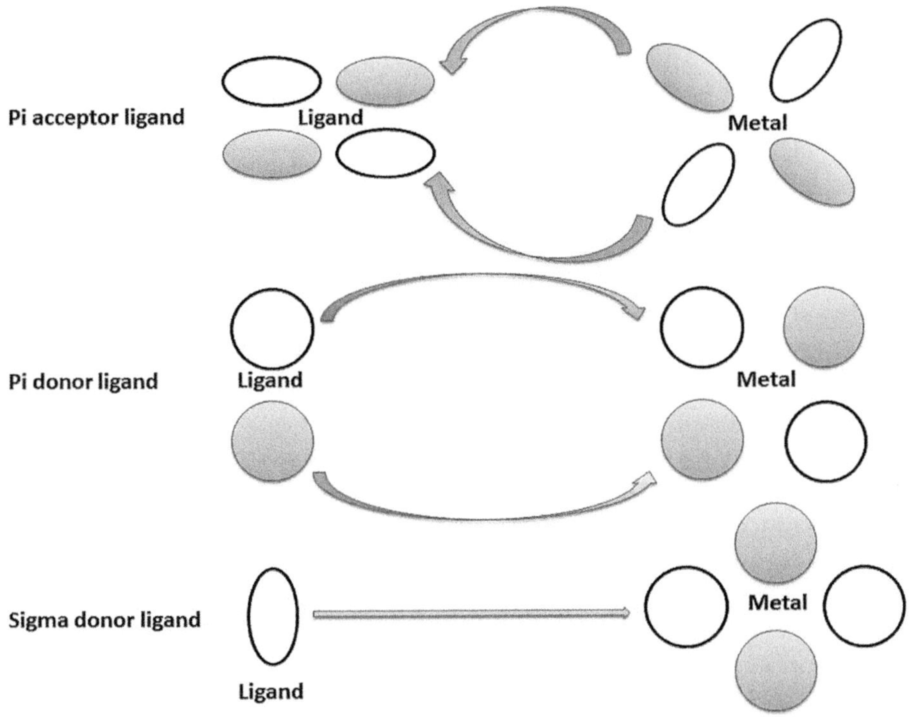

Fig. 1 Metal-ligand interactions. *(Modified from C. Erkey, Supercritical fluids and organometallic compounds: from recovery of trace metals to synthesis of nanostructured materials, Supercritical Fluid Science and Technology, Vol. 1, Series Editor, Erdogan Kiran, Elsevier, 2011, with permission from the Elsevier, license number: 4461730202464.)*

- Preparation of ultra-high molecular weight polyethylene (UHMWPE)/silver nanocomposites using SCF for biomedical applications [22];
- Polymer/metal nanocomposites synthesis through organometallic compounds impregnation into polymers using $SCCO_2$ [23];
- Synthesis of gold nanoparticles by the reduction of triphenylphosphine gold(I) perfluorooctanoate with dimethylamineborane in $SCCO_2$ phase [24];
- Organometallic precursor degradation in $SCCO_2$ for nanocrystals formation with controlled size distribution [25];
- Copolymerization of CO/4-*tert*-butylstyrene (TBS) and CO/styrene (ST) in $SCCO_2$ with palladium complexes as catalysts [26];
- Separation of heavy metals from liquid and solid matrices using SCF by formation of metal chelates [27];
- Mironization of titanocene dichloride by rapid expansion of supercritical solvent (RESS) and its ethylene polymerization [28]; and
- Synthesis of nanometer sized silver particles using $SCCO_2$ for impregnation of porous media with silver coordination compounds [29].

Erkey [12] and Wai and Wang [9] have reviewed SCF extraction of metals including organometallic complexes from aqueous media. An interesting review on solubility of organometallic complexes in $SCCO_2$ has been published by Teoh et al. [13].

3. Supercritical carbon dioxide ($SCCO_2$)

When a fluid is heated and compressed above its critical conditions (critical temperature and critical pressure) it is usually referred to as a supercritical fluid (SCF). Carbon dioxide (CO_2) is the most widely applied SCF. The critical conditions of CO_2 are relatively low ($T_c = 304\,K$ and $P_c = 73.7\,bar$), allowing it to be applied under mild conditions of temperatures without leaving harmful organic wastes. It is inert, non-toxic, non-flammable, non-explosive, and readily available at high purities and low costs [3, 9, 17]. The high diffusivity and low viscosity of $SCCO_2$ are also beneficial for direct separation and extraction of solutes from solid media, which can affect the rate, yield, design, and economy of the extraction process [9].

$SCCO_2$ has been utilized as a suitable and environmentally friendly solvent in different applications and industries including petroleum and petrochemical, pharmaceutical, food, material and polymer, and the separation of toxic pollutants from wastewater or soil [3, 4, 30].

4. Solubility

In general, solubility is a chemical property that refers to the ability for a given solute to dissolve in a solvent under specified conditions of temperature and pressure. In other words, the concentration or mole fraction of a solute in the SCF phase at equilibrium with the pure fluid is called solubility [31].

The solubility of the solute in the SCF can affect the rate, yield, design, and economy of the supercritical fluid processes and must be considered as an important factor in the SFT [32]. In reality, solubility data of the solutes in the SCFs are required for the feasibility study, establishing optimum operating conditions, and design of supercritical fluid processes [3, 10, 14–16, 32, 33]. In addition, the thermodynamic limit to a supercritical fluid process is provided by solubility data [13, 34].

4.1 Solubility measurement techniques

Solubility is usually determined as mole fraction or weight fraction of the solute (e.g., organometallic compound) in the SCF (e.g., $SCCO_2$) at equilibrium [32]. The dynamic and static techniques are the solubility measurement methods of the solutes in $SCCO_2$ [10, 16, 18, 31, 32, 35].

4.1.1 Dynamic technique

In the dynamic method, solute is constantly, but at slow flow rates, swept with $SCCO_2$. The $SCCO_2$ stream that leaves the sample vessel or cell becomes saturated with the solute (e.g., organometallic compound), and analyzed by chromatographic, spectroscopic, gravimetric, dielectric, and other methods [18, 32, 35]. A schematic diagram of the dynamic method experimental setup used by Aschenbrenner et al. [36] is depicted in Fig. 2. The system consists of a gas cylinder, a syringe pump (ISCO 100DX), a saturation vessel, an oven (ISCO SFX 2–10), a back pressure regulator (Jasco 880-81), and a sampling tube. Collection of the dissolved solute precipitated is done in a glass tube with a frit and the gravimetric method using a balance accurate to ± 0.0001 g (Sartorius Handy H110) is used to quantify the value of solute sampled in the glass tube [36]. In the dynamic technique, solubility can generally be computed as:

$$y_2 = \frac{n_2}{n_1 + n_2} = \frac{n_2}{Q_1 \rho_1 t + n_2} \tag{1}$$

where

n_1 and n_2 denote the moles of CO_2 and the solute sampled in time t, respectively;
Q_1 is volumetric flow rate of CO_2; and
ρ_1 denotes molar density of CO_2, both at the same conditions, which are usually in back-pressure regulator exit conditions [32].

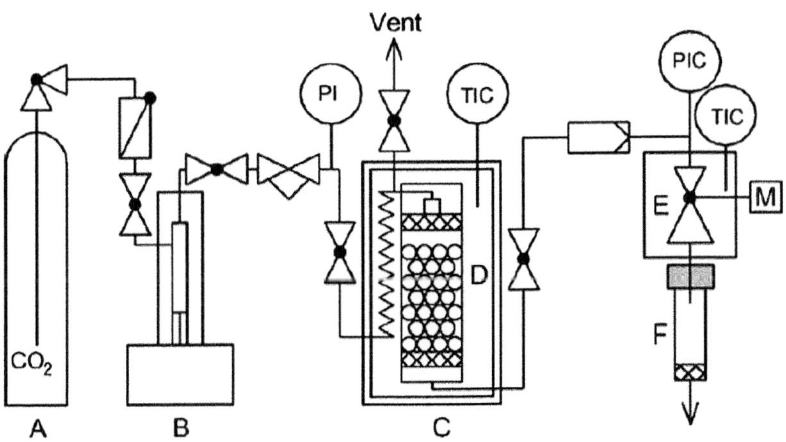

Fig. 2 A schematic diagram of the dynamic method experimental setup used by Aschenbrenner et al.: (A) Gas cylinder, (B) syringe pump, (C) oven, (D) saturation vessel, (E) back pressure regulator, and (F) sampling tube. *(From O. Aschenbrenner, S. Kemper, N. Dahmen, K. Schaber, E. Dinjus, Solubility of B-diketonates, cyclopentadienyls, and cyclooctadiene complexes with various metals in supercritical carbon dioxide, J. Supercrit. Fluids 41 (2007) 179–186, reproduced with permission from the Elsevier, license number: 4461730626721.)*

Achieving the saturation condition in the dynamic technique may be approximately fast even in the situation of low solubility [10, 16], which facilitates obtaining a lot of data. In addition, fractionation data may be acquired. The mass-transfer rate issues between $SCCO_2$ and solute and the restrictor and valve clogging by heavy solutes should be well-considered [32]. This technique requires a fully saturated flow and the technique for trapping and measuring the solute from the $SCCO_2$ phase may also become less precise at low solubility conditions (at lower pressures or higher temperatures) [10, 16].

4.1.2 Static techniques

In static techniques (batch mode), the solute can be in static contact with $SCCO_2$ for a long time in order to achieve equilibrium [18, 32, 35]. There are three different types of static techniques, including analytical, synthetic, and gravimetric, with respect to the sampling and the type of applied high-pressure cell. A schematic diagram of the static method experimental setup applied for the solubility measurement of $Ru(tmhd)_2(COD)$ and $Ru(acac)_3$ in $SCCO_2$ [37] is depicted in Fig. 3. In the static/analytical technique, a constant-volume vessel is applied. This vessel containing solute is charged with a known amount of $SCCO_2$. After equilibration, a small amount of the fluid phase containing the solute is removed and analyzed. In the static/synthetic technique, a variable-volume view vessel provided with a sapphire window for visual observations is used for adjusting operating conditions. A determined amount of solute is put in the cell and the cell is charged

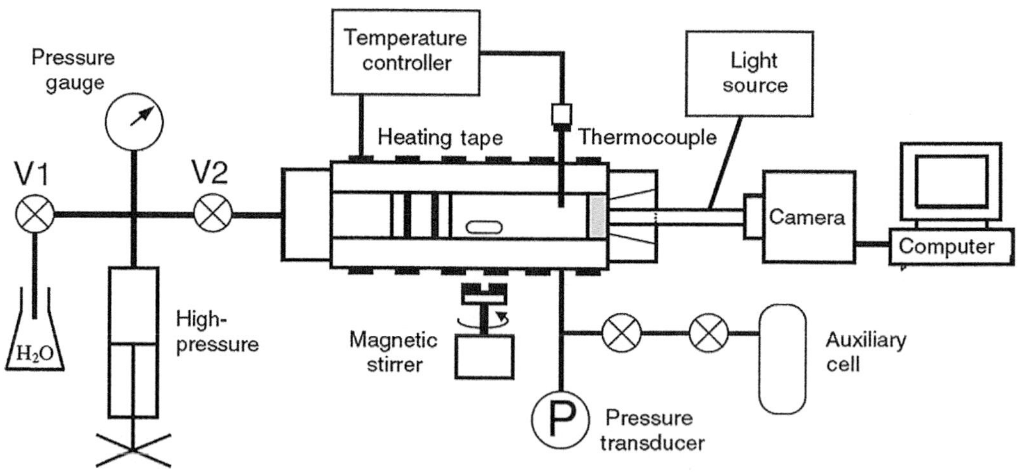

Fig. 3 A schematic diagram of the static method experimental setup applied for the solubility measurement of $Ru(tmhd)_2(COD)$ and $Ru(acac)_3$ in $SCCO_2$. *(From J. Morère, M.J. Tenorio, C. Pando, J.A.R. Renuncio, A. Cabañas, Solubility of two metal-organic ruthenium precursors in supercritical CO_2 and their application in supercritical fluid technology, J. Chem. Thermodyn. 58 (2013) 55–61, reproduced with permission from the Elsevier, license number: 4461730929888.)*

with a known amount of $SCCO_2$ until the entire solute dissolves in the $SCCO_2$. Then, pressure is reduced by gradually increasing the volume of vessel to slowly decrease the solubility of the mixture and cause precipitation. In the beginning of the precipitation (the cloud point), the solubility at this condition is computed using the value of solute and the value of $SCCO_2$ charged into the cell.

In the static/gravimetric technique, a specified amount of solute is charged in a vial and the vessel is charged with $SCCO_2$. The $SCCO_2$ reaches the vial, dissolves the solute, and brings it out. After equilibration, solubility can be obtained by depressurizing the vessel and gravimetrically measuring the mass of remaining solute in the vial [18, 32]. According to Gupta and Shim [32], the solubility of the solute can be computed by:

$$y_2 = \frac{\Delta n_2}{\Delta n_1 + \Delta n_2} \approx \frac{\Delta n_2}{\rho_1 \Delta V + \Delta n_2} \tag{2}$$

where

Δn_2 denotes the difference between initial and final moles of solute in the vial;

Δn_1 is the difference between total moles of CO_2 and that in the vial;

ΔV denotes the vessel minus vial volume; and

ρ_1 is the molar density of CO_2.

In comparison to the dynamic technique, remarkably less solute is applied in the static techniques [18] and accurate solubility data can be acquired. However, a long time is needed to reach the equilibrium in the static techniques and it can be difficult for low solubility solutes [10, 16].

4.2 Variables affecting organometallic compounds solubility in $SCCO_2$

Physical and chemical factors generally contribute to the solubility of an organometallic compound in $SCCO_2$. Temperature, pressure, intermolecular forces, free volume difference, clustering of $SCCO_2$ molecules around organometallic compounds, metal and ligand side-chains, and the crystal structure of metal complex can be the variables to affect the solubility of organometallic compounds in $SCCO_2$ [13, 38].

The solubility of organometallic compounds in $SCCO_2$ is increased with increasing pressure because of the density enhancement of $SCCO_2$ [9, 39–41]. The temperature offers significant effects on the organometallic compounds in $SCCO_2$ as it alters the vapor pressure of the solute as well as the $SCCO_2$ density. The effects can be different since the vapor pressure of a solid solute is usually favored by increased temperatures, while the density of $SCCO_2$ reduces with the rise in temperature [9]. In other words, the solubility can be enhanced with rising temperature at a constant pressure above the cross-over pressure (the vapor pressure effect dominates) and it can be reduced with enhancing temperature at a constant pressure below the cross-over pressure in the highly compressible region (the density effect dominates) [18]. Morere et al. [37] studied the solubility of ruthenium(III) acetylacetonate [Ru(acac)$_3$] and bis(2,2,6,6-tetramethyl-3,5-heptanedionato)(1,5-cyclooctadiene) ruthenium(II) [Ru(tmhd)$_2$(COD)] in $SCCO_2$ between the temperatures 40°C and 80°C and pressures up to 20 MPa. The solubilities of Ru(tmhd)$_2$(COD) were reported to vary from 2×10^{-4} to 9×10^{-4} whilst those for Ru(acac)$_3$ were reported to change from 2×10^{-5} to 13×10^{-5}. At constant temperature, the solubility is enhanced with pressure because of the higher density of $SCCO_2$. For a given pressure, it is reduced with temperature because of the reduction in density of $SCCO_2$. The cross-over phenomenon was not reported in the pressure range investigated in both complexes [37].

Intermolecular forces (IMFs) are attraction or repulsion forces that act between molecules and other types of neighboring matter. Different investigations have demonstrated that the presence of inter- or intra-molecular hydrogen bonds in an SCF such as $SCCO_2$ can increase solubility [13, 42–49]. In other words, addition of a modifier such as ethanol to $SCCO_2$, which can form hydrogen bonds with metal complex, usually can enhance its solubility [13, 49].

The empty space attainable to the molecule after deducting the hard-core volume of the molecule itself is called free volume [13, 50], which relates to the compressibility or expansivity of matter. The free volume difference between an organometallic compound and $SCCO_2$ can be related to the $SCCO_2$ density as the interaction level between molecules in supercritical phase is affected by it [38]. $SCCO_2$ density increase can result in the ability of $SCCO_2$ to solvate an organometallic compound [13, 51].

The clustering of $SCCO_2$ molecules around organometallic compounds can enhance their solubilities in $SCCO_2$. In addition, the metal and ligand side-chains and the crystal structure of a metal complex can affect the solubility of metal complexes in $SCCO_2$. Further details on these factors can be found in [12, 13, 36, 52–56]. Teoh et al. [13] reviewed the organometallic compounds solubility in $SCCO_2$ and reported that:

(i) the solubilities of metal chelates containing fluorinated ligands > aliphatic-substituted systems > phenyl-substituted systems;
(ii) the solubility of organometallic compounds with identical valence electron configuration usually reduces with enhancing atomic numbers; and
(iii) metal chelates with higher oxidation states usually exhibit higher solubility in $SCCO_2$ [13].

5. Literature review of organometallic compounds solubility in $SCCO_2$

Knowledge of how the organometallic compounds solubility changes with operating conditions is an important factor to design the $SCCO_2$ related processes of organometallic compounds [37]. In recent decades, different investigations have been performed around the world to measure and model the solubility of organometallic compounds in $SCCO_2$. Erkey [12, 18], Wai and Wang [9], Gupta and Shim [32], and Teoh et al. [13] have reviewed their solubilities in $SCCO_2$. A summary of these investigations is presented in Table 1. Teoh et al. [13] reported that fluorinated ligands have generally been

Table 1 Published solubility data for organometallic compounds in $SCCO_2$

Organometallic compound	Experimental conditions	Solubility (in mole fraction unless stated otherwise)	Reference
Ga(acac)$_3$	T: 60°C, P: 98–294 bar	≥3.01 mg/L	[56]
In(acac)$_3$	T: 60°C, P: 98–294 bar	≥2.63 mg/L	[56]
Mn(acac)$_3$	T: 60°C, P: 98–294 bar	≥1.26 mg/L	[56]
Li(acac)	T: 60°C, P: 98–294 bar	≥0.01 mg/L	[56]
Rh(acac)$_3$	T: 40°C, P: 100–298 bar	$(3.32–10.30) \times 10^{-5}$	[57]
Ru(acac)$_3$	T: 40°C, P: 100–300 bar	$(1.70–9.50) \times 10^{-5}$	[57]
Ag(acac)	T: 40°C, P: 103–300 bar	$(0.89–3.46) \times 10^{-7}$	[57]
UO$_2$(acac)	T: 40°C, P: 100–250 bar	$(4.18–26.7) \times 10^{-4}$ M	[58]
Co(acac)$_2$	T: 40°C, P: 160 bar	0.53×10^{-5} kg/m^3	[59]
	T: 40°C, P: 161–300 bar	$(3.91–8.59) \times 10^{-5}$	[59]
Cu(acac)$_2$	T: 60°C, P: 100–300 bar	$(0–3) \times 10^{-5}$ mol/mol	[36]
	T: 150–170°C, P: 120–220 bar	$(6.20–155) \times 10^{-4}$	[60]
	T: 40°C, P: 103.4–344.7 bar	0.750–2.307	[53]
	T: 35–55°C, ρ: 652–902.3 bar	$(0.508–9.126) \times 10^{-5}$	[34]
	T: 60°C, P: 98–294 bar	≥0.21 mg/L	[56]
	T: 40°C, P: 160 bar	7.3×10^{-5} kg/m^3	[59]
	T: 40–70°C, P: 143–300 bar	$(0.997–4.74) \times 10^{-5}$	[30]
Fe(acac)$_2$	T: 40–60°C, P: 100–201 bar	$(0.108–5.97) \times 10^{-4}$	[61]
Mn(acac)$_2$.2H$_2$O	T: 60°C, P: 98–294 bar	≥0.40 mg/L	[56]
Ni(acac)$_2$	T: 60°C, P: 100–300 bar	$(0–6) \times 10^{-5}$ mol/mol	[36]
	T: 40°C, P: 160 bar	3.4×10^{-5} kg/m^3	[59]
Pd(acac)$_2$	T: 60°C, P: 100–300 bar	$(0–6) \times 10^{-5}$ mol/mol	[36]
	T: 40°C, P: 100–300 bar	$(0.93–5.75) \times 10^{-5}$	[57]
Pt(acac)$_2$	T: 40°C, P: 105–290 bar	$(0.93–5.75) \times 10^{-5}$	[57]
Zn(acac)$_2$	T: 60°C, P: 98–294 bar	≥1.01 mg/L	[56]
	T: 60°C, P: 101–232 bar	$(9.5–90) \times 10^{-4}$ mol/L	[41]
	T: 40°C, P: 160 bar	7.2×10^{-5} kg/m^3	[59]
Co(acac)$_2$.2H$_2$O	T: 60°C, P: 98–294 bar	≥0.25 mg/L	[56]
Co(acac)$_3$	T: 60°C, P: 98–294 bar	≥0.62 mg/L	[56]
	T: 40–60°C, P: 40–97 bar	$(0.83–10.669) \times 10^{-5}$	[10]
Cr(acac)$_3$	T: 40°C, P: 103.4–344.7 bar	$(1.716–19.09) \times 10^{-5}$	[53]
	T: 60°C, P: 203–405 bar	$(2–3.5) \times 10^{-3}$ mol/L	[54]
Fe(acac)$_3$	T: 40–60°C, P: 98.2–353.7 bar	$(0.06–0.30) \times 10^{-3}$	[62]
	T: 40–60°C, P: 90.1–275.8 bar	$(8.75–576) \times 10^{-6}$	[63]
	T: 40–60°C, P: 101–176 bar	$(0.0509–2.89) \times 10^{-4}$	[61]
Y(acac)$_3$	T: 150–170°C, P: 120–220 bar	$(0.47–3.40) \times 10^{-5}$	[60]
Cu(bdc)$_2$	T: 60°C, P: 101–232 bar	$(1.3–72) \times 10^{-5}$ mol/L	[41]
Hg(bdc)$_2$	T: 60°C, P: 101–232 bar	$(5.6–56) \times 10^{-5}$ mol/L	[41]
Zn(bdc)$_2$	T: 60°C, P: 101–232 bar	$(8.2–69) \times 10^{-5}$ mol/L	[41]
	T: 55°C, P: 240.5 bar	12.42×10^{-6} g/mL	[64]
Co(cp)$_2$	T: 60°C, P: 100–175 bar	$(26–222) \times 10^{-5}$ mol/mol	[36]
Cr(cp)$_2$	T: 60°C, P: 100–175 bar	$(19–206) \times 10^{-5}$ mol/mol	[36]

Continued

Table 1 Published solubility data for organometallic compounds in $SCCO_2$—cont'd

Organometallic compound	Experimental conditions	Solubility (in mole fraction unless stated otherwise)	Reference
Fe(cp)$_2$	T: 60°C, P: 100–175 bar	$(24–250) \times 10^{-5}$ mol/mol	[36]
	T: 40–70°C, P: 97.5–366.3 bar	$(2.22–240) \times 10^{-4}$	[65]
	T: 35–50°C, P: 80.2–403.4 bar	$(0.098–3.986) \times 10^{-3}$	[10]
	T: 40–60°C, P: 101–232 bar	$(0.233–2.41) \times 10^{-3}$	[61]
Mn(cp)$_2$	T: 60°C, P: 100–300 bar	$(8–9) \times 10^{-5}$ mol/mol	[36]
Ni(cp)$_2$	T: 60°C, P: 100–175 bar	$(18–218) \times 10^{-5}$ mol/mol	[36]
Os(cp)$_2$	T: 60°C, P: 100–250 bar	$(1-13) \times 10^{-5}$ mol/mol	[36]
Ru(cp)$_2$	T: 60°C, P: 100–170 bar	$(4–29) \times 10^{-5}$ mol/mol	[36]
Ti(cp)$_2$*Cl*$_2$	T: 64°C, P: 179–300 bar	$(0.86–3.39) \times 10^{-5}$ mol/mol	[66]
Na(edc)	T: 50°C, P: 101 bar	1.5×10^{-4} mol/L	[52]
Ni(edc)$_2$	T: 50°C, P: 101 bar	8.5×10^{-7} mol/L	[52]
Cu(edc)$_2$	T: 50°C, P: 101 bar	1.1×10^{-6} mol/L	[52]
	T: 60°C, P: 101–232 bar	$(1.4–11) \times 10^{-6}$ mol/L	[41]
	T: 35–55°C, ρ: 759.93–898.45.3 bar	$(1.95–13.9) \times 10^{-7}$	[34]
Hg(edc)$_2$	T: 50°C, P: 152 bar	8.2×10^{-6} mol/L	[52]
	T: 60°C, P: 101–132 bar	$(6.8–53) \times 10^{-6}$ mol/L	[41]
	T: 35–55°C, P: 100–300 bar	4.086–7.181 mol/L	[67]
Bi(edc)$_3$	T: 50°C, P: 101 bar	6×10^{-6} mol/L	[52]
Co(edc)$_3$	T: 50°C, P: 101 bar	2.4×10^{-6} mol/L	[52]
Zn(edc)	T: 55°C, P: 240.5 bar	0.59×10^{-6} g/mL	[64]
Na(fddc)	T: 50°C, P: 101 bar	4.7×10^{-4} mol/L	[52]
Ni(fddc)$_2$	T: 50°C, P: 101 bar	7.2×10^{-4} mol/L	[52]
Cu(fddc)$_2$	T: 50°C, P: 101 bar	9.1×10^{-4} mol/L	[52]
	T: 60°C, P: 101–232 bar	$(9.1–10) \times 10^{-4}$ mol/L	[41]
Hg(fddc)$_2$	T: 50°C, P: 152 bar	5×10^{-3} mol/L	[52]
	T: 60°C, P: 101–232 bar	$(3.0–14) \times 10^{-3}$ mol/L	[41]
Bi(fddc)$_3$	T: 50°C, P: 101 bar	70.3×10^{-4} mol/L	[52]
Co(fddc)$_3$	T: 50°C, P: 101 bar	8.0×10^{-4} mol/L	[52]
Cu(hdc)$_2$	T: 60°C, P: 101–232 bar	$(9.1–10) \times 10^{-4}$ mol/L	[41]
Hg(hdc)$_2$	T: 60°C, P: 101–232 bar	$(1.6–38) \times 10^{-4}$ mol/L	[41]
Zn(hdc)$_2$	T: 60°C, P: 101–232 bar	$(3.2–58) \times 10^{-4}$ mol/L	[41]
UO$_2$*(hfa)*	T: 40°C, P: 100–250 bar	$(1.21–3.73) \times 10^{-2}$ M	[58]
Ni(hfa)$_2$	T: 60°C, P: 203–405 bar	$(8–9.9) \times 10^{-3}$ mol/L	[46]
Ni(hfa)$_2$*.2H*$_2$*O*	T: 40–60°C, P: 94–251 bar	$(2.95–20.23) \times 10^{-5}$	[49]
Ba(hfa)$_2$	T: 150–170°C, P: 120–220 bar	$(1.3–24.0) \times 10^{-5}$	[60]
Cu(hfa)$_2$	T: 60°C, P: 203–405 bar	$(8–9.9) \times 10^{-3}$ mol/L	[46]
	T: 40°C, P: 103.4–344.7 bar	$(6.173–74.15) \times 10^{-5}$	[53]
	T: 40°C, P: 100 bar	0.029 mol/dm^3	[68]
Cu(hfa)$_2$*.H*$_2$*O*	T: 40°C, P: 103.4–344.7 bar	$(152–414) \times 10^{-5}$	[53]
UO(hfa)$_2$*.H*$_2$*O*	T: 40°C, P: 100 bar	0.040 mol/dm^3	[68]

Table 1 Published solubility data for organometallic compounds in $SCCO_2$—cont'd

Organometallic compound	Experimental conditions	Solubility (in mole fraction unless stated otherwise)	Reference
$VO(hfa)_2.H_2O$	T: 40°C, P: 100 bar	0.025 mol/dm^3	[68]
$Cr(hfa)_3$	T: 60°C, P: 203–405 bar	$(8–9.9) \times 10^{-3}$ mol/L	[46]
	T: 40°C, P: 100 bar	0.03 mol/dm^3	[68]
$Y(hfa)_3$	T: 150–170°C, P: 120–220 bar	$(1.30–24.0) \times 10^{-5}$	[60]
$Zr(hfa)_4$	T: 40°C, P: 100 bar	0.044 mol/dm^3	[68]
$Cu(pdc)_2$	T: 60°C, P: 101–232 bar	$(4.1–40) \times 10^{-7}$ mol/L	[41]
$Hg(pdc)_2$	T: 60°C, P: 101–232 bar	$(3.5–34) \times 10^{-7}$ mol/L	[41]
$Zn(pdc)_2$	T: 60°C, P: 101–232 bar	$(3.2–90) \times 10^{-7}$ mol/L	[41]
$Cu(p3dc)_2$	T: 60°C, P: 101–232 bar	$(6.3–120) \times 10^{-6}$ mol/L	[41]
$Hg(p3dc)_2$	T: 60°C, P: 101–232 bar	$(1.2–23) \times 10^{-5}$ mol/L	[41]
$Zn(p3dc)_2$	T: 60°C, P: 101–232 bar	$(7.9–150) \times 10^{-6}$ mol/L	[41]
$Cu(p5dc)_2$	T: 60°C, P: 101–232 bar	$(0.90–18) \times 10^{-4}$ mol/L	[41]
$Hg(p5dc)_2$	T: 60°C, P: 101–232 bar	$(1.0–20) \times 10^{-4}$ mol/L	[41]
$Zn(p5dc)_2$	T: 60°C, P: 101–232 bar	$(1.6–32) \times 10^{-4}$ mol/L	[41]
Ag(thd)	T: 60°C, P: 100–300 bar	0	[36]
K(thd)	T: 60°C, P: 100–200 bar	0	[36]
Rb(thd)	T: 60°C, P: 100–200 bar	0	[36]
$Co(thd)_2$	T: 40–70°C, P: 100–159 bar	$(2.10–12.8) \times 10^{-4}$	[30]
$Cu(thd)_2$	T: 60°C, P: 203–405 bar	$(1–82.0) \times 10^{-5}$	[36]
	T: 60°C, P: 203–405 bar	$(6.173–74.15) \times 10^{-5}$	[53]
	T: 40–70°C, P: 119–280 bar	$(2.70–33.9) \times 10^{-4}$	[30]
$Ni(thd)_2$	T: 60°C, P: 100–250 bar	$(0–28.0) \times 10^{-5}$	[36]
$Zn(thd)_2$	T: 60°C, P: 100–150 bar	$(7–>500) \times 10^{-5}$	[36]
$Co(thd)_3$	T: 60°C, P: 100–170 bar	$(2-213) \times 10^{-5}$	[36]
	T: 40–70°C, P: 100–190 bar	$(0.368–2.83) \times 10^{-3}$	[69]
$Cr(thd)_3$	T: 60°C, P: 100–150 bar	$(4–>500) \times 10^{-5}$	[36]
	T: 40°C, P: 103.4–344.7 bar	$(400–604.9) \times 10^{-5}$	[53]
	T: 40–70°C, P: 101–163 bar	$(0.233–5.24) \times 10^{-3}$	[69]
$Fe(thd)_3$	T: 60°C, P: 100–150 bar	$(3–>500) \times 10^{-5}$	[36]
	T: 40–60°C, P: 94.8–306.7 bar	$(1.64–10.13) \times 10^{-3}$	[62]
$Mn(thd)_3$	T: 60°C, P: 100–150 bar	$(3–>500) \times 10^{-5}$	[36]
$Ru(thd)_3$	T: 60°C, P: 100–150 bar	$(1–174) \times 10^{-5}$	[36]
$Tb(thd)_3$	T: 40–60°C, P: 124.6–352.3 bar	$(0.07–2.34) \times 10^{-3}$	[62]
$Ti(thd)_3$	T: 60°C, P: 100 bar	>500	[36]
$Ce(thd)_4$	T: 40–60°C, P: 101.2–350.9 bar	$(0.05–0.62) \times 10^{-3}$	[62]
$Zn(thd)_4$	T: 60°C, P: 100–175 bar	$(1–109) \times 10^{-5}$	[36]
$Cu(tod)_2$	T: 40°C, P: 103.4–344.7 bar	$(26.03–269.7) \times 10^{-5}$	[53]
$Fe(tod)_3$	T: 40–60°C, P: 90.6–177.4 bar	$(1.39–12.67) \times 10^{-3}$	[62]
$Tb(tod)_3$	T: 40–60°C, P: 124.3–318.4 bar	$(0.53–3.45) \times 10^{-3}$	[62]
$Ce(tod)_4$	T: 40–60°C, P: 98–243 bar	$(1.53–7.84) \times 10^{-3}$	[62]

Continued

Table 1 Published solubility data for organometallic compounds in $SCCO_2$—cont'd

Organometallic compound	Experimental conditions	Solubility (in mole fraction unless stated otherwise)	Reference
Cu(acac-Br)$_3$	T: 40°C, P: 103.4–344.7 bar	$(1.7–9.5) \times 10^{-5}$	[53]
Rh(acac)(cod)	T: 60°C, P: 100–200 bar	$(0-23) \times 10^{-5}$	[36]
Cu(bzac)$_2$	T: 40°C, P: 103.4–344.7 bar	$(0.179–1.047) \times 10^{-5}$	[53]
Pt(cod)(me)$_2$	T: 60°C, P: 100–200 bar	$(3-132) \times 10^{-5}$	[36]
	T: 40–80°C, P: 125.9–296.4 bar	$(6.453–34.36) \times 10^{-4}$	[70]
Zn(dc)$_2$	T: 55°C, P: 240.5 bar	0.089×10^{-6} g/mL	[64]
Cu(dibm)$_2$	T: 40°C, P: 103.4–344.7 bar	$(9.17–88.41) \times 10^{-5}$	[53]
Cu(dmhd)$_2$	T: 40°C, P: 137.9–344.7 bar	$(3.696–36.21) \times 10^{-5}$	[53]
Cu(tfa)$_2$	T: 40°C, P: 103.4–344.7 bar	$(29.60–59.38) \times 10^{-5}$	[53]
UO(tfa)	T: 40°C, P: 100–250 bar	$(2.20–13.7) \times 10^{-3}$ M	[58]
Cu(tfbzm)$_2$	T: 40°C, P: 137.9–344.7 bar	$(0.702–4.269) \times 10^{-5}$	[53]
trans-Cr(tfa)3	T: 40°C, P: 103.4–344.7 bar	$(147.9–272.1) \times 10^{-5}$	[53]
Ru(thd)$_2$(cod)	T: 60°C, P: 100–150 bar	$(1–56) \times 10^{-5}$	[36]
cis-Cr(tfa)$_3$	T: 40°C, P: 103.4–344.7 bar	$(67.3–190.8) \times 10^{-5}$	[53]
Mo(CO)$_6$	T: 40–60°C, P: 76.5–113.9 bar	$(9.5–145.1) \times 10^{-4}$	[71]
Ni(dctp)$_2$Cl$_2$	T: 35–55°C, P: 108–279 bar	$(1.32–5.33) \times 10^{-6}$	[72]
Ti(Oi-Pr)$_2$(dpm)$_2$	T: 40–60°C, P: 60–200 bar	$(0.025–1.1) \times 10^{-2}$	[73]
UO$_2$(dfh)$_2$DMSO	T: 40°C, P: 100 bar	0.12 mol/dm^3	[68]
trans-$Co_2(CO)_6$ [3,5-bis(CF_3) $C_6H_3P(i\text{-}C_3H_7)_2$]	T: 40–70°C, P: 100–260 bar	$(0.8–16.7) \times 10^{-3}$ M	[74]

Notes: acac: Acetylacetonate, acac-Br: 3-Bromopentane-2,4-dionato, bzac: 1-Phenylpentane-1,3-dionato, bdc: Dibutyldithiocarbamate, CO: Carbonyl, cod: 1,5-Cyclooctadiene, cp: Cyclopentadienyl, dc: Dithiocarbamate, dctp: Dichlorobis(triphenylphosphine), dfh: 4H,4H-Devafluoroheptane-3,5-dionate, dibm: 2,6-Dimethylheptane-3,5-dionato, dmhd: 1,1-Dimethylhexane-3,5-dionato, DMSO: Dimethyl sulfoxide, edc: Diethyldithiocarbamate, hdc: Dihexyldithiocarbamate, hfa: Hexafluoroacetylacetonate, fddc: (Trifluoroethyl)dithiocarbamate, me: Methyl, pdc: Pyrrolidinedithiocarbamate, p3dc: Dipropyldithiocarbamate, p5dc: Dipentyldithiocarbamate, thd: 2,2,6,6-Tetramethyl-3,5-heptanedionato, tfa: 1,1,1-Trifluoropentane-2,4-dionato, tfbzm: 1,1,1-Trifluoro-4-phenylbutane-2,4-dionato, tod: 2,2,7-Trimethyl-3,5-octanedionato.

Modified from W.H. Teoh, R. Mammucari, N.R. Foster, Solubility of organometallic complexes in supercritical carbon dioxide: a review, J. Org. Chem. 724 (2013) 102–116 with permission from the Elsevier, license number: 4461731076776.

Modified from W.H. Teoh, R. Mammucari, N.R. Foster, Solubility of organometallic complexes in supercritical carbon dioxide: a review, J. Org. Chem. 724 (2013) 102–116 with permission from the Elsevier, license number: 4461731076776.

more CO_2-philic, representing higher solubility, while aromatic substituted ligands showed poor solubility in $SCCO_2$. In addition, metal chelates with higher oxidation states have demonstrated higher solubility in $SCCO_2$.

6. Thermodynamic modeling

The solubility of organometallic compounds in $SCCO_2$ has generally been modeled based on regular solution theory, empirical or semi-empirical models, and equations of states

(EOSs). Models derived from EOSs require difficult calculation methods, which do not exist in commonly applied commercial software. In addition, solubility predictions applying EOSs can be affected by the numerical values of the solute properties such as critical parameters, acentric factor, molar volumes, and vapor pressure; because these models apply these properties that frequently cannot be easily obtained experimentally [3, 75].

6.1 Model based on regular solution theory

Lagalante et al. [53] applied a model based on regular solution theory to predict an organometallic compound [Cr(acac)$_3$] solubility in SCCO$_2$. According to this model, solubility can be predicted from the following equation.

$$y_2 = \exp\left[-\frac{\Delta H_f}{RT_{m2}}\left(\frac{T_{m2}}{T}-1\right)-\frac{v_2\varphi_1^2}{RT}\left(\delta_2-\delta_1\right)^2\right] \tag{3}$$

where

y_2 denotes the mole fraction of the organometallic compound;
ΔH_f is the enthalpy of fusion of the organometallic compound at its melting point;
R denotes the gas constant;
T_{m2} is the melting point of the organometallic compound;
T is the temperature of the system;
v_2 denotes the molar volume of the organometallic compound;
φ_1 is the volume fraction of the CO_2; and
δ_1 and δ_2 are the solubility parameters for CO_2 and the organometallic compound, respectively.

The Benedict-Webb-Rubin EOS was utilized to calculate the solubility parameter of pure CO_2, while the solubility parameter of Cr(acac)$_3$ was estimated by the group contribution approach of Fedors [53, 76]. The Cr(acac)3 solubility at pressures higher than 30 MPa was adequately estimated. But, according to the high deviation from experimental data at pressures around 10 MPa, Lagalante et al. [53] concluded that "this theory is a poor model for the estimation of organometallic compounds solubility in SCCO$_2$ due to simplifying assumptions such as a zero mixing volume" [12, 13, 18, 53].

6.2 Empirical models

Different empirical or semi-empirical models such as Chrastil [77], Ziger and Eckert [78], Del Valle and Aguilera [79], Bartle et al. [80], Yu et al. [81], Bush and Eckert [82], Gordillo et al. [83], Mèndez-Santiago and Teja [84], Jouyban et al. [85], Jafarinejad et al. [3], Hozhabr et al. [86], Si-Moussa et al. [87], Ota et al. [88], etc. have widely been used to predict the solute solubility in SCCO$_2$.

Relating to organometallic compounds, Chrastil [77] as well as Mèndez-Santiago and Teja models [84] have been widely used in organometallic compounds-SCCO$_2$ systems.

The Chrastil model assumes that one molecule of solute, A, associates with a molecules of solvent, B, to create a solvate-complex AB_a, in equilibrium with the system. The Chrastil model is expressed with the following equation.

$$\ln y_2 = a \ln\rho + \frac{b}{T} + c \tag{4}$$

where

y_2 is the mole fraction of the solute (organometallic compound);

ρ (kg/m^3) is the density of the $SCCO_2$;

a is the association number;

b is a constant, defined as $\Delta H/R$ (where ΔH is the sum of the enthalpies of vaporization and solvation of the solute and R is the gas constant); and

c is another constant related to the molecular mass of the solute and solvent [3, 37, 77].

For the Chrastil model, the absolute standard deviations (ASDs) between calculated and experimental data have been reported to be 2%–20% [13, 89].

According to the theory of dilute solutions, Mèndez-Santiago and Teja [84] have proposed a model to correlate the solubility of solutes in a supercritical fluid:

$$T\ln(y_2 P) = A' + B'\rho + C'T \tag{5}$$

where

A', B', and C' are constants, considered as temperature independent;

T (K) is the temperature of the system;

P (MPa) is the pressure of the system; and.

ρ (mol/cm^3) is the density of the supercritical fluid ($SCCO_2$) [3, 37, 84].

For the Mèndez-Santiago and Teja model, the deviations (ASDs) between experimental and calculated data have been reported in the range of 2%–19% [13, 37, 49, 70].

6.3 Equations of state (EOSs)

For solid-supercritical fluid equilibria, the equilibrium condition is.

$$\hat{f}_i^{S} = \hat{f}_i^{SCF} \tag{6}$$

where

$\hat{f}_i^{SCF}$ is the fugacity of species i in the SCF phase; and

$\hat{f}_i^{S}$ is the fugacity of species i in the solid phase.

The fugacity of the solute (organometallic compound) in the SCF phase can be calculated from the following equation:

$$\hat{f}_2^{SCF} = y_2 P \hat{\varphi}_2^{SCF} \tag{7}$$

where

P is the pressure;

$\hat{\varphi}_2^{SCF}$ is the fugacity coefficient of the organometallic compound in SCF; and

y_2 is the solubility or mole fraction of the organometallic compound in the SCF phase. For phase equilibrium between a high-boiling compound and an SCF with low critical temperature, it may usually be assumed:

(i) the solid solute is pure;

(ii) the molar volume of solid solute is a very weak function of pressure and can be constant; and.

(iii) the saturated vapor of solid solute-vapor is considered as an ideal gas. Using the first assumption, the fugacity of solute in the solid state, $\hat{f}_2^S$, is identical to the pure solid fugacity, f_2^S, and it can be written as:

$$\hat{f}_2^S = f_2^S = P_2^{sat}\hat{\varphi}_2^{sbl}\exp\left(\int_{P_2^{sat}}^{P}\frac{v_2^S}{RT}dP\right) \tag{8}$$

where

R is the gas constant;

T is the temperature;

P is the pressure;

P_2^{sat} is the saturated vapor pressure of the solute;

$\hat{\varphi}_2^{sbl}$ is the fugacity coefficient of the solute in the solid phase at sublimation; and

v_2^S is the solid-state molar volume of the solute.

Using assumptions (ii) and (ii), and the thermodynamic equilibrium condition, Eq. (6), the solubility or mole fraction of the solute can be expressed by the following equation:

$$y_2 = \frac{P_2^{sat}\hat{\varphi}_2^{sbl}}{P\hat{\varphi}_2^{SCF}}\exp\left(\int_{P_2^{sat}}^{P}\frac{v_2^S}{RT}dP\right) = \frac{P_2^{sat}}{P\hat{\varphi}_2^{SCF}}\exp\left(\frac{v_2^S\left(P - P_2^{sat}\right)}{RT}\right) \tag{9}$$

where the fugacity coefficient of the solute in the solid phase at sublimation is considered to be unity.

The saturated vapor pressure and solid molar volume can be acquired from literature data or by applying a suitable correlation [90]. The fugacity coefficient of the solute in SCF phase ($\hat{\varphi}_2^{SCF}$) can be computed from the following Eq. [91]:

$$\ln\hat{\varphi}_2^{SCF} = \frac{1}{RT}\int_{\infty}^{v}\left[\left(\frac{\partial P}{\partial n_2}\right)_{T,v,n_1} - \frac{RT}{v}\right]dV - \ln\left(\frac{Pv}{RT}\right) \tag{10}$$

where

n_2 is the number of moles of solute;

n_1 is the number of moles of CO_2; and

v is the molar volume of the mixture.

A proper equation of state can be utilized to perform the integration and evaluate v. Cubic EOSs such as van der Waals EOS [92], Soave-Redlick-Kwong EOS (SRK-EOS) [93], and

Peng-Robinson EOS (PR-EOS) [94] are usually applied for phase equilibria calculations. The general form of cubic EOS is represented by the following expression [95, 96]:

$$P = \frac{RT}{v-b} - \frac{a}{v(v+c)+b(3v+c)} \tag{11}$$

where a, b, and c are parameters that can be calculated from the critical properties and the acentric factors of the pure species.

The most widely applied cubic EOS is the PR-EOS given by

$$P = \frac{RT}{v-b} - \frac{a}{v(v+b)+b(v-b)} \tag{12}$$

where a and b are constants given by:

$$a = \sum_i \sum_j y_i y_j a_{ij} \tag{13}$$

$$a_{ij} = \sqrt{a_i a_j}\left(1-k_{ij}\right) \tag{14}$$

$$b = \sum_i y_i b_i \tag{15}$$

where k_{ij} denotes the characteristic parameter between unlike molecules i and j. Note that for evaluation of parameters, different nonquadratic mixing rules are also available [91].

$$a_i = 0.45724\alpha \frac{R^2 T_{ci}^2}{P_{ci}} \tag{16}$$

$$\alpha = \left[1 + \left(0.37464 + 1.54226w - 0.26992w^2\right)\left(1 - T_r^{0.5}\right)^2\right] \tag{17}$$

$$w = -1 - \log\left(P_r^{sat}\right)_{T_r=0.7} \tag{18}$$

$$b_i = 0.7780 \frac{RT_{ci}}{P_{ci}} \tag{19}$$

Using these equations, the fugacity coefficient of the solute in SCF phase ($\hat{\varphi}_2^{SCF}$) can be calculated from the following equation:

$$\ln \hat{\varphi}_2^{SCF} = \frac{B_2}{B}(Z-1) - \ln(Z-B) + \frac{A}{2\sqrt{2}B}\left[\frac{B_2}{B}\right] \ln\left(\frac{Z+2.414B}{Z-0.41B}\right) \tag{20}$$

where

$$B = \frac{bP}{RT} \tag{21}$$

$$B_2 = \frac{b_2 P^2}{RT} \tag{22}$$

$$A = \frac{a\alpha P}{(RT)^2} \tag{23}$$

and Z can be calculated from the following equation:

$$Z^3 - (1 - B)Z^2 + \left(A - 3B^2 - 2B\right)Z - \left(AB - B^2 - B^3\right) = 0 \tag{24}$$

The following approach can be applied to regress the critical temperature and the critical pressure of the solute and the binary interaction parameter, k_{ij}:

$$f\left(T, P, P_c, T_c, k_{ij}\right) = \sum_i \left(\frac{y_{i,\text{model}} - y_{i,\text{experimental}}}{y_{i,\text{experimental}}}\right) \tag{25}$$

where $y_{i,\text{model}}$ and $y_{i,\text{experimental}}$ are the mole fractions estimated by the model and experimental mole fractions, respectively.

The lack of sublimation pressure data for organometallic compounds is the main drawback in the application of EOSs. In addition, because most of the organometallic compounds decompose even before reaching their boiling points, the critical temperature of the organometallic compound is not a physically realistic parameter [18].

Roggeman et al. [63] investigated the modeling of $Fe(acac)_3$ solubility in $SCCO_2$ with the PR-EOS applying the van der Waals mixing rule. Due to the lack of critical property data, they used the Joback group contribution method [97] to predict the critical properties. The boiling point of the complex and its acentric factor were predicted by its experimental heat of vaporization and sublimation pressure near the melting point. The average absolute relative deviations (AARDs) between the calculated and the experimental data were reported 11.3% and 15.3% at 40°C and 60°C, respectively [63].

7. Conclusions

The dynamic and static techniques are the solubility measurement methods of the solutes in $SCCO_2$. Solubility data of the organometallic compounds in supercritical fluids (SCFs) are required for establishing optimum operating conditions and designing SCF processes. In addition, better knowledge of the dissolution phenomenon can be provided by mathematical modeling of organometallic compounds solubility in SCFs. Furthermore, solubility modeling can provide an approach to estimate solubility at different operating conditions with a minimum of experimental data. All this can enhance the development and progress rate of an SCF process. Studies have demonstrated that:

(i) the solubilities of metal chelates containing fluorinated ligands > aliphatic-substituted systems > phenyl-substituted systems;

(ii) the solubility of organometallic compounds with identical valence electron configuration usually reduces with enhancing atomic numbers; and

(iii) metal chelates with higher oxidation states usually exhibit higher solubility in $SCCO_2$.

The solubility of organometallic compounds in $SCCO_2$ has generally been modeled based on regular solution theory, empirical or semi-empirical models, and equations of states. Lagalante et al. concluded that "regular solution theory is a poor model for the estimation of organometallic compounds solubility in $SCCO_2$ due to simplifying assumptions such as a zero mixing volume." For empirical models, for example, the Mèndez-Santiago and Teja model, the deviations (ASDs) between experimental and calculated data have been reported in the range of 2%–19%. Solubility predictions using equations of states can be affected by the numerical values of the solute properties such as critical parameters, acentric factor, molar volumes, and vapor pressure; because these models apply these properties that frequently cannot be easily obtained experimentally.

References

[1] P. Dondeti, Y. Desai, Supercritical fluid technology in pharmaceutical research, Encyclopedia Pharmaceut. Technol. 18 (1999) 219–248.

[2] P. Coimbra, M.H. Gil, C.M.M. Duarte, B.M. Heron, H.C. de Sousa, Solubility of a spiroindolinonaphthoxazine photochromic dye in supercritical carbon dioxide: experimental determination and correlation, Fluid Phase Equilibr. 238 (2005) 120–128.

[3] S. Jafarinejad, H. Abolghasemi, M.A. Moosavian, M.G. Maragheh, Prediction of solute solubility in supercritical carbon dioxide: A novel semi-empirical model, Chem. Eng. Res. Des. 88 (2010) 893–898.

[4] S. Jafarinejad, H. Abolghasemi, A. Golzary, M.A. Moosavian, M.G. Maragheh, Fractional factorial design for the optimization of hydrothermal synthesis of lanthanum oxide under supercritical water condition, J. Supercrit. Fluids 52 (2010) 292–297.

[5] S. Jafarinejad, M.A. Khansary, F. Amiri, Solubility prediction of some disperse azo dyes in supercritical carbon dioxide using equation of states (EOSs), Inter. J. Thermodyn. 15 (2) (2012) 103–110.

[6] S. Jafarinejad, M.A. Khansary, F. Amiri, Modeling of solubility of disperse blue dyes in supercritical carbon dioxide using equation of states (EOSs), inter, J. Chem. Eng. App. 2 (4) (2011) 279–284.

[7] J.K. Sabet, S. Jafarinejad, A. Golzary, Supercritical water oxidation for the recovery of dysprosium ion from aqueous solutions, Inter. Res. J. Appl. Basic Sci. 8 (8) (2014) 1079–1083.

[8] S. Jafarinejad, H. Abolghasemi, M.A. Moosavian, M.G. Maragheh, Fractional factorial design for the optimization of supercritical carbon dioxide extraction of La^{3+}, Ce^{3+} and Sm^{3+} ions from a solid matrix using bis(2,4,4-trimethylpentyl)dithiophosphinic acid + tributylphosphate, Chem. Eng. Res. Des. 89 (2011) 827–835.

[9] C.M. Wai, S. Wang, Supercritical fluid extraction: Metals as complexes, J. Chromatography A. 785 (1997) 369–383.

[10] C.Y. Kong, K. Sone, T. Sako, T. Funazukuri, S. Kagei, Solubility determination of organometallic complexes in supercritical carbon dioxide by chromatographic impulse response method, Fluid Phase Equilib. 302 (2011) 347–353.

[11] J.W. King, K. Srinivas, Multiple unit processing using sub- and supercritical fluids, J. Supercrit. Fluids 47 (2009) 598–610.

[12] C. Erkey, Supercritical carbon dioxide extraction of metals from aqueous solutions: A review, J. Supercrit. Fluids 17 (2000) 259–287.

[13] W.H. Teoh, R. Mammucari, N.R. Foster, Solubility of organometallic complexes in supercritical carbon dioxide: a review, J. Org. Chem. 724 (2013) 102–116.

[14] R. Dohrn, G. Brunner, High-pressure fluid-phase equilibria: experimental methods and systems investigated (1988-1993), Fluid Phase Equilib. 106 (1995) 213–282.

[15] J. Mendez-Santiago, A.S. Teja, The solubility of solids in supercritical sluids, Fluid Phase Equilib. 158-160 (1999) 501–510.

[16] D.E. Knox, Solubilities in supercritical fluids, Pure Appl. Chem. 77 (2005) 513–530.

[17] F.P. Lucien, N.R. Foster, Solubilities of solid mixtures in supercritical carbon dioxide: a review, J. Supercrit. Fluids 17 (2000) 111–134.

[18] C. Erkey, Supercritical fluids and organometallic compounds: from recovery of trace metals to synthesis of nanostructured materials, in: E. Kiran (Ed.), Supercritical Fluid Science and Technology, Vol. 1, Elsevier, 2011.

[19] J.W. Rathke, R.J. Klingler, T.R. Krause, Propylene hydroformylation in supercritical carbon dioxide, Organometallics 10 (1991) 1350–1355.

[20] A. Cabañas, J.M. Blackburn, J.J. Watkins, Deposition of cu films from supercritical fluids using cu(I) beta-diketonate precursors, Microelectron. Eng. 64 (2002) 53–61.

[21] A.I. Cooper, S.G. Kazarian, M. Poliakoff, Supercritical fluid impregnation of polyethylene films, a new approach to studying equilibria in matrices; the hydrogen bonding of fluoroalcohols to (Z5-C5Me5)Ir (CO) and the effect on C-H activation, Chem. Phys. Lett. 206 (1993) 175.

[22] K.S. Morley, P.B. Webb, N.V. Tokareva, A.P. Krasnov, V.K. Popov, J. Zhang, C.J. Roberts, S.M. Howdle, Synthesis and characterisation of advanced UHMWPE/silver nanocomposites for biomedical applications, European Poly. J. 43 (2007) 307–314.

[23] J.J. Watkins, T.J. McCarthy, Polymer/metal nanocomposite synthesis in supercritical CO_2, Chem. Mater. 7 (1995) 1991–1994.

[24] K. Esumi, S. Sarashina, T. Yoshimura, Synthesis of gold nanoparticles from an organometallic compound in supercritical carbon dioxide, Langmuir 20 (2004) 5189–5191.

[25] P.S. Shah, S. Husain, K.P. Johnston, B.A. Korgel, Nanocrystal arrested precipitation in supercritical carbon dioxide, J. Phys. Chem. B 105 (2001) 9433–9440.

[26] A. Campos-Carrasco, C.T. Estorach, A. Bastero, M. Reguero, A.M. Masdeu- Bultó, G. Franciò, W. Leitner, A. D'Amora, B. Milani, Cationic palladium(II) complexes for CO/vinyl Arene copolymerization in the presence of carbon dioxide, Organometallics 30 (2011) 6572–6586.

[27] K.E. Laintz, C.M. Wai, C.R. Yonker, R.D. Smith, Extraction of metal ions from liquid and solid materials by supercritical carbon dioxide, Anal. Chem. 64 (1992) 2875–2878.

[28] J. Wang, J. Chen, Y. Yang, Micronization of titanocene dichloride by rapid expansion of supercritical solution and its ethylene polymerization, J. Supercrit. Fluids 33 (2005) 159–172.

[29] S.K. Morley, P.C. Marr, P.B. Webb, A.R. Berry, F.J. Allison, G. Moldovan, P.D. Brown, S.M. Howdle, Clean preparation of nanoparticulate metals in porous supports: a supercritical route, J. Mater. Chem. 12 (2002) 1898–1905.

[30] M. Haruki, F. Kobayashi, S. Kihara, S. Takishima, Solubility of β-diketonate complexes of copper(II) and cobalt(II) in supercritical carbon dioxide, J. Chem. Eng. Data 56 (2011) 2230–2235.

[31] S.A. Abdullah, Solubility in Supercritical Carbon Dioxide, MS Thesis, New Jersey Institute of Technology, 2007.

[32] R.B. Gupta, J.J. Shim, Solubility in Supercritical Carbon Dioxide, CRC Press, Taylor & Francis Group, USA, 2007.

[33] M. Sauceau, J. Letourneau, D. Richon, J. Fages, Enhanced densitybased models for solid compound solubilities in supercritical carbon dioxide with cosolvents, Fluid Phase Equilib. 208 (1–2) (2003) 99–113.

[34] W. Cross, A. Akgerman, C. Erkey, Determination of metal-chelate complex solubilities in supercritical carbon dioxide, Indust. Eng. Chemistry Res. 35 (1996) 1765–1770.

[35] M. Ashraf-Khorassani, M.T. Combs, L.T. Taylor, Supercritical fluid extraction of metal ions and metal chelates from different environments, J. Chromatography A 774 (1997) 37–49.

[36] O. Aschenbrenner, S. Kemper, N. Dahmen, K. Schaber, E. Dinjus, Solubility of B-diketonates, cyclopentadienyls, and cyclooctadiene complexes with various metals in supercritical carbon dioxide, J. Supercrit. Fluids 41 (2007) 179–186.

[37] J. Morère, M.J. Tenorio, C. Pando, J.A.R. Renuncio, A. Cabañas, Solubility of two metal-organic ruthenium precursors in supercritical CO_2 and their application in supercritical fluid technology, J. Chem. Thermodyn. 58 (2013) 55–61.

[38] M.A. McHugh, V.J. Krukonis, Supercritical Fluid Extraction: Principles and Practice, Butterworth, Stoneham, MA, 1986.

[39] K.E. Laintz, C.M. Wai, C.R. Yonker, R.D. Smith, Solubility of fluorinated metal dithiocarbamates in supercritical carbon dioxide, J. Supercrit. Fluids 4 (1991) 194–198.

[40] C.M. Wai, Y. Lin, R. Brauer, S.F. Wang, W.F. Beckert, Talanta 40 (1993) 1325.
[41] C.M. Wai, S.F. Wang, J.J. Yu, Solubility parameters and solubilities of metal dithiocarbamates in supercritical carbon dioxide, Anal. Chem. 68 (1996) 3516–3519.
[42] S.L. Draper, G.A. Montero, B. Smith, K. Beck, Solubility relations for disperse dyes in supercritical carbon dioxide, Dyes Pigments 45 (2000) 177–183.
[43] K.E. Anderson, J.I. Siepmann, Solubility in supercritical carbon dioxide: importance of the poynting correction and entrainer effects, J. Phys. Chem. B 112 (2008) 11374–11380.
[44] D. Tuma, B. Wagner, G.M. Schneider, Comparative solubility investigations of anthraquinone disperse dyes in near- and supercritical fluids, Fluid Phase Equilibria 182 (2001) 133–143.
[45] J. Fasihi, Y. Yamini, F. Nourmohammadian, N. Bahramifar, Investigations on the solubilities of some disperse azo dyes in supercritical carbon dioxide, Dyes Pigments 63 (2004) 161–168.
[46] B. Guzel, A. Akgerman, Solubility of disperse and mordant dyes in supercritical CO2, J. Chem. Eng. Data 44 (1999) 83–85.
[47] H.D. Sung, J.J. Shim, Solubility of C.I. disperse red 60 and C.I. disperse blue 60 in supercritical carbon dioxide, J. Chem. Eng. Data 44 (1999) 985–989.
[48] S.G. Kazarian, P.A. Hamley, M. Poliakoff, Is intermolecular hydrogen-bonding to uncharged metal centers of organometallic compounds widespread in solution? A spectroscopic investigation in hydrocarbon, noble gas, and supercritical fluid solutions of the interaction between fluoro alcohols and (.Eta.5-C5R5)ML2 (R = H, me; M = CO, Rh, Ir; L = CO, C2H4, N2, PMe3) and its relevance to protonation, J. Am. Chem. Soc. 115 (1993) 9069–9079.
[49] M.J. Tenorio, A. Cabañas, C. Pando, J.A.R. Renuncio, Solubility of $Pd(hfac)_2$ and $Ni(hfac)_2{\cdot}2H_2O$ in supercritical carbon dioxide pure and modified with ethanol, J. Supercrit. Fluids 70 (2012) 106–111.
[50] G.M. Kontogeorgis, in: C.M. Hensen (Ed.), Hansen Solubility Parameters: A User's Handbook, CRC Press, Boca Raton, 2007, pp. 77–94.
[51] S.K. Kumar, K.P. Johnston, Modeling the solubility of solids in supercritical fluids with density as the independent variable, J. Supercrit. Fluids 1 (1988) 15–22.
[52] K.E. Laintz, C.M. Wai, C.R. Yonker, R.D. Smith, Solubility of fluorinated metal diethyldithiocarbamates in supercritical carbon dioxide, J. Supercrit. Fluids 4 (1991) 194–198.
[53] A.F. Lagalante, B.N. Hansen, T.J. Bruno, R.E. Sievers, Solubilities of copper(II) and chromium(III) β-diketonates in supercritical carbon dioxide, Inorg. Chem. 34 (1995) 5781–5785.
[54] M. Ashraf-Khorassani, M.T. Combs, L.T. Taylor, Solubility of metal chelates and their extraction from an aqueous environment via supercritical CO_2, Talanta 44 (1997) 755–763.
[55] A. Wolf, Bestimmung derĽoslichkeit vonMetallchelaten in¨uberkritischem Kohlendioxid, Ph.D. Thesis, University of Siegen, Germany, 1999.
[56] N. Saito, Y. Ikushima, T. Goto, Liquid/solid extraction of acetylacetone chelates with supercritical carbon dioxide, Bulletin Chem. Socie. Japan 63 (1990) 1532–1534.
[57] S. Yoda, Y. Mizuno, T. Furuya, Y. Takebayashi, K. Otake, T. Tsuji, T. Hiaki, Solubility measurements of noble metal acetylacetonates in supercritical carbon dioxide by high performance liquid chromatography (HPLC), J. Supercrit. Fluids 44 (2008) 139–147.
[58] T. Tsukahara, Y. Kachi, Y. Kayaki, T. Ikariya, Y. Ikeda, Spectrophotometric study on solubility of $UO_2(\beta\text{-diketonato})_2$dmso complexes (β-diketonate = acetylacetonate, trifluoroacetylacetonate, hexafluoroacetylacetonate; dmso = dimethyl sulfoxide) in supercritical carbon dioxide, J. Supercrit. Fluids 39 (2006) 6–12.
[59] C. Kersch, G.J. Witkamp, G.F. Woerlee, Supercritical fluid extraction: A study on metal recovery and regeneration of β-diketones and organophosphorus extractants, Sol. Extract. Ion Exchange 23 (2005) 189–212.
[60] R. M'Hamdi, J.F. Bocquet, K. Chhor, C. Pommier, Solubility and decomposition studies on metal chelates in supercritical fluids for ceramic precursor powders synthesis, J. Supercrit. Fluids 5 (1992) 55–59.
[61] M. Haruki, F. Kobayashi, S.-i. Kihara, S. Takishima, Effect of the chemical structures of iron complexes on the solubilities in supercritical carbon dioxide, Fluid Phase Equilib. 308 (2011) 1–7.
[62] W.C. Andersen, R.E. Sievers, A.F. Lagalante, T.J. Bruno, Solubilities of cerium(IV), terbium(III), and Iron(III) β-Diketonates in supercritical carbon dioxide, J. Chem. Eng. Data 46 (2001) 1045–1049.

[63] E.J. Roggeman, A.M. Scurto, J.F. Brennecke, Spectroscopy, solubility, and modeling of cosolvent effects on metal chelate complexes in supercritical carbon dioxide solutions, Indust. Eng. Chemistry Res. 40 (2000) 980–989.

[64] J. Wang, W.D. Marshall, Recovery of metals from aqueous media by extraction with supercritical carbon dioxide, Anal. Chem. 66 (1994) 1658–1663.

[65] C.M. Cowey, K.D. Bartle, M.D. Burford, A.A. Clifford, S. Zhu, N.G. Smart, N.D. Tinker, Solubility of ferrocene and a nickel complex in supercritical fluids, J. Chem. Eng. Data 40 (1995) 1217–1221.

[66] J. Wang, J. Chen, Y. Yang, Solubility of titanocene dichloride in supercritical propane, Fluid Phase Equilib. 220 (2004) 147–151.

[67] Y. Ding, H. Huang, Y. Ding, C. Yang, J. Ji, Study of metal complexes' solubility in supercritical carbon dioxide, Adv. Mat. Res. 236–238 (2011) 411–416.

[68] N. Miyamoto, T. Tsukahara, Y. Kachi, M. Harada, Y. Kayaki, T. Ikariya, Y. Ikeda, Studies on solubility of uranyl complexes in supercritical carbon dioxide and its controlling factors using UV-visible and ^{17}O- and ^{19}F-NMR spectroscopy, J. Nuclear Sci. Technol. 49 (2012) 37–46.

[69] M. Haruki, F. Kobayashi, M. Okamoto, S. Kihara, S. Takishima, Solubility of β-diketonate complexes for cobalt(III) and chromium(III) in supercritical carbon dioxide, Fluid Phase Equilib. 297 (2010) 155–161.

[70] M. Türk, M. Crone, G. Upper, Effect of gas pressure on the phase behaviour of organometallic compounds, J. Supercrit. Fluids 58 (2011) 1–6.

[71] R.P. Warzinski, C.-H. Lee, G.D. Holder, Supercritical-fluid solubilization of catalyst precursors: The solubility and phase behavior of molybdenum hexacarbonyl in supercritical carbon dioxide and application to the direct liquefaction of coal, J. Supercrit. Fluids 5 (1992) 60–71.

[72] D.R. Palo, C. Erkey, Solubility of dichlorobis(triphenylphosphine)nickel(II) in supercritical carbon dioxide, J. Chem. Eng. Data 43 (1998) 47–48.

[73] H. Uchida, K. Sekino, Y. Hayakawa, S. Koda, Solubility of titanium diisopropoxide bis(dipivaloylmethanate) complex in supercritical carbon dioxide and its effect on supercritical fluid deposition process, J. Supercrit. Fluids 66 (2012) 59–65.

[74] N. Dahmen, P. Griesheimer, P. Makarczyk, S. Pitter, O. Walter, Solubility of *trans*-$Co_2(CO)_6$ [3,5-bis $(CF_3)C_6H_3P(i\text{-}C_3H_7)_2$] in dense carbon dioxide, J. Org. Chem. 690 (2005) 1467–1473.

[75] S. Garnier, E. Neau, P. Alessi, A. Cortesi, I. Kikic, Modeling solubility of solids in supercritical fluids using fusion properties, Fluid Phase Equilibr. 158-162 (1999) 491.

[76] A.F.M. Barton, CRC Handbook of Solubility Parameters and Other Cohesion Parameters, CRC Press, Boca Raton, FL, 1983.

[77] J. Chrastil, Solubility of solids and liquids in supercritical gases, J. Phys. Chem. 86 (1982) 3016–3022.

[78] D.H. Ziger, C.A. Eckert, Correlation and prediction of solid supercritical fluid phase equilibria, Ind. Eng. Chem. Process. Des. Dev. 22 (1983) 582–593.

[79] J.M. Del Valle, J.M. Aguilera, An improved equation for predicting the solubility of vegetable oils in supercritical CO_2, Ind. Eng. Chem. Res. 27 (1988) 1551–1559.

[80] K.D. Bartle, A.A. Clifford, S.A. Jafar, G.F. Shilstone, Solubilities of solids and liquids of low volatility in supercritical carbon dioxide, J. Phys. Chem. Ref. Data Monogr. 20 (1991) 713–725.

[81] Z. Yu, B. Singh, S.S.H. Rizvi, J.A. Zollewg, Solubilities of fatty acids, fatty acid esters, and fats and oils in supercritical carbon dioxide, J. Supercrit. Fluids 7 (1994) 51–61.

[82] D. Bush, C.A. Eckert, Estimation of solid solubilities in supercritical carbon dioxide from solute solvatochromic parameters, J. Supercrit. Fluid (1997) 37–50.

[83] M.D. Gordillo, M.A. Blanco, A. Molero, E. Martinez de la Ossa, Solubility of the antibiotic penicillin G in supercritical carbon dioxide, J. Supercrit. Fluids 15 (1999) 183–189.

[84] J. Mèndez-Santiago, A.S. Teja, Solubility of solids in supercritical fluids: consistency of data and a new model for cosolvent systems, Ind. Eng. Chem. Res. 39 (2000) 4767–4771.

[85] A. Jouyban, H.K. Chan, N.R. Foster, Mathematical representation of solute solubility in supercritical carbon dioxide using empirical expressions, J. Supercrit. Fluids 24 (2002) 19–35.

[86] S.B. Hozhabr, S.H. Mazloumi, J. Sargolzaei, Correlation of solute solubility in supercritical carbon dioxide using a new empirical equation, Chem. Eng. Res. Des. 92 (11) (2014) 2734–2739.

[87] C. Si-Moussa, A. Belghait, L. Khaouane, S. Hanini, A. Halilali, Novel density-based model for the correlation of solid drugs solubility in supercritical carbon dioxide, C. R. Chim. 20 (5) (2017) 559–572.
[88] M. Ota, Y. Sato, R.L. SmithJr, H. Inomata, Predictive dimensionless solubility (*p*DS) model for solid solutes in supercritical CO_2 that requires only pure-component physical properties, Chem. Eng. Res. Des. 136 (2018) 251–261.
[89] M. Skerget, Z.e. Knez, M.a. Knez-Hrncic, Solubility of solids in sub- and supercritical fluids: A review, J. Chem. Eng. Data 56 (2011) 694–719.
[90] H. Higashi, Y. Iwai, Y. Arai, Solubilities and diffusion coefficients of high boiling compounds in supercritical carbon dioxide, Chem. Eng. Sci. 56 (2001) 3027–3044.
[91] J.M. Prausnitz, R.N. Lichtenthaler, E.D. de Azevedo, Molecular Thermodynamics of Fluid-Phase Equilibria, third ed., Prentice-Hall, Upper Saddle River, NJ, 1999.
[92] A.I. Kipnis, B.E. Efimovich, J.S. Rowlinson, Van der Waals and Molecular Sciences, Clarendon Press, Oxford, 1996.
[93] G. Soave, Equilibrium constants from a modi8ed Redlich-Kwong equation of state, Chem. Eng. Sci. 27 (1972) 1197–1203.
[94] D.Y. Peng, D.B. Robinson, A new two-constant equation of state, Indust. Eng. Chem. Fundament. 15 (1976) 59–64.
[95] Y. Yu, B.C.Y. Lu, A three-parameter cubic equation of state for asymmetric mixture density calculations, Fluid Phase Equilib. 34 (1987) 1–19.
[96] L.M. Yu, B.C.Y. Lu, Y. Iwai, Simultaneous calculations of VLE and standard liquid and vapor volumes by means of a 3P1T cubic EOS, Fluid Phase Equilib. 37 (1987) 207–222.
[97] K.G. Joback, R.C. Reid, Estimation of pure-component properties from group-contributions, Chem. Eng. Communic. 57 (1987) 233–243.

Index

Note: Page numbers followed by *f* indicate figures, *t* indicate tables, and *s* indicate schemes.

D

E

F

G

P

CPI Antony Rowe
Eastbourne, UK
November 20, 2019